Cancer Stem Cells

Cancer Stem Cells

Targeting the Roots of Cancer, Seeds of Metastasis, and Sources of Therapy Resistance

Huiping Liu

Case Western Reserve University, Cleveland, OH, United States; Northwestern University Feinberg School of Medicine, Chicago, IL, United States; National Center for Regenerative Medicine, Cleveland, OH, United States

Justin D. Lathia

Cleveland Clinic, Cleveland, OH, United States; Cleveland Clinic Lerner College of Medicine of Case Western Reserve University, Cleveland, OH, United States; Case Comprehensive Cancer Center, Cleveland, OH, United States

ELSEVIER

AMSTERDAM • BOSTON • HEIDELBERG • LONDON • NEW YORK • OXFORD
PARIS • SAN DIEGO • SAN FRANCISCO • SINGAPORE • SYDNEY • TOKYO

Academic Press is an imprint of Elsevier

Academic Press is an imprint of Elsevier
125 London Wall, London EC2Y 5AS, United Kingdom
525 B Street, Suite 1800, San Diego, CA 92101-4495, United States
50 Hampshire Street, 5th Floor, Cambridge, MA 02139, United States
The Boulevard, Langford Lane, Kidlington, Oxford OX5 1GB, United Kingdom

Library of Congress Cataloguing-in-Publication Data
A catalog record for this book is available from the Library of Congress

British Library Cataloguing-in-Publication Data
A catalogue record for this book is available from the British Library

ISBN: 978-0-12-803892-5

For information on all Academic Press publications
visit our website at https://www.elsevier.com/

Publisher: Mica Haley
Acquisition Editor: Rafael Teixeira
Editorial Project Manager: Lisa Eppich
Production Project Manager: Laura Jackson
Designer: Maria Inês Cruz

Typeset by TNQ Books and Journals

Dedication

We would like to dedicate this book to a beloved middle school teacher Ms. Aiying Peng and others who lost their lives to cancer, the ones inspired by stem cells, and all committed to save lives and fight against cancer.

Contents

List of Contributors

Lauren Agro University Health Network, Toronto, ON, Canada; University of Toronto, Toronto, ON, Canada
Shideng Bao Lerner Research Institute, Cleveland, OH, United States
Chi-Hsuan Chang Baylor College of Medicine, Houston, TX, United States
Keith Syson Chan Baylor College of Medicine, Houston, TX, United States
Samuel H. Cheshier Stanford University, Stanford, CA, United States
Avijeet S. Chopra University of Connecticut, Storrs, CT, United States
Anastasia Chumakova Cleveland Clinic, Cleveland, OH, United States
Michael F. Clarke Stanford University, Stanford, CA, United States
Salvatore Condello Indiana University School of Medicine, Indianapolis, IN, United States
Leanne R. Donahue Cornell University, Ithaca, NY, United States
Rogelio Esparza Stanford University, Stanford, CA, United States
Fang Fang Indiana University, Bloomington, IN, United States
Christine Fillmore Brainson Stem Cell Program, Children's Hospital Boston, Boston, MA, United States; Harvard Stem Cell Institute, Cambridge, MA, United States; Harvard Medical School, Boston, MA, United States
Austin Gurney OncoMed Pharmaceuticals, Inc., Redwood City, CA, United States
Andrew Haller University Health Network, Toronto, ON, Canada; University of Toronto, Toronto, ON, Canada
Jennifer Haynes University Health Network, Toronto, ON, Canada; University of Toronto, Toronto, ON, Canada
Diane Heiser Stanford University, Stanford, CA, United States
Joy Q. He Stanford University, Stanford, CA, United States
Masahiro Hitomi Cleveland Clinic, Cleveland, OH, United States; Cleveland Clinic Lerner College of Medicine of Case Western Reserve University, Cleveland, OH, United States
Timothy Hoey OncoMed Pharmaceuticals, Inc., Redwood City, CA, United States
Jason K. Hseih University Hospitals-Case Medical Center & Case Comprehensive Cancer Center, Cleveland, OH, United States
Gregor Hutter Stanford University, Stanford, CA, United States
Awad Jarrar Cleveland Clinic, Cleveland, OH, United States

Carla F. Kim Stem Cell Program, Children's Hospital Boston, Boston, MA, United States; Harvard Stem Cell Institute, Cambridge, MA, United States; Harvard Medical School, Boston, MA, United States

Molly Kozminsky University of Michigan, Ann Arbor, MI, United States

Antonina V. Kurtova Baylor College of Medicine, Houston, TX, United States

Justin D. Lathia Cleveland Clinic, Cleveland, OH, United States; Cleveland Clinic Lerner College of Medicine of Case Western Reserve University, Cleveland, OH, United States; Case Comprehensive Cancer Center, Cleveland, OH, United States

Cherry Leung University Health Network, Toronto, ON, Canada; University of Toronto, Toronto, ON, Canada

Evelyne Lima-Fernandes Structural Genomics Consortium, Toronto, ON, Canada

Huiping Liu Case Western Reserve University, Cleveland, OH, United States; Northwestern University, Chicago, IL, United States

Xia Liu Case Western Reserve University, Cleveland, OH, United States

Neethan A. Lobo Stanford University, Stanford, CA, United States; Celgene Quanticel Research, San Francisco, CA, United States

Daniela Matei Indiana University School of Medicine, Indianapolis, IN, United States; Indiana University Melvin and Bren Simon Cancer Center, Indianapolis, IN, United States; VA Roudebush Hospital, Indianapolis, IN, United States

Siddhartha S. Mitra Stanford University, Stanford, CA, United States

Hyeongsun Moon Cornell University, Ithaca, NY, United States

Sunitha Nagrath University of Michigan, Ann Arbor, MI, United States

Kenneth P. Nephew Indiana University, Bloomington, IN, United States; Indiana University School of Medicine, Indianapolis, IN, United States; Indiana University Melvin and Bren Simon Cancer Center, Indianapolis, IN, United States

Catherine A. O'Brien University Health Network, Toronto, ON, Canada; University of Toronto, Toronto, ON, Canada; Toronto General Hospital, Toronto, ON, Canada

Claudia Petritsch University of California San Francisco, San Francisco, CA, United States

Akshita Puri University Health Network, Toronto, ON, Canada; University of Toronto, Toronto, ON, Canada

Dalong Qian Stanford University, Stanford, CA, United States

Sumaiyah Rehman University Health Network, Toronto, ON, Canada; University of Toronto, Toronto, ON, Canada

Ofer Reizes Cleveland Clinic, Cleveland, OH, United States; Cleveland Clinic Lerner College of Medicine of Case Western Reserve University, Cleveland, OH, United States; Case Western Reserve University, Cleveland, OH, United States

Jeffrey M. Rosen Baylor College of Medicine, Houston, TX, United States

Kiera Rycaj University of Texas MD Anderson Cancer Center, Smithville, TX, United States
Yogen Saunthararajah Taussig Cancer Institute, Cleveland Clinic, Cleveland, OH, United States
Xiling Shen Duke University, Durham, NC, United States
Andrew E. Sloan University Hospitals-Case Medical Center & Case Comprehensive Cancer Center, Cleveland, OH, United States
Benjamin T. Spike University of Utah, Salt Lake City, UT, United States
Swetha J. Sundar University Hospitals-Case Medical Center & Case Comprehensive Cancer Center, Cleveland, OH, United States
Dean G. Tang University of Texas MD Anderson Cancer Center, Smithville, TX, United States; Tongji University School of Medicine, Shanghai, China; University of Texas MD Anderson Cancer Center, Houston, TX, United States
Praveena S. Thiagarajan Cleveland Clinic, Cleveland, OH, United States
Linda J. van Weele Stanford University, Stanford, CA, United States
Yadong Wang University Health Network, Toronto, ON, Canada; University of Toronto, Toronto, ON, Canada
Yinu Wang Indiana University, Bloomington, IN, United States
Irving Weissman Stanford University, Stanford, CA, United States
Andrew C. White Cornell University, Ithaca, NY, United States
James Wright University Hospitals-Case Medical Center & Case Comprehensive Cancer Center, Cleveland, OH, United States
Maider Zabala Stanford University, Stanford, CA, United States
Wenchao Zhou Lerner Research Institute, Cleveland, OH, United States

Foreword by Jane Visvader

Since the declaration of war on cancer in 1971, we have witnessed an exponential increase in our understanding of cancer at the molecular and cellular level. Yet, mortality from cancer remains a leading cause of death, and the incidence of most cancers continues to rise, in part due to an aging global population. Despite the development of revolutionary anticancer drugs over the past decades, the range of therapeutic armaments remains limited. The immense heterogeneity between cancer subtypes and within individual tumors poses a great challenge for biologists and clinicians. The cancer stem cell (CSC) hypothesis has stimulated great interest as a mechanism to explain intratumoral heterogeneity and therapeutic resistance. The lessons learned from stem cell regeneration and the emergence of the CSC field offer an opportunity to unravel the complexities of cancer and to develop more effective therapies. This book provides a comprehensive summary of advances in the CSC field over the past 20 years, encompassing concepts, definitions, methods and cutting-edge technologies to study CSCs, the emerging features of these cells in tumors of diverse organs, and their therapeutic potential. Moreover, it demonstrates how stem cell biology and traditional cancer models can be integrated.

Tumors show dramatic heterogeneity in their cellular morphology, signaling pathways, genetic/epigenetic lesions, and therapeutic response. How does intratumoral heterogeneity arise? On the snowy mountains of Breckenridge, Colorado, I was approached by Huiping to read this book, which addresses and seeks to answer some of the most poignant questions in the cancer field. The book boasts a list of prominent authors and provides an extraordinarily comprehensive account of CSCs, spanning both basic and translational aspects. The field has been lacking such a broad overview.

The clinical relevance of CSCs has been a crucial question. This book not only helps us to appreciate the features of CSCs in many representative systems, but also illustrates the importance of CSCs to tumor relapse and predicting patient outcome. By identifying molecular commonalities (such as self-renewal pathways) between CSC populations in diverse tumors, it has become clear that it is possible to target CSCs. Indeed, the targeting of developmental (stem cell) pathways usurped by CSCs is showing considerable promise in the clinic. The equally important issue of how CSCs evade radiation and chemotherapy is also discussed in this book. The dynamic nature of CSCs and the presence of metastatic cells with properties similar to CSCs continue to underscore the importance of studying these cells. Cancer patients usually succumb to death through metastasis. An improved understanding of the relationship between CSCs and metastatic cells will enhance the development of metastasis-targeting therapies. Despite the emerging evidence for tumors being evolving entities, the chapters herein reveal the considerable potential for targeting the unique features of CSCs in tumor development and metastasis.

This book also reviews recent developments in the field on the interplay between CSCs, their niches, and the immune microenvironment. Insights into how CSCs evade the innate and adaptive immune responses are discussed, as well as the potential of CSC-directed immunotherapy. Such therapies in synergy with conventional therapies hold the promise of affording new therapeutic strategies against cancer. The book provides a valuable source of information for trainees and scientific investigators at all levels, as well as physicians and industrial partners. Finally, this book is timely, presenting a summary of the state of the field and highlighting key priorities for future investigation. I very much enjoyed reading this book and highly commend it as a resource for the cancer biology field.

Sincerely,

Jane Visvader, PhD, Professor
Fellow of the Australian Academy of Science
Joint Division Head, Stem Cells and Cancer
The Walter and Eliza Hall Institute of Medical Research
Melbourne, Australia

Foreword by Stanton L. Gerson

To provide a foreword for the definitive treatise on the state of the art of cancer stem cells by all of the authors I would like to hear from is rather overwhelming.

This text will become a reference of the field because it has assembled a timely review of critical issues in the field and will catapult the efforts forward to take what is learned about cancer stem cells into diagnostic, prognostic, therapeutic, and ultimately preventive fields of clinical oncology. In this regard, this effort is very special.

The field of stem cell characterization starts and persists to this day, in all fields, with confusion and contention about functional characteristics, isolation techniques, identification strategies, and purification to homogeneity. Every stem cell field struggles with the efforts to outwit the cells through characterization, whether they be mesenchymal, hematopoietic, neural, or cancer stem cells. To me, it is part of the endearing scientific quest for perfection, that sometimes sidelines itself in the bigger, faster, and better method of today only to be outgunned tomorrow.

Looking through these chapters, all attempt a definition of the cancer stem cell: no two definitions are the same, but the quest remains one clear focus—to identify the most problematic cell in the cancer lineage, characterized by its pleiotropic lineage potential, responsiveness to signals similar but never identical to normal stem cells, and persistence through tumor evolution and treatment. No one wants these cells in a tumor yet every tumor seems to have them. And, eliminating them is, for the moment, elusive.

One thing is clear, the outstanding compilation herein will teach about the complexity of the subject, the links between approaches, the consensus on surface markers and gene expression characterization, and the compelling approaches to isolation and targeted therapeutics. I would love five pages to critique every chapter with a provocative question and a suggestion for experimentation. But alas, my duty is to engage and encourage you to review, compare, and reflect for yourself why the field has yet to come to closure on critical issues about cancer stem cells: When do they originate in a tumor—is a single cell sufficient to set the clone and is that cell the cancer stem cell? Is the treatment resistance intrinsic to cancer and instability of gene expression and mutation profile, or a static property selected for among cancer stem cells? Will a common Achilles' heel emerge given that there are so many similarities why should not a treatment be effective across tumor types? Do all cancer stem cells circulate and if so, as single cells covered in accessory cells and exosomes, or do they break away as clusters? What is the best clinical test of the cancer stem cell hypothesis for diagnosis, treatment, or prevention, and what is the best clinical trial design to address the role and clinical functionality of cancer stem cells? The good news is that the authors of these pages know better than anyone.

Have a good read.

Stanton L. Gerson, MD
Case Western Reserve University School of Medicine
Case Comprehensive Cancer Center and Seidman Cancer Center
National Center for Regenerative Medicine
Cleveland, OH, United States

Preface

Cancer Stem Cells—Targeting the Roots of Cancer, Seeds of Metastasis, and Sources of Therapy Resistance is intended for scientific researchers from students to investigators as well as clinicians and industry partners who can impact daily and future cancer medicine.

Cancer has been a pathology that has continued to vex physicians and scientists, despite our ever evolving understanding of this class of diseases. In our attempts to understand its complexity and generate paradigms to guide investigation, we have turned to the elegance of development processes. Tumors have long been considered aberrant organs, and many key developmental signals reemerge during tumor initiation and progression. Another key feature shared by tumors and developing organs is cellular heterogeneity, which in normal organs, is organized in a cellular hierarchy with a self-renewing stem cell at apex. Stem cells are essential for normal development and homeostasis, and their dysfunction is linked to many degenerative diseases. Stem cells are tightly controlled by the integration of intrinsic and extrinsic regulatory mechanisms that ensure that at each cell division, at least one daughter cell retains stem cell function.

To fully illustrate the cellular heterogeneity present within tumors, investigators have been inspired by the hierarchal organization present in development to complement the stochastic model. Using transplantation assays, it was recognized that a fraction of tumor cells in a given cancer could give rise to a tumor that could serially be passaged. These tumor-initiating cells first identified in leukemias gave rise to heterogeneous tumors and had enhanced self-renewal capabilities, similar to that of normal stem cells. Out of these observations, the cancer stem cell hypothesis was born, and now cancer stem cells have been identified in a variety of advanced cancers including leukemias, breast cancer, colorectal cancer, brain tumors, pancreatic cancer, cutaneous tumors, lung cancer, bladder cancer, and other cancers. The cancer stem cell hypothesis has provided an additional model to help appreciate the complexity of cancer and complements traditional models based on genetic aberrations. By integrating the cancer stem cell hypothesis with genetic models and regulatory immune microenvironment, additional regulatory mechanisms have been identified that drive tumor growth, therapy resistance, and metastasis that claim patient lives. The first generation cancer stem cell pathway inhibitors are in clinical trials, highlighting the future potential of cancer stem cell targeting.

In this book, we have provided an overview of the current state of understanding within the field and highlight key areas that require additional investigation. In the first section, cancer stem cells are introduced along with self-renewal, a fundamental property of these cells. A synopsis on cancer stem cell enrichment methods is also included. In the second section, cancer stem cells in representative organs systems are discussed and include leukemia, breast cancer, lung cancer, colorectal cancer, bladder cancer, and ovarian cancer. In the third section, key cancer stem cell features are highlighted and include understanding the cell of origin, the role of asymmetric cell division and cell fate choice, the role of cancer stem cells in metastasis, the interaction between cancer stem cells and the immune system, and the mechanisms of therapeutic resistance. In the final section, a clinical perspective is provided for cancer stem cell–based therapies and detection, including circulating tumor cells.

We have assembled this book to engage readers of varying levels within the field and hope that you find the concepts within this book engaging and a source of inspiration. We look forward to receiving feedback and suggestions to improve the book in future editions.

Huiping Liu, MD, PhD
Case Western Reserve University

Justin D. Lathia, PhD
Cleveland Clinic Foundation

Acknowledgments

This book represents a true team effort, and we are extremely grateful to numerous individuals who were integral to the production of this book. First off, we would like to acknowledge our partners at Elsevier for reaching out to us and assisting with putting this book together, an exercise neither of us had previously participated in. We would especially like to thank Catherine (Cassie) Van Der Laan who assisted at the initial stages of the book conception as well as Lisa Eppich who worked with us through the entire process and was invaluable in getting the book published. We are deeply indebted to Drs. Jane Visvader and Stanton Gerson for the foreword as well as each and every one of the authors who spent time putting together high-quality chapters. Along with writing the chapters, the authors also served as peer reviewers (Dr. Jeffrey Rosen served twice). Without them, the book would not be available. We would also like to credit the following additional peer reviewers for their valuable and insightful comments: Dr. Anita Hjelmeland (University of Alabama, Birmingham), Dr. Masahiro Hitomi (Cleveland Clinic), Dr. Monica Venere (Ohio State University), Dr. Jialiang Wang (Vanderbilt University), Dr. Stefani Spranger (The University of Chicago), and Dr. Golam Kibria and Valery Adorno-Cruz (Case Western Reserve University). We are thankful to our respective laboratory members who informally contributed to the book by providing ideas and interesting points of discussion. Finally, we would like to acknowledge the many investigators in the cancer stem cell field for providing a strong foundation upon which this book was generated and sincerely apologize to the investigators whose work we were unable to include or cite due to space limitations.

With love, we really appreciate our respective family members who provided tremendous support and inspiration for the book.

Huiping Liu, MD, PhD **Justin D. Lathia, PhD**

SECTION 1

CSC Overview and Methodology

CHAPTER 1

Introduction: Cancer Stem Cells

S.S. Mitra, J.Q. He, R. Esparza, G. Hutter, S.H. Cheshier, I. Weissman
Stanford University, Stanford, CA, United States

INTRODUCTION

Somatic stem cells have been known to exist since at least the 19th century through the study of lower organisms such as planarians and salamanders those were capable of remarkable tissue regeneration [1]. The subsequent study of human teratomas, germ cell tumors that could grow tissues such as bone and teeth in ectopic locations, brought the concept of stem cells—a population of cells which could remain poised and uncommitted to specific lineages even after adulthood—to human biology [2]. By the late 1950s, Conrad Waddington [3] had postulated that by regulating gene expression, cells become more specialized the more they divide—eventually becoming committed to a specific cell fate [3]. The earliest cells in the division hierarchy, termed stem cells, have the capacity not only to divide and proliferate into committed cells, but also to self-renew and create more cells with the same uncommitted identity. These asymmetric divisions thus result in the maintenance of a self-renewing pool of stem cells with a life-long responsibility to provide differentiated cell types. Of note, at each subsequent differentiation step in the cell division hierarchy, self-renewal potential is lost. Committed cells lose their ability to regenerate and therefore generally have limited cellular life spans.

Cancer stem cells (CSCs) are a subset of tumor cells which have escaped cell cycle regulatory mechanisms, cell death, and yet have retained the immense self-renewing and proliferative potential of stem cells [4]. In this manner, they have the capacity to regenerate entire tumors from a limited number of cells. Their existence was first documented by Bonnet and Dick [5] in transplantation studies of human acute myeloid leukemia (AML) in mice with severe combined immunodeficiency disease (SCID). Of the transplanted leukemic cells, only an estimated population of between 0.01 and 1% of the total cell population was capable of initiating AML in the immunocompromised mice. Termed SCID leukemia–initiating cells, they were found to undergo rapid clonal expansion and appeared to be at the top of a cancer cell hierarchy. Later,

CONTENTS

Cancer Stem Cells. http://dx.doi.org/10.1016/B978-0-12-803892-5.00001-2

it was shown that most AML tumor-initiating cells were at the stage of the multipotent progenitors (MPPs), and not the hematopoietic stem cells (HSCs) [6], opening the question as to how does a normally non-self-renewing cells gain both self-renewal and unlimited expansion. The introduction of the CSC hypothesis has led to new developments and insights into the nature of tumor propagation and possible therapeutic targets. Yet, much is still unknown about this important tumor cell subtype in each of the human and experimental cancers. In this chapter, we will discuss the biological properties of CSCs, current research on their identification and function, and future goals for CSC-directed therapies.

CANCER STEM CELL ORIGINS

Cancer Stem Cell Generation From a Single Cell of Origin

The cells within a tumor are derived from tissues and organs which contain normal stem cells, progenitors, and lineage committed cells, eg, the lineage hierarchy of the blood system. Identifying the complete roadmap of transitions from HSC through MPPs lacking self-renewal (short term-HSC, MPPs [7,8]) to common myeloid progenitor in mice [9] and humans [10], and Common lymphoid progenitor (CLP) [11], and downstream from them ever more committed progenitors (eg, granulocyte–macrophage progenitor (GMP) and megakaryocyte/erythroid progenitor) [9,12] allowed Weissman and colleagues to phenotypically isolate the cell types from which gave rise to leukemias. The critical development of a strain of mice where two pathways important in programmed cell death was blocked in hematopoetic cells [13]. Serial transplantation of leukemias from mice that developed AML could only be achieved with GMP cells. They further inferred that the progression to leukemia required at least five to seven rare events, either genetic or epigenetic; however, most of these events could not confer self-renewal, and so must have occurred in self-renewing cells to persist sufficiently to form a clone that was leukemic [14]. The pathways from the cell type that acquired the first oncogenic mutation, also known as the cell of origin to CSC, are varied and complex. However, it has been shown that in in vitro models of hematopoietic malignancy, certain oncoproteins may activate genetic programs involved in self-renewal, thereby conferring "stemness" to committed malignant cells and leading to the creation of CSCs [15–17]. In vivo studies using an MLL1–AF9 mouse model of AML confirmed this finding but with an additional caveat: the amount of translocation product expressed determined the efficiency of CSC generation [18]. Below a certain threshold, oncogenic capacity was limited, demonstrating the importance not only of oncoprotein presence, but also of gene dosage, in tumor formation. It is important to note, however, that the CSC population of a tumor is almost always genetically distinct from the cell of origin. Two models have been proposed to explain why this may be the case.

Two Models for Tumor Growth and Expansion

Two models exist to explain the heterogeneity of tumor growth. The first, which emphasizes a population hierarchy in which all stemness derives from rare CSCs at the top and the tumor bulk is comprised of differentiated CSCs, is referred to as the hierarchical model. By contrast, the second, which emphasizes the functional heterogeneity within the tumor population and posits that all tumor cells contribute to tumor maintenance, is referred to as the clonal evolution (CE) model. It is important to note that these two pathways are not mutually exclusive, and that both pathways may very well occur in the same tumor. In both cases, it is postulated that an acquired mutation confers a growth and proliferative advantage on a certain malignant cell type, allowing it to outcompete surrounding cell types and thus forms a tumor [19]. Yet the CE model postulates that the heterogeneity of tumor cells—specifically, the sheer number of different mutations often found in a single tumor—make it likely that tumors form from a multitude of genetic and epigenetic errors [20]. A dominant CSC may later give way to a cell downstream that has acquired an even more explosive growth advantage [21]. While the hierarchical model works well to characterize somatic stem cells, in the CE model, such structure is often obliterated by *the rapid accumulation of mutation* process that accelerates as essential DNA repair and cell cycle checkpoint processes are increasingly disrupted [20].

IDENTIFICATION OF THE CANCER STEM CELL POPULATION

Features of Cancer Stem Cell

CSCs are robust cells which may have acquired characteristics similar to their normal tissue stem cell. The expression of ABC transporters and telomerase and glutathione synthetase are properties of normal tissue stem cells that are extended into their CSC progeny [22–27] that allow for cell survival and proliferation even after exposure to anticancer therapeutics. For example, certain gastrointestinal cancer cell lines show increased resistance to oxidative stress via interactions between CD44 and cell surface cystine–glutamate exchange transporters which result in increased synthesis of reduced glutathione, a key molecule involved in the neutralization of reactive oxygen species [15]. In this manner, they are able to gain a survival advantage in inflammatory environments. Other studies have shown that CSCs are also capable of extensive metabolic reprogramming [16,17], rapid DNA damage repair [18,28], as well as enhanced drug excretion through ATP-binding cassette (ABC) transporters [29] of particular concern in the context of chemotherapeutic agents and other anticancer drugs.

Given the aggressively proliferative nature of CSCs, it is no wonder that they have been found to contribute significantly to the formation of minimal residual disease (MRD) [30,31]. These cells, which can escape targeting by anticancer therapy and remain undetected even as the tumor bulk regresses, are capable of regenerating entire tumors from an astonishingly small starting population. Moreover, because of the selection pressures introduced by chemotherapy, cancer recurrence after drug administration often grows more aggressively than the original tumor. Kurtova et al. found that the administration of chemotherapeutic agents triggered apoptosis in healthy cells which subsequently released prostaglandin E2, inducing previously dormant keratin14+ bladder CSCs to proliferate again [32]. Overcoming these hyperaggressive characteristics of CSC, especially the paradoxical enrichment often found after incomplete prior therapy, is of particular interest in the current development of novel therapeutics.

Biomarkers for Cancer Stem Cell Identification

No specific set of markers have yet been identified that both sensitively and specifically detect the CSC population. For example, while cell surface protein CD133 has been widely implicated in multiple cancer types and is currently used as part of the standard practice for CSC identification from a heterogeneous cell population, it has been shown that separation of $CD133^+$ and $CD133^-$ populations and subsequent functional assays reveal two genetically distinct cell populations that have different growth dynamics but nevertheless both retain tumorigenic potential [33,34]. Moreover, in the setting of pancreatic cancer, it was noted that $CD44^+/CD24^-$ cells, which expressed CSC-specific markers first isolated in breast cancer, exhibited variable overlap with $CD133^+$ cells [35]. These two studies indicated that although these markers were apparently specific to the CSCs described in these studies, different CSCs have different markers; thus these markers were not universal, just as normal stem cell markers for one tissue stem cell do not tell us much about other tissue stem cells, were not necessarily sensitive, and were unable to identify large portions of the CSC population. Nevertheless, the combination of the various surface markers along with high ALDH1 levels did yield a functionally enriched CSC population with a remarkably low cellular threshold needed to initiate tumor formation [36]. Table 1.1 illustrates which tumors currently have a verified CSC population and includes their specific markers.

The gold standard assay for identifying a CSC population is the in vivo limiting-dilution assay, in which primary tumor cells are mechanically disassociated, cellular fractions are separated, and the cellular mixtures are injected into the homologous organ or tissue in immune-deficient mice to form xenografts in recipient mice [73]. By monitoring tumor development in these mice, it is possible to set a lower limit for the frequency of CSCs in a given cell population [74]; the in situ CSC, which do not need to have genes expressed to protect them between isolation and transplantation, neither of which are physiological. The observation of tumors in recipient mice supports the existence of a

Table 1.1 Verified Cancer Stem Cell Populations and Their Corresponding Markers in Various Cancers

Cancer Class	Markers (eg)	References
Leukemia/Lymphoma	$CD34^+$, $CD47^+$, $CD96^+$, $CD25^+$, $CCL\text{-}1^+$ $CD38^-$	[5,37–41]
Head and neck squamous cell carcinoma	$CD44^+$, BMI 1^+, $CD24^+$, $CD133^+$	[42–44]
Glioblastoma multiforme	$CD133^+$, $CD49f^+$, JAM-A, $HER2^+$, $EGFRvIII^+$	[45–49]
Lung	$CD44^+$ $CD133^+$	[50,51]
Breast	$ESA^+CD44^+/CD24^{-/low}$ Lin^-, $ALDH1^+$, $CD133^+$, $CD61^+$	[33,35,36,49–54]
Ovarian	$CD44^+$, $CD117^+$	[55]
Pancreas	$CD44^+$, CD24, ESA^+	[56,57]
Gastric	$CD44^+$, $CD133^+$, $ABCB1^+$, $ABCG2^+$	[58,59]
Colorectal	$CD44^+$, $CD133^+$, $CD166^+$, $CD24^+$	[60–63]
Prostate	$CD44^+$, $CD133^+$, $ALDH^+$	[64–66]
Bladder	$CD44^+$, $CD90^+$, $CD49f^+$	[67–69]
Melanoma	$CD20^+$, $CD271^+$, $ABCB5^+$	[70–72]

CSC population that is proliferative, capable of self-renewal, and able to make the bulk of the differentiated tumor cells that lack stemness. Moreover, because it is possible to separate tumor cells into distinct fractions through techniques such as fluorescence-activated cell sorting (FACS), one can also use this assay to probe the existence of a hierarchical cellular structure within different tumor compartments [73].

An alternative to the in vivo limiting-dilution assay is the in vitro limiting-dilution assay, which utilizes cell culture techniques to measure tumorigenicity rather than transplantation. In this approach, tumor cells are again mechanically dissociated and cellular fractions separated, with the difference being that culture then takes place in microtiter wells [75]. Stemness is assayed by staining for stem cell associated markers, as well as by functional assays such as the formation of cellular progenitor structures (eg, tumorspheres) and the ability to maintain self-renewal after serial passages [47,75–77]. However, any interpretation from in vitro experiments is tempered by the hyperoxic in vitro environment and the lack of a stem cell niche.

In vitro measures are often used as a convenient and cost-effective alternative to animal studies. However, whether cell culture is representative of in vivo work is an ongoing concern [78]. For example, the receptor EGFRvIII, overexpressed in up to 70% of glioblastoma (GBM) tumors, is a promising therapeutic target with multiple ongoing clinical trials selectively targeting EGFRvIII-amplifying cells [79,80]. Use of $EGFRvIII^+$ cells in murine xenografts accurately model CSC attributes, including preferential residence in the perivascular niche and the ability to differentiate into endothelial cells [49]. However, normal in vitro culture conditions select against EGFRvIII expression, and cultured human-derived cell lineages quickly lose the marker after serial passaging [81,82]. EGFRvIII expression can be retained by culturing cells in niche-specific serum-free neurosphere media [49]. Thus, whether in vitro studies may be representative of in vivo studies may ultimately depend on how well culture conditions reflect the composition of the native CSC niche.

Key Signaling Pathways Involved in Cancer Stem Cell Maintenance

CSCs maintain their tumorigenicity and stemness through several key signaling pathways, including PI3K/AKT, NF-kB, Notch, Hedgehog, Wnt/β-catenin, and others [83–87]. Aberrant signaling in one or more of these pathways results in the formation of cells that are hyperproliferative, yet resistant to normal cellular checkpoint mechanisms. For example, Notch signaling is implicated in cell self-renewal and plasticity, and Notch4 has been found to be upregulated in breast CSCs [87,88]. Similarly, Wnt has signaling also been associated with maintenance of stemness in CSC populations, with abnormal Wnt/β-catenin signaling contributing to leukemias as well as solid organ malignancies such as colon and breast carcinoma [89,90]. In colon carcinoma specifically, abnormal Wnt expression has been found to inappropriately activate an epithelial–mesenchymal transition in gut cells, leading the cells to gain invasive properties and increasing potential for cancer metastasis [91]. PI3K/AKT affects Bcl-2 and other apoptotic inhibitory proteins, thereby conferring increased resistance to apoptosis in leukemias and other cancers such as melanoma, prostate, and endometrial [65,92]. NF-kB also affects Bcl-2 but also regulates the cellular inflammatory response to cytokines [86]. Aberrant NF-kB expression in CSCs thus promotes a hyperinflammatory state while increasing CSC resistance to inflammation, resulting in cancer progression.

MOUSE MODELS OF CANCER STEM CELL FUNCTION

Mouse models have contributed significantly to the understanding of CSC biology. In preclinical CSC research, there is now a multitude of transplantable patient-derived or animal-derived models that can be placed orthotopically (in the same anatomical location where the tumors arose from) or heterotopically (in a different anatomical location where the tumors arose from) into new syn-, allo-, or xenogeneic hosts. Additionally, organ- and cancer-specific genetically engineered mouse (GEM) models, further stratified into either germline or somatic (non-germline), serve as an important tool to study cancer initiation, progression, and metastasis. Recently, optimized humanized mouse models have also been developed. We will give a brief overview of each.

Patient-Derived Xenograft Models

The current gold standard for tumor induction is the transplantation of either low-passage CSC lines from patient-derived primary or metastatic tumors, or of directly FACS-sorted subsets of potential CSC candidates, either heterotopically or orthotopically into immunocompromised host mice. These tumors, called patient-derived xenografts (PDXs), can also be serially transplanted into subsequent hosts to confirm the tumorigenic potential of the putative CSC candidate, although this presents the tumors with selections that may not occur in humans, and within a few serial transplants tumors are commonly selected that grow more rapidly and have increased instances of new genetic and epigenetic alterations.

NSG, NOG mice (NOD.Cg-Prkdc^{scid}Il2rg^{tm1Wjl}/SzJ: NSG and NOD.Cg-Prkdc^{scid} Il2rg^{tm1Sug}/Jic: NOG) and Rag1—Il2Rg—mice, which are B- and T-cell-deficient and lack natural killer (NK) cell activity represent a well-established model system to study the engraftment and mobilization of human peripheral blood stem cells and are excellent hosts for growing experimental leukemias and solid tumors [9,10,93,94]. The understanding and identification of CSC in hematopoietic malignancies, especially in leukemias [95] and lymphomas [5] has largely profited from this approach, but recently, CSCs in solid cancers such as breast [36], colon [14,96], skin [72,97], melanoma [71], and brain [47] emerged using the SCID repopulation approach as proof of tumorigenicity. However, the degree of compromise of the host immune system can be permissive for the tumorigenic potential of unselected patient-derived cells as shown by Quintana et al. in melanoma-derived xenografts and indicated that the tumor-initiating potential of cells within the tumor bulk may vary significantly depending on the tumor type [78]. However, Quintana et al. transplanted small numbers of melanoma cells in high-protein matrigel, a basement membrane preparation extracted from Engelbreth-Holm-Swarm (EHS) mouse sarcoma, a tumor rich in ECM proteins used by very few laboratories. While the use of high-protein matrigel does bring up interesting questions of using the correct niche for xenotransplant models it also brings in the added possibility of the contribution from the matrigel components in increased tumor initiation.

While heterotopic models, eg, those created via subcutaneous injection of CSCs or cancer organoids, are simpler to implement versus orthotopic models, their most prominent disadvantage is the absence of the original supporting stem cell niche or microenvironment. This problem could partially be solved by co-transplanting tumor cells and human stroma cells or basement membrane matrix proteins (eg, Matrigel) [98]. However, most stromal cell compartments such as endothelial cells and fibroblasts will rapidly be replaced by mouse tissue [99]. While heterotopic PDX models may be used in high throughput testing of anti-CSC drugs, the microenvironment, nontumor cell stroma, lymphatic drainage, and vascularization of heterotopic tumors are not the same as from where the tumor originated, thus limiting the external validity of this model [100].

In contrast, orthotopic models possess tissue-specific stromal cell types and extracellular matrix components that support the cancer type inherent CSC niche [101]. Due to the lack of certain components of the immune system, however, PDX tumors tend to progress faster than tumors otherwise would in patients, which might render the model more sensitive to targeted anti-CSC drugs than the true physiologic response [102]. Using fluorescent and luminescent labeling techniques, tumor growth and effect of anti-CSC drug candidates can be monitored noninvasively or using intravital microscopy to specifically visualize the interaction of the cancer cells within the tumor microenvironment. For example, Lee et al. showed that the homing of hematopoietic cancer cells and normal stem cells to their niche in the bone marrow could be modeled using implantable marrowlike microenvironments and monitored by intravital microscopy and fluorescence computed tomography [103].

Germline Genetically Engineered Mouse Models

Of all murine cancer models, GEM models provide the most complete representation of cancer development; cancers develop from initiation through progression, coevolve with intrinsic stroma, and possess an intact immune system [104]. Techniques to engineer the germline of mice include constitutive and conditional knock-in/knock-out strategies, targeted transgene expression, and inducible transgene expression. In the last several decades, numerous organ-specific germline deletions of tumor suppressor genes or mutations of oncogenes have been created that shed light on tumorigenesis [105]. In some instances, spatial and temporal induction or deletion of oncogenes/tumor suppressor genes, mainly by exploiting countless variations of the powerful organ-specific recombination tools, Cre/LoxP and FLP/FRT, resulted in a sequence of oncogenic events mirroring human tumorigenesis to a great extent [105]. Bacterial Cre and yeast FLP enzymes are site-specific recombinases that catalyze specific recombination between defined *loxP* and *FRT* sites, respectively [24,106,107]. Therefore, in the presence of Cre or FLP protein expression, homologous recombination is induced between *loxP* or *FRT* sites that flank the gene of interest thus recombining out the flanked genetic sequence and deleting the gene of interest. If a STOP codon is flanked by *loxP* or *FRT* site, potentially oncogenic mutations can be driven as well. The fact that Cre recombinases have been engineered to be inducible and driven by tissue-specific promoters renders this approach an invaluable tool to simulate oncogenesis [108].

Based on dissociation of germline GEM-derived tumors, it was possible to isolate tumor-initiating cells and propagate and characterize them further in allograft or xenograft models; exemplified by $CD15^+$ cells sorted from *Ptch* mutant mice with medulloblastoma [109]. Yet, GEM cancer models provide the only opportunity to evaluate drug delivery, therapeutic response, and biomarker expression for cancers evolving within their natural microenvironment (autochthonous cancers) [104]. Recently, the discovery of the CRISPR/Cas9 technique has decreased the time to achieve germline genome editing significantly [110].

Somatic Genetically Engineered Mouse Models

Given the low penetrance rate and high heterogeneity of traditional germline GEM models, other approaches considering direct manipulation of somatic cells have been developed. By utilizing genetically engineered murine embryonic stem cells for chimera-formation and transplantation, these models have the additional benefit of requiring less time and fewer resources to develop, two major obstacles to preclinical trials [111].

Chimeric mouse models take advantage of genetic mutations that predispose toward tumor formation. They are created by transplanting genetically engineered embryonic stem cells into the inner cell mass of preimplantation-stage embryos, commonly at the blastocyst stage. These blastocytes are then implanted into pseudopregnant mice, eventually generating chimeric mice whose cells are a mosaic of engineered and original host cells. One such chimeric model by Zhou et al. employed ESCs engineered to contain tissue-specific, inducible oncogenes and a luciferase marker, and deletions of prominent tumor suppressors to generate chimeras which were thus predisposed to developing specific tumor types [112].

Transplantation mouse models harness the ability of stem and tumor cells to proliferate and compete with the native cells in a host mouse. The models often involve the implantation of genetically modified stem or tumor cells into adult tissue, and rely on the ability of the implanted cells to hone to their respective stem cell niches in the host animal and act as new cells of origin. One such example is a transplantation model by Alcantara Llaguno et al., that described the localization of GBM cells-of-origin to neural stem cells in the subventricular zone [113].

Germline-Derived Allograft Models

GEM-derived allograft (GDA) models combine the human cancer similarity of GEM models with the relative simplicity of transplantation approaches. Importantly, the host mouse is a genetic-background matched, immunocompetent animal that enables researchers to study the immune-related microenvironment of the CSC niche as well. GDA models represent a better alternative to assess immunomodulatory therapies against CSCs, since a fully functional immune system is in many instances necessary to validate the effect of immunomodulatory agents [114].

Humanized Mouse Models

To overcome the immune-related limitations of PDX models, humanized mouse models (eg, those created by transferring functional human immune cells or purified HSCs into immunocompromised mice) have been developed and optimized in recent years. Novel humanized mouse model with much better engraftment of human myeloid cells will prove very useful in monitoring the full potential of human immune cells against CSC xenografts [115].

STRATEGIES FOR TARGETING CANCER STEM CELL

As previously discussed, the stemness of CSCs confers undesirable traits that promote cancer progression and metastasis, and contributes to the formation of cancers that are refractory to chemo- and radiotherapy. Here, we will describe several strategies to target CSCs based on these acquired characteristics.

Protective Signaling Pathways

Protective signaling pathways such as antiapoptotic pathways have generated much interest among researchers attempting to halt the proliferation of CSCs. For example, inhibitors of NF-kB have been shown to improve cancer response to chemotherapy. Combined treatment with NF-kB and chemotherapeutic agents such as the DNA intercalating antibiotic doxorubicin leads to a marked reduction in multidrug resistance. Moreover, inhibitors of the Hedgehog pathway such as cyclopamine have shown the ability to inhibit metastasis in orthotopic mouse models of pancreatic cancer and in in vitro adherent glioma models of GBM [116,117].

Also of great research interest is the development of antagonists to the Notch and Wnt/β-catenin pathways, which have been implicated in metastasis and tumorigenicity in breast and colon cancer [118,119]. Notch signaling, mediated by γ-secretase, was found to be important in the development of brain metastases from breast cancer in mouse models. Inhibition of Notch1 in these mice reduced the $CD44^+$ and $CD24^-$ CSC subpopulation and lowered the incidence of brain metastases [119].

CSC quiescence—the activation of a hypometabolic state of dormancy—when faced with unfavorable cellular environments, is also under consideration as a potential therapeutic target. Since dormant CSCs are largely refractory to cytotoxic agents, which often target actively replicating cells through disruption of DNA replication, strategies to "lock out" these cells from their resting G0 cell cycle phase have gained prominence. One such strategy involves Fbw7, a component of the E3 ubiquitin ligase which is involved in the degradation of proto-oncogenic proteins such as Notch and c-Myc [120–122]. Fbw7 inactivation promotes the transition of CSCs from dormancy into a proliferative state, allowing for their targeting by conventional chemotherapeutics. For example, in chronic myeloid leukemia CSCs which had acquired resistance to the tyrosine kinase inhibitor imatinib, Fbw7 inhibition improved drug response by stimulating CSCs to proliferate, leaving them vulnerable once again to antiproliferative therapeutics [123].

Finally, as mentioned previously, the protective role of the CD44-xCT-GSH axis against oxidative stress has been exploited extensively by $CD44^+$ CSCs to gain a competitive advantage in hyperinflammatory environments [15]. This feature of CSCs is currently being targeted by anti-xCT therapies, including the drugs sulfasalazine and auranofin, conventionally used to treat rheumatoid arthritis and other autoimmune disorders [124]. Interestingly, while sulfasalazine treatment preferentially targets $CD44^+$ cells, the epidermal growth factor receptor-targeted drug cetuximab selectively targets the $CD44^-$ population [125,126]. Combination therapy with sulfasalazine and cetuximab therefore might be an effective strategy to treat tumors that may otherwise develop marked drug resistance.

Immunotherapies Against Cell Surface Markers

Another key strategy to target CSCs has been the development of monoclonal antibodies to surface markers found specifically on the CSC population. As we have discussed, no "unifying" surface antigen has been found to represent the entire CSC population, but antibody therapy against specific markers have been associated with depletion of the CSC population in certain cancers. For example, in the case of AML, in which markers such as CD33, CD44, and IL-3R are expressed predominantly in the CSC population, drugs such as gemtuzumab ozogamicin (an anti-CD33 humanized mouse monoclonal antibody conjugated to the cytotoxic drug calicheamicin) have been developed for use in clinical treatment [127]. Anti-CD44 and anti-IL-3R therapies have shown to be promising in mice; however, it appears that the bone marrow niche may be in part shielding CSCs from the brunt of the cytotoxic damage via activation of the PI3K/Akt pathway in CSCs [128,129].

Tumor Microenvironment—The Complicit Cells of the Cancer Stem Cell Niche

The role of the stem cell microenvironment, or niche, in maintaining stemness is well known. For HSCs, the bone marrow niche comprising cellular and noncellular elements provide a "fertile ground" for stem cell maintenance [130]. The cellular elements include specialized compartments of the bone marrow such as osteoblastic [131] and vascular niche [132]. Similar to HSC, leukemic stem cells (LSC) require the marrow niche to sustain their perpetual, malignant self-renewal state. The niche promotes angiogenesis to nourish tumors and secretes important growth factors and chemoresistance signals, thus serving as a shield from chemotherapeutic agents and other proapoptotic factors [133]. For example, bone marrow stromal cells which serve as important niche cells for leukemia CSCs have been found to secrete stromal cell-derived factor-1 (SDF-1/CXCR12). CXCR12 could allow tumor cells, which have the associated receptor CXCR4, to home to the niche for nourishment and replenishment. While it is known that leukemic cells co-opt the bone marrow niche and create an abnormal microenviroment to sequester transplanted human CD34^{+} cells [134], it is possible that the receptor–ligand interaction ensures close associate of LSC with stromal cells to manipulate the stromal cells into secreting additional growth and antiapoptotic signals.

Similarly, solid tissue stem cells also reside within stem cell niches; highlighting the fundamental importance of these specialized microenvironments for normal stem cell biology [135]. It was shown as far back as in 1940s that tumor cells migrate into the normal brain around blood vessels (perivascular satellitosis), suggesting that glioblastoma cells might develop a special relationship with the surrounding vasculature [136]. However, only in the last decade that a functional relationship between the tumor vasculature and glioblastoma CSC has emerged where CD133^{+} GBM CSC produced high levels of vascular endothelial growth factor (VEGF) which lead to highly vascular and hemorrhagic tumors in immunocompromised mice. We now know that CSC from GBM and other brain tumors are maintained within vascular niches that mimic the neural stem cell niche [137].

Tumor angiogenesis through the secretion of VEGF is another characteristic of the niche that promotes tumor growth and survival. Studies in mice have shown that administration of bevacizumab, a recombinant antibody against VEGF-A, decreases the number of GBM CSCs by disrupting tumor vasculature, thereby compromising the CSC niche [137,138]. Moreover, combination therapy of VEGF antibodies with cyclophosphamide, a cytotoxic drug, was more effective against glioma xenografts in vivo than either therapy alone, providing more support for simultaneously targeting CSCs via multiple pathways to achieve maximal antitumor effect [139].

ATP-Binding Cassette Transporters: Multidrug and Redox Stress Resistance

In addition to tumor angiogenesis and activation of antiapoptotic pathways, CSCs also appear to express high levels of ABC transporters that can pump a variety of structurally unique drugs out of the cell, thus conferring an additional degree of resistance to CSCs and contributing to MRD [114,140,141]. The ABC superfamily includes multidrug resistance proteins, breast cancer resistance protein, and P-glycoprotein [142]. Currently, Chen et al. are attempting to synthesize inhibitors of ABC function using 20(S)-protopanoxadiol (PPD) derivatives. Of note, PPD12 was found to increase the concentration of cytotoxic drugs adriamycin and rhodamine123 in chemoresistant cells, presumably by directly binding to an inhibitor of the transporter itself [143]. Because of their extensive capacity to contribute to MRD and tumor recurrence and metastasis, ABC transporters are a highly promising target for future therapeutics.

FUTURE OF CANCER STEM CELL-DIRECTED THERAPIES

CSC-directed immunotherapies are among the most promising novel agents, yet also have the greatest potential to conflict with current standards of care. Immunotherapies are dependent on a controlled activation of the immune system to selectively target and attack cells carrying a preselected antigen. An often highlighted advantage of active immunotherapies is their long-lasting efficacy, but such success is critically dependent on eliciting a sustained immune response. This necessity of an inflammatory response becomes a complicating factor when analyzed in the context of current practices in medicine. Given the diversity of immune system functionality among individuals, inflammation in a postimmunotherapy patient is something that must be carefully evaluated, especially in the context of their specific therapy. For example, corticosteroids, which have immunosuppressive and antiinflammatory effects, are mainstays of the postoperative and postradiotherapeutic management of cancer patients but their obvious potentially counterproductive effects in the setting of immunotherapy-induced inflammation brings into question their use in these unique circumstances.

The understanding of immune responses in the setting of CSC-targeted immunotherapy and the ability to gauge efficacy in the clinic will be a major obstacle in the evaluation of immunotherapies. Unlike conventional immunotherapies in which pseudo-progression, the radiographic finding of an apparent increase in tumor size following treatment, is a major potential confounding factor in assessing the efficacy of the therapeutic, the low prevalence of CSCs within a tumor population will instead pose the inverse challenge when assessing their efficacy. As shown in multiple clinical trials, immunotherapies often demonstrate a delayed response time, at times demonstrating efficacy years after initial treatment [144]. A delayed response time in conjunction with the inherent rarity of CSCs following tumor resection will make it increasingly difficult for clinicians to assess whether a therapeutic is acutely active. An efficient method to overcome this obstacle is the identification and stratification of patients who would benefit from an immunotherapy prior to administration. As shown by Snyder et al., therapeutic response to an immunotherapy could potentially be assessed through genetic means. In their study, they utilized melanoma tissue from patients who received anti-CTLA-4 therapies and through whole-exome sequencing, discovered a neo-antigen repertoire that was expressed specifically in responders to the treatment [145]. This preemptive screening for efficacy will not only allow us to have a better idea of who may benefit from a novel therapy but may also help us identify those who may pose an inappropriately aggressive autoimmune response.

Where CSC-targeted immunotherapies lie within the landscape of current cancer treatment options is a question that is ultimately dependent on technologic bottlenecks and our ability to create truly personalized medicine. As mentioned previously, the successful removal of only a subpopulation of CSCs does not fully prevent tumor recurrence and characterization of these subpopulations and how they react to current therapeutic strategies is necessary to achieve prolonged remissions with the goal of eventual cures. CSC heterogeneity is a multifaceted issue; tumors are evolving entities constantly responding to their local environments. Boisgerault et al. highlighted the impact immunotherapies and their subsequent environmental change can have on therapeutic efficacy [146]. In their work, they characterize a subpopulation of CSCs which were able to escape initial rounds of treatment and eventually lead to aggressive tumor recurrence. They subsequently demonstrated that this subpopulation of cells had the ability to change their surface antigen signature in response to immune pressure and thus evade targeted single-agent immunotherapies. This molecular adaptation did allow the tumor cells to evade the systemic immune response, but, as a tradeoff, also made them newly vulnerable to chemotherapeutic agents. This finding holds major implications for the techniques and strategies that must be employed when designing treatment regimens for individuals and exemplifies the potential nuances we will encounter in the age of personalized medicine.

CONCLUSION

The world of CSC research and CSC-targeted immunotherapy holds much promise in allowing us to achieve significant advancements in what appears to have become a stagnant battle against many common cancers. Although this goal is continually coming closer to a reality, there are still many aspects of employing these new therapeutics which remain unclear. The arrival of CSC-targeted therapies will present challenges that may not be adequately addressed in current medical infrastructure or easily incorporated into the current standard of care. The success of new therapeutics developed from this new CSC-centric paradigm rely not only their inherent levels of efficacy, but on accurate methods by which to appropriately assess their success in addition to identifying those individuals that would benefit most from these landmark achievements in the new age of individualized medicine.

List of Acronyms and Abbreviations

ABC ATP-binding cassette
ALDH1 Aldehyde dehydrogenase 1
AML Acute myeloid leukemia

CE Clonal evolution (model)
CRISPR/Cas9 Clustered regularly interspaced short palindromic repeats/CRISPR-associated protein 9
CSC Cancer stem cell
CTLA4 Cytotoxic T-lymphocyte-associated protein 4
EGFRvIII Epidermal growth factor receptor variant 3
FACS Fluorescence-activated cell sorting
Fbw7 F-box and WD40 repeat domain-containing protein 7
GBM Glioblastoma multiforme
GDA Germline-derived allograft
GEM Genetically engineered mouse
MLL1–AF9 ALL1-fused gene from chromosome 9 or MLLT3
MRD Minimal residual disease
NF-kB Nuclear factor kappa-light-chain-enhancer of activated B cells
PDX Patient-derived xenograft
PI3K/AKT Phosphoinositide 3-kinase/protein kinase B, also Akt
Ptch Protein patched homolog
SCID Severe combined immunodeficiency disease
SDF-1/CXCR12 Stromal cell-derived factor-1

References

[1] Dinsmore CE. Animal regeneration: from fact to concept. BioScience 1995;45:484–92.

[2] Pierce B, Verney EL, Dixon FJ. The biology of testicular cancer: I. Behavior after transplantation. Cancer Res 1957;17:134–8.

[3] Waddington CH. Genetic assimilation of the bithorax phenotype. Evolution 1956;10:1–13.

[4] Reya T, Morrison SJ, Clarke MF, Weissman IL. Stem cells, cancer, and cancer stem cells. Nature 2001;414:105–11.

[5] Bonnet D, Dick JE. Human acute myeloid leukemia is organized as a hierarchy that originates from a primitive hematopoietic cell. Nat Med 1997;3:730–7.

[6] Miyamoto T, Weissman IL, Akashi K. AML1/ETO-expressing nonleukemic stem cells in acute myelogenous leukemia with 8;21 chromosomal translocation. Proc Natl Acad Sci USA 2000;97(13):7521–6.

[7] Morrison SJ, Weissman IL. The long-term repopulating subset of hematopoietic stem cells is deterministic and isolatable by phenotype. Immunity 1994;1(8):661–73.

[8] Morrison SJ, Wandycz AM, Hemmati HD, Wright DE, Weissman IL. Identification of a lineage of multipotent hematopoietic progenitors. Development 1997;124(10):1929–39.

[9] Akashi K, Traver D, Miyamoto T, Weissman IL. A clonogenic common myeloid progenitor that gives rise to all myeloid lineages. Nature 2000;404(6774):193–7.

[10] Manz MG, Miyamoto T, Akashi K, Weissman IL. Prospective isolation of human clonogenic common myeloid progenitors. Proc Natl Acad Sci USA 2002;99(18):11872–7.

[11] Kondo M, Weissman IL, Akashi K. Identification of clonogenic common lymphoid progenitors in mouse bone marrow. Cell 1997;91(5):661–72.

[12] Na Nakorn T, Traver D, Weissman IL, Akashi K. Myeloerythroid-restricted progenitors are sufficient to confer radioprotection and provide the majority of day 8 CFU-S. J Clin Investig 2002;109(12):1579–85.

[13] Traver D, Akashi K, Weissman IL, Lagasse E. Mice defective in two apoptosis pathways in the myeloid lineage develop acute myeloblastic leukemia. Immunity 1998;9(1):47–57.

[14] Weissman I. Stem cell research: paths to cancer therapies and regenerative medicine. JAMA 2005;294(11):1359–66.

[15] Ishimoto T, Nagano O, Yae T, Tamada M, Motohara T, Oshima H, et al. CD44 variant regulates redox status in Cancer cells by stabilizing the xCT subunit of system xc- and thereby promotes tumor growth. Cancer Cell 2011;19:387–400.

[16] Wu Z, Wei D, Gao W, Xu Y, Hu Z, Ma Z, et al. Tpo-induced metabolic reprogramming drives liver metastasis of colorectal cancer $CD110^{+}$ tumor-initiating cells. Cell Stem Cell 2015;17:47–59.

[17] Shen YA, Wang CY, Hsieh YT, Chen YJ, Wei YH. Metabolic reprogramming orchestrates cancer stem cell properties in nasopharyngeal carcinoma. Cell Cycle 2015;14:86–98. http://www.tandfonline.com/doi/pdf/10.4161/15384101.2014.974419.

[18] Crosstalk between DNA repair and cancer stem cell (CSC) associated intracellular pathways. 2015.

[19] Nowell PC. The clonal evolution of tumor cell populations. Sci (New York, NY) 1976;194:23–8.

[20] Barabé F, Kennedy JA, Hope KJ, Dick JE. Modeling the initiation and progression of human acute leukemia in mice. Sci (New York, NY) 2007;316:600–4.

[21] Metildi CA, Kaushal S, Hoffman RM, Bouvet M. In vivo serial selection of human pancreatic cancer cells in orthotopic mouse models produces high metastatic variants irrespective of Kras status. J Surg Res 2013;184:290–8.

[22] Chaudhary PM, Roninson IB. Induction of multidrug resistance in human cells by transient exposure to different chemotherapeutic drugs. J Natl Cancer Inst 1993;85(8):632–9.

[23] Chaudhary PM, Mechetner EB, Roninson IB. Expression and activity of the multidrug resistance P-glycoprotein in human peripheral blood lymphocytes. Blood 1992;80(11):2735–9.

[24] Allsopp RC, Cheshier S, Weissman IL. Telomerase activation and rejuvenation of telomere length in stimulated T cells derived from serially transplanted hematopoietic stem cells. J Exp Med 2002;196(11):1427–33.

[25] Allsopp RC, Weissman IL. Replicative senescence of hematopoietic stem cells during serial transplantation: does telomere shortening play a role? Oncogene 2002;21(21):3270–3.

[26] Allsopp RC, Cheshier S, Weissman IL. Telomere shortening accompanies increased cell cycle activity during serial transplantation of hematopoietic stem cells. J Exp Med 2001;193(8):917–24.

[27] Diehn M, Cho RW, Lobo NA, Kalisky T, Dorie MJ, Kulp AN, et al. Association of reactive oxygen species levels and radioresistance in cancer stem cells. Nature 2009;458(7239):780–3.

[28] Maugeri-Saccà M, Bartucci M, De Maria R. DNA damage repair pathways in cancer stem cells. Mol Cancer Ther 2012;11:1627–36.

[29] Dean M. ABC transporters, drug resistance, and cancer stem cells. J Mammary Gland Biol Neoplasia 2009;14:3–9.

[30] Creighton CJ, Li X, Landis M, Dixon JM, Neumeister VM, Sjolund A, et al. Residual breast cancers after conventional therapy display mesenchymal as well as tumor-initiating features. Proc Natl Acad Sci USA 2009;106:13820–5.

[31] Meads MB, Gatenby R, Dalton WS. Environment-mediated drug resistance: a major contributor to minimal residual disease. Nat Rev Cancer 2009;9:665–74.

[32] Kurtova AV, Xiao J, Mo Q, Pazhanisamy S, Krasnow R, Lerner SP, et al. Blocking PGE2-induced tumour repopulation abrogates bladder cancer chemoresistance. Nature 2014;517:209–13.

[33] Grosse-Gehling P, Fargeas CA, Dittfeld C, Garbe Y, Alison MR, Corbeil D, et al. CD133 as a biomarker for putative cancer stem cells in solid tumours: limitations, problems and challenges. J Pathol 2013;229:355–78.

[34] Beier D, Hau P, Proescholdt M, Lohmeier A, Wischhusen J, Oefner PJ, et al. CD133^{+} and CD133− glioblastoma-derived cancer stem cells show differential growth characteristics and molecular profiles. Cancer Res 2007;67:4010–5.

[35] Hermann PC, Huber SL, Herrler T, Aicher A, Ellwart JW, Guba M, et al. Distinct populations of cancer stem cells determine tumor growth and metastatic activity in human pancreatic cancer. Cell Stem Cell 2007;1:313–23.

[36] Al-Hajj M, Wicha MS, Benito-Hernandez A, Morrison SJ, Clarke MF. Prospective identification of tumorigenic breast cancer cells. Proc Natl Acad Sci USA 2003;100(7):3983–8.

[37] Majeti R, Chao MP, Alizadeh AA, Pang WW, Jaiswal S, Gibbs KD, et al. CD47 is an adverse prognostic factor and therapeutic antibody target on human acute myeloid leukemia stem cells. Cell 2009;138:286–99.

[38] Van Rhenen A, Van Dongen GAMS, Kelder AL, Rombouts EJ, Feller N, Moshaver B, et al. The novel AML stem cell-associated antigen CLL-1 aids in discrimination between normal and leukemic stem cells. Blood 2007;110:2659–66.

[39] Hosen N, Park CY, Tatsumi N, Oji Y, Sugiyama H, Gramatzki M, et al. CD96 is a leukemic stem cell-specific marker in human acute myeloid leukemia. Proc Natl Acad Sci USA 2007;104:11008–13.

[40] Jan M, Chao MP, Cha AC, Alizadeh AA, Gentles AJ, Weissman IL, et al. Prospective separation of normal and leukemic stem cells based on differential expression of TIM3, a human acute myeloid leukemia stem cell marker. Proc Natl Acad Sci USA 2011;108:5009–14.

[41] Saito Y, Kitamura H, Hijikata A, Tomizawa-Murasawa M, Tanaka S, Takagi S, et al. Identification of therapeutic targets for quiescent, chemotherapy-resistant human leukemia stem cells. Sci Transl Med 2010;2:17ra9.

[42] Han J, Fujisawa T, Husain SR, Puri RK. Identification and characterization of cancer stem cells in human head and neck squamous cell carcinoma. BMC Cancer 2014;14:173.

[43] Prince ME, Sivanandan R, Kaczorowski A, Wolf GT, Kaplan MJ, Dalerba P, et al. Identification of a subpopulation of cells with cancer stem cell properties in head and neck squamous cell carcinoma. Proc Natl Acad Sci USA 2007;104:973–8.

[44] Wei XD, Zhou L, Cheng L, Tian J, Jiang JJ, MacCallum J. In vivo investigation of CD133 as a putative marker of cancer stem cells in hep-2 cell line. Head Neck 2009;31:94–101.

[45] Lathia JD, Gallagher J, Heddleston JM, Wang J, Eyler CE, Macswords J, et al. Integrin alpha 6 regulates glioblastoma stem cells. Cell Stem Cell 2010;6(5):421–32.

[46] Lathia JD, Li M, Sinyuk M, Alvarado AG, Flavahan WA, Stoltz K, et al. High-throughput flow cytometry screening reveals a role for junctional adhesion molecule a as a cancer stem cell maintenance factor. Cell Rep 2014;6(1):117–29.

[47] Singh SK, Clarke ID, Terasaki M, Bonn VE, Hawkins C, Squire J, et al. Identification of a cancer stem cell in human brain tumors. Cancer Res 2003;63:5821–8.

[48] Ahmed N, Salsman VS, Kew Y, Shaffer D, Powell S, Zhang YJ, et al. HER2-specific T cells target primary glioblastoma stem cells and induce regression of autologous experimental tumors. Clin Cancer Res 2010;16:474–85.

[49] Emlet DR, Gupta P, Holgado-Madruga M, Del Vecchio CA, Mitra SS, Han SY, et al. Targeting a glioblastoma cancer stem-cell population defined by EGF receptor variant III. Cancer Res 2014;74:1238–49.

[50] Leung EL-H, Fiscus RR, Tung JW, Tin VP-C, Cheng LC, Sihoe AD-L, et al. Non-small cell lung cancer cells expressing CD44 are enriched for stem cell-like properties. PLoS One 2010;5:e14062.

[51] Eramo A, Lotti F, Sette G, Pilozzi E, Biffoni M, Di Virgilio A, et al. Identification and expansion of the tumorigenic lung cancer stem cell population. Cell Death Differ 2008;15:504–14.

[52] Ginestier C, Hur MH, Charafe-Jauffret E, Monville F, Dutcher J, Brown M, et al. ALDH1 is a marker of normal and malignant human mammary stem cells and a predictor of poor clinical outcome. Cell Stem Cell 2007;1(5):555–67.

[53] Meyer MJ, Fleming JM, Lin AF, Hussnain SA, Ginsburg E, Vonderhaar BK. CD44posCD49f-hiCD133/2hi defines xenograft-initiating cells in estrogen receptor-negative breast cancer. Cancer Res 2010;70(11):4624–33.

[54] Vaillant F, Asselin-Labat ML, Shackleton M, Forrest NC, Lindeman GJ, Visvader JE. The mammary progenitor marker CD61/beta3 integrin identifies cancer stem cells in mouse models of mammary tumorigenesis. Cancer Res 2008;68(19):7711–7.

[55] Zhang S, Balch C, Chan MW, Lai H-C, Matei D, Schilder JM, et al. Identification and characterization of ovarian cancer-initiating cells from primary human tumors. Cancer Res 2008;68:4311–20.

[56] Li L, Hao X, Qin J, Tang W, He F, Smith A, et al. Antibody against CD44s inhibits pancreatic tumor initiation and postradiation recurrence in mice. Gastroenterology 2014;146:1108–18.

[57] Li C, Heidt DG, Dalerba P, Burant CF, Zhang L, Adsay V, et al. Identification of pancreatic cancer stem cells. Cancer Res 2007;67:1030–7.

[58] Takaishi S, Okumura T, Tu S, Wang SSW, Shibata W, Vigneshwaran R, et al. Identification of gastric cancer stem cells using the cell surface marker CD44. Stem Cells (Dayton, Ohio) 2009;27:1006–20.

[59] Jiang Y, He Y, Li H, Li HN, Zhang L, Hu W, et al. Expressions of putative cancer stem cell markers ABCB1, ABCG2, and CD133 are correlated with the degree of differentiation of gastric cancer. Gastric Cancer 2012;15:440–50.

[60] Fanali C, Lucchetti D, Farina M, Corbi M, Cufino V, Cittadini A, et al. Cancer stem cells in colorectal cancer from pathogenesis to therapy: controversies and perspectives. World J Gastroenterol 2014;20:923–42.

[61] Ricci-Vitiani L, Lombardi DG, Pilozzi E, Biffoni M, Todaro M, Peschle C, et al. Identification and expansion of human colon-cancer-initiating cells. Nature 2007;445:111–5.

[62] Kemper K, Grandela C, Medema JP. Molecular identification and targeting of colorectal cancer stem cells. Oncotarget 2010;1:387–95.

[63] Choi D, Lee HW, Hur KY, Kim JJ, Park GS, Jang SH, et al. Cancer stem cell markers CD133 and CD24 correlate with invasiveness and differentiation in colorectal adenocarinoma. World J Gastroenterol 2009;15:2258–64.

[64] Patrawala L, Calhoun T, Schneider-Broussard R, Li H, Bhatia B, Tang S, et al. Highly purified $CD44^+$ prostate cancer cells from xenograft human tumors are enriched in tumorigenic and metastatic progenitor cells. Oncogene 2006;25:1696–708.

[65] Dubrovska A, Kim S, Salamone RJ, Walker JR, Maira S-M, García-Echeverría C, et al. The role of PTEN/Akt/PI3K signaling in the maintenance and viability of prostate cancer stem-like cell populations. Proc Natl Acad Sci USA 2009;106:268–73.

[66] Domingo-Domenech J, Vidal SJ, Rodriguez-Bravo V, Castillo-Martin M, Quinn SA, Rodriguez-Barrueco R, et al. Suppression of acquired docetaxel resistance in prostate cancer through depletion of notch- and hedgehog-dependent tumor-initiating cells. Cancer Cell 2012;22:373–88.

[67] Chan KS, Espinosa I, Chao M, Wong D, Ailles L, Diehn M, et al. Identification, molecular characterization, clinical prognosis, and therapeutic targeting of human bladder tumor-initiating cells. Proc Natl Acad Sci USA 2009;106:14016–21.

[68] Chan KS, Volkmer JP, Weissman I. Cancer stem cells in bladder cancer: a revisited and evolving concept. Curr Opin Urol 2010;20(5):393–7.

[69] Volkmer JP, Sahoo D, Chin RK, Ho PL, Tang C, Kurtova AV, et al. Three differentiation states risk-stratify bladder cancer into distinct subtypes. Proc Natl Acad Sci USA 2012;109(6):2078–83.

[70] Fang D, Nguyen TK, Leishear K, Finko R, Kulp AN, Hotz S, et al. A tumorigenic subpopulation with stem cell properties in melanomas. Cancer Res 2005;65(20):9328–37.

[71] Boiko AD, Razorenova OV, van de Rijn M, Swetter SM, Johnson DL, Ly DP, et al. Human melanoma-initiating cells express neural crest nerve growth factor receptor CD271. Nature 2010;466(7302):133–7.

[72] Schatton T, Murphy GF, Frank NY, Yamaura K, Waaga-Gasser AM, Gasser M, et al. Identification of cells initiating human melanomas. Nature 2008;451:345–9.

[73] Hu Y, Smyth GK. ELDA: extreme limiting dilution analysis for comparing depleted and enriched populations in stem cell and other assays. J Immunol Methods 2009;347: 70–8.

[74] Breivik H. Haematopoietic stem cell content of murine bone marrow, spleen, and blood. Limiting dilution analysis of diffusion chamber cultures. J Cell Physiol 1971;78:73–8.

[75] Bellows CG, Aubin JE. Determination of numbers of osteoprogenitors present in isolated fetal rat calvaria cells in vitro. Dev Biol 1989;133:8–13.

[76] Tropepe V, Sibilia M, Ciruna BG, Rossant J, Wagner EF, van der Kooy D. Distinct neural stem cells proliferate in response to EGF and FGF in the developing mouse telencephalon. Dev Biol 1999;208:166–88.

[77] Zimmerer RM, Korn P, Demougin P, Kampmann A, Kokemüller H, Eckardt AM, et al. Functional features of cancer stem cells in melanoma cell lines. Cancer Cell Int 2013;13:78.

[78] Quintana E, Shackleton M, Sabel MS, Fullen DR, Johnson TM, Morrison SJ. Efficient tumour formation by single human melanoma cells. Nature 2008;456:593–8.

[79] Sampson JH, Heimberger AB, Archer GE, Aldape KD, Friedman AH, Friedman HS, et al. Immunologic escape after prolonged progression-free survival with epidermal growth factor receptor variant III peptide vaccination in patients with newly diagnosed glioblastoma. J Clin Oncol Off J Am Soc Clin Oncol 2010:1–9.

[80] Del Vecchio CA, Wong AJ. Rindopepimut, a 14-mer injectable peptide vaccine against EGFRvIII for the potential treatment of glioblastoma multiforme. Curr Opin Mol Ther 2010;12:741–54.

[81] Pandita A, Aldape KD, Zadeh G, Guha A, James CD. Contrasting in vivo and in vitro fates of glioblastoma cell subpopulations with amplified EGFR. Genes Chromosomes Cancer 2004;39:29–36.

[82] Saikali S, Avril T, Collet B, Hamlat A, Bansard JY, Drenou B, et al. Expression of nine tumour antigens in a series of human glioblastoma multiforme: interest of EGFRvIII, IL-13Rα2, gp100 and TRP-2 for immunotherapy. J Neuro Oncol 2007;81:139–48.

[83] Gutierrez A, Sanda T, Grebliunaite R, Carracedo A, Salmena L, Ahn Y, et al. High frequency of PTEN, PI3K, and AKT abnormalities in T-cell acute lymphoblastic leukemia. Blood 2009;114:647–50.

[84] Merchant AA, Matsui W. Targeting Hedgehog–a cancer stem cell pathway. Clin Cancer Res Off J Am Assoc Cancer Res 2010;16:3130–40.

[85] Reya T, Clevers H. Wnt signalling in stem cells and cancer. Nature 2005;434:843–50.

[86] Chaturvedi MM, Sung B, Yadav VR, Kannappan R, Aggarwal BB. NF-kappaB addiction and its role in cancer: one size does not fit all. Oncogene 2011;30:1615–30.

[87] Dontu G, Jackson KW, McNicholas E, Kawamura MJ, Abdallah WM, Wicha MS. Role of notch signaling in cell-fate determination of human mammary stem/progenitor cells. Breast Cancer Res BCR 2004;6:R605–15.

[88] Harrison H, Farnie G, Howell SJ, Rock RE, Stylianou S, Brennan KR, et al. Regulation of breast cancer stem cell activity by signaling through the Notch4 receptor. Cancer Res 2010;70:709–18.

[89] Zeng YA, Nusse R. Wnt proteins are self-renewal factors for mammary stem cells and promote their long-term expansion in culture. Cell Stem Cell 2010;6:568–77.

[90] Reya T, Duncan AW, Ailles L, Domen J, Scherer DC, Willert K, et al. A role for Wnt signalling in self-renewal of haematopoietic stem cells. Nature 2003;423:409–14.

[91] Chaffer CL, Weinberg R. A perspective on cancer cell metastasis. Sci (New York, NY) 2011;331:1559–64.

[92] Roy SK, Srivastava RK, Shankar S. Inhibition of PI3K/AKT and MAPK/ERK pathways causes activation of FOXO transcription factor, leading to cell cycle arrest and apoptosis in pancreatic cancer. J Mol Signal 2010;5:10.

[93] Shultz LD, Lyons BL, Burzenski LM, Gott B, Chen X, Chaleff S, et al. Human lymphoid and myeloid cell development in NOD/LtSz-scid IL2R gamma null mice engrafted with mobilized human hemopoietic stem cells. J Immunol 2005;174:6477–89.

[94] Ito M, Hiramatsu H, Kobayashi K, Suzue K, Kawahata M, Hioki K, et al. NOD/SCID/γc null mouse: an excellent recipient mouse model for engraftment of human cells. Blood 2002;100:3175–82.

[95] Abrahamsson AE, Geron I, Gotlib J, Dao KH, Barroga CF, Newton IG, et al. Glycogen synthase kinase 3beta missplicing contributes to leukemia stem cell generation. Proc Natl Acad Sci USA 2009;106(10):3925–9.

[96] O'Brien CA, Pollett A, Gallinger S, Dick JE. A human colon cancer cell capable of initiating tumour growth in immunodeficient mice. Nature 2007;445:106–10.

[97] Colmont CS, Benketah A, Reed SH, Hawk NV, Telford WG, Ohyama M, et al. CD200-expressing human basal cell carcinoma cells initiate tumor growth. Proc Natl Acad Sci USA 2013;110:1434–9.

[98] Fridman R, Benton G, Aranoutova I, Kleinman HK, Bonfil RD. Increased initiation and growth of tumor cell lines, cancer stem cells and biopsy material in mice using basement membrane matrix protein (Cultrex or Matrigel) co-injection. Nat Protoc 2012;7:1138–44.

[99] Alonso-Camino V, Santos-Valle P, Ispizua MC, Sanz L, Alvarez-Vallina L. Engineered human tumor xenografts with functional human vascular networks. Microvasc Res 2011;81:18–25.

[100] Kim S, Park Y-W, Schiff BA, Doan DD, Yazici Y, Jasser S, et al. An orthotopic model of anaplastic thyroid carcinoma in athymic nude mice. Clin Cancer Res Off J Am Assoc Cancer Res 2005;11:1713–21.

[101] Talmadge JE, Singh RK, Fidler IJ, Raz A. Murine models to evaluate novel and conventional therapeutic strategies for cancer. Am J Pathol 2007;170:793–804.

[102] Ruggeri BA, Camp F, Miknyoczki S. Animal models of disease: pre-clinical animal models of cancer and their applications and utility in drug discovery. Biochem Pharmacol 2014;87:150–61.

[103] Lee J, Li M, Milwid J, Dunham J, Vinegoni C, Gorbatov R, et al. Implantable microenvironments to attract hematopoietic stem/cancer cells. Proc Natl Acad Sci USA 2012;109:19638–43.

[104] Day C-P, Merlino G, Van Dyke T. Preclinical mouse Cancer models: a maze of opportunities and challenges. Cell 2016;163:39–53.

[105] Deng C-X. Conditional knockout mouse models of cancer. Cold Spring Harb Protoc 2014;2014. pdb.top074393.

[106] Talking about a revolution: the impact of site-specific recombinases on genetic analyses in mice. 2004.

[107] Branda CS, Dymecki SM. Talking about a revolution: the impact of site-specific recombinases on genetic analyses in mice. Dev Cell 2004;6(1):7–28.

[108] Jackson EL, Willis N, Mercer K, Bronson RT, Crowley D, Montoya R, et al. Analysis of lung tumor initiation and progression using conditional expression of oncogenic K-ras. Genes Dev 2001;15:3243–8.

[109] Read T-A, Fogarty MP, Markant SL, McLendon RE, Wei Z, Ellison DW, et al. Identification of CD15 as a marker for tumor-propagating cells in a mouse model of medulloblastoma. Cancer Cell 2009;15:135–47.

[110] Sánchez-Rivera FJ, Jacks T. Applications of the CRISPR-Cas9 system in cancer biology. Nat Rev Cancer 2015;15:387–95.

[111] Heyer J, Kwong LN, Lowe SW, Chin L. Non-germline genetically engineered mouse models for translational cancer research. Nat Rev Cancer 2010;10:470–80.

[112] Zhou Y, Rideout WM, Zi T, Bressel A, Reddypalli S, Rancourt R, et al. Chimeric mouse tumor models reveal differences in pathway activation between ERBB family- and KRAS-dependent lung adenocarcinomas. Nat Biotechnol 2010;28:71–8.

[113] Alcantara Llaguno SR, Chen J, Parada LF. Signaling in malignant astrocytomas: role of neural stem cells and its therapeutic implications. Clin Cancer Res Off J Am Assoc Cancer Res 2009;15:7124–9.

[114] Kruger JA, Kaplan CD, Luo Y, Zhou H, Markowitz D, Xiang R, et al. Characterization of stem cell-like cancer cells in immune-competent mice. Blood 2006;108:3906–12.

[115] Rongvaux A, Willinger T, Martinek J, Strowig T, Gearty SV, Teichmann LL, et al. Development and function of human innate immune cells in a humanized mouse model. Nat Biotechnol 2014;32:364–72.

[116] Feldmann G, Habbe N, Dhara S, Bisht S, Alvarez H, Fendrich V, et al. Hedgehog inhibition prolongs survival in a genetically engineered mouse model of pancreatic cancer. Gut 2008;57:1420–30.

[117] Bar EE, Chaudhry A, Lin A, Fan X, Schreck K, Matsui W, et al. Cyclopamine-mediated hedgehog pathway inhibition depletes stem-like cancer cells in glioblastoma. Stem Cells 2007;25:2524–33.

[118] Vermeulen L, De Sousa EMF, van der Heijden M, Cameron K, de Jong JH, Borovski T, et al. Wnt activity defines colon cancer stem cells and is regulated by the microenvironment. Nat Cell Biol 2010;12:468–76.

[119] McGowan PM, Simedrea C, Ribot EJ, Foster PJ, Palmieri D, Steeg PS, et al. Notch1 inhibition alters the CD44hi/CD24lo population and reduces the formation of brain metastases from breast cancer. Mol Cancer Res MCR 2011;9:834–44.

[120] The CDK inhibitors in cancer research and therapy. 2011.

[121] Welcker M, Clurman BE. FBW7 ubiquitin ligase: a tumour suppressor at the crossroads of cell division, growth and differentiation. Nat Rev Cancer 2008;8:83–93.

[122] Yada M, Hatakeyama S, Kamura T, Nishiyama M, Tsunematsu R, Imaki H, et al. Phosphorylation-dependent degradation of c-Myc is mediated by the F-box protein Fbw7. EMBO J 2004;23:2116–25.

[123] Takeishi S, Matsumoto A, Onoyama I, Naka K, Hirao A, Nakayama KI. Ablation of Fbxw7 eliminates leukemia-initiating cells by preventing quiescence. Cancer Cell 2013;23:347–61.

[124] Sulfasalazine: a review of its use in the management of rheumatoid arthritis. 2005.

[125] Harris IS, Treloar AE, Inoue S, Sasaki M, Gorrini C, Lee K, et al. Glutathione and thioredoxin antioxidant pathways synergize to drive cancer initiation and progression. Cancer Cell 2015;27:211–22.

[126] Yoshikawa M, Tsuchihashi K, Ishimoto T, Yae T, Motohara T, Sugihara E, et al. XCT inhibition depletes CD44v-expressing tumor cells that are resistant to EGFR-targeted therapy in head and neck squamous cell carcinoma. Cancer Res 2013;73:1855–66.

[127] Immunotherapy: a useful strategy to help combat multidrug resistance. 2012.

[128] Jin L, Hope KJ, Zhai Q, Smadja-Joffe F, Dick JE. Targeting of CD44 eradicates human acute myeloid leukemic stem cells. Nat Med 2006;12:1167–74.

[129] Jin L, Lee EM, Ramshaw HS, Busfield SJ, Peoppl AG, Wilkinson L, et al. Monoclonal antibody-mediated targeting of CD123, IL-3 receptor alpha chain, eliminates human acute myeloid leukemic stem cells. Cell Stem Cell 2009;5:31–42.

[130] Morrison SJ, Spradling AC. Stem cells and niches: mechanisms that promote stem cell maintenance throughout life. Cell 2008;132(4):598–611.

[131] Calvi LM, Adams GB, Weibrecht KW, Weber JM, Olson DP, Knight MC, et al. Osteoblastic cells regulate the haematopoietic stem cell niche. Nature 2003;425(6960):841–6.

[132] Arai F, Hirao A, Ohmura M, Sato H, Matsuoka S, Takubo K, et al. Tie2/angiopoietin-1 signaling regulates hematopoietic stem cell quiescence in the bone marrow niche. Cell 2004;118(2):149–61.

[133] Ishikawa F, Yoshida S, Saito Y, Hijikata A, Kitamura H, Tanaka S, et al. Chemotherapy-resistant human AML stem cells home to and engraft within the bone-marrow endosteal region. Nat Biotechnol 2007;25(11):1315–21.

[134] Colmone A, Amorim M, Pontier AL, Wang S, Jablonski E, Sipkins DA. Leukemic cells create bone marrow niches that disrupt the behavior of normal hematopoietic progenitor cells. Science 2008;322(5909):1861–5.

[135] Scheres B. Stem-cell niches: nursery rhymes across kingdoms. Nat Rev Mol Cell Biol 2007;8(5):345–54.

[136] Scherer HJ. A critical review: the pathology of cerebral gliomas. J Neurol Psychiatry 1940;3(2):147–77.

[137] Calabrese C, Poppleton H, Kocak M, Hogg TL, Fuller C, Hamner B, et al. A perivascular niche for brain tumor stem cells. Cancer Cell 2007;11(1):69–82.

[138] Burkhardt JK, Hofstetter CP, Santillan A, Shin BJ, Foley CP, Ballon DJ, et al. Orthotopic glioblastoma stem-like cell xenograft model in mice to evaluate intra-arterial delivery of bevacizumab: from bedside to bench. J Clin Neurosci 2012;19:1568–72.

[139] Folkins C, Man S, Xu P, Shaked Y, Hicklin DJ, Kerbel RS. Anticancer therapies combining antiangiogenic and tumor cell cytotoxic effects reduce the tumor stem-like cell fraction in glioma xenograft tumors. Cancer Res 2007;67:3560–4.

[140] Ho MM, Ng AV, Lam S, Hung JY. Side population in human lung cancer cell lines and tumors is enriched with stem-like cancer cells. Cancer Res 2007;67:4827–33.

[141] Matsui W, Wang Q, Barber JP, Brennan S, Smith BD, Borrello I, et al. Clonogenic multiple myeloma progenitors, stem cell properties, and drug resistance. Cancer Res 2008;68:190–7.

[142] Deeley RG, Westlake C, Cole SPC. Transmembrane transport of endo- and xenobiotics by mammalian ATP-binding cassette multidrug resistance proteins. Physiol Rev 2006;86:849–99.

[143] Chen G, Liu J, Chen W, Xu Q, Xiao M, Hu L, et al. A 20(S)-protopanoxadiol derivative overcomes multi-drug resistance by antagonizing ATP-binding cassette subfamily B member 1 transporter function. Oncotarget 2016;7(8).

[144] Prieto PA, Yang JC, Sherry RM, Hughes MS, Kammula US, White DE, et al. CTLA-4 blockade with ipilimumab: long-term follow-up of 177 patients with metastatic melanoma. Clin Cancer Res 2012;18:2039–47.

[145] Snyder A, Makarov V, Merghoub T, Yuan J, Zaretsky JM, Desrichard A, et al. Genetic basis for clinical response to CTLA-4 blockade in melanoma. N Engl J Med 2014:2189–99.

[146] Boisgerault N, Kottke T, Pulido J, Thompson J, Diaz RM, Rommelfanger-Konkol D, et al. Functional cloning of recurrence-specific antigens identifies molecular targets to treat tumor relapse. Mol Ther J Am Soc Gene Ther 2013;21:1507–16.

CHAPTER 2

Overview: Cancer Stem Cell Self-Renewal

M. Zabala[1], N.A. Lobo[1,2], D. Qian[1], L.J. van Weele[1], D. Heiser[1], M.F. Clarke[1]
[1]*Stanford University, Stanford, CA, United States;* [2]*Celgene Quanticel Research, San Francisco, CA, United States*

CONTENTS

DEFINITION OF SELF-RENEWAL IN CANCER STEM CELLS

Long-lived organisms must properly maintain their organs and tissues to replace damaged or aging cells in a tightly controlled and balanced way [1]. In adult tissues, the stem cell population is crucial to maintaining tissue integrity because this is the only cell type endowed with the capacity to self-renew and differentiate into heterogeneous, more specialized mature cells that constitute the full cellular repertoire of the adult organ [1,2]. Self-renewal is the key biological process that ultimately defines a stem cell, where upon a specialized mitotic cell division, a stem cell produces one (asymmetric) or two (symmetric) daughter cells that retain the ability to self-renew, ensuring that the stem cell population is maintained and expanded for long-term clonal growth [3]. As a consequence of asymmetric division, or symmetric differentiation cell division, the stem cells create progenitor cells that differentiate as they divide, further specializing into terminally differentiated cells that eventually become senescent or undergo apoptosis. In addition, the balance between self-renewal and differentiation or so-called homeostatic control is critical for proper maintenance of the tissue. Due to all of these special qualities that are necessary to function as a stem cell, it is not surprising that stem cells are a rare population of cells. Furthermore, it has been suggested that stem cells rarely divide, meaning most are in a state of dormancy, which may play an important role in maintaining the long-term pool of stem cells [4]. In tissues such as the blood, skin, and small intestine, where stem cell populations have been defined, more than one pool of stem cells can coexist: one population of stem cells may be more important to the homeostasis of the regenerative tissue, whereas the other, more slowly cycling or even dormant population, may serve as a reserve pool that activates in situations of injury or stress.

Cancer Stem Cells. http://dx.doi.org/10.1016/B978-0-12-803892-5.00002-4

Cancer is widely understood as a heterogeneous disease composed of diverse populations of cells that differ phenotypically. This was first described by pathology and later by function, when it was found that not all myeloma cells have the potential to form tumors in vitro or in vivo [5]. One interpretation of these findings is explained by the stochastic theory (Fig. 2.1), which hypothesizes that although all cancer cells are biologically identical, they show random behavior in their potential to form tumors due to stochastic differences in intrinsic and extrinsic cues [1]. The hierarchical model (Fig. 2.1), which was first considered as an alternative to the stochastic model, postulates that a fraction of cancer cells—the cancer stem cells (CSCs)—are responsible for growth and maintenance of the tumor. The CSC theory proposes that tumors have a cellular hierarchy that is a caricature of their normal tissue counterpart. As such, the CSCs are the only cells in the tumor that possess the ability to self-renew, which enables them to initiate and propagate long-term clonal growth

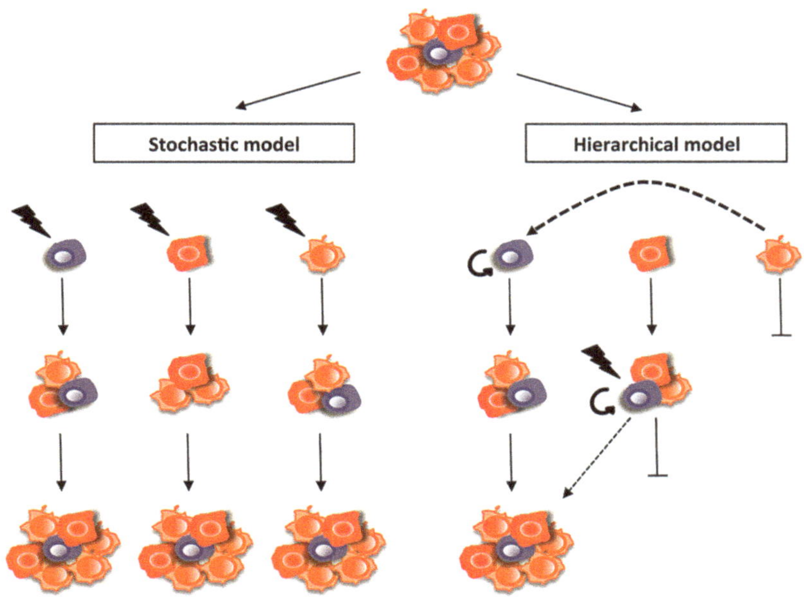

FIGURE 2.1 **Different models of tumorigenesis.**
Tumors are formed by heterogeneous subpopulations of cells. In stochastic models, all the cells that form the tumor have the same potential to become tumorigenic due to external cues or insults (indicated by the lightning bolts). In the hierarchical model, only a fraction of tumor cells, the cancer stem cells (CSCs), are able to form the tumor and perpetually maintain its growth. This is because the CSCs are the only cell type endowed with self-renewal (indicated by *curved arrows*) potential. Also, progenitor-like cancer cells can acquire self-renewal potential due to aberrant cues. In recent years, a modified CSC model called the "dedifferentiation" model has been proposed. In this model, mature cells can dedifferentiate or regenerate stem cells (indicated by *dotted-curved arrow*).

of tumors. In comparison to somatic stem cells, the processes of self-renewal and differentiation are poorly regulated in their malignant counterparts.

Recently, a modified CSC model, called the "plasticity" or "dedifferentiation" model, has been proposed. In this model, the mature cells can regenerate stem cells (Fig. 2.1). We favor the deterministic model because, in our hands, "dedifferentiation" may be explained by two stem cell populations transitioning back and forth between phenotypic stem cell states (quiescent, epithelial stem cells transitioning to proliferating mesenchymal cells or vice versa). Note that the "plasticity" models are based on the assumption that epithelial-to-mesenchymal transition (EMT) defines stem cells and therefore stem cells are defined as EMT cells, and not based on functional assays. Thus far, it has not been proven that all stem cells are undergoing EMT. There remains the possibility that there are distinct, nonmesenchymal CSCs as well as mesenchymal stem cells, which can both be prospectively isolated and distinguished from terminally differentiated cells. However, one can never completely rule out dedifferentiation because there are just too many possible combinations of cell signals to test in laboratory assays. Regardless of which model is correct, it is clear that the stem cell state and expression of stem cell genes is obligated for the maintenance of a tumor. It follows that targeting stem cell pathways should have therapeutic benefit regardless of which model is correct.

It has been suggested incorrectly that CSCs are so named because they originate from their normal counterparts. Although normal stem cells can acquire genetic and epigenetic mutations that lead to their transformation into CSCs, it has also been shown that progenitor cells can aberrantly upregulate self-renewal pathways, revealing them as the responsible population behind maintenance of tumor growth. In the progression of chronic myelogenous leukemia (CML) to blast crisis, the leukemic stem cells (LSCs) arise from the granulocyte-macrophage progenitor cells that have aberrantly activated the β-catenin pathway [6]. In *BRCA1*-associated breast basal tumors, it has been suggested that the cells of origin are aberrant luminal progenitor cells [7]. In mouse glioma models, it has been shown that oligodendrocyte progenitors with defective transmembrane proteoglycan NG2 asymmetry aberrantly self-renew, fail to differentiate and turn into glioma-initiating cells [8].

CSCs are defined by their ability to self-renew and differentiate, not by their phenotype (Fig. 2.2). In fact, at the present time, none of the sorting strategies for stem cells isolates a pure population of cells; it is always a mixture of stem cells and more differentiated cells. Different avenues have been used to test the identity of CSCs, from the prospective isolation of different populations of cancer cells based on differential expression of cell surface markers to lineage-tracing the CSCs that express specific genes/reporters, or a mixture of both. In solid tumors, it is necessary to create single-cell suspensions using

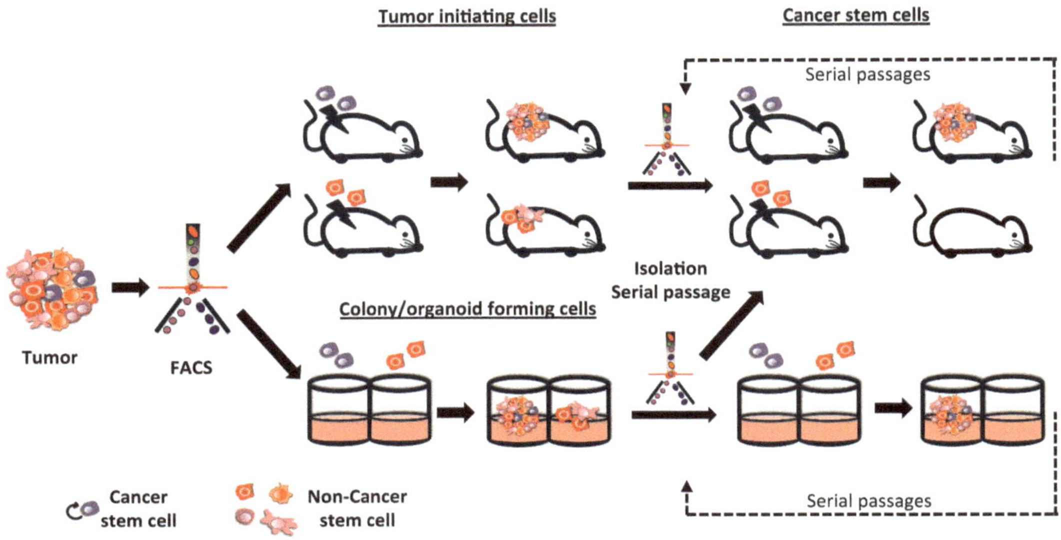

FIGURE 2.2 **Characterization of cancer stem cells (CSCs).**
Tumors are formed by heterogeneous subpopulations of cells. Using cell surface markers, different subpopulations can be isolated by flow cytometry. The subpopulations are then isolated and injected into mice and/or plated in vitro, to assess their tumor or colony formation potential. The newly arising tumors/colonies can be tested for their capacity to form new tumors/colonies after serial passages in vitro and in vivo, and the composition of the tumors/colonies can be studied. Only the CSCs will be able to form a tumor with the same heterogeneous repertoire of cancer cells and perpetually propagate tumor growth.

partial enzymatic digestion. The single-cell suspension is stained with a cocktail of antibodies, which will be used to distinguish between cells that compose the tumor and cells that support tumor growth. Using fluorescence-activated cell sorting, different populations are separated and then injected into immunocompromised or syngeneic mice depending on the origin of the tumor. Limiting dilution injection strategies are recommended to compare the frequency of engrafting cells within the different populations (Fig. 2.2). These experiments reveal the frequency of tumor-initiating cells (TICs) in each population, but the self-renewal capability and differentiation potential of the candidate CSCs still need to be tested. To do so, the resulting xenograft tumors are processed to analyze whether the complete repertoire of tumor cells are present (confirming CSC differentiation). This whole process must be repeated again to determine whether the TICs are able to perpetually proliferate and maintain tumor growth on serial passaging (confirming self-renewal) (Fig. 2.2). Only after these questions have been answered can the TICs be called CSCs. These laborious experiments come with some caveats: the identity and frequency of self-renewing cancer cells can be obscured by technical problems with the cocktail of antibodies, the harshness of the methods used to process and isolate the cells, or the model and media used to inject the cells. The frequency of TICs improved when more immunocompromised mice were used as recipients, as well as when different types of Matrigel were utilized for injections [9]. In mouse models of cancer, in vivo imaging of tumors [10] and lineage tracing experiments in mice have been used to further validate the CSCs model. Using lineage tracing mouse models to study squamous skin tumors, a recent study identified the clonal growth of different tumors populations, some of which resemble those of skin stem cells [11]. In mouse intestinal adenomas, another study found that expression of Lrg5 also marks a subpopulation of adenoma cells that fuel the growth of established intestinal adenoma [12]. In glioblastoma (GBM) multiforme, using transgenic mice where the suicide gene thymidine kinase was expressed under Nestin, a neural stem cell–specific promoter, the authors elegantly demonstrated that a subset of endogenous tumor cells that are the source of new tumor cells after the drug temozolomide is administered to transiently arrest tumor growth [13].

Another approach to studying self-renewal uses the in vitro culture of CSCs (Fig. 2.1). Several media formulations have been developed to maintain the stemness of CSCs isolated from different tumor types. Some advantages that these assays include a simpler setup, faster results, and easier readouts [2]. They allow tracking of single cells and as they are easier to manipulate, they can be scaled up for high-throughput assays. Usually, these media formulations consist of a cocktail of growth factors and secreted proteins that are able to maintain the pool of self-renewing cancer cells while also allowing differentiation patterns to mimic tumor growth in vivo. These assays interrogate the ability of the cancer cells to form colonies. However, their capacity for differentiation and long-term proliferation potential should be tested by performing serial passages in vitro and, ideally, in vivo.

ASSAYS TO MEASURE SELF-RENEWAL POTENTIAL OF CANCER STEM CELLS

Immune Deficient Mouse Strains

During the last decade, converging of concepts in both stem cell biology and oncology has led to vigorous research effort in the field of designing adequate models to assay CSCs. Validation of the capacity of long-term self-renewal and differentiation in animal models of cancer is essential. In vivo xenograftment modeling consists of adoptive transplantation of a specific subset of cells of a human cancer into a heterogeneic host, immunodeficient mice, the predominant in vivo animal recipient, aiming to definitively confirm the self-renewing and differentiating capacity of prospective CSCs. This model demonstrated the first experimental evidence of a functional hierarchy of stem cell system in a solid tumor, breast epithelial carcinoma [14]. Since then, this model has been used extensively for many other human solid tumors [15–21].

Immunodeficient mice, as recipients, are crucial to the success in growing patient-derived xenografts or PDXs (Table 2.1). In alignment with that, it has been observed that the frequency of engrafting cells and CSCs can be directly altered by the immunodeficient mice used in the study [9]. Some of the reasons encompass residual immune system cells from the host as well as differences in the tumor microenvironment (TME) of mouse to human [22]. Even though, it is relevant to name the advances done in the different immunodeficient mice. The increasing successful application of immunodeficient mice for in vivo mouse xenograftment is paralleled with improvements in the immunodeficient murine hosts in four phases [23]. The model used may directly affect the engraftment, self-renewal, and differentiation potential of CSCs.

Athymic *Foxn1*^{nu} (nude) mice. The pilot use of immunodeficient mice in the study of human cancer was an early model for engraftment of human cell lines derived from solid tumors nearly 40 years ago [24], following the discovery of congenitally athymic nude mice in the 1960s. The nude mice allows limited engraftment with most primary solid human tumors, but prevents engraftment of human normal or malignant hematopoietic cells, due to the presence of both an intact humoral adaptive immune system and an intact innate immune system, including high natural killer (NK) cell activity, even if the nude mutation impairs the development of functional T cells (Table 2.1).

C.B17- *Prkdc*^{scid} (*scid*) Mice. The next advance in immunodeficient modeling was the discovery of a spontaneous mutation in C.B17 mice, named "*scid*" (*Prkdc*^{scid}, protein kinase DNA-activated catalytic polypeptide) [25]. C.B17-*scid* mice, as compared with nude mice, permit the engraftment of human solid tumors. Moreover, it is the first immunodeficient mouse model that can be engrafted with human hematopoietic cells. Although the *scid* mutation prevents the development of mature T and B lymphocytes of the adaptive immune system, C.B17-*scid* mouse maintains an intact innate immune system, including moderate NK cell activity. In addition, the toxicity of NK cells for human hematopoietic stem cells (HSCs) and TICs [26], and the possible development of mature T and B cells upon aging [27] further restricts C.B17-*scid* mouse for the study of tumor growth.

NOD.C.B17-*Prkdc*^{scid} (NOD-*scid*) Mice. NOD-*scid* strain bears NOD/Lt background introduced to C.B-17-*scid* mice by backcrossing [28]. The NOD background features intrinsic defects in innate immunity, including lowered activity of NK cell and macrophage, abnormal dendritic cell, hemolytic complement. Combined with the loss of adaptive immunity from the *Prkdc*^{scid}, *Rag1*^{null}, or *Rag2*^{null} mutations, NOD-*scid* mice are more receptive for engraftment of human primary solid tumors and hematopoietic malignant cells, as compared to C.B17-*scid* strain. However, because of remaining NK cell activity and other residual innate immune function, many cancers and hematological malignancies still fail to engraft efficiently and grow in NOD-*scid* mice.

Table 2.1 Immune Deficient Mouse Strains [23]

Mouse Strains	Mutant Genes	Immune Deficiency			Xenograftment Acceptance			
		B Cell Functions	**T Cell Functions**	**NK Cell Activity**	**Solid Tumor Cell Lines**	**Solid Tumor Primary Cells**	**Hematopoietic Normal Cells**	**Hematopoietic Malignancies**
Athymic *Foxn1^{nu}* (nude) mice [23]	*Foxn1^{nu}*	Intact	Compromised	High	Accepted	Most not accepted	Not accepted	Not accepted
C.B17-*Prkd^{scid}* (*scid*) mice [24–26]	*Parkdc^{scid}*	Compromised	Compromised	Moderate	Accepted	Accepted	Some cells accepted at low level	Some cells accepted at low level
NOD.C.B17-*Prkdc^{scid}* (NOD-*scid*) mice [27]	*Parkdc^{scid}* *Rag1^{null}* *Rag2^{null}*	Completely deficient	Intrinsic T cell defect with lowered macrophage activity	Low	More accepted	More accepted	Some cells accepted	Some cells accepted
NOD-*Prkdc^{scid}* *IL2rg^{null}* (NSG) mice [22,26,28,29]	*Parkdc^{scid}* *Rag1^{null}* *Rag2^{null}* *IL2rg^{null}*	Completely deficient	Lack of T cell and macrophage activity	Low	Most accepted	Most accepted	Many cells accepted	Many cells accepted

NOD-*Prkdc*^{scid} *IL2rg*^{null} (NSG) Mice. Development of NSG mice was a hallmark and leap-forward in the immunodeficient model for engraftment of primary human cells, tissues, and tumors. The NSG mice carry triple mutations, combining, in addition to the mutation of its predecessor NOD-*scid* mice, with IL-2 receptor common γ chain (*IL2rg*^{null}) mutation, and *Rag1*^{null} or *Rag2*^{null} mutations. The IL2rg is responsible for high-affinity signaling for the receptors for IL-2, IL-4, IL-7, IL-9, IL-15, and IL-21. The *IL-2rg*^{null} mutation compromises both adaptive and innate immune system [29]. NSG is completely deficient in adaptive immunity and severely deficient in innate immunity, therefore, highly receptive to engraftment of human cells, tissues, and primary tumors [27,30,31].

There are three major strains of immunodeficient *IL2rg*^{null} mice [27,30,31]. They differ in strain background (NOD vs BALB/c vs BALB/c × 129 mixed background), the mutation used to induce their immunodeficiency in the adaptive immune system (*scid* or *Rag1*^{null} or *Rag2*^{null}) and in the mutation of the *IL2rg* gene itself (complete null or truncated). These variables result in differences in their ability to be engrafted with human cells and tissues. For the purpose of engraftment of primary human tumors, NSG strain is by far the most receptive immunodeficient mouse model [32]. Many more human cancers, particularly hematological malignancies, can be engrafted successfully in NSG mice than in its predecessor NOD-*scid* mice [32,33].

In Vitro Three-Dimensional Culture

As stated earlier, in vitro culture of CSCs could bring great advantages to study the different characteristics of CSCs as their self-renewal potential. Three-dimensional (3D) culture refers to a spectrum of in vitro culture systems for different applications, with various terms describing a spherelike structure in well-rounded shape. For numerous sphere structures that contain cancer cells and can be maintained as free-floating cultures, a 4-group classification was proposed [33] to divide them based on their differences in the method of culture preparation and subsequent spheres formed in liquid culture: (1) multicellular tumor spheroids—cancer cell lines cultured in serum-containing media; (2) tumorispheres—suspension of dissociated cancer cells cultured in serum-free media with growth factors; (3) tissue-derived tumor spheres—partially dissociated cancer tissues cultured in serum-containing media; (4) organotypic multicellular spheroids—minced tumor fragments cultured in serum-containing media.

In vitro culture of primary organoids in nonliquid culture has gained momentum during recent years. It has evolved into a third human-derived cancer model, intermediate between the other two typical human-derived animal cancer xenograft models—in vitro culture of cancer cell line and in vivo growth of primary patient-derived tumor xenograft [34]. Gut organoid is the most systemically studied model. The establishment of long-term culture of gut organoids was first reported in 2009 by Clevers and Kuo groups, respectively [35,36]. Clevers and colleagues reported that mouse intestine crypt-villus structure can be grown from a single-sorted LGR5+ in vitro without the presence of nonepithelial niche. Lineage tracing indicates that the LGR5+ stem cell hierarchy is maintained in the structure. The long-term (more than 8 months) culture condition requires suspension of base membrane matrix (Matrigel), recombinant EGF, RSPO1, and NOGGIN. Kuo and colleagues described their long-term culture method for either mouse small or large intestine. The culture is initiated with minced epithelium tissue fractions in a matrix of agarose gel of collagen with an air-liquid interface, within a Wnt-dependent condition. The prolonged intestinal epithelial expansion as organoids with proliferation and multilineage differentiation can sustain as long as 350 days. Later, long-term 3D culture of gut epithelium organoids from normal mouse intestine cells was extended to culture from human colon, including normal colon, adenoma, adenocarcinoma, and Barrett's epithelium [37].

Culturing tissue in organotypic assays represents an acknowledgment that traditional cell line models, although useful in many contexts, may lack biological features that are crucial to understanding in vivo mechanisms of growth and survival. Organotypic modeling of normal and malignant tissue generally retains at least some of the tissue architecture and cellular heterogeneity found in the original specimen. The addition of signaling ligands for pathways such as Wnt and bone morphogenic protein (BMP) encourage the propagation of self-renewing cells in vitro, enabling detailed dissection of signal transduction pathways that translate to in vivo models. However, organotypic models typically rely on extracellular matrix–based cultures that make manipulation challenging and expensive. In addition, the limited amount of cells available from organotypic assays present challenges when trying to use traditional analysis tools such as Western blotting. The lack of environmental cell types also limits interpretation of experiments when translating to in vivo settings. Despite these challenges, it is generally accepted that organotypic assays are more appropriate systems to test self-renewal effectors than cell lines.

Single Cell–Based Techniques

It has been asserted that tumor heterogeneity contributes to failure of treatment and leads to dissemination of disease. Increasing evidence points to CSCs as the responsible subpopulation with malignant properties that are spared by many therapies. A gene expression signature composed of self-renewal-associated genes has been described that is specific to CSCs isolated from a diverse range of tumors. This signature can be used to predict prognosis, relapse and/or survival of patients with breast, colon, leukemia, and bladder tumors [38–42].

It is important to understand that multiple independent mutations must occur to initiate a cancer [43]. To be sustained, each of these mutations must happen in a self-renewing population given that mutations occurring in non-self-renewing cells will be lost over time. Thus, oncogenic mutations must occur in a normal stem cell or, alternatively, a mutation, or more likely a combination of mutations, must confer self-renewal capacity on a downstream cell.

One important strategy of understanding how particular mutations affect the self-renewal of cancer cells is the investigation of both normal stem and progenitor cells as well as malignant stem cells. This enables one to determine normal self-renewal pathways in normal cells and then to understand how a particular mutation perturbs the self-renewal pathway in cancer. In cancers, the self-renewing cells are a result of activation of genes necessary for self-renewal and the lack of expression of genes that drive differentiation. In cancer, the CSCs can utilize normal regulatory pathways to express self-renewal genes and repress expression of differentiation/senescence genes; or oncogenic mutations can activate self-renewal pathways and/or inactivate differentiation/senescence pathway genes. By understanding the differences in how self-renewal is regulated in normal and malignant cells, one can hopefully identify self-renewal regulation pathways that are unique to the malignant cells [43].

To identify and study CSCs, it has been critical to develop new tools and models. The advancement in sequencing and PCR techniques together with the development of biotechnologies able to analyze low or even single cells have brought advances to the fields of normal stem cell and CSC biology (Fig. 2.3). Furthermore, mouse models and more robust in vitro culture assays that allow us to study and track CSCs at clonal levels have also significantly contributed to the advancement of the field. In this section we will focus on single-cell sequencing and PCR strategies used to further interrogate CSCs.

Why is it so important to perform these studies at the single-cell level? Tumors are organized in a heterogeneous population of cells, which also implies heterogeneity in their genomes and transcription profiles. If the tumor bulk population is analyzed, it is nearly impossible to deconvolute the findings to

used. In this figure, single-cell real-time polymerase chain reaction (RT-PCR) is described. Different populations are isolated and enriched based on cell surface markers and, using flow cytometry, single cells are sorted. Single-cell RT, multiplexed amplification and PCR are performed in microfluidic devices. In this case, the analysis of mRNA expression results is displayed as a hierarchical heat map. The stem cells are characterized by high expression of self-renewing genes (self-renewal genes 1–3) and low expression of differentiated genes (differentiation genes 1–5). The differentiated cells will express an opposite gene expression pattern from the stem cells, whereas progenitor cells will show intermediate expression profiles. CSCs can also be isolated and analyzed at the single-cell level. This technique allows the identification of novel genes that are highly expressed in CSCs. Those genes that are highly expressed or unique to CSCs are interesting candidates to further characterize the CSC population or provide possible therapeutic targets. Furthermore, by comparing the gene expression profiles, some clues to the cell of origin behind the CSCs emerge, in determining whether they come from a somatic stem cell or more differentiated cell.

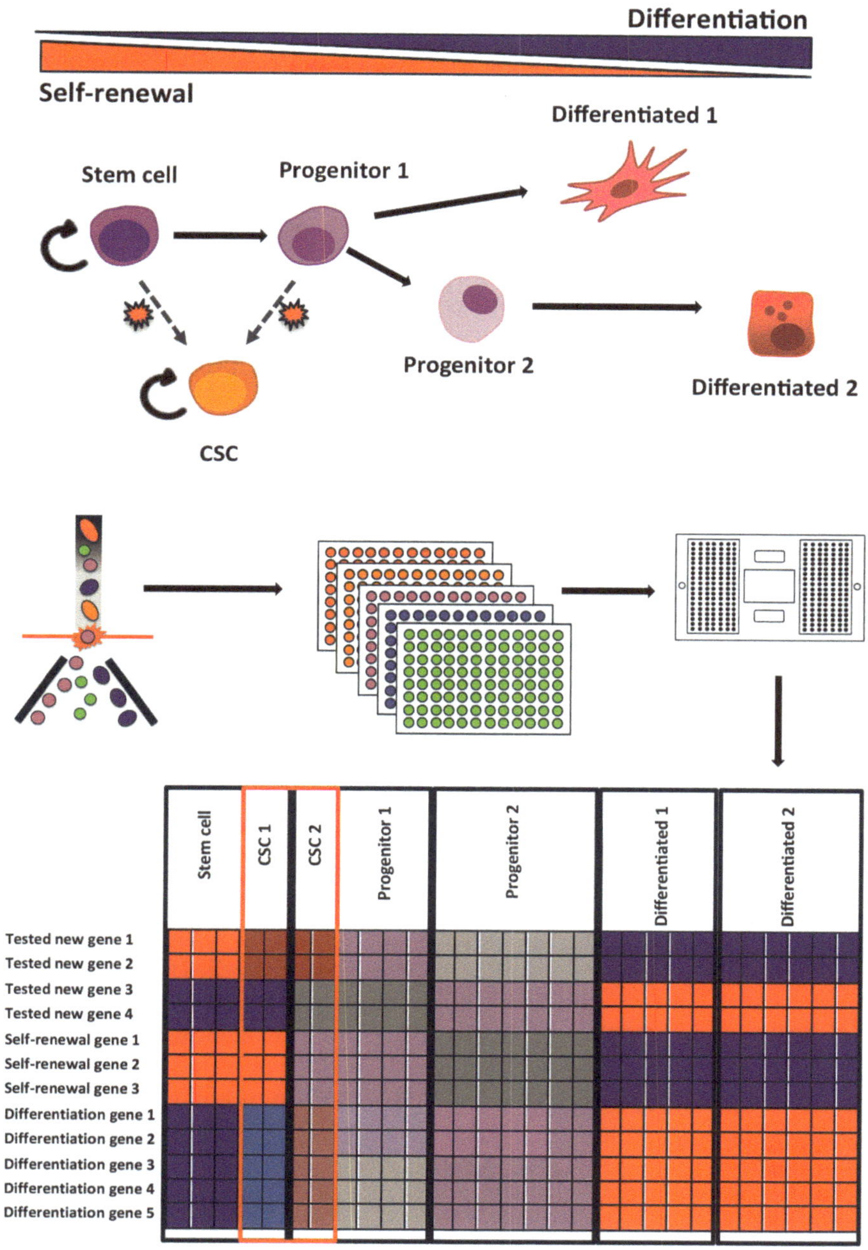

FIGURE 2.3 **Analysis of genes/transcripts to characterize cancer stem cells (CSCs) at the single-cell level.**

Some tissues are characterized by their hierarchical organization where the stem cells are the cells at the top of the hierarchy because they give rise to the more differentiated cells that constitute the organ. In order to identify novel self-renewal or differentiation genes, single-cell analysis approaches have been each clonal population, which can occur as a result of either ongoing mutations or due to epigenetic differences as a result of normal differentiation programs [39]. Targeted single-cell analysis allows finer details on populations to be revealed. Furthermore, it must be taken into consideration that the CSCs in many tumors are a minority cell population. Thus, if the analysis is not skewed toward increasing the number of CSCs to be analyzed, it is likely the information is more informative of the gene expression profile of the non-CSC population than the CSCs, and interpretation of the results and conclusions made may be compromised. Several important issues are made possible by using these new genomic tools, such as identification of the cell of origin of the CSCs, determination of the cellular hierarchy of normal and malignant tissues, and identification of modifications responsible for the acquisition of self-renewal and resistance to therapy (Fig. 2.3).

In some tissues and organs, the presence of stem cells or the full hierarchy remains to be fully elucidated. This may be in part due to the lack of cell surface markers able to separate the stem from differentiated cells. There have been several examples where single-cell RNA-sequencing and single-cell PCR have helped to further identify populations in the hierarchy of normal tissues. In the mouse lung epithelium, microfluidic single-cell RNA-sequencing technology performed on 198 individual cells was able to further define the developmental and cellular hierarchy on this organ [44]. This strategy led to the discovery of many previously unknown cell-type markers, including transcriptional regulators that discriminate between the different populations. In small intestine where heterogeneity of the intestinal stem cell pool remains controversial, single-cell RNA-sequencing was recently performed [45]. Individual cells were isolated from organoids arising from LGR5+ cells as well as *Lgr5* reporter mice. In this study, the authors concluded that LGR5+ cells represent an abundant population of stem cells mixed with a rare population of LGR5+ secretory cells. Furthermore, they identified Reg4 as a novel marker for enteroendocrine cells. In the intestine, Paneth cells are a key cell type that compose the niche of the undifferentiated LGR5+ cells. In the colon, while the presence of undifferentiated LGR5+ cells was known, the equivalent to the Paneth cell remained undetermined. In this case, single-cell PCR of sorted cells was able to identify those cells as a cKIT/CD117+ subpopulation of goblet cells [46]. This new population responded to Notch signaling.

Similar approaches have been used to identify the hierarchy of tumors as well as novel populations. Single-cell PCR gene expression analysis demonstrated that the cellular composition and transcriptional profiles of primary colon cancer epithelia mirrored those found in normal human colon [39]. In this work, it was also demonstrated that the transcriptional diversity of cancer tissues is largely explained by in vivo multilineage differentiation and not solely by clonal genetic heterogeneity.

Single-cell analyses of GBM have led to novel insights into this uniformly lethal cancer. GBM contains a primitive subpopulation of stemlike cells with preferential resistance to existing therapies [47]. Single cell–based transcriptomic analysis of five different human GBM tumors revealed their transcriptional, functional, and genetic diversity [48]. Comparison of the transcriptomes of spherogenic cultures arising from GBM stem cells or more differentiated cells allowed the identification of a stemness signature able to identify a fraction of the GBM cells with high expression of transcription factors related to stemness such as *POU3F2*, *NFIA*, and *NFIB*. This stemness signature was able to differentiate subtypes of GBM tumors as well as predict patient prognosis.

In blood, similar approaches have been used to determine the hierarchy of normal and malignant blood samples. Of all adult tissues, the differentiation hierarchy of blood (hematopoietic) cells is the best understood. Novel bar-coding approaches have been used to track the differentiation of single HSCs [49]. This technique genetically labels single HSCs using a viral vector and using high-throughput sequencing techniques, enabling one to track the hierarchy of their descendants. Using the same technology, a novel descendant of NK cells was found in monkeys [50]. Single-cell RNA-sequencing and single-cell PCR approaches were also used to determine the hierarchy of blood maturation [51,52]. In the latter study, the computational lineage progression analysis of 1500 single cells isolated from different hematopoietic populations allowed the mapping of the mouse hematopoietic stem cell differentiation hierarchy. Furthermore, they were able to identify two independent self-renewing lineages with different clonal activities in a mouse model of acute myeloid leukemia (AML). Using this approach, they were able to indirectly assess the purity of HSCs from different enrichment protocols, once more highlighting the importance of using the correct dissociation, staining, and isolation protocols to perform an accurate study of stem cells or CSCs. Also, the combination of exome sequencing and single-cell genetic characterization of residual preleukemic HSCs isolated from six different AML patient samples, established a sequence of accumulative mutations occurring in AML [53]. They also found out that some of these samples follow the classic clonal evolution model of cancer in which it is postulated that the more genetically altered clones are increasingly dominant, while other AML samples did not follow this model.

Single-cell transcriptomic analysis has also been used to study circulating tumor cells (CTCs) and metastatic cells. 221 individual candidate prostate CTCs were analyzed to address their heterogeneity within and across different patients and their differences from primary tumor specimens, as well as to identify clinically relevant mechanisms of drug resistance [54]. Sixty percent of CTCs robustly expressed transcripts described as putative stem cell markers. Furthermore, they identified *WNT5B* and *WNT7A* noncanonical Wnt pathway members providing survival signals after treatment with an androgen receptor inhibitor. Current theories hypothesize that CSCs may be responsible for disease dissemination. Recently, the identity of metastasis-initiating cells in human breast cancer was interrogated by single-cell PCR analysis [55]. The authors observed that early stage metastatic cells possess a distinct stemlike gene expression signature. These results along with others [56] confirm a hierarchical model for metastasis in breast cancer, in which stemlike cells initiate metastases.

One of the exciting directions in the field of stem cell and CSCs research is the genetic and transcriptomic analysis of single cells or a bulk of clonally derived cells. Although this type of approach can answer existing questions about the existence and origin of CSCs, it is not applicable to all systems, nor is it available universally to all investigators. These problems are largely due to the cost of this technology, bias from the method used to amplify the starting material, and complicated analysis of the results. Similar to what has occurred with other novel techniques, hopefully the price will decrease as the amount of researchers using the technology increases. Also, as novel cell surface markers and transcriptomic signatures that define self-renewing cells or differentiated cells are discovered to enrich for the population to be studied, fewer cells and analysis of a small number of genes will be needed to get robust results.

MODULATORS OF SELF-RENEWAL IN CANCER STEM CELLS

CSCs share their self-renewal and differentiation properties with normal stem cells. Thus, the molecular players that regulate self-renewal on normal stem cells is, in most cases, shared by the CSCs. Furthermore, stem cells are not only defined just by high expression of genes that give them stemness but also for low expression of differentiation genes. This concept opens the door to a wide list of different stemness-related transcriptomes that gets anything but more complex when talking about CSCs. Genes and pathways that have been related with self-renewal are diverse and encompass transcription factors, cell signaling pathways, and epigenetic regulators. In the next section we will focus our attention in these master self-renewal pathways of stem cells that have been corrupted at different extent in CSCs (Table 2.2).

Cell Signaling and Microenvironment

As in normal tissue, the niche of the CSCs can play a pivotal role in regulating their stemness and differentiation. For CSCs, the TME—other tumor cells that lack stemness potential and normal counterparts—could be considered their niche. These different niches may modulate the self-renewal of CSCs in different ways since the TME provides a source of cytokines and growth factors that support CSCs to maintain their stemness [57]. There are several

Table 2.2 Key Players in Cancer Stem Cell (CSC) Self-Renewal

Tissue/Cell Type	Key Players in CSC Self-Renewal	Citations
LSCs	DNM	[112,114]
	LSD1	[142]
	Bmi-1	[138]
	β-Catenin	[6]
	Smo	[88,89,91]
	Shh	[91]
	CD47	[96]
	HIF-1α-Wnt	[99]
	miR-22 via Tet2	[151]
Colon CSCs	DNMTs	[117]
	HIF-1α	[98]
	miR-451	[153]
	miR-34a	[154]
Breast CSCs	Dnmt1	[115]
	RAF1-B-catenin signaling	[121]
	Snail and twist	[103]
	Taz	[107]
	miR-200c via Bmi-1	[104]
Lung CSCs	Bmi-1	[136]
	TET1	[118]
	Smo	[87]
	Gli	[87]
Brain CSCs	EZH2, BMI-1	[134]
	MLL activation of HOXA10	[140]
	MLL activation of HIF2α	[141]
	FoxM1-β-catenin	[70]
	Notch	[77,78]
	Shh	[85]
	Gli	[86]
	miR-128 via Bmi-1	[152]
Head and neck CSCs	TWIST regulation of Bmi-1	[139]
Epidermal CSCs	β-Catenin	[74]
	PTCH1	[84]
Liver CSCs	BMI-1	[75]
	β-Catenin	[75]
Pancreatic CSCs	Notch	[79]
Prostate CSCs	miR-34a	[155]

pathways that are responsive to extrinsic signals and have been described as important for self-renewal in both normal stem cells and CSCs. For example, cancer-associated fibroblasts (CAFs) can stimulate the secretion of Wnt and Notch stemness factors as well as BMP and hedgehog differentiation factors to modulate their stemness and lineage fate.

Wnt: The Wnt pathway is critical for normal stem cell self-renewal, since Wnt activation leads to the expansion of the stem cell compartment [58]. Wnt family members are examples of molecules secreted by the microenvironment that regulate stemness. Wnt molecules are short-range signals that bind to Frizzled and LRP5/6 coreceptors to form a complex. Upon Wnt binding, β-catenin is stabilized and can relocate to the nucleus to act as a coactivator for Wnt-regulated gene transcription such as *Axin2* and *Lgr5* [59]. Lineage tracing studies of *Axin2* and *Lgr5* have been successfully used to identify self-renewing cells in various tissues such as intestine and colon, stomach, skin, mammary gland, kidney, prostate, and liver [60–67]. Stem cells can be activated by paracrine or autocrine signaling. For instance, in the colon *Lgr5*+ stem cells receive paracrine Wnt signaling from interspersed Paneth cells and surrounding stromal cells to maintain stemness [37].

The first evidence for a direct link between the Wnt pathway and cancer came with the discovery that the adenomatous polyposis coli (*APC*) tumor suppressor gene directly interacts with β-catenin [68,69]. There are various examples of Wnt activity in stem/progenitor cells and cancer. In glioma, WNT3A increases the levels of FOXM1, which binds to β-catenin to enhance nuclear localization and is essential for glioma formation [70]. In *mixed lineage leukemia* (MLL), β-catenin is activated during the development of MLL CSCs. Suppression of β-catenin lead to a significant drop in growth of MLL cells and the deletion of β-catenin completely abolished tumorigenesis [71]. In the progression of CML to blast crisis, the origin of the CSCs is a granulocyte-macrophage progenitor cell with an aberrantly activated β-catenin pathway [6]. The *MMTV-Wnt1* mouse model has been extensively used to study the role of Wnt in breast cancer. In the premalignant mammary gland of these mice there is an enrichment of the stem cell subpopulation [72]. Furthermore, murine breast CSCs were first prospectively isolated from the tumors developed in this model, which legitimized the use of this mouse model to study CSCs [73]. In early epidermal tumors a CSC population was found to share similarities to normal bulge skin stem cells. The ablation of the β-catenin gene lead to depletion of those CSCs, and subsequently to complete tumor regression [74]. In another study, forced expression of *BMI-1* and constitutively active β-catenin induced self-renewal in both hepatocytes and transplanted cells form tumors [75]. In 2012, several studies using lineage tracing mouse models were published providing in vivo evidence for the relevance of Wnt signaling in CSCs. One of these studies showed that in primary intestinal adenomas

5–10% of the tumor cells were *Lgr5+* cells and those cells were the population responsible for the growth of established intestinal adenomas [12].

Notch: Notch signaling is an evolutionarily conserved pathway that regulates cell fate determination and tissue homeostasis. The canonical pathway is made up of secreted ligands (DLL1, DLL3, DLL4, JAG1, and JAG2) that bind to cognate cell surface receptors (Notchs 1–4). Binding triggers cleavage of the intracellular domain of the receptor, which translocates to the nucleus where it interacts with the CBF-1 protein and initiates transcription of target genes. Deregulated expression of Notch pathway components has been described in a wide variety of solid and blood cancers [76]. CSCs in medulloblastomas often express high levels of Notch and are sensitive to Notch pathway inhibitors [77]. Notch receptors and ligands are also expressed in some human GBM, as well as in a *Kras*-induced mouse GBM model. In GBM, Notch signaling promotes Nestin expression, a hallmark of GBM stem cells. Enforced expression of Notch and *Kras* signaling in vivo generated GBM-like lesions in the subventricular zone of mice, which is known to harbor normal stem cells [78]. In some pancreatic cancers, Notch pathway components are upregulated. Furthermore, in certain pancreatic cancers inhibition of the Notch pathway using a gamma-secretase inhibitor or Hes1 shRNA decreased the CSC population both in vitro and in vivo [79]. Enforced activation of Notch signaling increased the CSC population. Currently, Notch inhibitors are being evaluated in early stage clinical trials for their efficacy against cancers such as head and neck squamous cell carcinoma [80]. Although these inhibitors reduce the CSC population in preclinical studies, early clinical data suggests that they are unable to completely eliminate the CSC compartment.

Hedgehog: Hedgehog (Hh) signaling is a highly conserved signaling pathway that regulates embryogenesis, tissue regeneration, and maintenance [81]. Hh ligands (Sonic, Desert, or Indian) are posttranslationally lipidated and secreted by cells to affect their neighbors. In the canonical pathway, these secreted molecules bind to Patched (PTCH) membrane receptors, which trigger the depression and subsequent activation of Smoothened (SMO), a GPCR-like protein. SMO is then able to activate GLI transcription factors to effect gene transcription. A role for Hh signaling in CSCs has been shown in certain subsets of patients form many cancers most notably basal cell carcinomas (BCCs) of the skin and a subset of medulloblastomas [82]. It may also play a role in certain patients with glioma, small cell lung cancer (SCLC), non-small cell lung cancer (NSCLC), metastatic melanoma, some blood malignancies, and prostate cancer.

Hh signaling has been shown to be a powerful regulator of CSC proliferation in solid tumors. BCC, which is the most common type of skin cancer in the United States, was the first group of cancers in which the deregulation of

Hh was linked to tumorigenic potential based on studies drawn from Gorlin's syndrome [83]. BCC is characterized by the frequent loss of *PTCH1*, which results in constitutive Hh signaling. Recently, multiple hair follicle stem cell populations were found to develop BCC-like tumors when *Ptch1* was deleted [84]. These studies demonstrated that Hh signaling drives CSC development in BCC. Gorlin's syndrome patients also have a predisposition for developing medulloblastoma. Mutations in various components of the Hh pathway, including *PTCH1* and *SMO*, occur in approximately 30% of sporadic medulloblastomas, resulting in activated Sonic hedgehog signaling. Quiescent, therapy-resistant SOX2+ CSCs drive the sonic Hh subgroup of medulloblastomas and are enriched by antimitotic chemotherapy and Smoothen inhibition [85]. In addition, Hh signaling has been shown to be necessary for the growth and maintenance of human GBM CSCs [86]. In a mouse model of SCLC driven by deletion of *Rb1* and *Trp53* genes, Hh signaling is activated in an autocrine fashion. Deletion of *Smo* in suppressed SCLC initiation and progression in mice and pharmacological abrogation of Hh signaling inhibited the growth of mouse and human SCLC [87].

Hh signaling may also play a crucial role in the maintenance of some types of LSCs. Hh signaling is activated in LSCs through upregulation of *SMO* [88]. Loss of *SMO* inhibits the induction of CML by the BCR-ABL1 oncoprotein through the depletion of CML LSCs [89]. However, Hh signaling driving AML LSC growth is likely to be dependent on the genetics of the disease. In the *MLL-AF9* mouse model of AML, which has a defined C-KIT+ LSC population, abrogation of Hh signaling through conditional deletion of *Smo* did not effect the development of the disease [90]. In B-ALL, Hh pathway components are highly expressed. Hh pathway inhibitors such as cyclopamine and IPI-926 limited immature B-ALL cells' proliferation in vitro and in vivo [91]. The hedgehog inhibitors vismodegib, itraconazole, and sonidegib are currently being evaluated in clinical trials in cancers such as adult recurrent medulloblastoma and advanced BCC [92–94]. These inhibitors are able to reduce the CSC population but it is unclear if they are able to completely eliminate the compartment.

Immune system: Immune cells play a pivotal role in shaping the microenvironment that surrounds and infiltrates the tumor. The immune cells can orchestrate a cascade of cellular signaling and secretion of cytokines and chemokines that can both impair or promote tumor growth [95]. The role of CSCs in immune modulation is still in its infancy, although seminal works have been published. The CD47 molecule was first discovered because it is highly expressed in a subset of self-renewing LSCs, allowing them to avoid phagocytosis by innate immune cells [96]. Several studies in blood and solid tumors have found that upon inhibition of the CD47:SIPR-α interaction there is an increase in tumor cell phagocytosis [97]. Hypoxic niches can also regulate the function of stem and precursor cells; hypoxia inducible factor or HIF-1α

modulates the Wnt/β-catenin pathway in colon CSCs [98] and LSCs [99]. In some head and neck cancers, the CD44+ cells, which are enriched for CSCs, express the immune checkpoint modulator PD-L1 which is immunosuppressive [100]. Taken together, these data suggest that stem cells, including CSCs, may routinely express immunosuppressors to escape elimination by both the innate and acquired immune system.

TGF-β: The tumor niches can also play a critical role in the dissemination of the disease. CTCs need an appropriate niche to engraft, survive, and activate their cell cycle if they have been in a dormant state [57]. This new environment also must be able to maintain the self-renewal and differentiation capacities of the metastatic self-renewing cancer cells. EMT is an essential process of development that in cancer has been linked to migration and metastasis formation. Some of the pathways able to modulate self-renewal have been associated with the EMT process such as TGF-β, hedgehog, and both canonical and noncanonical Wnt signaling. The TGF-β signaling pathway is comprised on many structurally related proteins that include the BMPs, glial cell–derived neurotrophic factor proteins (GDNFs), activins/inhibins and growth and differentiation factors (GDFs). This powerful protein superfamily regulates a diverse set of biological processes such as tissue homeostasis, development, and immune system regulation [101].

In a seminal paper, the invading cells at the periphery of many colon cancer tumors had a gene expression pattern that suggested that they were undergoing EMT and also resembled CSCs [102]. In breast cancer, at least some of the stem cells were also shown to resemble cells undergoing EMT [103]. Additionally, some CSCs and normal stem cells express low levels of miR-200 family, which can regulate both EMT pathway genes as well as self-renewal genes [102,104]. Note that although it is clear that some CSCs are also EMT cells, it has not been proven that all stem cells are also EMT cells, ie, that there are non-EMT stem cells, and not all cells undergoing EMT are stem cells.

Hippo: Hippo signaling is a highly conserved pathway that controls organ size through regulation of self-renewal, apoptosis, and proliferation [105]. Hippo signaling relies on a kinase cascade that is triggered by cell surface proteins that segregate to the apico-basolateral axis of cells in their native tissue contexts. Hippo pathway members are regulated by canonical and noncanonical Wnt pathway members [106]. Dysregulation of cell polarity highly correlates with the malignancy of cancer. The abilities of breast cancer cells to initiate tumors in vivo and to self-renew depend on the activity of TAZ, a transducer protein in the Hippo signaling cascade [107]. Importantly, ectopic expression of *TAZ* confers self-renewal activity to non-CSCs. Interestingly, the induction of EMT disrupts the inhibitory association of TAZ with other core Hippo kinases mammalian sterile 20-like kinase (MST) and large tumor suppressor (LATS). These studies clearly link cell polarity with CSC proliferation.

Epigenetic Regulators

Epigenetic regulation of the transcription of genes is defined as the covalent modification of DNA, or RNA, or protein components of the chromatin associated with chromosomes that result in the alteration of the molecules' function without changing the nucleic or amino acid sequence itself. These chemical modifications mediate a wide range of physical effects, from altering the topographical structure of chromatin to transforming nucleic acid bases. Epigenetic modifications can be classified into those that are established and passed along to daughter cells and others whose presence is dynamically controlled by the cell's intrinsic needs and extrinsic signals. Stable epigenetic modifications that are passed along to daughter cells (called imprinted genes) constitute an inheritance system that accompanies the information encoded in the DNA itself. Epigenetic inheritance is common to all multicellular organisms and a number of unicellular organisms as well. Dynamic epigenetic modifications have the ability to impact gene transcription and protein function in predictable ways. Both types of epigenetic modifications can run askew during cellular transformation either in a causative fashion or as a result of mutations that activate or repress the transcription of enzymes that establish the marks themselves. An example of epigenetic inheritance is the stable programming of a specialized, terminally differentiated tissue cells such as granulocytes which cannot alter their cell fate to become a red blood cell because the genes responsible for that lineage commitment have been epigenetically and stably silenced. An example of dynamic epigenetic modification is the changing of epithelial cells to a motile mesenchymal phenotype by removal and establishment of epigenetic marks on or near the genes that control those processes.

DNA methylation: This is the most well-studied epigenetic modification. DNA methylation is defined as the covalent addition of a methyl group, usually to cytosines in CpG dinucleotides. These CpGs are often clustered together into genomic regions called CpG islands. In addition, DNA methylation tends to associate with other histone modifications that together regulate the expression and accessibility of a genomic region. Typically, sites in the genome with hypomethylation tend to be expressed, while methylated regions tend to be repressed. Global hypomethylation induces genomic instability and contributes to cellular transformation. Hypermethylation typically inactivates tumor suppressor genes within the promoter regions. DNA hypermethylation also occurs at repetitive DNA sequences, leading to regulation of noncoding RNA expression, such as microRNAs (miRs). DNA methylation is a dynamic type of chromatin modification, and thus is reversible and may be affected by a CSCs' phenotypic state in addition to environmental factors. There are three major DNA methyltransferase (DNMT) enzymes that drive DNA methylation. DNMT1 recognizes hemimethylated DNA and maintains DNA methylation patterns after cellular replication [108]. DNMT3A and DNMT3B are enzymes that are largely responsible for establishing the methyl moieties, also known as de novo methylation [109].

DNMTs are overexpressed in a wide variety of cancers [110]. CSCs were first identified in acute leukemias [111]. DNMT1 is crucial for the maintenance of LSCs in bilinear myeloid-B lymphoid leukemia induced by transduction of *c-myc* and *Bcl-2* [112]. The MLL-AF9 AML mouse model contains LSCs that sustain, expand, and regenerate the disease [113]. *Dnmt1*, but not *Dnmt3a* or *Dnmt3b*, was more highly expressed in LSCs compared to differentiated cancer progeny. Conditional deletion of *Dnmt1* abrogated leukemic transformation, delayed progression of preexisting leukemias, and suppressed LSC self-renewal in the *Mll-Af9* model [114]. In addition, DNMT1 is highly expressed in mammary tumors, a tumor type that has been shown to contain CSCs at least in some tumors [14]. In a mouse model of breast cancer, mammary gland–specific *Dnmt1* deletion protected mice from mammary tumorigenesis by reducing the numbers of CSCs [115]. This study identified ISL1 as a DNMT1-target gene that acts as a tumor suppressor. In colon, genetic deletion of *Apc* in normal colon stem cells results in tumor progression [62]. DNMT enzymes share about half of their target genes with Polycomb complexes PRC1 and PRC2 in colorectal cancer [116]. Interestingly, treatment of the APC$^{(min/+)}$ colon cancer mouse model with the demethylating agent 5-aza-2′-deoxycytidine reduces colon tumor formation [117]. Biologically mediated DNA demethylation is a recently identified phenomenon with the discovery of The *ten-eleven-translocation* (*TET*) family genes as a set of enzymes that perform the opposite function of DNMTs. Members of the Tet family of proteins (*TET1/2/3*) are dioxygenase enzymes that convert cytosine-5-methylation (5mC) to 5-hydroxymethyl-cytosine (5hmC). The removal of the 5hmC moiety contributes to the DNA demethylation process. A recent study demonstrated that TET1-mediated production of 5hmV was required for the tumorigenicity of GBM stem cells through a protein methyltransferase-mediated activation of target genes [118].

Polycomb proteins: These proteins were initially identified in *Drosophila melanogaster* and found to silence genes by localizing to Polycomb-responsive elements. In mammalian cells, Polycomb-mediated gene repression is conducted through two complexes, PRC1 and PRC2. PRC2 binds to chromatin and trimethylates H3K27 through the enzymatic activity of EZH2. This repressive chromatin mark is then recognized by a CBX protein that is part of PRC1. Classically, binding of the mark by a CBX protein localizes the PRC1 complex to specific chromatin regions and directs the PRC1 E3 ligases RING1/2 to monoubiquitinate histone H2A on K119. There are multiple PRC1 complexes, and some may ubiquitinate target genes in the absence of the H3K27 trimethyl mark [119]. The additional H2AK119 ubiquitin mark inhibits RNA polymerase II by inducing chromatin condensation around the histone, leaving the DNA inaccessible for transcription [120]. As a silencing mechanism, PRC complexes generally target tumor suppressor genes and suppress their transcription. Individual PRC complex members have been shown to be important for the self-renewal of CSCs. Recent studies indicate that the Wnt/β-catenin pathway can also be regulated by histone modifications in cancer. EZH2, for example, has been shown to enhance the number of breast TICs by activating RAF1-β-catenin signaling [121]. High levels of EZH2 led to growth of Ewing tumors and inhibition of differentiation [122].

BMI-1, a member of PRC1, was the first gene associated with the unique property of self-renewal. The phenotype of the *Bmi-1* knockout (KO) mouse first suggested a link between *BMI-1* and self-renewal. It was shown to be essential for the self-renewal of HSCs and subsequently neuronal, mammary, and prostate stem cells [123–129]. The effect is, at least in part, mediated through the repression of the *Cdkn2a* locus, which encodes the *Arf* and *Ink4a* genes that activate cellular apoptosis and senescence [126]. This effect can be reversed through the enzymatic action of USP16, which deubiquitinates the H2AK119 mark [130]. Furthermore, in the murine hematopoietic system, knocking out the downstream targets of *Bmi-1* (*Trp53*, $p16^{ink4a}$, and $p19^{arf}$) led multipotent progenitors to acquire long-term reconstitution capability [131].

Although *Bmi-1* was first identified as an oncogene [132,133], its link to CSCs has only recently been described. In CSCs isolated from liver, GBM multiforme, lung, colon, intestine, and head and neck tumors, *Bmi-1* is preferentially expressed in the CSC fraction [20,39,75,134,135]. In some glioma, AML, prostate, lung, and breast tumors, where CSCs have been described, it has been shown that *Bmi-1* is involved in pathogenesis and maintenance of tumor growth [104,124,129,136]. As in normal stem cells, the inhibition of $p16^{ink4a}$ and $p19^{arf}$ by *Bmi-1* is partially responsible for the underlying mechanism of stemness in these CSCs, although $p16^{ink4a}$ and $p19^{arf}$ independent pathways have been described [137]. Loss of *Bmi-1* results in reduced self-renewal and expansion of the CSC compartment in a *Kras* lung cancer model [136]. In GBM, knockdown of either *EZH2* or *BMI-1* expression results in diminished self-renewal and proliferation of tumors [134]. Bmi-1 was shown to promote LSC expansion in CD34+ human AML cells [138]. KO of *Bmi-1* in AML LSCs prevented initiation of disease, indicating a critical role for these cells in tumor initiation. In addition, the *BMI-1* gene itself may be regulated by TWIST1, an EMT regulator, affecting the tumor-initiating capacity of head and neck cancer cells [139]. Interestingly, the group of proteins belonging to the Trithorax protein family, which activates genes by establishing H3K4 methyl marks, countering the Polycomb family's repression, has been shown to mediate GBM CSC tumorigenicity [140]. The histone methyltransferase MLL1 activates HIF2a protein and regulates the tumorigenicity of glioma stem cells [141]. The inhibition of *Bmi-1* in some of these tumor types using small molecule inhibitors, miRs or knockdown strategies has effectively impaired tumor growth, further demonstrating the important role of *Bmi-1* in the self-renewal of CSCs [104,135].

Other histone modifications: LSD1, or KDM1A, is a histone demethylase that inhibits gene expression by enzymatically removing the H3K4me1 and H3K4me2 marks and converting them to unmethylated states. LSD1 inhibition was associated with a decrease in CSC frequency in a mouse model of *Mll-Af9* leukemia, which is driven by the proliferation of cKIT+ CSCs [113,142].

miRs: miRs comprise another epigenetic class of regulators involved in both normal and CSC self-renewal. These small 21–25 bp noncoding RNA molecules bind to transcripts and alter mRNA levels and protein expression of target genes posttranscriptionally [143–145]. miRs often work together in global networks, with multiple miRs regulating a single transcript, and multiple transcripts regulated by a single miR [146]. This allows miRs as a class to broadly regulate both physiologic and pathophysiologic cellular processes, including self-renewal and proliferation.

miRs have been implicated as important players in both pluripotent and somatic stem cells. Several miR families are necessary for human embryonic stem cell self-renewal, proliferation, and differentiation, including the miR-302/367 cluster and the miR-371 cluster [147]. Ablating canonical miR expression in a mouse model by knocking out the miR-processing enzyme DGCR8 leads to embryonic lethality and a loss of stem cell self-renewal [148]. Many miR with roles in stem cell biology regulation also play a role in tumorigenesis and progression.

miRs are well-known players in cancer, with several "oncomiRs" and "tumor suppressor miRs" being described in different systems. Like protein-coding oncogenes and tumor suppressors, oncomiRs are often amplified or activated in cancers, while tumor suppressor miRs are silenced [149]. miR expression signatures in cancers are so unique they can often differentiate between different subtypes and stages more completely than can gene expression signatures [150]. As miRs play critical roles in both cancer and stem cells, it is not surprising that recent evidence suggests a role for these small RNA in CSC biology. In particular, several miRs have been implicated in aberrant self-renewal processes across a broad range of CSC systems.

miR-22 promotes LSC self-renewal and transformation through inhibition of the TET2 tumor suppressor [151]. *Bmi-1* is targeted by miR-128 in glioma stem cells and miR-200 in breast cancer [104]. Depletion of miR-128 allows both maintenance of normal neural stem cells as well as aberrant self-renewal of glioma CSCs [125,152]. miR-200 is downregulated in breast CSCs (as well as normal mammary stem cells), in part by relieving inhibition of its target and self-renewal regulator *Bmi-1* [104]. At the same time, miRs from the miR-200 family (miR-200c, miR-203, and miR-183) are regulated by the EMT player ZEB1 [102]. The authors associated EMT activation, metastatis, and stemness-maintenance by suppressing stemness-inhibiting miRs in colon and pancreatic cancer. Similarly, in colon CSCs, downregulation of miR-451 leads to acquisition of self-renewal, while downregulation of miR-34a biases CSCs toward symmetric rather than asymmetric cell division, leading to maintenance of the self-renewing CSC population [153,154]. miR-34a plays a similar role in prostate CSCs, suggesting a widely ranging tumor suppressor role for this miR [155]. These studies across a broad range of systems demonstrate an integral role for miRs in both positive and negative regulation of self-renewal in both normal and CSCs.

CONCLUSION

The presence of self-renewing cancer cells in tumors has been described in a wide variety of human and mouse tumors. The existence of CSCs has been independently validated by a number of assays including xenograft transplants and in vitro assays of human tumors, as well as syngeneic transplantation, lineage tracing, and in vivo imaging assays of murine tumors. More importantly, their presence has substantial clinical implications. The unique characteristics of CSCs has made them relevant in designing new therapeutic strategies since it has been proven that CSCs are often resistant to current radiation treatments and chemotherapies, allowing the disease to recur at the original lesion and/or distant metastatic sites after initial treatment. Thus, targeting the CSC's unique characteristic of self-renewal is a promising therapeutic strategy. Next generation techniques are also being used to further characterize and identify novel genes unique to CSCs. These newer tools such as single-cell genomic analysis should not only reveal promising therapeutic targets but also novel self-renewing/differentiation genes. The former could lead to the development of new drugs, bringing advances in cancer therapy, while the latter may aid in treating degenerative diseases.

Glossary

Cancer stem cells Tumorigenic cells able to self-renew and has the capacity to give rise to every cell type within a tumor.

Differentiation A process of specialization that a cell can undergo to support a certain tissue.

Epigenetics Modulations that affect gene expression without altering the genetic code.

Immunodeficient mice Mouse model that lacks certain immune cells to enable the implantation of nonsyngeneic cells.

microRNAs Small, noncoding endogenous RNAs that regulate gene expression by directing their target mRNAs for degradation or translational repression.

Patient-derived xenograft PDX models are created when a cancerous tissue from a patient's primary tumor is implanted directly into an immunodeficient mouse.

RNA-sequencing Sequencing of the RNA pool in a cell or a population to determine their transcriptome.

Single-cell PCR A qRT-PCR performed in one cell to check expression for a set amount of genes in a set number of individual cells.

Stem cells Cells that upon division can give rise to both the same cell type as well as to a more differentiated daughter cell.

Three-dimensional culture In vitro cell culture systems that allow cell growth in all three dimensions, allowing for a more accurate model than 2D cultures.

List of Acronyms and Abbreviations

3D Three-dimensional
5hmC 5-Hydroxymethyl-cytosine
5mC Cytosine-5-methylation
AML Acute myeloid leukemia
APC Adenomatous polyposis coli (protein)
Arf Alternative reading frame
BCC Basal cell carcinoma
BMI1 B cell–specific Moloney murine leukemia virus integration site 1
BMP Bone morphogenic protein
BRCA1 Breast cancer 1
CAF Cancer-associated fibroblast
CD Cluster of differentiation
cKit V-kit Hardy-Zuckerman 4 feline sarcoma viral oncogene homolog
CML Chronic myelogenous leukemia
CSC Cancer stem cell
CTC Circulating tumor cells
DGCR8 DiGeorge syndrome critical region gene 8
DNMT DNA methyltransferase
EMT Epithelial–mesenchymal transition
EZH2 Enhancer of zeste 2 Polycomb repressive complex 2 subunit
FGF Fibroblast growth factor
FOXM1 Forkhead box M1
GBM Human glioblastoma
GDF Activing/inhibins and growth and differentiation factor
GDNF Glial cell–derived neurotrophic factor protein
Hh Hedgehog
HIF Hypoxia inducible factor
HMLE Human mammary epithelial cell
ISL1 Islet
KDM1A Lysine (K)-specific demethylase 1A
KO Knockout
KRAS Kirsten rat sarcoma viral oncogene homolog
LATS Large tumor suppressor
LGR5 Leucine-rich repeat containing g protein-coupled receptor 5
LRP Low-density lipoprotein receptor-related protein
LSC Leukemic stem cell
miRs MicroRNAs
MLL Mixed lineage leukemia
mRNA Messenger RNA

MST Mammalian sterile 20-like kinase
NFI Nuclear factor I
NK cell Natural killer cell
Nod Nonobese diabetic
Nod-scid mouse NOD.C.B17-Prkdcscid mouse
NSCLC Non-small cell lung cancer
NSG mouse NOD-Prkdcscid IL2rgnull mouse
Nude mouse Athymic Foxn1nu mouse
PCR Polymerase chain reaction
PDX Patient-derived xenograft
POU3F2 POU domain, class 3, transcription factor 2
PRC Polycomb repressive complex
Prkdcscid Protein kinase DNA-activated catalytic polypeptide
PTCH Patched
qRT-PCR Quantitative real time polymerase chain reaction
RAF1 Raf-1 proto-oncogene
Reg4 Regenerating islet-derived family, member 4
RSPO R-spondin
Scid mouse C.B17-Prkdcscid mouse
SCLC Small cell lung cancer
SMO Smoothened
TAZ Transcriptional coactivator with PDZ-binding motif
TET Ten-eleven-translocation
TGFB1 Transforming growth factor, beta 1
TIC Tumor-initiating cell
TME Tumor microenvironment
TWIST1 Twist family BHLH transcription factor 1
USP16 Ubiquitin-specific peptidase 16

Acknowledgments

We thank present and past members of the Clarke lab for enriching discussions as well as colleagues of the Institute of Stem Cell Biology and Regenerative Medicine at Stanford University.

Conflict of Interest

M.F. Clarke is a founder, shareholder of Oncomed Pharmaceuticals, a company that focuses on the development of therapeutic methods to target tumor-initiating cells.

References

[1] Reya T, Morrison SJ, Clarke MF, Weissman IL. Stem cells, cancer, and cancer stem cells. Nature 2001;414:105–11.

[2] Clarke MF, Dick JE, Dirks PB, Eaves CJ, Jamieson CH, Jones DL, et al. Cancer stem cells–perspectives on current status and future directions: AACR Workshop on cancer stem cells. Cancer Res 2006;66:9339–44.

[3] Lobo NA, Shimono Y, Qian D, Clarke MF. The biology of cancer stem cells. Annu Rev Cell Dev Biol 2007;23:675–99.

[4] Schillert A, Trumpp A, Sprick MR. Label retaining cells in cancer–the dormant root of evil? Cancer Lett 2013;341:73–9.

[5] Park CH, Bergsagel DE, McCulloch EA. Mouse myeloma tumor stem cells: a primary cell culture assay. J Natl Cancer Inst 1971;46:411–22.

[6] Jamieson CH, Ailles LE, Dylla SJ, Muijtjens M, Jones C, Zehnder JL, et al. Granulocyte-macrophage progenitors as candidate leukemic stem cells in blast-crisis CML. N Engl J Med 2004;351:657–67.

[7] Lim E, Vaillant F, Wu D, Forrest NC, Pal B, Hart AH, et al. Aberrant luminal progenitors as the candidate target population for basal tumor development in BRCA1 mutation carriers. Nat Med 2009;15:907–13.

[8] Sugiarto S, Persson AI, Munoz EG, Waldhuber M, Lamagna C, Andor N, et al. Asymmetry-defective oligodendrocyte progenitors are glioma precursors. Cancer Cell 2011;20:328–40.

[9] Quintana E, Shackleton M, Sabel MS, Fullen DR, Johnson TM, Morrison SJ. Efficient tumour formation by single human melanoma cells. Nature 2008;456:593–8.

[10] Zomer A, Ellenbroek SI, Ritsma L, Beerling E, Vrisekoop N, Van Rheenen J. Intravital imaging of cancer stem cell plasticity in mammary tumors. Stem Cells 2013;31:602–6.

[11] Driessens G, Beck B, Caauwe A, Simons BD, Blanpain C. Defining the mode of tumour growth by clonal analysis. Nature 2012;488:527–30.

[12] Schepers AG, Snippert HJ, Stange DE, van den Born M, van Es JH, van de Wetering M, et al. Lineage tracing reveals $Lgr5^+$ stem cell activity in mouse intestinal adenomas. Science 2012;337:730–5.

[13] Chen J, Li Y, Yu TS, McKay RM, Burns DK, Kernie SG, et al. A restricted cell population propagates glioblastoma growth after chemotherapy. Nature 2012;488:522–6.

[14] Al-Hajj M, Wicha MS, Benito-Hernandez A, Morrison SJ, Clarke MF. Prospective identification of tumorigenic breast cancer cells. Proc Natl Acad Sci USA 2003;100:3983–8.

[15] Galli R, Binda E, Orfanelli U, Cipelletti B, Gritti A, De Vitis S, et al. Isolation and characterization of tumorigenic, stem-like neural precursors from human glioblastoma. Cancer Res 2004;64:7011–21.

[16] Singh SK, Hawkins C, Clarke ID, Squire JA, Bayani J, Hide T, et al. Identification of human brain tumour initiating cells. Nature 2004;432:396–401.

[17] Ricci-Vitiani L, Lombardi DG, Pilozzi E, Biffoni M, Todaro M, Peschle C, et al. Identification and expansion of human colon-cancer-initiating cells. Nature 2007;445:111–5.

[18] Dalerba P, Dylla SJ, Park IK, Liu R, Wang X, Cho RW, et al. Phenotypic characterization of human colorectal cancer stem cells. Proc Natl Acad Sci USA 2007;104:10158–63.

[19] O'Brien CA, Pollett A, Gallinger S, Dick JE. A human colon cancer cell capable of initiating tumour growth in immunodeficient mice. Nature 2007;445:106–10.

[20] Prince ME, Sivanandan R, Kaczorowski A, Wolf GT, Kaplan MJ, Dalerba P, et al. Identification of a subpopulation of cells with cancer stem cell properties in head and neck squamous cell carcinoma. Proc Natl Acad Sci USA 2007;104:973–8.

[21] Li C, Heidt DG, Dalerba P, Burant CF, Zhang L, Adsay V, et al. Identification of pancreatic cancer stem cells. Cancer Res 2007;67:1030–7.

[22] Meacham CE, Morrison SJ. Tumour heterogeneity and cancer cell plasticity. Nature 2013;501:328–37.

[23] Shultz LD, Goodwin N, Ishikawa F, Hosur V, Lyons BL, Greiner DL. Human cancer growth and therapy in immunodeficient mouse models. Cold Spring Harb Protoc 2014;2014:694–708.

[24] Fogh J, Fogh JM, Orfeo T. One hundred and twenty-seven cultured human tumor cell lines producing tumors in nude mice. J Natl Cancer Inst 1977;59:221–6.

[25] Blunt T, Gell D, Fox M, Taccioli GE, Lehmann AR, Jackson SP, et al. Identification of a nonsense mutation in the carboxyl-terminal region of DNA-dependent protein kinase catalytic subunit in the scid mouse. Proc Natl Acad Sci USA 1996;93:10285–90.

[26] Greiner DL, Hesselton RA, Shultz LD. SCID mouse models of human stem cell engraftment. Stem Cells 1998;16:166–77.

[27] Shultz LD, Ishikawa F, Greiner DL. Humanized mice in translational biomedical research. Nat Rev Immunol 2007;7:118–30.

[28] Shultz LD, Schweitzer PA, Christianson SW, Gott B, Schweitzer IB, Tennent B, McKenna S, Mobraaten L, Rajan TV, Greiner DL, et al. Multiple defects in innate and adaptive immunologic function in NOD/LtSz-scid mice. J Immunol 1995;154:180–91.

[29] DiSanto JP, Muller W, Guy-Grand D, Fischer A, Rajewsky K. Lymphoid development in mice with a targeted deletion of the interleukin 2 receptor gamma chain. Proc Natl Acad Sci USA 1995;92:377–81.

[30] Rongvaux A, Takizawa H, Strowig T, Willinger T, Eynon EE, Flavell RA, et al. Human hemato-lymphoid system mice: current use and future potential for medicine. Annu Rev Immunol 2013;31:635–74.

[31] Shultz LD, Brehm MA, Garcia-Martinez JV, Greiner DL. Humanized mice for immune system investigation: progress, promise and challenges. Nat Rev Immunol 2012;12:786–98.

[32] Agliano A, Martin-Padura I, Mancuso P, Marighetti P, Rabascio C, Pruneri G, et al. Human acute leukemia cells injected in NOD/LtSz-scid/IL-2Rgamma null mice generate a faster and more efficient disease compared to other NOD/scid-related strains. Int J Cancer 2008;123:2222–7.

[33] Weiswald LB, Bellet D, Dangles-Marie V. Spherical cancer models in tumor biology. Neoplasia 2015;17:1–15.

[34] Sachs N, Clevers H. Organoid cultures for the analysis of cancer phenotypes. Curr Opin Genet Dev 2014;24:68–73.

[35] Ootani A, Li X, Sangiorgi E, Ho QT, Ueno H, Toda S, et al. Sustained in vitro intestinal epithelial culture within a Wnt-dependent stem cell niche. Nat Med 2009;15:701–6.

[36] Sato T, Vries RG, Snippert HJ, van de Wetering M, Barker N, Stange DE, et al. Single Lgr5 stem cells build crypt-villus structures in vitro without a mesenchymal niche. Nature 2009;459:262–5.

[37] Sato T, van Es JH, Snippert HJ, Stange DE, Vries RG, van den Born M, et al. Paneth cells constitute the niche for Lgr5 stem cells in intestinal crypts. Nature 2011;469:415–8.

[38] Liu R, Wang X, Chen GY, Dalerba P, Gurney A, Hoey T, et al. The prognostic role of a gene signature from tumorigenic breast-cancer cells. N Engl J Med 2007;356:217–26.

[39] Dalerba P, Kalisky T, Sahoo D, Rajendran PS, Rothenberg ME, Leyrat AA, et al. Single-cell dissection of transcriptional heterogeneity in human colon tumors. Nat Biotechnol 2011;29:1120–7.

[40] Eppert K, Takenaka K, Lechman ER, Waldron L, Nilsson B, van Galen P, et al. Stem cell gene expression programs influence clinical outcome in human leukemia. Nat Med 2011;17:1086–93.

[41] Volkmer JP, Sahoo D, Chin RK, Ho PL, Tang C, Kurtova AV, et al. Three differentiation states risk-stratify bladder cancer into distinct subtypes. Proc Natl Acad Sci USA 2012;109:2078–83.

[42] Merlos-Suarez A, Barriga FM, Jung P, Iglesias M, Cespedes MV, Rossell D, et al. The intestinal stem cell signature identifies colorectal cancer stem cells and predicts disease relapse. Cell Stem Cell 2011;8:511–24.

[43] Weissman I. Stem cell research: paths to cancer therapies and regenerative medicine. JAMA 2005;294:1359–66.

[44] Treutlein B, Brownfield DG, Wu AR, Neff NF, Mantalas GL, Espinoza FH, et al. Reconstructing lineage hierarchies of the distal lung epithelium using single-cell RNA-seq. Nature 2014;509:371–5.

[45] Grun D, Lyubimova A, Kester L, Wiebrands K, Basak O, Sasaki N, et al. Single-cell messenger RNA sequencing reveals rare intestinal cell types. Nature 2015;525:251–5.

[46] Rothenberg ME, Nusse Y, Kalisky T, Lee JJ, Dalerba P, Scheeren F, et al. Identification of a cKit(+) colonic crypt base secretory cell that supports Lgr5(+) stem cells in mice. Gastroenterology 2012;142:1195–205.e6.

[47] Bao S, Wu Q, McLendon RE, Hao Y, Shi Q, Hjelmeland AB, et al. Glioma stem cells promote radioresistance by preferential activation of the DNA damage response. Nature 2006;444:756–60.

[48] Patel AP, Tirosh I, Trombetta JJ, Shalek AK, Gillespie SM, Wakimoto H, et al. Single-cell RNA-seq highlights intratumoral heterogeneity in primary glioblastoma. Science 2014; 344:1396–401.

[49] Lu R, Neff NF, Quake SR, Weissman IL. Tracking single hematopoietic stem cells in vivo using high-throughput sequencing in conjunction with viral genetic barcoding. Nat Biotechnol 2011;29:928–33.

[50] Wu C, Li B, Lu R, Koelle SJ, Yang Y, Jares A, et al. Clonal tracking of rhesus macaque hematopoiesis highlights a distinct lineage origin for natural killer cells. Cell Stem Cell 2014;14:486–99.

[51] Jaitin DA, Kenigsberg E, Keren-Shaul H, Elefant N, Paul F, Zaretsky I, et al. Massively parallel single-cell RNA-seq for marker-free decomposition of tissues into cell types. Science 2014;343:776–9.

[52] Guo G, Luc S, Marco E, Lin TW, Peng C, Kerenyi MA, et al. Mapping cellular hierarchy by single-cell analysis of the cell surface repertoire. Cell Stem Cell 2013;13:492–505.

[53] Jan M, Snyder TM, Corces-Zimmerman MR, Vyas P, Weissman IL, Quake SR, et al. Clonal evolution of preleukemic hematopoietic stem cells precedes human acute myeloid leukemia. Sci Transl Med 2012;4:149ra118.

[54] Miyamoto DT, Zheng Y, Wittner BS, Lee RJ, Zhu H, Broderick KT, et al. RNA-Seq of single prostate CTCs implicates noncanonical Wnt signaling in antiandrogen resistance. Science 2015;349:1351–6.

[55] Lawson DA, Bhakta NR, Kessenbrock K, Prummel KD, Yu Y, Takai K, et al. Single-cell analysis reveals a stem-cell program in human metastatic breast cancer cells. Nature 2015;526: 131–5.

[56] Liu H, Patel MR, Prescher JA, Patsialou A, Qian D, Lin J, et al. Cancer stem cells from human breast tumors are involved in spontaneous metastases in orthotopic mouse models. Proc Natl Acad Sci USA 2010;107:18115–20.

[57] Plaks V, Kong N, Werb Z. The cancer stem cell niche: how essential is the niche in regulating stemness of tumor cells? Cell Stem Cell 2015;16:225–38.

[58] Willert K, Brown JD, Danenberg E, Duncan AW, Weissman IL, Reya T, et al. Wnt proteins are lipid-modified and can act as stem cell growth factors. Nature 2003;423:448–52.

[59] Clevers H, Loh KM, Nusse R. Stem cell signaling. An integral program for tissue renewal and regeneration: Wnt signaling and stem cell control. Science 2014;346:1248012.

[60] Barker N, Huch M, Kujala P, van de Wetering M, Snippert HJ, van Es JH, et al. Lgr5(+ve) stem cells drive self-renewal in the stomach and build long-lived gastric units in vitro. Cell Stem Cell 2010;6:25–36.

[61] Barker N, Rookmaaker MB, Kujala P, Ng A, Leushacke M, Snippert H, et al. Lgr5(+ve) stem/progenitor cells contribute to nephron formation during kidney development. Cell Rep 2012;2:540–52.

[62] Barker N, van Es JH, Kuipers J, Kujala P, van den Born M, Cozijnsen M, et al. Identification of stem cells in small intestine and colon by marker gene Lgr5. Nature 2007;449:1003–7.

[63] Jaks V, Barker N, Kasper M, van Es JH, Snippert HJ, Clevers H, et al. Lgr5 marks cycling, yet long-lived, hair follicle stem cells. Nat Genet 2008;40:1291–9.

[64] Lim X, Tan SH, Koh WL, Chau RM, Yan KS, Kuo CJ, et al. Interfollicular epidermal stem cells self-renew via autocrine Wnt signaling. Science 2013;342:1226–30.

[65] van Amerongen R, Bowman AN, Nusse R. Developmental stage and time dictate the fate of Wnt/β-catenin-responsive stem cells in the mammary gland. Cell Stem Cell 2012;11:387–400.

[66] Wang BE, Wang X, Long JE, Eastham-Anderson J, Firestein R, Junttila MR. Castration-resistant Lgr5(+) cells are long-lived stem cells required for prostatic regeneration. Stem Cell Rep 2015;4:768–79.

[67] Zeng YA, Nusse R. Wnt proteins are self-renewal factors for mammary stem cells and promote their long-term expansion in culture. Cell Stem Cell 2010;6:568–77.

[68] Rubinfeld B, Souza B, Albert I, Muller O, Chamberlain SH, Masiarz FR, et al. Association of the APC gene product with beta-catenin. Science 1993;262:1731–4.

[69] Su LK, Vogelstein B, Kinzler KW. Association of the APC tumor suppressor protein with catenins. Science 1993;262:1734–7.

[70] Zhang N, Wei P, Gong A, Chiu WT, Lee HT, Colman H, et al. FoxM1 promotes β-catenin nuclear localization and controls Wnt target-gene expression and glioma tumorigenesis. Cancer Cell 2011;20:427–42.

[71] Yeung J, Esposito MT, Gandillet A, Zeisig BB, Griessinger E, Bonnet D, et al. β-Catenin mediates the establishment and drug resistance of MLL leukemic stem cells. Cancer Cell 2010;18:606–18.

[72] Shackleton M, Vaillant F, Simpson KJ, Stingl J, Smyth GK, Asselin-Labat ML, et al. Generation of a functional mammary gland from a single stem cell. Nature 2006;439:84–8.

[73] Cho RW, Wang X, Diehn M, Shedden K, Chen GY, Sherlock G, et al. Isolation and molecular characterization of cancer stem cells in MMTV-Wnt-1 murine breast tumors. Stem Cells 2008;26:364–71.

[74] Malanchi I, Peinado H, Kassen D, Hussenet T, Metzger D, Chambon P, et al. Cutaneous cancer stem cell maintenance is dependent on beta-catenin signalling. Nature 2008;452:650–3.

[75] Chiba T, Zheng YW, Kita K, Yokosuka O, Saisho H, Onodera M, et al. Enhanced self-renewal capability in hepatic stem/progenitor cells drives cancer initiation. Gastroenterology 2007;133:937–50.

[76] Espinoza I, Pochampally R, Xing F, Watabe K, Miele L. Notch signaling: targeting cancer stem cells and epithelial-to-mesenchymal transition. Onco Targets Ther 2013;6:1249–59.

[77] Fan X, Matsui W, Khaki L, Stearns D, Chun J, Li YM, et al. Notch pathway inhibition depletes stem-like cells and blocks engraftment in embryonal brain tumors. Cancer Res 2006;66:7445–52.

[78] Shih AH, Holland EC. Notch signaling enhances nestin expression in gliomas. Neoplasia 2006;8:1072–82.

[79] Abel EV, Kim EJ, Wu J, Hynes M, Bednar F, Proctor E, et al. The Notch pathway is important in maintaining the cancer stem cell population in pancreatic cancer. PLoS One 2014;9:e91983.

[80] Piha-Paul SA, Munster PN, Hollebecque A, Argiles G, Dajani O, Cheng JD, et al. Results of a phase 1 trial combining ridaforolimus and MK-0752 in patients with advanced solid tumours. Eur J Cancer 2015;51:1865–73.

[81] Petrova R, Joyner AL. Roles for Hedgehog signaling in adult organ homeostasis and repair. Development 2014;141:3445–57.

[82] Li Y, Maitah MY, Ahmad A, Kong D, Bao B, Sarkar FH. Targeting the Hedgehog signaling pathway for cancer therapy. Expert Opin Ther Targets 2012;16:49–66.

[83] Hahn H, Wicking C, Zaphiropoulous PG, Gailani MR, Shanley S, Chidambaram A, et al. Mutations of the human homolog of Drosophila patched in the nevoid basal cell carcinoma syndrome. Cell 1996;85:841–51.

[84] Peterson SC, Eberl M, Vagnozzi AN, Belkadi A, Veniaminova NA, Verhaegen ME, et al. Basal cell carcinoma preferentially arises from stem cells within hair follicle and mechanosensory niches. Cell Stem Cell 2015;16:400–12.

[85] Vanner RJ, Remke M, Gallo M, Selvadurai HJ, Coutinho F, Lee L, et al. Quiescent sox2(+) cells drive hierarchical growth and relapse in sonic hedgehog subgroup medulloblastoma. Cancer Cell 2014;26:33–47.

[86] Clement V, Sanchez P, de Tribolet N, Radovanovic I, Ruiz i Altaba A. HEDGEHOG-GLI1 signaling regulates human glioma growth, cancer stem cell self-renewal, and tumorigenicity. Curr Biol 2007;17:165–72.

[87] Park KS, Martelotto LG, Peifer M, Sos ML, Karnezis AN, Mahjoub MR, et al. A crucial requirement for Hedgehog signaling in small cell lung cancer. Nat Med 2011;17:1504–8.

[88] Dierks C, Beigi R, Guo GR, Zirlik K, Stegert MR, Manley P, et al. Expansion of Bcr-Abl-positive leukemic stem cells is dependent on Hedgehog pathway activation. Cancer Cell 2008;14:238–49.

[89] Zhao C, Chen A, Jamieson CH, Fereshteh M, Abrahamsson A, Blum J, et al. Hedgehog signalling is essential for maintenance of cancer stem cells in myeloid leukaemia. Nature 2009;458:776–9.

[90] Hofmann I, Stover EH, Cullen DE, Mao J, Morgan KJ, Lee BH, et al. Hedgehog signaling is dispensable for adult murine hematopoietic stem cell function and hematopoiesis. Cell Stem Cell 2009;4:559–67.

[91] Lin TL, Wang QH, Brown P, Peacock C, Merchant AA, Brennan S, et al. Self-renewal of acute lymphocytic leukemia cells is limited by the Hedgehog pathway inhibitors cyclopamine and IPI-926. PLoS One 2010;5:e15262.

[92] Kim DJ, Kim J, Spaunhurst K, Montoya J, Khodosh R, Chandra K, et al. Open-label, exploratory phase II trial of oral itraconazole for the treatment of basal cell carcinoma. J Clin Oncol 2014;32:745–51.

[93] Migden MR, Guminski A, Gutzmer R, Dirix L, Lewis KD, Combemale P, et al. Treatment with two different doses of sonidegib in patients with locally advanced or metastatic basal cell carcinoma (BOLT): a multicentre, randomised, double-blind phase 2 trial. Lancet Oncol 2015;16:716–28.

[94] Robinson GW, Orr BA, Wu G, Gururangan S, Lin T, Qaddoumi I, et al. Vismodegib exerts targeted efficacy against recurrent sonic hedgehog-subgroup medulloblastoma: results from phase II pediatric brain tumor consortium studies PBTC-025B and PBTC-032. J Clin Oncol 2015;33:2646–54.

[95] Dunn GP, Bruce AT, Ikeda H, Old LJ, Schreiber RD. Cancer immunoediting: from immunosurveillance to tumor escape. Nat Immunol 2002;3:991–8.

[96] Majeti R, Chao MP, Alizadeh AA, Pang WW, Jaiswal S, Gibbs Jr KD, et al. CD47 is an adverse prognostic factor and therapeutic antibody target on human acute myeloid leukemia stem cells. Cell 2009;138:286–99.

[97] Willingham SB, Volkmer JP, Gentles AJ, Sahoo D, Dalerba P, Mitra SS, et al. The CD47-signal regulatory protein alpha (SIRPa) interaction is a therapeutic target for human solid tumors. Proc Natl Acad Sci USA 2012;109:6662–7.

[98] Santoyo-Ramos P, Likhatcheva M, Garcia-Zepeda EA, Castaneda-Patlan MC, Robles-Flores M. Hypoxia-inducible factors modulate the stemness and malignancy of colon cancer cells by playing opposite roles in canonical Wnt signaling. PLoS One 2014;9:e112580.

[99] Giambra V, Jenkins CE, Lam SH, Hoofd C, Belmonte M, Wang X, et al. Leukemia stem cells in T-ALL require active Hif1alpha and Wnt signaling. Blood 2015;125:3917–27.

[100] Lee Y, S JH, Longmire M, Wang H, Kohrt HE, Chang HY, et al. CD44+ cells in head and neck squamous cell carcinoma suppress T cell-mediated immunity by selective constitutive and inducible expression of PD-L1. Clin Cancer Res 2016. http://dx.doi.org/10.1158/1078-0432.CCR-15-2665.

[101] Shi Y, Massague J. Mechanisms of TGF-beta signaling from cell membrane to the nucleus. Cell 2003;113:685–700.

[102] Wellner U, Schubert J, Burk UC, Schmalhofer O, Zhu F, Sonntag A, et al. The EMT-activator ZEB1 promotes tumorigenicity by repressing stemness-inhibiting microRNAs. Nat Cell Biol 2009;11:1487–95.

[103] Mani SA, Guo W, Liao MJ, Eaton EN, Ayyanan A, Zhou AY, et al. The epithelial-mesenchymal transition generates cells with properties of stem cells. Cell 2008;133:704–15.

[104] Shimono Y, Zabala M, Cho RW, Lobo N, Dalerba P, Qian D, et al. Downregulation of miRNA-200c links breast cancer stem cells with normal stem cells. Cell 2009;138:592–603.

[105] Harvey KF, Zhang X, Thomas DM. The Hippo pathway and human cancer. Nat Rev Cancer 2013;13:246–57.

[106] Azzolin L, Panciera T, Soligo S, Enzo E, Bicciato S, Dupont S, et al. YAP/TAZ incorporation in the beta-catenin destruction complex orchestrates the Wnt response. Cell 2014;158:157–70.

[107] Cordenonsi M, Zanconato F, Azzolin L, Forcato M, Rosato A, Frasson C, et al. The Hippo transducer TAZ confers cancer stem cell-related traits on breast cancer cells. Cell 2011;147:759–72.

[108] Eden A, Gaudet F, Waghmare A, Jaenisch R. Chromosomal instability and tumors promoted by DNA hypomethylation. Science 2003;300:455.

[109] Okano M, Bell DW, Haber DA, Li E. DNA methyltransferases Dnmt3a and Dnmt3b are essential for de novo methylation and mammalian development. Cell 1999;99:247–57.

[110] Gravina GL, Festuccia C, Marampon F, Popov VM, Pestell RG, Zani BM, et al. Biological rationale for the use of DNA methyltransferase inhibitors as new strategy for modulation of tumor response to chemotherapy and radiation. Mol Cancer 2010;9:305.

[111] Bonnet D, Dick JE. Human acute myeloid leukemia is organized as a hierarchy that originates from a primitive hematopoietic cell. Nat Med 1997;3:730–7.

[112] Broske AM, Vockentanz L, Kharazi S, Huska MR, Mancini E, Scheller M, et al. DNA methylation protects hematopoietic stem cell multipotency from myeloerythroid restriction. Nat Genet 2009;41:1207–15.

[113] Somervaille TC, Cleary ML. Identification and characterization of leukemia stem cells in murine MLL-AF9 acute myeloid leukemia. Cancer Cell 2006;10:257–68.

[114] Trowbridge JJ, Sinha AU, Zhu N, Li M, Armstrong SA, Orkin SH. Haploinsufficiency of Dnmt1 impairs leukemia stem cell function through derepression of bivalent chromatin domains. Genes Dev 2012;26:344–9.

[115] Pathania R, Ramachandran S, Elangovan S, Padia R, Yang P, Cinghu S, et al. DNMT1 is essential for mammary and cancer stem cell maintenance and tumorigenesis. Nat Commun 2015;6:6910.

[116] Jin B, Yao B, Li JL, Fields CR, Delmas AL, Liu C, et al. DNMT1 and DNMT3B modulate distinct polycomb-mediated histone modifications in colon cancer. Cancer Res 2009;69:7412–21.

[117] Laird PW, Jackson-Grusby L, Fazeli A, Dickinson SL, Jung WE, Li E, et al. Suppression of intestinal neoplasia by DNA hypomethylation. Cell 1995;81:197–205.

[118] Takai H, Masuda K, Sato T, Sakaguchi Y, Suzuki T, Suzuki T, et al. 5-Hydroxymethylcytosine plays a critical role in glioblastomagenesis by recruiting the CHTOP-methylosome complex. Cell Rep 2014;9:48–60.

[119] Ismail IH, Andrin C, McDonald D, Hendzel MJ. BMI1-mediated histone ubiquitylation promotes DNA double-strand break repair. J Cell Biol 2010;191:45–60.

[120] Francis NJ, Kingston RE, Woodcock CL. Chromatin compaction by a polycomb group protein complex. Science 2004;306:1574–7.

[121] Chang CJ, Yang JY, Xia W, Chen CT, Xie X, Chao CH, et al. EZH2 promotes expansion of breast tumor initiating cells through activation of RAF1-beta-catenin signaling. Cancer Cell 2011;19:86–100.

[122] Richter GH, Plehm S, Fasan A, Rossler S, Unland R, Bennani-Baiti IM, et al. EZH2 is a mediator of EWS/FLI1 driven tumor growth and metastasis blocking endothelial and neuro-ectodermal differentiation. Proc Natl Acad Sci USA 2009;106:5324–9.

[123] Sangiorgi E, Capecchi MR. Bmi1 is expressed in vivo in intestinal stem cells. Nat Genet 2008;40:915–20.

[124] Lessard J, Sauvageau G. Bmi-1 determines the proliferative capacity of normal and leukaemic stem cells. Nature 2003;423:255–60.

[125] Molofsky AV, Pardal R, Iwashita T, Park IK, Clarke MF, Morrison SJ. Bmi-1 dependence distinguishes neural stem cell self-renewal from progenitor proliferation. Nature 2003;425: 962–7.

[126] Park IK, Qian D, Kiel M, Becker MW, Pihalja M, Weissman IL, et al. Bmi-1 is required for maintenance of adult self-renewing haematopoietic stem cells. Nature 2003;423:302–5.

[127] Pietersen AM, Evers B, Prasad AA, Tanger E, Cornelissen-Steijger P, Jonkers J, et al. Bmi1 regulates stem cells and proliferation and differentiation of committed cells in mammary epithelium. Curr Biol 2008;18:1094–9.

[128] Zencak D, Lingbeek M, Kostic C, Tekaya M, Tanger E, Hornfeld D, et al. Bmi1 loss produces an increase in astroglial cells and a decrease in neural stem cell population and proliferation. J Neurosci 2005;25:5774–83.

[129] Lukacs RU, Memarzadeh S, Wu H, Witte ON. Bmi-1 is a crucial regulator of prostate stem cell self-renewal and malignant transformation. Cell Stem Cell 2010;7:682–93.

[130] Adorno M, Sikandar S, Mitra SS, Kuo A, Nicolis Di Robilant B, Haro-Acosta V, et al. Usp16 contributes to somatic stem-cell defects in Down's syndrome. Nature 2013;501:380–4.

[131] Akala OO, Park IK, Qian D, Pihalja M, Becker MW, Clarke MF. Long-term haematopoietic reconstitution by Trp53-/-p16Ink4a-/-p19Arf-/- multipotent progenitors. Nature 2008;453:228–32.

[132] Haupt Y, Alexander WS, Barri G, Klinken SP, Adams JM. Novel zinc finger gene implicated as myc collaborator by retrovirally accelerated lymphomagenesis in E mu-myc transgenic mice. Cell 1991;65:753–63.

[133] van Lohuizen M, Verbeek S, Scheijen B, Wientjens E, van der Gulden H, Berns A. Identification of cooperating oncogenes in E mu-myc transgenic mice by provirus tagging. Cell 1991;65:737–52.

[134] Abdouh M, Chatoo W, El Hajjar J, David J, Ferreira J, Bernier G. Bmi1 is down-regulated in the aging brain and displays antioxidant and protective activities in neurons. PLoS One 2012;7:e31870.

[135] Kreso A, van Galen P, Pedley NM, Lima-Fernandes E, Frelin C, Davis T, et al. Self-renewal as a therapeutic target in human colorectal cancer. Nat Med 2014;20:29–36.

[136] Dovey JS, Zacharek SJ, Kim CF, Lees JA. Bmi1 is critical for lung tumorigenesis and bronchioalveolar stem cell expansion. Proc Natl Acad Sci USA 2008;105:11857–62.

[137] Bruggeman SW, Hulsman D, Tanger E, Buckle T, Blom M, Zevenhoven J, et al. Bmi1 controls tumor development in an Ink4a/Arf-independent manner in a mouse model for glioma. Cancer Cell 2007;12:328–41.

[138] Rizo A, Olthof S, Han L, Vellenga E, de Haan G, Schuringa JJ. Repression of BMI1 in normal and leukemic human CD34(+) cells impairs self-renewal and induces apoptosis. Blood 2009;114:1498–505.

[139] Yang MH, Hsu DS, Wang HW, Wang HJ, Lan HY, Yang WH, et al. Bmi1 is essential in Twist1-induced epithelial-mesenchymal transition. Nat Cell Biol 2010;12:982–92.

[140] Gallo M, Ho J, Coutinho FJ, Vanner R, Lee L, Head R, et al. A tumorigenic MLL-homeobox network in human glioblastoma stem cells. Cancer Res 2013;73:417–27.

[141] Heddleston JM, Hitomi M, Venere M, Flavahan WA, Yang K, Kim Y, et al. Glioma stem cell maintenance: the role of the microenvironment. Curr Pharm Des 2011;17:2386–401.

[142] Harris WJ, Huang X, Lynch JT, Spencer GJ, Hitchin JR, Li Y, et al. The histone demethylase KDM1A sustains the oncogenic potential of MLL-AF9 leukemia stem cells. Cancer Cell 2012;21:473–87.

[143] Lagos-Quintana M, Rauhut R, Lendeckel W, Tuschl T. Identification of novel genes coding for small expressed RNAs. Science 2001;294:853–8.

[144] Lau NC, Lim LP, Weinstein EG, Bartel DP. An abundant class of tiny RNAs with probable regulatory roles in *Caenorhabditis elegans*. Science 2001;294:858–62.

[145] Lee RC, Ambros V. An extensive class of small RNAs in *Caenorhabditis elegans*. Science 2001;294:862–4.

[146] Baek D, Villen J, Shin C, Camargo FD, Gygi SP, Bartel DP. The impact of microRNAs on protein output. Nature 2008;455:64–71.

[147] Stadler B, Ivanovska I, Mehta K, Song S, Nelson A, Tan Y, et al. Characterization of microRNAs involved in embryonic stem cell states. Stem Cells Dev 2010;19:935–50.

[148] Wang Y, Medvid R, Melton C, Jaenisch R, Blelloch R. DGCR8 is essential for microRNA biogenesis and silencing of embryonic stem cell self-renewal. Nat Genet 2007;39:380–5.

[149] Esquela-Kerscher A, Slack FJ. Oncomirs – microRNAs with a role in cancer. Nat Rev Cancer 2006;6:259–69.

[150] Lu J, Getz G, Miska EA, Alvarez-Saavedra E, Lamb J, Peck D, et al. MicroRNA expression profiles classify human cancers. Nature 2005;435:834–8.

[151] Song SJ, Ito K, Ala U, Kats L, Webster K, Sun SM, et al. The oncogenic microRNA miR-22 targets the TET2 tumor suppressor to promote hematopoietic stem cell self-renewal and transformation. Cell Stem Cell 2013;13:87–101.

[152] Godlewski J, Nowicki MO, Bronisz A, Williams S, Otsuki A, Nuovo G, et al. Targeting of the Bmi-1 oncogene/stem cell renewal factor by microRNA-128 inhibits glioma proliferation and self-renewal. Cancer Res 2008;68:9125–30.

[153] Bu P, Chen KY, Chen JH, Wang L, Walters J, Shin YJ, et al. A microRNA miR-34a-regulated bimodal switch targets Notch in colon cancer stem cells. Cell Stem Cell 2013;12:602–15.

[154] Bitarte N, Bandres E, Boni V, Zarate R, Rodriguez J, Gonzalez-Huarriz M, et al. MicroRNA-451 is involved in the self-renewal, tumorigenicity, and chemoresistance of colorectal cancer stem cells. Stem Cells 2011;29:1661–71.

[155] Liu C, Kelnar K, Liu B, Chen X, Calhoun-Davis T, Li H, et al. The microRNA miR-34a inhibits prostate cancer stem cells and metastasis by directly repressing CD44. Nat Med 2011;17:211–5.

CHAPTER 3

Enrichment and Interrogation of Cancer Stem Cells

A. Jarrar[1], A. Chumakova[1], M. Hitomi[1,2], J.D. Lathia[1,2,3]

[1]Cleveland Clinic, Cleveland, OH, United States; [2]Cleveland Clinic Lerner College of Medicine of Case Western Reserve University, Cleveland, OH, United States; [3]Case Comprehensive Cancer Center, Cleveland, OH, United States

CONTENTS

INTRODUCTION

It is well appreciated that tumors consist of heterogeneous neoplastic cells that display a spectrum of various degrees of differentiation similar to those found in the normal tissues from which the tumors originate [1–10]. The transplantation of patient-derived tumor cells into immunodeficient mice revealed different variation in their abilities to form tumors [11]. Upon secondary and tertiary transplantation, the resulting tumors maintain the same cellular heterogeneity similar to that of the original tumors. These observations highlight the possibility that intratumoral cellular heterogeneity may originate from a fraction of tumor cells with many characteristics similar to those of normal stem cells, namely the cells that give rise to all heterogeneous types of cell populations in the tissue, including stem cells themselves [12].

Similar observations made in multiple types of tumors led to the concept of the cancer stem cell (CSC), which was defined in the AACR Cancer Stem Cell Workshop as a cell in a tumor able to self-renew and give rise to the heterogeneous lineages of cancer cells that constitute the tumor [13]. This workshop group carefully pointed out that "self-renewal," the key defining characteristic of CSC, is not just simple proliferation but rather is a cell division that phenocopies the parental CSC in terms of tumorigenicity and the generation of differentiated cell lineages.

The CSC hypothesis has served to produce novel findings that provide new perspectives to understand cancer biology [14,15]. For example, many cell surface epitopes have been identified as CSC markers by virtue of their ability to enrich for cells with tumorigenic capacity confirmed by prospective xenograft tumor formation assays. Enrichment for CSCs via these markers has revealed targetable intrinsic and extrinsic factors required for maintaining the stem cell–like

Cancer Stem Cells. http://dx.doi.org/10.1016/B978-0-12-803892-5.00003-6

status. This can translate into improved patient survival, as CSCs are found to be more resistant to conventional chemo- and radiotherapy than other types of neoplastic cells in the tumors [16,17].

While the CSC hypothesis remains subject to scrutiny, studies on enriched tumor cells fulfilling the CSC criteria brought new perspectives for molecular mechanisms and dynamic cellular processes in self-renewal, tumor formation, therapeutic resistance, and tumor recurrence. Based on the accumulating knowledge from these studies, the recent in vivo lineage-tracing studies demonstrated compelling findings that CSCs can repopulate the tumor with a heterogeneous population of tumor cells in situ without transplantation assays in multiple mouse genetic tumor models [18–20]. These studies have illuminated the contribution of CSCs to intratumoral cellular heterogeneity in addition to or in parallel to the cellular heterogeneity caused by genetic instability of the tumor cells followed by clonal selection [21–23]. In this chapter, we describe the methods used to enrich for and analyze CSC populations and discuss the advantages and limitations of these approaches.

ENRICHMENT FOR CANCER STEM CELLS

Multiple approaches can be taken to enrich for CSCs (Fig. 3.1). Freshly resected patient tumor specimens or patient-derived xenografts (PDXs) maintained in immunocompromised mice are dissociated into a single-cell suspension using mechanical and enzymatic methods. Details of the dissociation protocols can be found in the following references [24,25]. CSC enrichment can be achieved by culturing bulk tumor cells in serum-free medium followed by introducing a reporter whose expression is driven by a stem cell–related gene promoter (Fig. 3.1a). Alternatively, CSCs can be enriched by culturing bulk tumor cells in serum-free medium containing basic fibroblast growth factor (bFGF) and epidermal growth factor (EGF). Such medium is permissible for CSC growth and will ensure genotypic and phenotypic resemblance to the original patient specimen (Fig. 3.1b) [26,27]. A third approach is to enrich for CSCs based on surface marker expression (Fig. 3.1c). Various cell surface markers have been reported to enrich for CSCs in different types of tumors. A compendium of CSC surface markers is summarized in the following section (Table 3.2). Each of the cell populations can then be interrogated further using functional studies, such as in vivo tumor formation analysis. These approaches will be discussed later.

Reporter Systems Used for Enrichment and Analysis of Cancer Stem Cells

Enrichment for CSCs based on cell surface marker expression, as discussed later, is the most commonly used approach, however, it has its limitations. The CSC state is plastic; under favorable conditions, non-CSCs can revert back to a CSC-like state [28–30]. Surface markers are limited in their ability to track

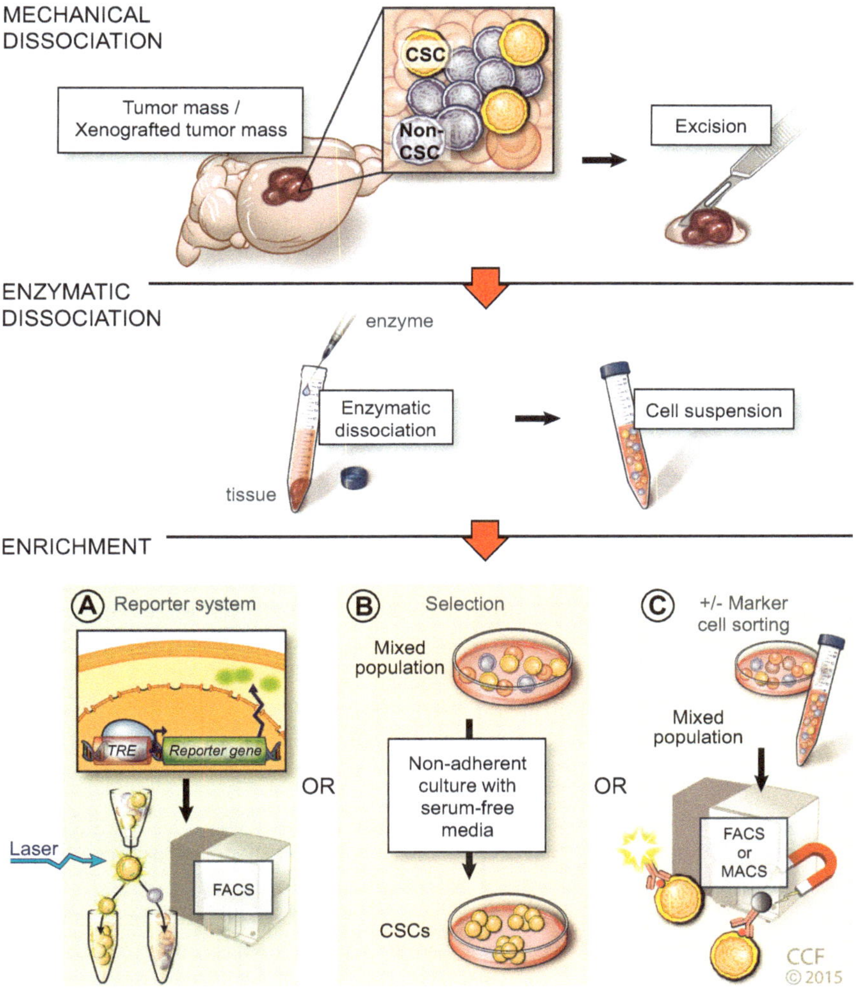

FIGURE 3.1

Multiple approaches are used to enrich for cancer stem cells (CSCs). Cancer specimens are dissociated into a single-cell suspension using mechanical (top panel) followed by enzymatic (middle panel) dissociation. The resultant single-cell tumor suspension can be: (a) expanded as bulk tumor cells in serum-free medium followed by introducing a reporter gene driven by a stem cell–related gene promoter, (b) enriched via culture in serum-free medium containing bFGF and EGF, and/or (c) sorted based on surface marker expression into marker-positive and marker-negative populations using immunofluorescently tagged antibodies via flow cytometry or magnetic beads tagged with antibodies via magnetic sorting. *Permission from The Cleveland Clinic Education Institute (CCEI), Cleveland, Ohio.*

changes in the CSC state in real time. Instead, the utilization of surface markers provides a surrogate snapshot of the stem cell status at the time of tumor dissociation. Additionally, the majority of the surface markers utilized lack a known function. These caveats raise the need to develop a reporter system that reflects

CSC functionality, that can be used to track cells in vitro as well as in vivo and can be combined with other modalities, such as the Cre recombinase, to lineage trace CSC progeny. This system has a wide range of applications, including but not limited to migration and invasion assays, time-lapse imaging to study CSC biology, and monitoring changes in the CSC state upon treatment with cytokines, drugs, or inhibitors.

The basic concept of reporter systems exploits the fact that pluripotent stem cell transcription factors such as NANOG, SOX2, OCT4, along with certain self-renewal pathways such as NOTCH and WNT, are upregulated in embryonic stem cells, induced pluripotent stem cells, and CSCs. Reporter systems enrich for the stem cell populations by indicating high activities of these molecules known to functionally drive a stem cell–like status. This type of system usually contains a fluorescent protein [green fluorescent protein (GFP), red fluorescent protein (RFP), enhanced green fluorescent protein (EGFP)] or a luminescent protein (luciferase) whose expression reports the activity of a gene of interest. There are two types of such reporter systems: promoter reporters and response reporters. The first directly monitors the activation of the promoter of a target gene by expressing a fluorescent or luminescent protein driven by the target promoter. The second monitors the binding of the transcription factor to its response element that then drives the expression of a fluorescent or luminescent protein. One disadvantage of the promoter reporters is that the relatively large promoter regions used in such constructs will invariably contain response elements for additional transcription factors, which may reduce reporter specificity.

Reporter systems are usually designed to express fluorescent proteins from the GFP family to enable real time tracing of promoter activity. To date, the family includes proteins with a rich palette of excitation and emission spectra (from blue to far red) and with different structural and stability properties. When designing a reporter system, two key characteristics of fluorescent proteins must be considered: maturation time and half-life in the cell. Traditional monomeric GFP matures within 40 min to 1 h [31]. Such a delay from the promoter activation to the appearance of fluorescence may not be acceptable for some studies. In this case, a quickly maturing tetrameric fluorescent protein with a maturation time as fast as several minutes can be used. The half-life of fluorescent proteins should be chosen depending on the purpose of the study. For the detection of transient expression or changes in promoter activation, short half-life fluorescent proteins are used. These proteins are fused with destabilization signals and are degraded within several hours [32,33]. For the analysis of accumulated signal, stable proteins with a half-life of more than 24 h (GFP) can be introduced into the reporter construct. Finally, for studies of temporal changes in gene expression, timer fluorescent proteins, such as DsRed, can be utilized. These proteins change their emission spectrum with time, and such aging of the fluorescent signal may help to determine the time of promoter activation. Timer fluorescent proteins can also have different maturation times and half-lives, which broadens their possible application [34].

In the following section, reporter systems used to enrich CSCs from a heterogeneous population of tumor cells along with daughter cell lineage-tracing techniques will be discussed. A summary of reporter systems discussed is provided in Table 3.1.

SOX2

Sex determining region Y box 2 (SOX2) is a transcription factor and a member of the HMG-box family that is expressed during embryogenesis [35,36] and in various cancers including glioblastoma (GBM), gastric cancer, small cell lung cancer, melanoma, esophageal squamous cell carcinoma, and breast cancer [37,38]. SOX2 is known to be an essential factor for inducing and maintaining stemness, namely through its effects on pluripotency and self-renewal [38]. The efficacy of SOX2 to enrich for CSCs has been shown in cervical cancer [39], GBM [40], and squamous cell carcinoma (SCC) [37]. In a mouse model

Table 3.1 Reporter Systems Validated for Cancer Stem Cell Enrichment

Reporter System	Gene Function	Cancers in Which the Reporter Has Been Validated
NANOG	Transcription factor, important for maintaining pluripotency in embryonic and adult stem cells, oncogene	Breast [57], prostate [199], nasopharynx [200], liver [37]
SOX2	Transcription factor, important for maintaining pluripotency in embryonic and adult stem cells, oncogene	Breast [38], glioma [40], skin [37], cervix [39]
OCT4	Transcription factor, important for maintaining pluripotency in embryonic and adult stem cells, oncogene	Liver [61], melanoma [60], sarcoma [62]
SORE6 "SOX2, OCT4"	Transcription factor, important for maintaining pluripotency in embryonic and adult stem cells, oncogene	Breast [65]
NOTCH pathway	Critical role in embryogenesis and fate decision, oncogene	Breast [71], lung [70]
Telomerase reverse transcriptase (TERT)	Catalytic subunit of the enzyme telomerase	Osteosarcoma [72]
Stem-SH2-domain-containing 5′-inositol phosphatase (s-SHIP)	Regulation of growth factor receptor–mediated signaling	Prostate [74]
TCF/LF	Wnt signaling pathway	Colorectal [201]
Alpha-fetoprotein (AFP)	Fetal plasma protein	Cholangiocarcinoma [75]
REX-1	Zinc finger gene	Glioblastoma [79] squamous cell carcinoma [80]

of GBM, a SOX2-EGFP reporter successfully isolated CSCs, which were phenotypically similar to patient-derived tumor-initiating cells with respect to modes of cell division, self-renewal, niche localization, and differentiation [40]. This transcription factor is also associated with invasion and chemoresistance [20,37,41]. Of interest, SOX2 has been shown to be upregulated in preneoplastic skin tumors. Using a SOX2-GFP knockin mouse, Boumahdi et al. demonstrated that SOX2-high SCC cells were greatly enriched for tumor-propagating cells and elucidated the role of SOX2 in the initiation and progression of primary skin tumors [37].

Taken together, SOX2 promoter-driven reporter constructs have been shown to enrich CSCs in different solid tumors and can be utilized as a powerful tool to track CSCs in real time using both in vitro and in vivo models.

NANOG

NANOG is a homeodomain-containing transcription factor involved in the self-renewal of undifferentiated embryonic stem cells and is thought to be a key factor in maintaining pluripotency [42,43]. The expression of NANOG has been detected by several groups in germ cell tumors as well as in other tumors, including breast, cervical, oral, kidney, prostate, lung, gastric, brain, and ovarian cancer [44–51]. High expression of NANOG is an indicator of poor prognosis for ovarian serous carcinoma, colorectal, and breast cancer patients [52–54]. In oral SCC and lung adenocarcinoma, higher expression of NANOG was associated with advanced cancer stage and shorter patient survival [45]. NANOG plays an important role in the maintenance of CSCs [55,56]. In a recent study, a NANOG promoter-driven GFP reporter was used to enrich for a CSC population in triple-negative breast cancer [57].

Moreover, increased levels of the NANOG protein have been demonstrated in CSCs from various tumor types that were isolated by differential expression of surface markers [58,59]. Interestingly, NANOG knockdown caused a decrease in tumorigenicity and clonogenic growth of several breast, prostate, and colon cancer cell lines [56]. In contrast, overexpression of NANOG alone was sufficient to induce self-renewal, tumor initiation and drug resistance in prostate and breast cancer cell lines. This change in cell behavior was also accompanied by a more stemlike phenotype with increased expression of CD133, CD44, and ABCG2 [56]. These findings indicate that NANOG is expressed in CSCs and that it confers pluripotency, similar to its effects in embryonic stem cells. Moreover, NANOG expression promotes malignant transformation and tumor progression.

Based on the abundant expression and importance of NANOG in CSCs, a novel CSC reporter system using a NANOG promoter-fluorescent reporter gene strategy has been employed in prostate and breast cancers [56,57].

OCT4

Similar to NANOG and SOX2, OCT4 is a key regulator of pluripotency and is upregulated in human solid tumors [60]. An OCT4-based reporter system has been used to enrich for CSCs in melanoma [60], hepatocellular carcinoma (HCC) [61], and soft tissue sarcomas [62]. This reporter system was used in two studies to investigate the preclinical activity of patient-derived cytokine-induced killer (CIK) cells against CSCs and non-CSCs from melanoma and soft tissue sarcoma [60,62]. In HCC, $OCT4^{+}$ cells demonstrated classical CSC features including higher expression of stemness-associated transcription factors, in vitro sphere formation, tumor initiation in immunocompromised mice, and resistance to sorafenib [61]. Therefore, an OCT4 promoter-driven reporter system is a promising novel way to isolate CSCs in solid tumors.

SORE6

To activate its target genes, SOX2 cooperatively binds to DNA with OCT4 [63,64], forming a stable tertiary structure with the DNA [43]. A fluorescent reporter system, SORE6, has been developed to monitor OCT4/SOX2 activity. This construct has six tandem repeats where composite OCT4/SOX2 response elements are used to drive expression of a fluorescent protein. Breast cancer cells enriched using expression of the SORE6 reporter exhibited higher self-renewal, tumor initiation, and resistance to chemotherapeutics [65]. The tumor cell subpopulation selected by the SORE6 reporter generated heterogeneous progeny, which is a cardinal characteristic of CSCs. The utility of this reporter in other solid tumors is under investigation.

NOTCH

The NOTCH signaling pathway is a highly conserved cell signaling system present in most multicellular organisms. Mammals possess four different NOTCH receptors, referred to as NOTCH 1–4 [66], which are involved in a downstream proteolysis pathway that eventually releases the NOTCH intracellular domain protein, which enhances gene transcription [67]. In addition to its critical role in embryogenesis and fate decision in various cellular systems, the NOTCH pathway has also been implicated in a wide variety of malignancies including non-small cell lung cancer (NSCLC) [68,69].

A NOTCH-GFP reporter construct was initially used to enrich a self-renewing cell population in NSCLC [70]. The expression of the NOTCH 2 and 4 receptors was higher in the GFP-high population, while the GFP-low population showed higher expression of the NOTCH 1 and NOTCH 3 receptors. GFP-high cells had higher tumor initiation capacity as assessed by serial in vivo xenotransplantation assays in non-obese diabetic (NOD)/severe combined immunodeficient (SCID) mice. Cells with high NOTCH activity were more resistant to chemotherapy, and NOTCH activity correlated with poor clinical outcome.

The NOTCH reporter system has also been used in breast cancer cell lines and successfully enriched for CSCs with higher tumor-initiating capacity [71].

TERT

Telomerase is known to be essential in maintaining CSCs/progenitor cells. Human telomerase reverse transcriptase (hTERT) is the catalytic subunit of the telomerase enzyme, and an hTERT reporter has been used to enrich CSCs in osteosarcoma [72]. It was found that a subpopulation of osteosarcoma cells expressed higher levels of telomerase, the readout of which was bright GFP expression linked to the hTERT promoter. Telomerase-positive cells had an enhanced ability to form sarcospheres and demonstrated higher levels of epithelial–mesenchymal transition markers.

s-SHIP

s-SHIP is a stem cell–specific isoform of the protein SH2 domain-containing 5′-inositol phosphatase (SHIP) and is expressed in both pluripotent embryonic stem (ES) cells and adult tissue-specific multipotent cells such as hematopoietic stem cells (HSCs). s-SHIP lacks an SH2 domain but contains a 5′-inositol phosphatase domain and several protein–protein interaction domains that potentially enable its participation in many different signaling pathways [73]. In the context of CSCs, the role of s-SHIP has been studied in human prostate cancer cell lines [74]. An s-SHIP promoter-linked GFP reporter has been used to isolate a stem cell–like subpopulation of cells. Cells with high s-SHIP activity demonstrated the ability to form spheres and had higher expression of pluripotency transcription factors (OCT4, SOX2) and stem cell surface markers (TROP2, CD44, CD166).

Alpha Fetoprotein

Alpha fetoprotein (AFP) is a well-known marker for hepatic progenitor cells, but it has also been studied as a marker for CSCs in cholangiocarcinoma [75]. Four cholangiocarcinoma cell lines were transfected with an AFP enhancer/promoter-driven EGFP gene, and one out of four cell lines showed a population of cells expressing both AFP and EGFP. EGFP-positive cells had higher tumorigenic potential and NOTCH 1 expression, which supported the role of the NOTCH pathway in maintaining CSC characteristics.

Transcription Factor/Lymphoid Enhancer Binding Factor

Wnt/beta-catenin signaling has been reported to be enhanced in CSCs. When Wnt binds to frizzled receptor it causes beta-catenin translocation to the nucleus. Beta-catenin binds to certain transcriptional factors such as transcription factor (TCF)/lymphoid enhancer binding factor (LEF). Adenomatous polyposis coli (APC) is a negative regulator of β-catenin concentrations. Colorectal cancers exhibit heterogeneity in β-catenin nuclear localization despite the high prevalence of a nonfunctional mutated version of the APC protein.

Using a TCF/LEF-driven GFP reporter system, it was shown that cancer cells with β-actin nuclear localization behave functionally like CSCs [76]. The luciferase or GFP β-catenin-based reporter systems identify binding of β-catenin to the tandem repeats of the TCF/LEF transcriptional response element present in the reporter. β-Catenin has been also shown to bind and regulate the activity of SOX2 and OCT4 [77]. Wnt reporter system identified a preferential localization of CSCs to specific niches such as peri-myofibroblasts [28,78] and perivascular [78] in colorectal and breast cancers.

REX-1

A less-studied pluripotency gene, retinoic acid-regulated zinc finger (REX-1) is another CSC marker. Rex-1 gene expression is increased in stem cells in chemoresistant GBM, SCC, pancreatic tumors, and other cancers [79,80]. A GFP-based promoter reporter system designed for normal human stem cells was described to be efficient for the detection of pluripotent cells reprogrammed from normal human fibroblasts [81].

Caveats of Using Reporters

One caveat to the use of reporters to enrich for stem cells is that long promoter sequences might contain binding sites for various transcription factors, which would adversely affect the specificity of the reporter. Therefore, meticulous validation of sorted cells using functional assays in vivo and in vitro is imperative. In addition, the validation of the endogenous expression of CSC markers (such as SOX2, OCT4, NANOG) in reporter high cells is essential. Second, experiments incorporating fluorescent dyes should be carefully designed to prevent cross talk between the GFP signal and other fluorophores. This is a common problem when comparing GFP^+ and GFP^- cells with commercially available assays (eg, annexin V-FITC staining, BCECF-based assays, aldefluor assays). Third, dead or dying cells emit autofluorescence and contribute a false positive signal. This potential complication can be eliminated by including cell death assays such as dye exclusion analysis.

Enrichment for Cancer Stem Cells Using Culture Media

A shared feature between CSCs and normal stem cells is their continuous self-renewal, and potential for full or partial differentiation—properties not found in differentiated cells or traditional commercially available cancer lines. One obvious difference in the in vitro propagation of normal stem cells compared to the commonly used cancer cell lines is the requirement for serum. Many established cell lines are grown in media containing serum, whereas the majority of normal stem cells are grown in serum-free media since serum induces their differentiation. Based on these analogies, a new method to select for CSCs was generated. Bulk cancer cells obtained from freshly resected specimens are cultured under conditions optimal for propagation and nondifferentiation of normal stem cells (serum-free media supplemented with bFGF and EGF), or conditions optimal for growth of more differentiated cells and most other cancer cell lines ("serum" conditions: Dulbecco's Modified Eagle Medium (DMEM) media containing 10% fetal bovine serum). The resultant cell populations in the former are cancer cells with CSC-like properties compared to a more differentiated one in the latter (Fig. 3.1b). Such method has been applied to multiple types of cancers including GBM and colorectal cancers [26,27]. One of the limitations of using culture conditions to enrich for CSCs is the lack of capability to enrich for and maintain non-CSCs.

Surface Markers Used for Enrichment and Analysis of Cancer Stem Cells

Using surface markers as a method of enrichment can be beneficial. Surface markers are easily accessible via fluorescence-activated cell sorting (FACS) or magnetic-activated cell sorting (MACS). Their utilization minimizes the genetic manipulation of the cells (as is the case with reporter systems), and acutely dissociated tumors can be sorted closer to the time of resection compared with other methods that necessitate longer periods in culture. However, enrichment via surface markers has certain limitations: solid tumors require protease digestion to obtain a single-cell suspension for FACS/MACS, which may impair some surface markers. Additionally, many of the markers utilized lack a known biological functionality, raising questions about their relevance and relation to the CSC state.

Following is a summary of the identified surface markers used most commonly to enrich for CSCs (Table 3.2). For a more comprehensive review, the reader can refer to additional review articles [82–84].

Acute Myelogenous Leukemia

$CD34^+/CD38^-$ is the earliest documented CSC marker expression pattern identified by Dick et al. in 1994. These markers are still widely used to identify acute myelogenous leukemia (AML) CSCs. $CD34^+/CD38^-$ cells were capable of repopulating AML in SCID mice. On the other hand, $CD34^+/CD38^+$ or $CD34^-$ cell fractions did not exhibit these properties [11]. Later studies provided evidence that $CD34^+/CD38^-$ was a common immune phenotype for leukemic CSCs in multiple AML subtypes [85]. Over time, different combinations and additions to $CD34^+/CD38^-$ were suggested. Examples of this are $CD34^+/CD90^-$ (Thy-1), $CD34^+/CD71^-/HLA\text{-}DR^-$, $CD34^+/CD117^-$ (c-kit), and $CD34^+/CD38^-/CD123^+$ (interleukin-3 receptor α chain) [86–88]. Recently, it was shown that $CD34^+/CD38^-/CD45^+/CD123^+$ is a global marker for AML CSCs [89]. $CD34^+/CD38^-/CD123^+$ CSCs have been noted for their upregulation of drug efflux pumps and drug resistance proteins [90].

Table 3.2 Cancer Stem Cells (CSCs) Surface Markers

Tumor Type	CSCs Markers
Acute myelogenous leukemia (AML)	($CD34^+/CD38^-$) [11], ($CD34^+/CD90^-$) [86], ($CD34^+/CD71^-/HLA\text{-}DR^-$) [87], ($CD34^+/CD117^-$) [88], ($CD34^+/CD38^-/CD123^+$) [202], ($CD34^+/CD38^-/CD45^+/CD123^+$) [90] CLL-1 [91], CD96 [92], ($CD25^+/CD34^+/CD38^-$), ($CD32^+/CD34^+/CD38^-$) [94]
Acute lymphoblastic leukemia (ALL)	($CD34^+/CD38^-$) [95], ($CD34^+/CD10^-$) [98], ($CD34^+/CD19^-$) [98], CD9 [99], CD90 (Thy-1), CD110 (c-Mpl) [100]
Chronic myelogenous leukemia (CML)	($CD34^+/Ph^+$) [102], ($CD34^+/CD38^-/CD123^+$) [103]
Breast cancer	($CD44^+/CD24^-$) [104], ($CD44^+/CD24^-/EpCAM^+$) [108], $ALDH1^{+a}$ [109], ($CD133^+/CD44^+$) [110–112], ($CD44^+/CD173^+$), ($CD44^+/CD174^+$), ($CD44^+/CD176^+$) [114], ($CD24^+/CD29^+$), ($CD24^+/CD49f^+$) [113]
Colorectal cancer	CD133 [27,116] EphB2high [120], ($EpCAM^+/CD44^+/CD166^+$) [121], $ALDH1^{+a}$ [122], $LGR5^+$ [123], and $CD44v6^+$ [124]
Liver cancer	$CD133^+$ [126], ($CD133^+/CD44^+$) [127], (CD176) [128], EpCAM [129,130], CD90 (Thy-1) [131], CD13 [132], OV6 [133]
Pancreatic cancer	($CD44^+/CD24^+/EpCAM^+$) [134], $ALDH1^{+a}$ [136], $CD133^+$ [135], c-met [137]
Glioblastoma	$CD133^+$ [140], integrin $\alpha6$ [25], L1CAM [144], SSEA-1, CD15 [145] $CD44^+$ [146,147], $A2B5^+$ [146,147]
Lung cancer	$CD133^+$ [150], CD176 [128]
Prostate cancer	$CD44^+/CD133^+$ [152,153], ($Sca\text{-}1^+/CD49f^+$), ($Sca\text{-}1^+/CD49f^-$) [157]

[a]*ALDH1 is not a surface marker.*

In addition to the markers mentioned earlier, C-type lectinlike molecule-1 (CLL-1) has also been found to be exclusively expressed on AML CSCs in 90% of studied cases. Isolation and transplantation of $CD34^+/CD38^-/CLL\text{-}1^+$ cells generated AML blasts in NOD/SCID mice [91]. Another important CSC marker in AML is CD96, which belongs to the immunoglobulin superfamily. It has been shown to be selectively overexpressed on $CD34^+/CD38^-/CD90^-/Lin^-$ phenotype compared with the basal expression level of CD96 on normal HSCs [92]. Several isoforms of the adhesion molecule CD44 are involved in the interaction between AML progenitor cells and the surrounding stromal cells (ie, the CSC niche) in primary human AML samples [93]. Another recent study identified CD25 or CD32 along with $CD34^+/CD38^-$ as potential CSC markers for AML. These two AML CSC populations ($CD25^+/CD34^+/CD38^-$ and $CD32^+/CD34^+/CD38^-$) were quiescent and resistant to chemotherapeutic drugs [94].

Acute Lymphoblastic Leukemia

Similar to AML, acute lymphoblastic leukemia (ALL) CSCs can be enriched using the marker pattern $CD34^+/CD38^-$; however, the addition of $CD33^-/CD19^-$ or the Philadelphia chromosome translocation (Ph) increases the robustness of defining CSCs [95]. Subsequent studies have shown that $CD34^+/$

$CD38^-/CD19^+$ cells carry the Philadelphia chromosome, while $CD19^-$ cells do not [96]. ALL can be classified as T lineage (T-ALL) or B lineage (B-ALL) [97]. ALL cells capable of long-term proliferation and differentiation into pre-B-ALL in vitro and in vivo (NOD/SCID transplant model) were derived from only $CD34^+/CD10^-$ or $CD34^+/CD19^-$ subfractions, indicating that the target cells for transformation in pre-B-ALL have a more immature phenotype than the bulk ALL population [98]. In childhood ALLs, $CD133^+/CD19^-$ marked cells with a CSC-like phenotype. CD9 has been recently suggested as a more reliable marker for CSCs in B-ALL compared with $CD34^+$ cells [99]. The same group also reported the importance of CD90 (Thy-1) and CD110 (c-Mpl) as markers for T-ALL CSCs [100].

Chronic Myelogenous Leukemia

Chronic myelogenous leukemia (CML) is characterized by the overproduction of mature myeloid cells. Like ALL, CML is associated with Philadelphia chromosomal translocation [101]. CML CSCs are also characterized as $CD34^+/Ph^+$ [102]. Additionally, CD123 has been characterized as a surface marker for CML CSCs (phenotype $CD34^+/CD38^-/CD123^+$) [103].

Breast Cancer

The first marker pattern identified for breast CSCs was $CD44^+/CD24^-$ [104], which correlated with therapeutic resistance [105], poor prognosis [106], and enhanced invasive properties [107]. The addition of epithelial cell adhesion molecule (EpCAM) enhanced the robustness for human breast CSC enrichment [108]. The cytoplasmic enzyme aldehyde dehydrogenase 1 (ALDH) and CD44 have been employed to identify putative breast CSCs and indicate poor prognosis [109]. Additional marker combinations proposed to enrich breast CSC, include $CD133^+/CD44^+$ [110–112], $CD24^+/CD29^+$ ($\beta 1$ integrin), or $CD24^+/CD49f^+$ (integrin $\alpha 6$) [113]. Coexpression of CD173, CD174, and CD176 with CD44 is another cell surface marker combination proposed to identify breast CSCs [114]. The molecular classification of breast cancer established four major subtypes: the luminal A and B, the HER2-overexpressing (HER2-OE), and basal-like tumors. Luminal A is the most prevalent subtype and is characterized by the expression of estrogen and progesterone receptors (ER and PR, respectively) in cancer cells, whereas the luminal B subtype is characterized by ER and PR expression together with HER2 overexpression. In contrast, HER2-OE tumors are negative for hormonal receptors and overexpress HER2 protein. Finally, the basal-like "triple-negative" tumors are characterized by the absence of ER, PR, and HER2 expression. The basal-like subtype still constitutes a heterogeneous group of tumors, expressing distinct basal markers. The described $CD44^+/CD24^-$ and $ALDH1^+$ stemlike phenotypes seem to identify CSC with distinct levels of differentiation based on the molecular subtype described earlier. $CD44^+/CD24^-$ being more related to basal-like carcinomas that most probably originate from the most primitive mammary stem cells, whereas ALDH1$^+$ is a marker of basal-like and HER2-overexpressing tumors, putatively originating from luminal committed progenitors [115]. It is worth mentioning that ALDH1 activity is commonly used to enrich for CSCs, although it is not a surface marker.

Colorectal Cancer

Human colorectal CSCs were first isolated using CD133 [27,116]. Several monoclonal antibodies have been developed against CD133. The most commonly used antibodies are AC133 (CD133/1) and 293C/AC141 (CD133/2), which are reported to recognize distinct epitopes [117]. AC133 is most frequently used to isolate CSCs and has been suggested to recognize a glycosylated epitope on CD133 [116]. Variations in the expression levels of AC133 are cell cycle dependent [118]. Cells in G0 (quiescent) or the G1 phase of the cell cycle have been reported to show the lowest expression. Once the same cells progress in the cell cycle, higher levels of AC133 were detected. As such, the association of AC133 with actively cycling cells may contribute to the basis for enrichment for CSCs [118]. The search for more specific surface markers of colorectal CSCs continued with the hope of finding a marker that is very specific to cancer and can be used as a biomarker to predict outcomes [119]. Markers that have been described to characterize colorectal CSCs include EphB2high [120], EpCAM$^+$/CD44$^+$/CD166$^+$ [121], ALDH$^+$ [122], LGR5$^+$ [123], and CD44v6$^+$ [124]. Many questions remain open in the field of colorectal CSCs. First, the consistency of CSC markers deserves further investigation, because the CSC phenotype itself has been shown to be plastic. It has been demonstrated that CSC populations that are LGR5 positive and negative can interconvert upon chemotherapy [125]. Moreover, cytokines produced by tumor-associated cells can induce increased CSC self-renewal [17,28] and reprogram transit-amplifying cells.

Liver Cancer

CD133$^+$ was the first reported putative marker for HCC CSCs. CD133$^+$ tumor cells showed an upregulation of stem-related pathways such as Wnt/β-catenin, NOTCH, and Hedgehog signaling [126]. Another well-established cell surface marker of HCC stem cells is CD44. Within the HCC CD133$^+$ population, CD44$^+$ cells were found to be more tumorigenic than CD44$^-$ cells in NOD/SCID mice [127]. Thomsen Friedenreich antigen (CD176) was also found to coexpress with CD44 in HCC [128]. Similar to colorectal cancer, EpCAM was found to further enrich for HCC CSCs [129,130]. CD90 (Thy-1) [131], CD13 (alanine aminopeptidase) [132], and OV6 [133] are other markers that have been reported to enrich for HCC CSCs.

Pancreatic Cancer

Pancreatic adenocarcinoma mortality approaches 100%. Identifying and characterizing a CSC compartment represents a window of opportunity in this highly lethal disease. A small population (0.2–0.8%) of pancreatic adenocarcinoma cells expressing CD44$^+$/CD24$^+$/EpCAM$^+$ showed a 100-fold increase in tumorigenicity when injected into immunocompromised mice [134]. In another study, CD133$^+$/CXCR4$^+$ markers identified another subgroup of pancreatic CSCs. Although CD133$^+$/CXCR4$^+$ cells and CD133$^+$/CXCR4$^-$ cells showed similar tumorigenicity when xenografted into nude mice, only the CD133$^+$/CXCR4$^+$ cells were capable of metastasis [135]. ALDH has also been implicated as a stem cell marker in pancreatic adenocarcinoma. ALDH$^+$ cells showed higher tumorigenicity compared with ALDH$^-$ cells independent of CD133 status [136]. c-Met is another promising marker of stemlike cells in pancreatic cancer. c-Met$^+$ cells have been shown to have a greater tumorigenic potential than CD44$^+$/CD24$^+$/ESA$^+$ and CD133$^+$ cells [137].

Side population cells are defined as cells capable of effluxing Hoechst dye via the adenosine triphosphate (ATP)-binding cassette (ABC) multidrug resistance transporter [138]. Side population isolation has been shown to enrich for pancreatic CSCs [139].

Glioblastoma

Singh et al. first identified CD133 as a CSC marker in GBM [140]. Subsequently, CD133 has been validated by multiple groups [141–143]. As stated earlier, an antibody against the glycosylated epitope of CD133 (AC133) is used to enrich for GBM CSCs. Another GBM CSC marker that has been identified is integrin α6. Integrin α6 is coexpressed with conventional CSCs markers such as CD133 and enriches for CSCs. Targeting integrin α6 in CSCs inhibits self-renewal, proliferation, and tumor formation [25] L1CAM, a transmembrane adhesion protein found in CD133$^+$ glioma cell populations, was also identified as a brain CSC marker [144]. Stage-specific embryonic antigen-1 (SSEA-1), or CD15, has also been identified as a potential marker of GBM CSCs [145]. CD44 and A2B5(+) have also been identified as potentials marker for GBM CSCs with CD44 being a preferential CSC marker in mesenchymal subtype of GBM [146–149].

Lung Cancer

CD133$^+$ enriched for CSC subpopulations in small cell and non-small cell lung cancers. The injection of 10,000 lung cancer CD133$^+$ cells into immunocompromised mice readily generated tumor xenografts phenotypically identical to the original tumor. Upon differentiation, lung cancer CD133$^+$ cells acquired specific lineage markers while losing both tumorigenic potential and CD133 expression [150]. CD176 has also been identified as a potential marker for lung CSC populations. Coexpression of CD176 and CD133 or CD44 has been observed, and CD176$^+$ cell populations were found to exhibit features of CSCs [128]. In lung cancer mouse models, the cell surface marker Sca1 has also been found to enrich for CSCs. Sca1 did not enrich for CSCs in tumors initiated with oncogenic Kras, and only Sca1-negative cells propagated EGFR mutant tumors. In contrast, Sca1-positive cells were enriched for tumor-propagating activity in Kras tumors with p53 deficiency suggesting a genotype-dependent CSC surface marker [151].

Prostate Cancer

It has been demonstrated that approximately 0.1% of prostate cancer cells are $CD44^+/CD133^+$ and possess stem cell features such as self-renewal and unlimited proliferation [152,153]. $CD44^+$ prostate cells exhibit sphere-forming ability and metastatic potential [154]. These cells show an upregulation of NANOG, Hedgehog signaling, and Bmi1 [155]. The side population method has also been used to isolate prostate CSCs [156]. Additional potential surface markers, Sca-1 and CD49f, were identified as surface markers of prostate CSCs. Sorted cells ($Sca\text{-}1^+/CD49f^+$ and $Sca\text{-}1^+/CD49f^-$) were then injected into SCID mice, and only $Sca\text{-}1^+/CD49f^+$ cells were able to form tumors [157].

FUNCTIONAL ANALYSIS

Lineage Tracing

In Vivo Lineage Tracing

In vivo lineage tracing using various technologies has been used in studies of normal stem cells to understand cellular dynamics during development, the maintenance of tissue homeostasis and regeneration. Lineage-tracing reagents must be able to mark the cells of interest, and the marking must be long-lasting and exclusively transferred to all successive progeny but not to the unrelated cells outside the progeny. Dye labeling, radioactive tracers, or stainable nucleotide analogs that are incorporated into DNA during S phase have been used successfully. Genetic marking is a preferred approach for a long-term stable and inheritable labeling, which is required for reliable lineage tracing in vivo [158].

Genetic Marking Methods in Transgenic Mouse Models

The site-specific recombination using Cre-Lox recombination technology is commonly used to introduce genetic marking for lineage tracing. By excising the LoxP flanked stop codon that prevents expression of a marker gene knocked into the Rosa26 locus, expressed Cre recombinase transgene product genetically marks all the progeny of the cells undergoing Cre-mediated recombination (Fig. 3.2A). Selecting the appropriate promoter to induce Cre recombinase expression allows for the genetic marking of specific stem/progenitor cells at their specific stage of differentiation and traces their progeny. A secondary layer of Cre recombinase activity control can be achieved using the CreERT system (Fig. 3.2B, top panel). The recombinase activity is inhibited by a fused modified ER domain, which can be reversed by an ER antagonist, Tamoxifen.

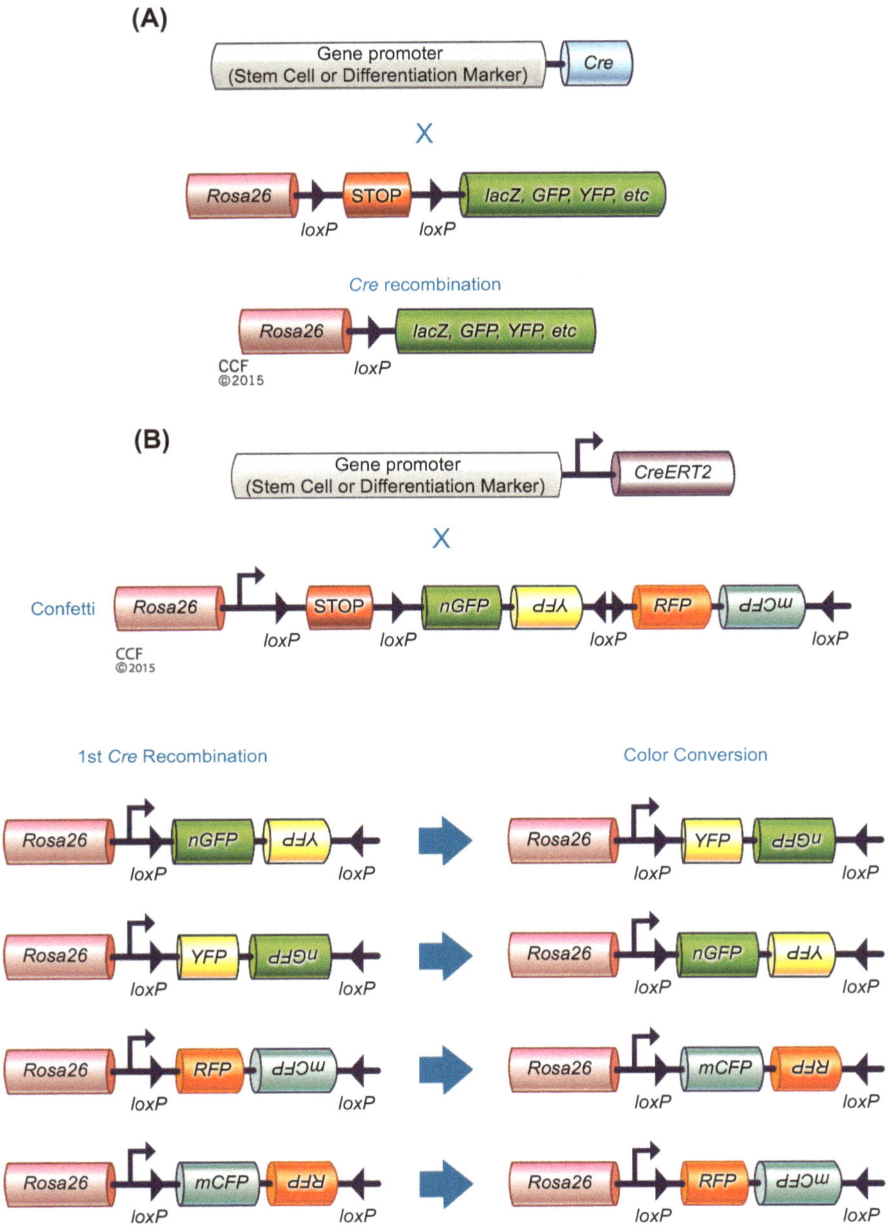

FIGURE 3.2

(A) Cre recombinase is driven by a specific promoter of interest, such as LGR5. Expressed Cre excises a stop codon sequence that prevents the expression of a marker gene knocked into the Rosa26 locus. Leading genetic marking of all progeny of the cells with activated promoter of interest. (B) A secondary layer of Cre recombinase activity control can be achieved in the CreERT system. The recombinase activity inhibited by the fused estrogen receptor (ER) domain can be reversed by an ER antagonist, Tamoxifen. Shown are the possible outcomes of the R26R-confetti locus after the first Cre recombination (tracing) and the second Cre recombination (color conversion).

Using this system, titrating the dosage of Tamoxifen can control the frequency of genetic marking of the cells, making this system suitable for lineage tracing at a single clonal resolution [18]. Alternatively, the activity of Cre in the cells can be quantitatively regulated by the doxycycline dosage when Cre expression is controlled by a reverse tetracycline-controlled transactivator (rtTA). In this system, the rtTA expression needs to be under the control of a defined promoter to ensure specific genetic marking of the cells of interest [159].

To simultaneously trace multiple clonal populations, a confetti construct in the Rosa26 locus has been used to mark multiple clones expressing fluorescent proteins of different colors. Subsequent activation of Cre recombinase can induce a color conversion to further trace the progeny (Fig. 3.2B) [19].

Lineage Tracing of Cancer Stem Cells In Vivo

These gene marking strategies have been applied together with knowledge of putative CSC promoter activity to examine hierarchical cellular behavior in mouse tumor models. Following are the examples of these applications.

Skin Papilloma

Keratin 14 is a keratin isoform strongly expressed by highly proliferative basal cells in chemically induced papilloma. The promoter of this keratin isoform was used to drive CreERT to mark papilloma cells at a frequency that allowed clonal linage tracing. A fraction of the resulting GFP-positive tumor cells initially detected at the basal layer of the papilloma gave rise to a clonal expansion encompassing the differentiation axis of the papilloma over time. The analysis of clonal expansion kinetics also detected GFP-positive cells with limited proliferative capacity, indicating that heterogeneity in proliferative capacity exists within the population of GFP-positive cells [18]. These data demonstrate that the presence of CSCs with the capacity of prolonged proliferation and differentiation contributes to the formation of papilloma consisting of heterogeneous tumor cell populations.

Intestinal Polyp

LGR5 marks an intestinal stem cell population at the basal area of crypts in mice. To study CSC biology in intestinal adenoma caused by deletion of the APC gene, LGR5 promoter-driven CreERT2-IRES-GFP was activated by administrating tamoxifen to genetically mark intestinal adenoma cells with various colors of fluorescent proteins by recombining and permanently activating confetti fluorescent proteins cloned in to the Rosa26 locus. At 4 weeks after the activation of LGR5-driven Cre, structural segments of adenoma similar to the crypt units of the intestine were uniformly labeled with a single fluorescent protein, whereas GFP expression reporting current LGR5 promoter activity was restricted to the basal part of the cryptlike structural segments of the adenoma.

When Cre was activated for the second time after clonal monocolored cryptlike segment establishment, cells with a color conversion were detected initially at the basal region of the cryptlike segments. Over time, these color-converted cells expanded from the basal to the apical direction. The cells undergoing color conversion were either positive for LGR5 reporter GFP or GFP-negative with the characteristics of Paneth cells or transit amplifying cells. Lineage tracing of the first Cre-mediated marking revealed that LGR5 cells have stem cell features because they can populate an entire cryptlike segment unit of the adenoma consisting of various types of cells, including those with an active LGR5 promoter. The second Cre-mediated marking with color conversion provided genuine experimental evidence of a self-renewing stem cell in vivo by demonstrating that progeny with an active LGR5 promoter generated from the first marking still possessed characteristics of stem cells and gave rise to all cell types, including LGR5-positive stem cells, that constitute the cryptlike unit of the adenoma [19].

Medulloblastoma

In a transgenic Patched (Ptc) mutant mouse model of medulloblastoma [160,161], kinetic analysis of incorporated 5-ethynyl-2′-deoxyuridine (EdU) positivity among the different marker-expressing tumor cell populations (CD15/SSEA-1) detected that SOX2-positive tumor cells retained the EdU label for a long time. This suggested that they were quiescent CSCs and gave rise to actively proliferating, double cortin-positive, amplifying tumor cells and eventually to differentiated, quiescent, NeuN-positive tumor cells. SOX2-positive medulloblastoma cells also showed other CSC characteristics such as coexpression of established medulloblastoma stem cell markers, self-renewal ability, and tumor propagation upon transplantation. To further confirm that SOX2-positive cells are genuine medulloblastoma stem cells, SOX2 promoter-driven CreERT2 was used to express a Rosa26 tdTomato fluorescent protein to trace the progeny of the SOX2-positive cells. As expected, tdTomato-positive tumor cells expanded in the tumor over time, as did the number of tdTomato-positive tumor cells that also expressed double cortin, NeuN, betaIII tubulin, glial fibrillary acidic protein (GFAP) and s100 B, indicating that SOX2-positive medulloblastoma cells give rise to various types of more differentiated tumor cells. Interestingly, when this lineage-tracing genetic marking was turned on prior to treatment with a Smoothened inhibitor, the recurrent tumor contained twice as many tdTomato-positive tumor cells, indicating CSC involvement in tumor recurrence. These results demonstrate the utility of genetic marking–based lineage tracing to delineate the therapeutic resistance of tumors [20].

Glioblastoma

To study the therapeutically resistant subpopulation of tumor cells in vivo, the activation of a similar stem cell gene promoter–driven transgene was utilized in a genetically engineered mouse model for GBM. In this model, the endogenous Nestin promoter was used to drive expression of HSV thymidine kinase (HSV-TK) and GFP (Nes-TK-GFP). GBM was induced by GFAP-Cre-dependent conditional inactivation of PTEN, p53, and NF1. In the resulting tumors, GFP-positive (Nes-TK-GFP reporter active) cells were rare and positive for SOX2 expression, a CSC marker. Lineage tracing using consecutive dual labeling with CldU and IdU revealed that Nes-Tk-GFP-positive cells are the origin of hierarchical tumor regrowth after antitumor treatment with Temozolomide; suggesting that these cells are the tumor repopulating cells during posttherapy tumor regrowth. To verify these lineage-tracing data, Nes-TK-GFP-positive cells were ablated by Ganciclovir treatment, which kills HSV-TK-expressing proliferating cells. As predicted, elimination of Nes-TK-GFP glioma cells inhibited tumor growth and increased the survival of tumor-bearing mice. This type of "lineage-erasing" approach is a good tool to validate "lineage-tracing" findings [41].

Potential Utility of Genetic Marking Based on the Zygosity State

Mosaic analysis with double marker (MADM) is a robust genetic marking system that allows labeling of cells according to the status of zygosity of the alleles of interest (Fig. 3.3). The activation of Cre during G2 phase of the cell cycle induces recombination between the introduced lox elements flanked by N- and C-terminal fragments of a fluorescent protein of a different color on the homologous chromosome loci. The cells with recombination undergoing X segregation will yield RFP or GFP expression with homozygosity for the alleles of interest (Fig. 3.3), whereas when undergoing Z segregation, heterozygous cells are either marked with both colors of fluorescent proteins or not marked with either of them [162]. This approach was successfully utilized to identify the cell of origin of GBM [163]. If combined with a mouse tumor model where Cre activation is not required to initiate tumors and CSC-specific promoters are well established, MADM utilizing such promoter activity could provide information on the gene dose (+/+, +/−, −/−; or wt/wt/, wt/mt, mt/mt) effect of the gene of interest on CSC biology. The previously described mouse papilloma model [18] may be a good candidate for such an approach because the papilloma is induced by chemical application and the keratin 10 promoter is active in tumor cells showing CSC hallmarks.

Lineage Tracing in Human Tumor Samples

Lineage tracing with gene marking is a powerful tool to understand CSC behavior in vivo, as described previously. Tumor cells derived from patients and maintained as xenografts in immunocompromised mice (PDX) have generated critical information for illuminating CSC biology. Ideally, these findings should be verified by CSC lineage tracing in vivo because the PDX approach may select for only the tumor cells that can survive the cell dissociation process

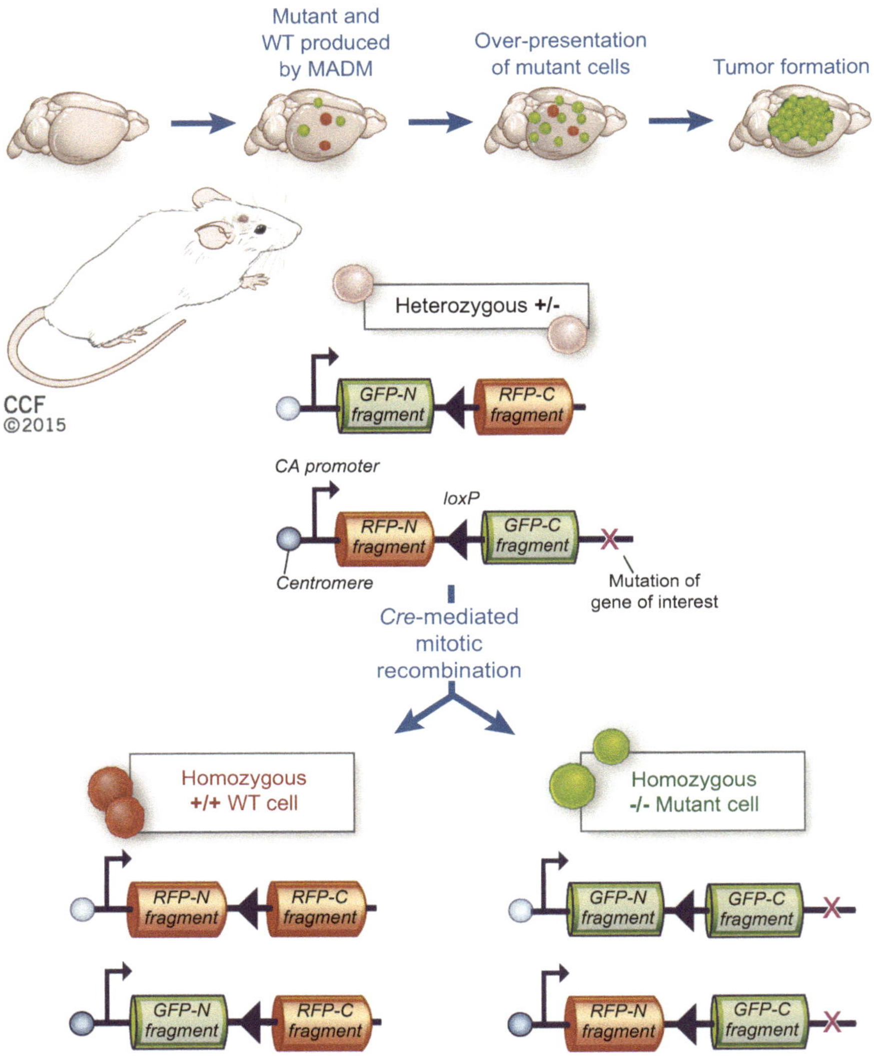

FIGURE 3.3

Schematic representation of mosaic analysis with double marker (MADM)-based system demonstrating how MADM concurrently mutates and labels cells.

and persist in the foreign environments of the recipient animal, which include the degree to which the immune system is compromised, depending on the recipient mouse model [164]. Any of the lineage-tracing approaches described previously are not applicable to human subjects to study CSC behaviors. The alternative is, however, to use naturally occurring endogenous mutations that distinguish different lineages as tracing markers. Histochemistry of human colon specimens is able to detect the loss of cytochrome c oxidase activity caused by rare somatic mutations in mitochondrial DNA. This naturally occurring genetic mutation was used as a lineage-tracing marker to analyze the stem cell clonality of single crypts and the frequency of crypt fission [165]. The same approach was used to illuminate tumor stem cell clonal loss and replacement in adenomatous crypts [166]. Recently, a report on the lineage tracing of neurons in the normal human brain took advantage of the somatic mutations caused by a retrotransposon element [167]. If such genetic mosaicism can be detected together with CSC properties, the cellular dynamics of CSCs during the tumorigenic processes can be traced in human tumor samples in situ.

In Vitro and In Vivo Tumor Formation and Clonogenic Assays

The ideal in vitro assay to assess CSC frequency would be rapid, quantitative, specific, and readily detectable even at a low frequency. CSCs share core properties: the ability to undergo self-renewal, the ability to differentiate, and the ability to establish tumors in vivo. To explore these dynamics, experimental assays must reliably distinguish stem cells and their progeny. The sphere formation assay and limiting dilution assay (LDA) are used for this purpose.

The ability to form spheres in vitro was described in neural precursor cells in 1992. A widely used assay to test stemlike characteristics for cells isolated from the neurogenic areas in the brain was derived from this original report [168,169]. In this assay, cells were cultured with EGF and prevented from adhering to the plate. These cells expressed nestin and other markers of precursor cells. When plated as an adherent culture, cells differentiated into glial and neural lineages. The association between the ability to form spheroids and stem cell–like characteristics has also been demonstrated for various pluripotent cells such as HSCs and prostate stem cells [170,171]. In that regard, CSCs are reported to behave in a similar manner. The approach of culturing cells in a special medium as nonadherent spheroids has become a standard for CSC propagation in various cancers [27,172–176]. In GBM, CSCs cultured as spheres in serum-free medium supplemented with EGF and bFGF have been shown to maintain relative genetic stability over time compared with culture in serum-containing medium [26,177]. Grown on nonadherent plastic, CSCs are able to form clonogenic spheres composed of daughter CSCs and differentiated daughter progenies, a property that relates to their tumorigenic capability [178,179]. Exploiting these characteristics, the sphere formation assay became a surrogate method to detect CSCs. When a sphere formation assay is performed in a limiting dilution format, the frequency of CSCs (sphere-forming cells) in a given cancer cell population can be determined.

The principle behind LDA is the culture of cancer cells at decreasing cell densities in liquid or semiliquid medium. Cells are suspended in CSC medium and plated at different densities in a 96-well plate (down to as low as 1 cell/well).

The number of wells without spheres at each dilution is scored after approximately 2 weeks. The results are presented as semilogarithmic plots as the number of wells negative for viable spheres [180].

Although LDA is a gold standard for assessing CSC frequency [181], it has potential pitfalls that complicate interpretation of the results. The rapidity of sphere development in many systems makes it unlikely that the sphere arose from a single cell solely through clonal expansion. The difficulty of distinguishing the relative contributions of aggregation and proliferation to sphere formation poses a major impediment to the use of this assay [182]. As described previously, culturing cells at decreasing densities can prevent this from happening and serve as a tool for isolating CSC clones. Another variation of the assay uses semiliquid medium to prevent active cell migration and fusion into pseudospheres. Multiple materials for preparing semiliquid medium for cell culture have been described and include the addition of collagen [183], agar [184], and methylcellulose [185].

FACS using a flow cytometer is the standard tool used to enrich for and plate CSCs for an LDA assay. Although FACS offers a sensitive, specific, and robust method of cell isolation and purification, some of its technical limitations make its application to stem cell purification challenging. For example, even when using advanced sorting techniques to distinguish single cells from aggregates, doublets (cells sticking to one another) can still occasionally sort together and need to be eliminated. Thus, microscopy is needed to show that single cells were indeed isolated. Most flow cytometers are typically set up to sort blood cells using small diameter streams at high pressures. These conditions are often not tolerated by the larger, more fragile cells found in many organs and tumors. Therefore, the diameter of the liquid stream and the sorting pressure frequently must be optimized for cells isolated from solid tissues. The interpretation of LDAs is another layer of complexity; a standard, agreed-upon definition of "a sphere" is required and should be reported in each study. It is highly preferable that the investigator quantifying the sphere formation assay be blinded from the experimental conditions. It is also preferable to have the quantification repeated by a third party to ensure the accuracy of the results. Performing serial secondary and tertiary in vitro LDAs is highly recommended; in these assays, the formed spheroids are dissociated into single cells and replated into a new LDA setting to ensure that the formed spheres adhere to the cardinal feature of CSCs: the ability to undergo self-renewal.

To prove that a particular marker or reporter system enriches for CSCs, in vivo transplantation LDAs must be performed with both the marker positive and marker negative populations; additional attention must be paid to the latter to ensure that the injected cells are viable tumor cells rather than from the stromal compartment. In vivo LDA remains the ultimate gold standard for identifying

CSCs, and any in vitro study should be supplemented with it. In transplantation assays, cells are xenografted subcutaneously or into an orthotopic site of immunocompromised mice (typically NOD/SCID) that are assayed at various time points for tumor formation. Herein lies the power of the prospective enrichment of CSCs by cell surface markers as such method is the only one enabling us to perform in vivo LDA from freshly dissociated tumors.

The purpose of in vivo LDA is to estimate CSC frequencies and to evaluate any cooperating effects between injected tumor cells in establishing a tumor. This is achieved by applying a goodness-of-fit test to the LDA data set. If a data set fails the goodness-of-fit test, it suggests that multiple cells, as opposed to a single CSC, are required to generate a tumor, thereby making it impossible to calculate an active cell frequency [181]. To determine the goodness of fit of the data, an LDA should include a wide range of dilutions, with a moderate to large number of replicates per dose. In addition, an LDA should ideally include doses that have both positive and negative results (ie, CSCs vs non-CSCs). Surprisingly, this type of thorough in vivo LDA testing has been performed for few solid tumors [181]. The field is becoming further complicated by the less rigorous use of sphere-forming in vitro LDAs as a surrogate for the gold standard in vivo LDA [186]. Notably, the sphere assay is informative but does not replace an in vivo LDA, a standard requirement in CSC research.

A recent article by Hu et al. described a web-based tool, extreme limiting dilution analysis (ELDA), which is available on the Walter and Eliza Hall Institute of Medical Research Website http://bioinf.wehi.edu.au/software/elda/, and can be used by researchers to calculate LDA frequencies [181]. Traditional LDA statistical programs were not designed to compare subpopulations that are depleted and enriched for active cells. In contrast, ELDA has the capacity to calculate frequencies for subpopulations that produce 0 or 100% positive results, a tool that is invaluable for CSC research [181,186]. ELDA makes these analyses available to all researchers and should be used as a standard in the field for both in vivo and in vitro LDAs.

Recent publications identified an important effect of the degree of immunodeficiency in the mouse model utilized for transplantation studies on the engraftment rate of CSCs [187]. The majority of early work in the CSC field was using SCID or NOD/SCID mice. Quintana et al. showed that at least in the case of melanoma, the usage of NOD/SCID mice can lead to an underestimation of the frequency of human cancer cells with tumorigenic potential and overestimation of the percentage of CSCs due to the additional lack of immune components [187].

Using in vivo LDA, 1 in 1,000,000 melanoma cells generated xenografts in NOD/SCID mice compared with 1 in 4 when injected together with Matrigel into NOD/SCID interleukin-2 receptor gamma chain null (IL2Rγnull) mice lacking T, B, and natural killer (NK) cells. This work leads to two important conclusions: (1) The degree of immune deficiency of the recipient mice can affect the frequency of CSCs estimation. (2) Not all types of cancer follow the CSC model. One method that researchers are using to address this ambiguity is the supplementation of human-derived tumor data with work performed in transgenic mouse models of cancer. These models allow for syngeneic transplantation of specific subsets of cells and therefore eliminate the cross-species barriers to engraftment. In addition, transgenic mouse models provide opportunities to trace lineage of CSC progeny, which has demonstrated the presence of CSCs without any complication associated with transplantation as discussed in linage-tracing section.

Another critique of the utilization of immunocompromised mouse models for CSC studies is the incompatibilities between mouse ligands and human receptors for certain growth factors and adhesion molecules that may impair the survival, proliferation, or migration of human cells in mice. Moreover, human cancers are often xenotransplanted in locations different to their native environment where they originally formed in patients. One possible alternative is the utilization of humanized mouse models which is still not very prevalent in the field.

The Utility of Time-Lapse Video Microscopy for Cancer Stem Cell Studies

Time-lapse video microscopy records the activity of individual cultured cells over time. Image capturing with phase contrast or differential interference contrast optics can visualize temporal morphological changes in live cells, whereas imaging with fluorescence microscopy can monitor temporal changes in the intensity and location of the fluorescently labeled molecules, such as GFP fusion proteins. When the imaging timing is optimized, the behavior of individual cells can be traced throughout the recorded period even if the cells are highly motile.

This individual cell traceability, a unique feature of time-lapse analysis, enabled studies on the cell state transition from CSC to non-CSC and vice versa. GFP expression driven by stem cell transcription factor response element(s) or by the promoter of the pluripotent transcription factor itself reports the activity of pluripotency transcription factors expressed to high levels in CSCs. Time-lapse analysis of CSCs or non-CSCs harboring such reporter genes revealed that the reporter-positive CSCs can give rise to the reporter-negative non-CSCs and that the reporter-negative non-CSCs can also give rise to the reporter-positive CSCs [188]. Similarly, a GFP reporter system for differentiation was successfully used to detect the conversion from the CSC to non-CSC and the non-CSC to CSC states [189]. Conversion from CSC to non-CSC or non-CSC to CSC can be also monitored by assessing changes in marker expression using enriched stem or nonstem populations by FACS or MACS. Criticisms of such approaches, however, are based on the argument that the observed conversion is not a genuine conversion from one state to the other, but rather an enrichment of a minor fraction of counterpart cells that contaminated the initial enriched population. The capability of time-lapse analysis to trace an individual cell can rule out such a possibility.

This individual cell traceability of time-lapse analysis also enables lineage tracing of CSCs after cell division. Time-lapse analysis of the GFP reporter systems described previously detected symmetric and asymmetric cell division of CSCs [189,190]. Without a GFP reporter system, time-lapse analysis combined with terminal immunostaining for stem cell or differentiation markers can be used to trace the lineage commitment of the progeny of CSCs [143]. These techniques will reveal the biological importance of the mode of cell division (symmetric vs asymmetric) in the maintenance of CSCs and the emergence of cellular heterogeneity in the tumor.

Intravital imaging technology utilizing imaging windows and multiphoton microscopy can visualize cells expressing fluorescent proteins at a single-cell resolution. Intravital imaging approaches have revealed many biological behaviors of cancer cells in vivo [191]. The proliferation of grafted CSCs and non-CSCs that had been enriched by a surface stem marker and differentially labeled with GFP or cyan fluorescent protein (CFP) was traced using intravital imaging at multiple time points. This real time in vivo time-lapse recording revealed that CSCs were more tumorigenic than non-CSCs [192]. When the lineages of cancer cells were traced by intravital imaging at multiple time points in a mouse genetic tumor model with the multicolor Cre recombinase-inducible confetti fluorescent protein lineage-tracing system, CSC clones, as defined as clonal populations with expansion capability, appeared and disappeared over time, indicating the plasticity of the CSC status [193]. Similar lineage-tracing studies utilizing the well-characterized promoters of CSC or differentiation markers as drivers for Cre expression are expected to provide vital informative to understand CSC biology in an unperturbed tumor. See the reporter systems and lineage-tracing section for in-depth details of the available reporter systems and lineage-tracing methods. Intravital time-lapse analysis of a reporter fluorescent protein for a critical CSC transcription factor would be useful to probe the CSC niche in vivo.

To obtain proper geometrical alignment throughout a time lapse is one of the challenges for intravital time-lapse imaging. In many cases, imaging the same areas relative to imaging windows with reference to blood vessels (visualized by intravenous injection of fluorescent dextran) or tissue structures detected by second harmonic signals can be difficult. Recently, the pattern of hair follicles was used as landmarks to successfully align images of an intravital time lapse for an extended period of time [194]. This approach would be helpful to study CSCs of skin cancers.

The advantage of using multiphoton microscopy in intravital imaging is that it can visualize fluorescent signals from a deeper area of the tissue because light with a longer wavelength compared with other types of fluorescence microscopies is used to activate fluorochromes. However, the depth that multiphoton microscopy can visualize is limited to 0.5–1 mm. The visualization of cellular events occurring in a deeper area of the tissue can be achieved through the use of imaging tubes to conduct the excitation light to deeper area of tissues [195].

Intravital time-lapse imaging with a good image registration technique is a powerful tool to investigate CSC biology in vivo when combined with lineage-tracing markers and fluorescent reporter systems.

CONCLUSION AND FUTURE DIRECTIONS

The accuracy of a measurement system is the degree of closeness of measurements to their true value. The precision of a measurement is related to the reproducibility and repeatability of the results. Although precision and accuracy can be synonymous in colloquial use, they are deliberately contrasted in the context of the scientific method [196]. Enrichment and analysis techniques discussed in this chapter should always be thought of within the context of these two concepts. How accurate are our methods to enrich for CSCs? How accurately do the cells obtained resemble their counterparts in patient tumors? How precise are our methods with respect to intertumor and intratumor variability? It has been said that no two cancers are identical, so how could a universal isolation method of CSCs be truly accurate and precise? Our accuracy and precision is limited by the measurement tools available to us and how precisely we apply the procedure for each analysis. The ideal enrichment method would be based on defining properties of CSCs (such as self-renewal, tumor initiation, etc) that is immediately lost upon differentiation and can be used with live cells. Such a digital readout is unattainable thus far because biological systems do not exhibit "all or none" phenomena [15]. To further complicate things, the analysis of human-derived CSCs is based on one snapshot in time when the tumor biopsy or resected specimen was attained. To generalize such findings to the natural history of cancer development starting from precancerous lesions to advanced-stage cancer is not possible given the limited tools available. These questions should be taken into consideration when evaluating various enrichment methods.

The term potential describes what cells are capable of doing under certain conditions, whereas fate describes what the cells actually do. The central tenet of the CSC theory has been that the vast majority of cancer cells has undergone differentiation and lost the potential to proliferate extensively. This is addressed experimentally by demonstrating the inability of these cells to propagate the tumor to immunocompromised mice and proliferate in culture. This does not necessarily reflect their biology in living patients. It is important to remember that simply because a cell has the potential to form a tumor does not mean that it actually does so within a patient. Cancer cells with tumorigenic potential might be held in check by environmental or immunological mechanisms that prevent them from actually contributing to disease. There are limited data in the field of cancer biology that address the question of whether many cancer cells or few cancer cells actually contribute to disease in human patients. This question is typically addressed via lineage-tracing and reporter systems in live animal models, as discussed previously in this chapter. Another layer of complexity is the concept of plasticity, where CSCs and non-CSCs are able to change states based on intrinsic and extrinsic stimuli. A fundamental premise of the CSC theory is the ability of CSCs to form tumors upon xenotransplantation in immunocompromised mice. A valid critique surrounds the question of whether immunocompromised mice are reliable models for studying human cancer. The answer may depend upon the aspect of cancer biology being studied. Transplantation into highly immunocompromised mice is the best, albeit imperfect, way of assessing which human cancer cells have the potential to form tumors.

It is highly recommended to supplement whatever method used by secondary validation in accordance with the rigorous application of the scientific method. An example of this is in vivo serial LDAs, in which CSCs are xenotransplanted into immunocompromised mice, the gold standard for the identification of CSCs. Self-renewal can be confirmed by this assay, in which prospectively reisolated CSC populations are placed into secondary and tertiary recipients. Multipotency is demonstrated by the ability of CSCs to generate tumor xenografts that reflect the cellular heterogeneity of the original tumor [13,197]. Many studies employ surrogate in vitro serial LDAs. However, only clonal serial in vivo repopulation assays can formally test self-renewal of stem cells [198].

A major disadvantage regarding all of the previously discussed techniques is the inherent difficulty of tracking cells over time in vivo in humans. Whether it is a tumor biopsy from patient-derived specimens or CSCs that may be followed via lineage-tracing techniques in mouse models of disease; both methods do not provide real-time imaging of CSCs in their natural niche during disease progression in humans. Our techniques today resemble the story of the blind man feeling the elephant, and as stated by the famous physicist Werner Heisenberg, "We have to remember that what we observe is not nature in itself, but nature exposed to our method of questioning."

We believe that major breakthroughs in the field of CSCs will result from improved high-resolution imaging techniques, enabling us to monitor the dynamics of CSCs during cancer treatment. However, such techniques require advancement in the field of imaging. In addition, it would be beneficial to find more specific CSC markers that can be tagged and traced. Advancements in the field of CSCs can also come from real-time in situ genetic analysis. With further development of fluorescent in situ sequencing, the relative significance and expression of each cancer cell will be evaluated in relation to its location in the tumor as well as its proximity to the CSC niches. The ideal theoretical tool to interrogate CSCs is to monitor CSCs in live patients and evaluate their genetic background and expression pattern in a quick and reliable manner which will reflect in personalized treatment seeking to eliminate such cells.

Additional advancements in the CSC field, in our opinion, will come from an increased appreciation of CSCs as a transition state that is subject to evolution. If we think of cancer cells as a wide spectrum with non-CSCs laying on one end of it and CSCs on the other end, with multiple transition states in between; our horizons will be widened beyond the dichotomous thinking of stochastic and clonal evolution models. It is important to remember that tumor cells are very plastic and highly evasive and, as such, taking this into consideration will help us tackle these various cancer states.

List of Acronyms and Abbreviations

AACR American Association for Cancer Research
ABCG2 ATP-binding cassette subfamily G member 2
AFP Alpha fetoprotein
ALDH1 Aldehyde dehydrogenase 1
ALL Acute lymphoblastic leukemia
AML Acute myelogenous leukemia
APC Adenomatous polyposis coli
B-ALL B-cell acute lymphoblastic leukemia
BCECF 2′,7′-bis-(2-carboxyethyl)-5-(and-6)-carboxyfluorescein acetoxymethyl ester
bFGF Basic fibroblast growth factor
Bmi1 Polycomb complex protein BMI-1
CFP Cyan fluorescent protein
CIK Cytokine-induced killer cells
CLL-1 C-type lectin-like molecule-1
CML Chronic myelogenous leukemia
CSC Cancer stem cells
DMEM Dulbecco's Modified Eagle Medium
EdU 5-Ethynyl-2′-deoxyuridine
EGF Epidermal growth factor
EGFP Enhanced green fluorescent protein
ELDA Extreme limiting dilution analysis

EpCAM Epithelial cell adhesion molecule
ER Estrogen receptor
ES Embryonic stem cells
FACS Fluorescence-activated cell sorting
FITC Fluorescein isothiocyanate
GBM Glioblastoma
GFAP Glial fibrillary acidic protein
GFP Green fluorescent protein
HCC Hepatocellular carcinoma
HER2 Human epidermal growth factor receptor 2
hTERT Human telomerase reverse transcriptase
IRES Internal ribosome entry site
KRAS V-Ki-ras2 Kirsten rat sarcoma viral oncogene homolog
L1CAM L1 cell adhesion molecule
LDA Limiting dilution analysis
LGR5 Leucine-rich repeat-containing G-protein coupled receptor 5
MAC Magnetic-activated cell sorting
MADM Mosaic analysis with double marker
Mt Mutant
NANOG Homeobox transcription factor Nanog
NK Natural killer
NOD Non-obese diabetic
NSCLC Non-small cell lung cancer
OE Overexpression
PDX Patient-derived xenograft
Ph Philadelphia chromosome
PR Progesterone receptor
REX-1 Retinoic acid-regulated zinc finger 1
RFP Red fluorescent protein
rtTA Reverse tetracycline-controlled transactivator
SCC Squamous cell carcinoma
SCID Severe combined immunodeficiency
SHIP SH2 domain-containing 5′-inositol phosphatase
SOX2 (Sex determining region Y)-box 2
SSEA-1 Stage-specific embryonic antigen-1
T-ALL T-cell acute lymphoblastic leukemia
TCF/LF Transcription factor/lymphoid enhancer binding factor
TK Thymidine kinase
WNT Wingless-type MMTV integration site family
Wt Wild type

Acknowledgment

We would like to thank Dr Erin Elizabeth Mulkearns-Hubert for her extensive help in editing the manuscript. We also thank Dr Daniel Sliver and Dr Caner Saygin for their very fruitful discussions that helped to shape this manuscript. We also would like to thank Amanda Mendelsohn for her help in formulating the artwork. We would like to apologize to the inability to cite all relevant references due to space limitations.

References

[1] Dexter DL, Kowalski HM, Blazar BA, Fligiel Z, Vogel R, Heppner GH. Heterogeneity of tumor cells from a single mouse mammary tumor. Cancer Res 1978;38(10):3174–81.

[2] Pertschuk LP, Tobin EH, Brigati DJ, Kim DS, Bloom ND, Gaetjens E, et al. Immunofluorescent detection of estrogen receptors in breast cancer. Comparison with dextran-coated charcoal and sucrose gradient assays. Cancer 1978;41(3):907–11.

[3] Poste G, Doll J, Hart IR, Fidler IJ. In vitro selection of murine B16 melanoma variants with enhanced tissue-invasive properties. Cancer Res 1980;40(5):1636–44.

[4] Mitelman F, Mark J, Levan G, Levan A. Tumor etiology and chromosome pattern. Science 1972;176(4041):1340–1.

[5] Shapiro JR, Yung WK, Shapiro WR. Isolation, karyotype, and clonal growth of heterogeneous subpopulations of human malignant gliomas. Cancer Res 1981;41(6):2349–59.

[6] Brocks D, Assenov Y, Minner S, Bogatyrova O, Simon R, Koop C, et al. Intratumor DNA methylation heterogeneity reflects clonal evolution in aggressive prostate cancer. Cell Rep 2014;8(3):798–806.

[7] Danielson KG, Anderson LW, Hosick HL. Selection and characterization in culture of mammary tumor cells with distinctive growth properties in vivo. Cancer Res 1980;40(6):1812–9.

[8] Gray JM, Pierce Jr GB. Relationship between growth rate and differentiation of melanoma in vivo. J Natl Cancer Inst 1964;32:1201–10.

[9] Heppner GH, Dexter DL, DeNucci T, Miller FR, Calabresi P. Heterogeneity in drug sensitivity among tumor cell subpopulations of a single mammary tumor. Cancer Res 1978;38(11 Pt 1):3758–63.

[10] Barranco SC, Ho DH, Drewinko B, Romsdahl MM, Humphrey RM. Differential sensitivities of human melanoma cells grown in vitro to arabinosylcytosine. Cancer Res 1972;32(12):2733–6.

[11] Lapidot T, Sirard C, Vormoor J, Murdoch B, Hoang T, Caceres-Cortes J, et al. A cell initiating human acute myeloid leukaemia after transplantation into SCID mice. Nature 1994;367(6464):645–8.

[12] Shackleton M, Quintana E, Fearon ER, Morrison SJ. Heterogeneity in cancer: cancer stem cells versus clonal evolution. Cell 2009;138(5):822–9.

[13] Clarke MF, Dick JE, Dirks PB, Eaves CJ, Jamieson CH, Jones DL, et al. Cancer stem cells–perspectives on current status and future directions: AACR workshop on cancer stem cells. Cancer Res 2006;66(19):9339–44.

[14] Clevers H. The cancer stem cell: premises, promises and challenges. Nat Med 2011;17(3):313–9.

[15] Lathia JD, Mack SC, Mulkearns-Hubert EE, Valentim CL, Rich JN. Cancer stem cells in glioblastoma. Genes Dev 2015;29(12):1203–17.

[16] Bao S, Wu Q, McLendon RE, Hao Y, Shi Q, Hjelmeland AB, et al. Glioma stem cells promote radioresistance by preferential activation of the DNA damage response. Nature 2006;444(7120):756–60.

[17] Lotti F, Jarrar AM, Pai RK, Hitomi M, Lathia J, Mace A, et al. Chemotherapy activates cancer-associated fibroblasts to maintain colorectal cancer-initiating cells by IL-17A. J Exp Med 2013;210(13):2851–72.

[18] Driessens G, Beck B, Caauwe A, Simons BD, Blanpain C. Defining the mode of tumour growth by clonal analysis. Nature 2012;488(7412):527–30.

[19] Schepers AG, Snippert HJ, Stange DE, van den Born M, van Es JH, van de Wetering M, et al. Lineage tracing reveals $Lgr5^+$ stem cell activity in mouse intestinal adenomas. Science (New York, NY) 2012;337(6095):730–5.

[20] Vanner RJ, Remke M, Gallo M, Selvadurai HJ, Coutinho F, Lee L, et al. Quiescent Sox2+ cells drive hierarchical growth and relapse in sonic hedgehog subgroup medulloblastoma. Cancer Cell 2014;26(1):33–47.

[21] Nowell PC. The clonal evolution of tumor cell populations. Science 1976;194(4260):23–8.

[22] Shah NP, Skaggs BJ, Branford S, Hughes TP, Nicoll JM, Paquette RL, et al. Sequential ABL kinase inhibitor therapy selects for compound drug-resistant BCR-ABL mutations with altered oncogenic potency. J Clin Invest 2007;117(9):2562–9.

[23] Barabe F, Kennedy JA, Hope KJ, Dick JE. Modeling the initiation and progression of human acute leukemia in mice. Science 2007;316(5824):600–4.

[24] Kreso A, O'Brien CA. Colon cancer stem cells. Curr Protoc Stem Cell Biol 2008. Chapter 3:Unit 3.1.

[25] Lathia JD, Gallagher J, Heddleston JM, Wang J, Eyler CE, Macswords J, et al. Integrin alpha 6 regulates glioblastoma stem cells. Cell Stem Cell 2010;6(5):421–32.

[26] Lee J, Kotliarova S, Kotliarov Y, Li A, Su Q, Donin NM, et al. Tumor stem cells derived from glioblastomas cultured in bFGF and EGF more closely mirror the phenotype and genotype of primary tumors than do serum-cultured cell lines. Cancer Cell 2006;9(5):391–403.

[27] Ricci-Vitiani L, Lombardi DG, Pilozzi E, Biffoni M, Todaro M, Peschle C, et al. Identification and expansion of human colon-cancer-initiating cells. Nature 2007;445(7123):111–5.

[28] Vermeulen L, De Sousa E, Melo F, van der Heijden M, Cameron K, de Jong JH, et al. Wnt activity defines colon cancer stem cells and is regulated by the microenvironment. Nat Cell Biol 2010;12(5):468–76.

[29] Heddleston JM, Wu Q, Rivera M, Minhas S, Lathia JD, Sloan AE, et al. Hypoxia-induced mixed-lineage leukemia 1 regulates glioma stem cell tumorigenic potential. Cell Death Differ 2012;19(3):428–39.

[30] Hjelmeland AB, Wu Q, Heddleston JM, Choudhary GS, MacSwords J, Lathia JD, et al. Acidic stress promotes a glioma stem cell phenotype. Cell Death Differ 2011;18(5):829–40.

[31] Sniegowski JA, Lappe JW, Patel HN, Huffman HA, Wachter RM. Base catalysis of chromophore formation in Arg96 and Glu222 variants of green fluorescent protein. J Biol Chem 2005;280(28):26248–55.

[32] Sakaue-Sawano A, Kurokawa H, Morimura T, Hanyu A, Hama H, Osawa H, et al. Visualizing spatiotemporal dynamics of multicellular cell-cycle progression. Cell 2008;132(3):487–98.

[33] Li X, Zhao X, Fang Y, Jiang X, Duong T, Fan C, et al. Generation of destabilized green fluorescent protein as a transcription reporter. J Biol Chem 1998;273(52):34970–5.

[34] Chudakov DM, Matz MV, Lukyanov S, Lukyanov KA. Fluorescent proteins and their applications in imaging living cells and tissues. Physiol Rev 2010;90(3):1103–63.

[35] Collignon J, Sockanathan S, Hacker A, Cohen-Tannoudji M, Norris D, Rastan S, et al. A comparison of the properties of Sox-3 with Sry and two related genes, Sox-1 and Sox-2. Development 1996;122(2):509–20.

[36] Takahashi K, Yamanaka S. Induction of pluripotent stem cells from mouse embryonic and adult fibroblast cultures by defined factors. Cell 2006;126(4):663–76.

[37] Boumahdi S, Driessens G, Lapouge G, Rorive S, Nassar D, Le Mercier M, et al. SOX2 controls tumour initiation and cancer stem-cell functions in squamous-cell carcinoma. Nature 2014;511(7508):246–50.

[38] Liang S, Furuhashi M, Nakane R, Nakazawa S, Goudarzi H, Hamada J, et al. Isolation and characterization of human breast cancer cells with SOX2 promoter activity. Biochem Biophys Res Commun 2013;437(2):205–11.

[39] Liu XF, Yang WT, Xu R, Liu JT, Zheng PS. Cervical cancer cells with positive Sox2 expression exhibit the properties of cancer stem cells. PLoS One 2014;9(1):e87092.

[40] Stoltz K, Sinyuk M, Hale JS, Wu Q, Otvos B, Walker K, et al. Development of a Sox2 reporter system modeling cellular heterogeneity in glioma. Neuro Oncol 2015;17(3):361–71.

[41] Chen J, Li Y, Yu T, McKay RM, Burns DK, Kernie SG, et al. A restricted cell population propagates glioblastoma growth after chemotherapy. Nature 2012;488(7412):522–6.

[42] Hastreiter S, Schroeder T. Nanog dynamics in single embryonic stem cells. Cell Cycle 2016;15(6):770–1.

[43] Chambers I, Colby D, Robertson M, Nichols J, Lee S, Tweedie S, et al. Functional expression cloning of Nanog, a pluripotency sustaining factor in embryonic stem cells. Cell 2003;113(5):643–55.

[44] Bussolati B, Bruno S, Grange C, Ferrando U, Camussi G. Identification of a tumor-initiating stem cell population in human renal carcinomas. FASEB J 2008;22(10):3696–705.

[45] Chiou SH, Wang ML, Chou YT, Chen CJ, Hong CF, Hsieh WJ, et al. Coexpression of Oct4 and Nanog enhances malignancy in lung adenocarcinoma by inducing cancer stem cell-like properties and epithelial-mesenchymal transdifferentiation. Cancer Res 2010;70(24):10433–44.

[46] Amsterdam A, Raanan C, Schreiber L, Freyhan O, Schechtman L, Givol D. Localization of the stem cell markers LGR5 and Nanog in the normal and the cancerous human ovary and their inter-relationship. Acta Histochem 2013;115(4):330–8.

[47] Ezeh UI, Turek PJ, Reijo RA, Clark AT. Human embryonic stem cell genes OCT4, NANOG, STELLAR, and GDF3 are expressed in both seminoma and breast carcinoma. Cancer 2005;104(10):2255–65.

[48] Hart LS, Dolloff NG, Dicker DT, Koumenis C, Christensen JG, Grimberg A, et al. Human colon cancer stem cells are enriched by insulin-like growth factor-1 and are sensitive to figitumumab. Cell Cycle 2011;10(14):2331–8.

[49] Hoei-Hansen CE, Almstrup K, Nielsen JE, Brask Sonne S, Graem N, Skakkebaek NE, et al. Stem cell pluripotency factor NANOG is expressed in human fetal gonocytes, testicular carcinoma in situ and germ cell tumours. Histopathology 2005;47(1):48–56.

[50] Guo Y, Liu S, Wang P, Zhao S, Wang F, Bing L, et al. Expression profile of embryonic stem cell-associated genes Oct4, Sox2 and Nanog in human gliomas. Histopathology 2011;59(4):763–75.

[51] Wen J, Park JY, Park KH, Chung HW, Bang S, Park SW, et al. Oct4 and Nanog expression is associated with early stages of pancreatic carcinogenesis. Pancreas 2010;39(5):622–6.

[52] Nagata T, Shimada Y, Sekine S, Hori R, Matsui K, Okumura T, et al. Prognostic significance of NANOG and KLF4 for breast cancer. Breast Cancer 2014;21(1):96–101.

[53] Lee M, Nam EJ, Kim SW, Kim S, Kim JH, Kim YT. Prognostic impact of the cancer stem cell-related marker NANOG in ovarian serous carcinoma. Int J Gynecol Cancer 2012;22(9):1489–96.

[54] Meng HM, Zheng P, Wang XY, Liu C, Sui HM, Wu SJ, et al. Over-expression of Nanog predicts tumor progression and poor prognosis in colorectal cancer. Cancer Biol Ther 2010;9(4):295–302.

[55] Iv Santaliz-Ruiz LE, Xie X, Old M, Teknos TN, Pan Q. Emerging role of Nanog in tumorigenesis and cancer stem cells. Int J Cancer 2014;135(12):2741–8.

[56] Jeter CR, Badeaux M, Choy G, Chandra D, Patrawala L, Liu C, et al. Functional evidence that the self-renewal gene NANOG regulates human tumor development. Stem Cells 2009;27(5):993–1005.

[57] Thiagarajan PS, Hitomi M, Hale JS, Alvarado AG, Otvos B, Sinyuk M, et al. Development of a fluorescent reporter system to delineate cancer stem cells in triple-negative breast cancer. Stem Cells 2015;33(7):2114–25.

[58] Cioffi M, D'Alterio C, Camerlingo R, Tirino V, Consales C, Riccio A, et al. Identification of a distinct population of $CD133^{+}CXCR4^{+}$ cancer stem cells in ovarian cancer. Sci Rep 2015;5:10357.

[59] Meyer MJ, Fleming JM, Lin AF, Hussnain SA, Ginsburg E, Vonderhaar BK. CD44posCD49f-hiCD133/2hi defines xenograft-initiating cells in estrogen receptor-negative breast cancer. Cancer Res 2010;70(11):4624–33.

[60] Gammaitoni L, Giraudo L, Leuci V, Todorovic M, Mesiano G, Picciotto F, et al. Effective activity of cytokine-induced killer cells against autologous metastatic melanoma including cells with stemness features. Clin Cancer Res 2013;19(16):4347–58.

[61] Wu G, Wilson G, Zhou G, Hebbard L, George J, Qiao L. Oct4 is a reliable marker of liver tumor propagating cells in hepatocellular carcinoma. Discov Med 2015;20(110):219–29.

[62] Sangiolo D, Mesiano G, Gammaitoni L, Leuci V, Todorovic M, Giraudo L, et al. Cytokine-induced killer cells eradicate bone and soft-tissue sarcomas. Cancer Res 2014;74(1):119–29.

[63] Yuan H, Corbi N, Basilico C, Dailey L. Developmental-specific activity of the FGF-4 enhancer requires the synergistic action of Sox2 and Oct-3. Genes Dev 1995;9(21):2635–45.

[64] Ambrosetti DC, Basilico C, Dailey L. Synergistic activation of the fibroblast growth factor 4 enhancer by Sox2 and Oct-3 depends on protein-protein interactions facilitated by a specific spatial arrangement of factor binding sites. Mol Cell Biol 1997;17(11):6321–9.

[65] Tang B, Raviv A, Esposito D, Flanders K, Daniel C, Nghiem B, et al. A flexible reporter system for direct observation and isolation of cancer stem cells. Stem Cell Rep 2015;4(1):155–69.

[66] Artavanis-Tsakonas S, Rand MD, Lake RJ. Notch signaling: cell fate control and signal integration in development. Science 1999;284(5415):770–6.

[67] Yahyanejad S, Theys J, Vooijs M. Targeting Notch to overcome radiation resistance. Oncotarget 2015.

[68] Ranganathan P, Weaver KL, Capobianco AJ. Notch signalling in solid tumours: a little bit of everything but not all the time. Nat Rev Cancer 2011;11(5):338–51.

[69] Dang TP, Gazdar AF, Virmani AK, Sepetavec T, Hande KR, Minna JD, et al. Chromosome 19 translocation, overexpression of Notch3, and human lung cancer. J Natl Cancer Inst 2000;92(16):1355–7.

[70] Hassan KA, Wang L, Korkaya H, Chen G, Maillard I, Beer DG, et al. Notch pathway activity identifies cells with cancer stem cell-like properties and correlates with worse survival in lung adenocarcinoma. Clin Cancer Res 2013;19(8):1972–80.

[71] D'Angelo RC, Ouzounova M, Davis A, Choi D, Tchuenkam SM, Kim G, et al. Notch reporter activity in breast cancer cell lines identifies a subset of cells with stem cell activity. Mol Cancer Ther 2015;14(3):779–87.

[72] Yu L, Liu S, Zhang C, Zhang B, Simoes BM, Eyre R, et al. Enrichment of human osteosarcoma stem cells based on hTERT transcriptional activity. Oncotarget 2013;4(12):2326–38.

[73] Desponts C, Ninos JM, Kerr WG. s-SHIP associates with receptor complexes essential for pluripotent stem cell growth and survival. Stem Cells Dev 2006;15(5):641–6.

[74] Bauderlique-Le Roy H, Vennin C, Brocqueville G, Spruyt N, Adriaenssens E, Bourette RP. Enrichment of human stem-like prostate cells with s-SHIP promoter activity uncovers a role in stemness for the long noncoding RNA H19. Stem Cells Dev 2015;24(10):1252–62.

[75] Ishii T, Yasuchika K, Suemori H, Nakatsuji N, Ikai I, Uemoto S. Alpha-fetoprotein producing cells act as cancer progenitor cells in human cholangiocarcinoma. Cancer Lett 2010;294(1):25–34.

[76] Vermeulen L, Morrissey E, van der Heijden M, Nicholson AM, Sottoriva A, Buczacki S, et al. Defining stem cell dynamics in models of intestinal tumor initiation. Science 2013;342(6161):995–8.

[77] Lee SH, Koo BS, Kim JM, Huang S, Rho YS, Bae WJ, et al. Wnt/β-catenin signalling maintains self-renewal and tumourigenicity of head and neck squamous cell carcinoma stem-like cells by activating Oct4. J Pathol 2014;234(1):99–107.

[78] Vadakkan TJ, Landua JD, Bu W, Wei W, Li F, Wong ST, et al. Wnt-responsive cancer stem cells are located close to distorted blood vessels and not in hypoxic regions in a p53-null mouse model of human breast cancer. Stem Cells Transl Med 2014;3(7):857–66.

[79] Kim B, Kang K, Choi J, Jung J, Im Y, Kang S. Knockdown of the potential cancer stem-like cell marker Rex-1 improves chemotherapeutic effects in gliomas. Hum Gene Ther 2011;22(12):1551–62.

[80] Chen Y, Huang AH, Lin L. Sphere-forming-like cells (squamospheres) with cancer stem-like cell traits from VX2 rabbit buccal squamous cell carcinoma. Int J Oral Sci 2014;6(4):212–8.

[81] Zhong JF, Weiner L, Jin Y, Lu W, Taylor CR. A real-time pluripotency reporter for human stem cells. Stem Cells Dev 2010;19(1):47–52.

[82] O'Brien CA, Kreso A, Dick JE. Cancer stem cells in solid tumors: an overview. Semin Radiat Oncol 2009;19(2):71–7.

[83] Moghbeli M, Moghbeli F, Forghanifard MM, Abbaszadegan MR. Cancer stem cell detection and isolation. Med Oncol 2014;31(9). 69-014-0069-6. Epub July 27, 2014.

[84] Andrews TE, Wang D, Harki DA. Cell surface markers of cancer stem cells: diagnostic macromolecules and targets for drug delivery. Drug Deliv Transl Res 2013;3(2):121–42.

[85] Bonnet D, Dick JE. Human acute myeloid leukemia is organized as a hierarchy that originates from a primitive hematopoietic cell. Nat Med 1997;3(7):730–7.

[86] Blair A, Hogge DE, Ailles LE, Lansdorp PM, Sutherland HJ. Lack of expression of Thy-1 (CD90) on acute myeloid leukemia cells with long-term proliferative ability in vitro and in vivo. Blood 1997;89(9):3104–12.

[87] Blair A, Hogge DE, Sutherland HJ. Most acute myeloid leukemia progenitor cells with long-term proliferative ability in vitro and in vivo have the phenotype $CD34^+/CD71^-/HLA\text{-}DR^-$. Blood 1998;92(11):4325–35.

[88] Blair A, Sutherland HJ. Primitive acute myeloid leukemia cells with long-term proliferative ability in vitro and in vivo lack surface expression of c-kit (CD117). Exp Hematol 2000;28(6):660–71.

[89] Florian S, Sonneck K, Hauswirth AW, Krauth MT, Schernthaner GH, Sperr WR, et al. Detection of molecular targets on the surface of $CD34^+/CD38^-$ stem cells in various myeloid malignancies. Leuk Lymphoma 2006;47(2):207–22.

[90] de Figueiredo-Pontes LL, Pintão MC, Oliveira LC, Dalmazzo LF, Jácomo RH, Garcia AB, et al. Determination of P-glycoprotein, MDR-related protein 1, breast cancer resistance protein, and lung-resistance protein expression in leukemic stem cells of acute myeloid leukemia. Cytom B Clin Cytom 2008;74B(3):163–8.

[91] van Rhenen A, van Dongen GA, Kelder A, Rombouts EJ, Feller N, Moshaver B, et al. The novel AML stem cell associated antigen CLL-1 aids in discrimination between normal and leukemic stem cells. Blood 2007;110(7):2659–66.

[92] Hosen N, Park CY, Tatsumi N, Oji Y, Sugiyama H, Gramatzki M, et al. CD96 is a leukemic stem cell-specific marker in human acute myeloid leukemia. Proc Natl Acad Sci 2007;104(26):11008–13.

[93] Bendall LJ, Bradstock KF, Gottlieb DJ. Expression of CD44 variant exons in acute myeloid leukemia is more common and more complex than that observed in normal blood, bone marrow or $CD34^+$ cells. Leukemia 2000;14(7):1239–46.

[94] Saito Y, Kitamura H, Hijikata A, Tomizawa-Murasawa M, Tanaka S, Takagi S, et al. Identification of therapeutic targets for quiescent, chemotherapy-resistant human leukemia stem cells. Sci Transl Med 2010;2(17):17ra9.

[95] Quijano CA, Moore D, Arthur D, Feusner J, Winter SS, Pallavicini MG. Cytogenetically aberrant cells are present in the $CD34^{+}CD33^{-}38^{-}19^{-}$ marrow compartment in children with acute lymphoblastic leukemia. Leukemia 1997;11(9):1508–15.

[96] Castor A, Nilsson L, Astrand-Grundström I, Buitenhuis M, Ramirez C, Anderson K, et al. Distinct patterns of hematopoietic stem cell involvement in acute lymphoblastic leukemia. Nat Med 2005;11(6):630–7.

[97] Yeoh E, Ross ME, Shurtleff SA, Williams WK, Patel D, Mahfouz R, et al. Classification, subtype discovery, and prediction of outcome in pediatric acute lymphoblastic leukemia by gene expression profiling. Cancer Cell 2002;1(2):133–43.

[98] Cox CV. Characterization of acute lymphoblastic leukemia progenitor cells. Blood 2004;104(9):2919–25.

[99] Nishida H, Yamazaki H, Yamada T, Iwata S, Dang NH, Inukai T, et al. CD9 correlates with cancer stem cell potentials in human B-acute lymphoblastic leukemia cells. Biochem Biophys Res Commun 2009;382(1):57–62.

[100] Yamazaki H, Nishida H, Iwata S, Dang NH, Morimoto C. CD90 and CD110 correlate with cancer stem cell potentials in human T-acute lymphoblastic leukemia cells. Biochem Biophys Res Commun 2009;383(2):172–7.

[101] Testa U. Leukemia stem cells. Ann Hematol 2010;90(3):245–71.

[102] Bocchia M, Ippoliti M, Gozzetti A, Abruzzese E, Calabrese S, Amabile M, et al. $CD34^{+}$/Ph^{+} cells are still detectable in chronic myeloid leukemia patients with sustained and prolonged complete cytogenetic remission during treatment with imatinib mesylate. Leukemia 2008;22(2):426–8.

[103] Nievergall E, Ramshaw HS, Yong AS, Biondo M, Busfield SJ, Vairo G, et al. Monoclonal antibody targeting of IL-3 receptor alpha with CSL362 effectively depletes CML progenitor and stem cells. Blood 2014;123(8):1218–28.

[104] Al-Hajj M. Cancer stem cells and oncology therapeutics. Curr Opin Intern Med 2007;6(2):178–81.

[105] Phillips TM, McBride WH, Pajonk F. The response of $CD24^{-}$/low/$CD44^{+}$ breast cancer-initiating cells to radiation. J Natl Cancer Inst 2006;98(24):1777–85.

[106] Buess M, Rajski M, Vogel-Durrer B, Herrmann R, Rochlitz C. Tumor-endothelial interaction links the $CD44^{+}$/$CD24^{-}$ phenotype with poor prognosis in early-stage breast cancer. Neoplasia 2009;11(10):987–1002.

[107] Sheridan C, Kishimoto H, Fuchs RK, Mehrotra S, Bhat-Nakshatri P, Turner CH, et al. $CD44^{+}$/$CD24^{-}$ breast cancer cells exhibit enhanced invasive properties: an early step necessary for metastasis. Breast Cancer Res 2006;8(5):R59.

[108] Al-Hajj M, Wicha MS, Benito-Hernandez A, Morrison SJ, Clarke MF. Prospective identification of tumorigenic breast cancer cells. Proc Natl Acad Sci 2003;100(7):3983–8.

[109] Neumeister V, Agarwal S, Bordeaux J, Camp RL, Rimm DL. In situ identification of putative cancer stem cells by multiplexing ALDH1, CD44, and cytokeratin identifies breast cancer patients with poor prognosis. Am J Pathol 2010;176(5):2131–8.

[110] Wright MH, Calcagno A, Salcido CD, Carlson MD, Ambudkar SV, Varticovski L. Brca1 breast tumors contain distinct $CD44^{+}$/$CD24^{-}$ and $CD133^{+}$ cells with cancer stem cell characteristics. Breast Cancer Res 2008;10(1):R10.

[111] Zhao P, Lu Y, Jiang X, Li X. Clinicopathological significance and prognostic value of CD133 expression in triple-negative breast carcinoma. Cancer Sci 2011;102(5):1107–11.

[112] Liu YL, Yang YM, Xu H, Dong XS. Increased expression of ubiquitin-specific protease 22 can promote cancer progression and predict therapy failure in human colorectal cancer. J Gastroenterol Hepatol 2010;25(11):1800–5.

[113] Vassilopoulos A. Identification and characterization of cancer initiating cells from BRCA1 related mammary tumors using markers for normal mammary stem cells. Int J Biol Sci 2008:133–42.

[114] Lin W, Karsten U, Goletz S, Cheng R, Cao Y. Co-expression of CD173 (H2) and CD174 (Lewis Y) with CD44 suggests that fucosylated histo-blood group antigens are markers of breast cancer-initiating cells. Virchows Arch 2010;456(4):403–9.

[115] Ricardo S, Vieira AF, Gerhard R, Leitao D, Pinto R, Cameselle-Teijeiro JF, et al. Breast cancer stem cell markers CD44, CD24 and ALDH1: expression distribution within intrinsic molecular subtype. J Clin Pathol 2011;64(11):937–46.

[116] O'Brien CA, Pollett A, Gallinger S, Dick JE. A human colon cancer cell capable of initiating tumour growth in immunodeficient mice. Nature 2007;445(7123):106–10.

[117] Green CL, Loken M, Buck D, Deeg HJ. Discordant expression of AC133 and AC141 in patients with myelodysplastic syndrome (MDS) and acute myelogeneous leukemia (AML). Leukemia 2000;14(4):770–2.

[118] Jaksch M, Munera J, Bajpai R, Terskikh A, Oshima RG. Cell cycle-dependent variation of a CD133 epitope in human embryonic stem cell, colon cancer, and melanoma cell lines. Cancer Res 2008;68(19):7882–6.

[119] Zeuner A, Todaro M, Stassi G, De Maria R. Colorectal cancer stem cells: from the crypt to the clinic. Cell Stem Cell 2014;15(6):692–705.

[120] Jung P, Sato T, Merlos-Suarez A, Barriga FM, Iglesias M, Rossell D, et al. Isolation and in vitro expansion of human colonic stem cells. Nat Med 2011;17(10):1225–7.

[121] Dalerba P, Clarke MF. Cancer stem cells and tumor metastasis: first steps into uncharted territory. Cell Stem Cell 2007;1(3):241–2.

[122] Huang EH, Hynes MJ, Zhang T, Ginestier C, Dontu G, Appelman H, et al. Aldehyde dehydrogenase 1 is a marker for normal and malignant human colonic stem cells (SC) and tracks SC overpopulation during colon tumorigenesis. Cancer Res 2009;69(8):3382–9.

[123] Kemper K, Prasetyanti PR, De Lau W, Rodermond H, Clevers H, Medema JP. Monoclonal antibodies against Lgr5 identify human colorectal cancer stem cells. Stem Cells 2012;30(11):2378–86.

[124] Todaro M, Gaggianesi M, Catalano V, Benfante A, Iovino F, Biffoni M, et al. CD44v6 is a marker of constitutive and reprogrammed cancer stem cells driving colon cancer metastasis. Cell Stem Cell 2014;14(3):342–56.

[125] Kobayashi S, Yamada-Okabe H, Suzuki M, Natori O, Kato A, Matsubara K, et al. LGR5-positive colon cancer stem cells interconvert with drug-resistant LGR5-negative cells and are capable of tumor reconstitution. Stem Cells 2012;30(12):2631–44.

[126] Ma S, Chan KW, Hu L, Lee TK, Wo JY, Ng IO, et al. Identification and characterization of tumorigenic liver cancer stem/progenitor cells. Gastroenterology 2007;132(7):2542–56.

[127] Zhu Z, Hao X, Yan M, Yao M, Ge C, Gu J, et al. Cancer stem/progenitor cells are highly enriched in $CD133^{+}CD44^{+}$ population in hepatocellular carcinoma. Int J Cancer 2010;126(9):2067–78.

[128] Lin W, Karsten U, Goletz S, Cheng R, Cao Y. Expression of CD176 (Thomsen-Friedenreich antigen) on lung, breast and liver cancer-initiating cells. Int J Exp Pathol 2010;92(2):97–105.

[129] Kimura O, Takahashi T, Ishii N, Inoue Y, Ueno Y, Kogure T, et al. Characterization of the epithelial cell adhesion molecule $(EpCAM)^{+}$ cell population in hepatocellular carcinoma cell lines. Cancer Sci 2010;101(10):2145–55.

[130] Yamashita T, Ji J, Budhu A, Forgues M, Yang W, Wang H, et al. EpCAM-positive hepatocellular carcinoma cells are tumor-initiating cells with stem/progenitor cell features. Gastroenterology 2009;136(3):1012–24. e4.

[131] Yang ZF, Ho DW, Ng MN, Lau CK, Yu WC, Ngai P, et al. Significance of $CD90^{+}$ cancer stem cells in human liver cancer. Cancer Cell 2008;13(2):153–66.

[132] Haraguchi N, Ishii H, Mimori K, Tanaka F, Ohkuma M, Kim HM, et al. CD13 is a therapeutic target in human liver cancer stem cells. J Clin Invest 2010;120(9):3326–39.

[133] Yang W, Yan HX, Chen L, Liu Q, He YQ, Yu LX, et al. Wnt/beta-catenin signaling contributes to activation of normal and tumorigenic liver progenitor cells. Cancer Res 2008;68(11):4287–95.

[134] Li C, Heidt DG, Dalerba P, Burant CF, Zhang L, Adsay V, et al. Identification of pancreatic cancer stem cells. Cancer Res 2007;67(3):1030–7.

[135] Hermann PC, Huber SL, Herrler T, Aicher A, Ellwart JW, Guba M, et al. Distinct populations of cancer stem cells determine tumor growth and metastatic activity in human pancreatic cancer. Cell Stem Cell 2007;1(3):313–23.

[136] Kim MP, Fleming JB, Wang H, Abbruzzese JL, Choi W, Kopetz S, et al. ALDH activity selectively defines an enhanced tumor-initiating cell population relative to CD133 expression in human pancreatic adenocarcinoma. PLoS One 2011;6(6):e20636.

[137] Li C, Wu JJ, Hynes M, Dosch J, Sarkar B, Welling TH, et al. c-Met is a marker of pancreatic cancer stem cells and therapeutic target. Gastroenterology 2011;141(6):2218–27. e5.

[138] Richard V, Nair MG, Santhosh Kumar TR, Pillai MR. Side population cells as prototype of chemoresistant, tumor-initiating cells. Biomed Res Int 2013;2013:517237.

[139] Asuthkar S, Stepanova V, Lebedeva T, Holterman AL, Estes N, Cines DB, et al. Multifunctional roles of urokinase plasminogen activator (uPA) in cancer stemness and chemoresistance of pancreatic cancer. Mol Biol Cell 2013;24(17):2620–32.

[140] Singh SK, Hawkins C, Clarke ID, Squire JA, Bayani J, Hide T, et al. Identification of human brain tumour initiating cells. Nature 2004;432(7015):396–401.

[141] Shmelkov SV, Butler JM, Hooper AT, Hormigo A, Kushner J, Milde T, et al. CD133 expression is not restricted to stem cells, and both $CD133^{+}$ and $CD133^{-}$ metastatic colon cancer cells initiate tumors. J Clin Invest 2008;118(6):2111–20.

[142] Saigusa S, Tanaka K, Toiyama Y, Yokoe T, Okugawa Y, Koike Y, et al. Clinical significance of CD133 and hypoxia inducible factor-1alpha gene expression in rectal cancer after preoperative chemoradiotherapy. Clin Oncol R Coll Radiol 2011;23(5):323–32.

[143] Lathia JD, Hitomi M, Gallagher J, Gadani SP, Adkins J, Vasanji A, et al. Distribution of CD133 reveals glioma stem cells self-renew through symmetric and asymmetric cell divisions. Cell Death Dis 2011;2:e200.

[144] Bao S, Wu Q, Li Z, Sathornsumetee S, Wang H, McLendon RE, et al. Targeting cancer stem cells through L1CAM suppresses glioma growth. Cancer Res 2008;68(15):6043–8.

[145] Son MJ, Woolard K, Nam D, Lee J, Fine HA. SSEA-1 is an enrichment marker for tumor-initiating cells in human glioblastoma. Cell Stem Cell 2009;4(5):440–52.

[146] Mao P, Joshi K, Li J, Kim SH, Li P, Santana-Santos L, et al. Mesenchymal glioma stem cells are maintained by activated glycolytic metabolism involving aldehyde dehydrogenase 1A3. Proc Natl Acad Sci USA 2013;110(21):8644–9.

[147] Tchoghandjian A, Baeza N, Colin C, Cayre M, Metellus P, Beclin C, et al. A2B5 cells from human glioblastoma have cancer stem cell properties. Brain Pathol 2010;20(1):211–21.

[148] Anido J, Saez-Borderias A, Gonzalez-Junca A, Rodon L, Folch G, Carmona MA, et al. TGF-beta receptor inhibitors target the CD44(high)/Id1(high) glioma-initiating cell population in human glioblastoma. Cancer Cell 2010;18(6):655–68.

[149] Ogden AT, Waziri AE, Lochhead RA, Fusco D, Lopez K, Ellis JA, et al. Identification of $A2B5^{+}CD133^{-}$ tumor-initiating cells in adult human gliomas. Neurosurgery 2008;62(2):505–14. discussion 514-5.

[150] Eramo A, Lotti F, Sette G, Pilozzi E, Biffoni M, Di Virgilio A, et al. Identification and expansion of the tumorigenic lung cancer stem cell population. Cell Death Differ 2007;15(3):504–14.

[151] Curtis SJ, Sinkevicius KW, Li D, Lau AN, Roach RR, Zamponi R, et al. Primary tumor genotype is an important determinant in identification of lung cancer propagating cells. Cell Stem Cell 2010;7(1):127–33.

[152] Collins AT. Prospective identification of tumorigenic prostate cancer stem cells. Cancer Res 2005;65(23):10946–51.

[153] Maitland NJ, Collins AT. Prostate cancer stem cells: a new target for therapy. J Clin Oncol 2008;26(17):2862–70.

[154] Patrawala L, Calhoun T, Schneider-Broussard R, Li H, Bhatia B, Tang S, et al. Highly purified $CD44^+$ prostate cancer cells from xenograft human tumors are enriched in tumorigenic and metastatic progenitor cells. Oncogene 2006;25(12):1696–708.

[155] Klarmann GJ, Hurt EM, Mathews LA, Zhang X, Duhagon MA, Mistree T, et al. Invasive prostate cancer cells are tumor initiating cells that have a stem cell-like genomic signature. Clin Exp Metastasis 2009;26(5):433–46.

[156] Patrawala L, Calhoun T, Schneider-Broussard R, Zhou J, Claypool K, Tang DG. Side population is enriched in tumorigenic, stem-like cancer cells, whereas $ABCG2^+$ and $ABCG2^-$ cancer cells are similarly tumorigenic. Cancer Res 2005;65(14):6207–19.

[157] Mulholland DJ, Xin L, Morim A, Lawson D, Witte O, Wu H. $Lin^-Sca\text{-}1^+CD49f^{high}$ stem/progenitors are tumor-initiating cells in the Pten-null prostate cancer model. Cancer Res 2009;69(22):8555–62.

[158] Kretzschmar K, Watt FM. Lineage tracing. Cell 2012;148(1–2):33–45.

[159] Romagnani P, Rinkevich Y, Dekel B. The use of lineage tracing to study kidney injury and regeneration. Nat Rev Nephrol 2015;11(7):420–31.

[160] Read TA, Fogarty MP, Markant SL, McLendon RE, Wei Z, Ellison DW, et al. Identification of CD15 as a marker for tumor-propagating cells in a mouse model of medulloblastoma. Cancer Cell 2009;15(2):135–47.

[161] Ward RJ, Lee L, Graham K, Satkunendran T, Yoshikawa K, Ling E, et al. Multipotent $CD15^+$ cancer stem cells in patched-1-deficient mouse medulloblastoma. Cancer Res 2009;69(11):4682–90.

[162] Zong H, Espinosa JS, Su HH, Muzumdar MD, Luo L. Mosaic analysis with double markers in mice. Cell 2005;121(3):479–92. 2016/01/19.

[163] Liu C, Sage JC, Miller MR, Verhaak RGW, Hippenmeyer S, Vogel H, et al. Mosaic analysis with double markers reveals tumor cell of origin in glioma. Cell 2011;146(2):209–21.

[164] Medema JP. Cancer stem cells: the challenges ahead. Nat Cell Biol 2013;15(4):338–44.

[165] Taylor RW, Barron MJ, Borthwick GM, Gospel A, Chinnery PF, Samuels DC, et al. Mitochondrial DNA mutations in human colonic crypt stem cells. J Clin Invest 2003;112(9):1351–60. 2016/01/21.

[166] Baker A, Cereser B, Melton S, Fletcher A, Rodriguez-Justo M, Tadrous P, et al. Quantification of crypt and stem cell evolution in the normal and neoplastic human colon. Cell Rep 2014;8(4):940–7. 2016/01/22.

[167] Evrony G, Lee E, Mehta B, Benjamini Y, Johnson R, Cai X, et al. Cell lineage analysis in human brain using endogenous retroelements. Neuron 2015;85(1):49–59. 2016/01/22.

[168] Reynolds BA, Weiss S. Generation of neurons and astrocytes from isolated cells of the adult mammalian central nervous system. Science (New York, NY) 1992;255(5052):1707–10.

[169] Pastrana E, Silva-Vargas V, Doetsch F. Eyes wide open: a critical review of sphere-formation as an assay for stem cells. Cell Stem Cell 2011;8(5):486–98.

[170] Acuff BR, Cohen JJ. An in vitro assay for T lymphocyte progenitors (CFU-pre T). J Supramol Struct 1980;14(2):215–22.

[171] Lawson DA, Xin L, Lukacs RU, Cheng D, Witte ON. Isolation and functional characterization of murine prostate stem cells. Proc Natl Acad Sci USA 2007;104(1):181–6.

[172] Ignatova TN, Kukekov VG, Laywell ED, Suslov ON, Vrionis FD, Steindler DA. Human cortical glial tumors contain neural stem-like cells expressing astroglial and neuronal markers in vitro. Glia 2002;39(3):193–206.

[173] Galli R, Binda E, Orfanelli U, Cipelletti B, Gritti A, De Vitis S, et al. Isolation and characterization of tumorigenic, stem-like neural precursors from human glioblastoma. Cancer Res 2004;64(19):7011–21.

[174] Hirschhaeuser F, Menne H, Dittfeld C, West J, Mueller-Klieser W, Kunz-Schughart L. Multicellular tumor spheroids: an underestimated tool is catching up again. J Biotechnol 2010;148(1):3–15.

[175] Chaichana K, Zamora-Berridi G, Camara-Quintana J, Quinones-Hinojosa A. Neurosphere assays: growth factors and hormone differences in tumor and nontumor studies. Stem Cells 2006;24(12):2851–7.

[176] Yan DW, Li DW, Yang YX, Xia J, Wang XL, Zhou CZ, et al. Ubiquitin D is correlated with colon cancer progression and predicts recurrence for stage II-III disease after curative surgery. Br J Cancer 2010;103(7):961–9.

[177] De Witt Hamer PC, Van Tilborg AAG, Eijk PP, Sminia P, Troost D, Van Noorden CJF, et al. The genomic profile of human malignant glioma is altered early in primary cell culture and preserved in spheroids. Oncogene 2008;27(14):2091–6.

[178] Visvader JE, Lindeman GJ. Cancer stem cells in solid tumours: accumulating evidence and unresolved questions. Nat Rev Cancer 2008;8(10):755–68.

[179] Duan J, Qiu W, Xu S, Wang B, Ye X, Ping Y, et al. Strategies for isolating and enriching cancer stem cells: well begun is half done. Stem Cells Dev 2013;22(16):2221–39.

[180] Leo O, Slaoui M, Marvel J, Milner EC, Hiernaux J, Moser M, et al. Idiotypic analysis of polyclonal and monoclonal anti-p-azophenylarsonate antibodies of BALB/c mice expressing the major cross-reactive idiotype of the A/J strain. J Immunol 1985;134(3):1734–9.

[181] Hu Y, Smyth GK. ELDA: extreme limiting dilution analysis for comparing depleted and enriched populations in stem cell and other assays. J Immunol Methods 2009;347(1–2):70–8.

[182] Singec I, Knoth R, Meyer RP, Maciaczyk J, Volk B, Nikkhah G, et al. Defining the actual sensitivity and specificity of the neurosphere assay in stem cell biology. Nat Methods 2006;3(10):801–6.

[183] Louis SA, Rietze RL, Deleyrolle L, Wagey RE, Thomas TE, Eaves AC, et al. Enumeration of neural stem and progenitor cells in the neural colony-forming cell assay. Stem Cells 2008;26(4):988–96.

[184] Meyskens FLJ, Thomson SP, Moon TE. Quantitation of the number of cells within tumor colonies in semisolid medium and their growth as oblate spheroids. Cancer Res 1984;44(1):271–7.

[185] Allieri MA, Douay L, Deloux J, Smadja N, Najman A, Gorin NC. The role of methylcellulose on colony growth of human myeloid leukemic progenitors (AML-CFU). Exp Hematol 1990;18(8):911–5.

[186] O'Brien CA, Kreso A, Jamieson CH. Cancer stem cells and self-renewal. Clin Cancer Res 2010;16(12):3113–20.

[187] Quintana E, Shackleton M, Sabel MS, Fullen DR, Johnson TM, Morrison SJ. Efficient tumour formation by single human melanoma cells. Nature 2008;456(7222):593–8.

[188] Flavahan WA, Wu Q, Hitomi M, Rahim N, Kim Y, Sloan AE, et al. Brain tumor initiating cells adapt to restricted nutrition through preferential glucose uptake. Nat Neurosci 2013;16(10):1373–82.

[189] Qin J, Liu X, Laffin B, Chen X, Choy G, Jeter CR, et al. The PSA(-/lo) prostate cancer cell population harbors self-renewing long-term tumor-propagating cells that resist castration. Cell Stem Cell 2012;10(5):556–69.

[190] Zheng Q, Banaszak L, Fracci S, Basali D, Dunlap SM, Hursting SD, et al. Leptin receptor maintains cancer stem-like properties in triple negative breast cancer cells. Endocr Relat Cancer 2013;20(6):797–808.

[191] Ellenbroek SI, van Rheenen J. Imaging hallmarks of cancer in living mice. Nat Rev Cancer 2014;14(6):406–18.

[192] Lathia JD, Gallagher J, Myers JT, Li M, Vasanji A, McLendon RE, et al. Direct in vivo evidence for tumor propagation by glioblastoma cancer stem cells. PLoS One 2011;6(9):e24807.

[193] Zomer A, Ellenbroek SI, Ritsma L, Beerling E, Vrisekoop N, Van Rheenen J. Intravital imaging of cancer stem cell plasticity in mammary tumors. Stem Cells 2013;31(3):602–6.

[194] Graf BW, Chaney EJ, Marjanovic M, Adie SG, De Lisio M, Valero MC, et al. Long-term time-lapse multimodal intravital imaging of regeneration and bone-marrow-derived cell dynamics in skin. Technology 2013;1(1):8–19.

[195] Barretto RP, Ko TH, Jung JC, Wang TJ, Capps G, Waters AC, et al. Time-lapse imaging of disease progression in deep brain areas using fluorescence microendoscopy. Nat Med 2011;17(2):223–8.

[196] Ferard G, Pontet F, Giroud C, Dumontet M. Introducing the new international vocabulary of metrology. Ann Biol Clin (Paris) 2009;67(5):591–4.

[197] Clarke MF. A self-renewal assay for cancer stem cells. Cancer Chemother Pharmacol 2005;56(Suppl. 1):64–8.

[198] Kreso A, Dick JE. Evolution of the cancer stem cell model. Cell Stem Cell 2014;14(3):275–91.

[199] Jeter CR, Liu B, Liu X, Chen X, Liu C, Calhoun-Davis T, et al. NANOG promotes cancer stem cell characteristics and prostate cancer resistance to androgen deprivation. Oncogene 2011;30(36):3833–45.

[200] Wei F, Rong XX, Xie RY, Jia LT, Wang HY, Qin YJ, et al. Cytokine-induced killer cells efficiently kill stem-like cancer cells of nasopharyngeal carcinoma via the NKG2D-ligands recognition. Oncotarget 2015;6(33):35023–39.

[201] Hirsch D, Hu Y, Ried T, Moll R, Gaiser T. Transcriptome profiling of LGR5 positive colorectal cancer cells. Genom Data 2014;2:212–5.

[202] Jordan CT, Upchurch D, Szilvassy SJ, Guzman ML, Howard DS, Pettigrew AL, et al. The interleukin-3 receptor alpha chain is a unique marker for human acute myelogenous leukemia stem cells. Leukemia 2000;14(10):1777–84.

2

SECTION

CSCs in Representative Organ Systems

CHAPTER 4

Critical Updates to the Leukemia Stem Cell Model

Y. Saunthararajah

Taussig Cancer Institute, Cleveland Clinic, Cleveland, OH, United States

INTRODUCTION

The terabytes of data generated in leukemia and cancer research require an organizing principle to facilitate interpretation. For some decades, serving this purpose has been the model that malignant cell populations recapitulate the hierarchical structure of normal hematopoiesis or tissue ontogeny, with apex stem cells (leukemia stem cells, cancer stem cells—LSCs, CSCs, respectively) responsible for generating and sustaining the overall malignant cell population [1]. This concept was solidified with a study of acute myeloid leukemia (AML) around two decades ago [2], and billions of public and private dollars have been invested in the concept and its implications [1]. Although controversy around the model has been acknowledged [1,3], that multiple, highly reproduced, and empiric lines of evidence directly contradict basic tenets of the original model is insufficiently appreciated, wasteful misconceptions persist [1]. Described in this report are the several strands of new evidence that taken together, synthesize a revised model with practical, actionable implications.

THE ORIGINAL LEUKEMIA STEM CELL MODEL

Surface phenotype was used to sort leukemia cell populations into subsets that resemble hematopoietic stem cells (HSCs) ($CD34^{+}CD38^{-}$) and committed progenitors ($CD34^{+}CD38^{+}$) of the normal hematopoietic hierarchy [2]. The capacity of these leukemia cell subsets to initiate leukemia in severe combined immune-deficient (SCID) mice was then analyzed. This initial study indicated that AML cells with leukemia initiating capacity were rare cells exclusively with a surface phenotype of normal HSC [2,4]. In other words, leukemia recapitulated the hierarchical structure of normal hematopoiesis from which it is derived, in that stem cell–like cells, possessing intrinsic, indefinite self-renewal capacity, created and sustained the overall malignant cell population [2,4].

CONTENTS

Cancer Stem Cells. http://dx.doi.org/10.1016/B978-0-12-803892-5.00004-8

This model promised mechanistic insights into clonal expansion and the enduring problem of resistance. It implied that the key to clonal advantage was deregulation of pathways that drive self-renewal of stem cells, or alternatively, the endowment onto committed progenitors of components of a stem cell program [5,6]; and that relapse was because distinct, self-renewing LSC escaped treatment using stem cell qualities such as quiescence [1,4].

WITH BETTER FLOW-SORTING PROTOCOLS, LEUKEMIA STEM CELLS ARE EXCLUSIVELY PROGENITORS, NOT HEMATOPOIETIC STEM CELLS, BY SURFACE PHENOTYPE

Contradicting the original report, several groups found that LSCs were much more frequent cells with surface phenotype features of lineage-committed progenitors, eg, $CD34^+38^+$, $CLL\text{-}1^+$, $CD71^+$, $CD90^-$, and $c\text{-}kit^-$ [7–18]. In fact, with the incorporation of additional parameters into the sorting strategy (eg, CD90), to better discriminate HSC from downstream committed progenitors, leukemia-initiating capacity was found to be absent from HSC-like cells, but present in committed progenitors [18–20]. That is, LSCs were exclusively lineage committed, not HSCs, by surface phenotype [18–20].

Why did the original and subsequent reports have contradictory results? One identified reason was that antibodies used to sort for $CD38^+$ cells have antiproliferative effects, thereby impeding leukemia-initiating capacity of $CD34^+CD38^+$ cells [7]. Another critical reason is that CD34 and CD38 as sole parameters do not sufficiently discriminate between HSCs and downstream early committed progenitors—hence, the strikingly different results when CD90 and other parameters were incorporated into cell-sorting protocols [18–20]. Other reasons relate to technical limitations of the immune-deficient mouse assay system used to identify LSC: with use of more immune-deficient mice (eg, NOD/Lt-Scid/IL2Rγ^{null}—NSG) as recipients, the frequency and phenotypes of LSC expanded [14–17]. Engraftment has been further enhanced by transgenic expression of human cytokines (SCF, GM-CSF, IL-3) which cross-react poorly with their murine counterparts [14]. Even so, primary AML cells from a large proportion of patients still fail to engraft, underscoring that technical and cross-species barriers continue to impact on the study of LSCs and the interpretation of their frequency.

EVIDENCE FOR LEUKEMIA STEM CELLS AND RESISTANCE

The original LSC model implied that AMLs with poor outcomes should be enriched for stem cell-like LSC, that is, contain greater number or characteristics of $CD34^+CD38^-$ LSCs, that evade treatment effects using stem cell–like properties like quiescence [1,4]. To evaluate this possibility, a 52-gene expression signature was generated, that distinguished leukemia cells with a multipotent progenitor (MPP)-phenotype ($Lin^-CD34^+CD38^-CD90^-$), defined as LSCs, from leukemia cells with downstream progenitor-phenotypes ($Lin^-CD34^+CD38^+$) or bulk leukemia blasts (Lin^-CD34^-) [21]. Then, this MPP/LSC signature, applied to primary bone marrow cells, was correlated with outcomes in four independent AML patient cohorts [21]. High expression of the MPP/LSC signature was associated with worse overall, event-free, and relapse-free survival. Significance was retained in multivariate incorporating age, mutations in *FLT3* and *NPM1*, and cytogenetic abnormalities [21]. In a separate study, a 42-gene gene expression signature was generated that distinguished leukemia cells that engrafted immune-deficient mice, defined as LSCs, from nonengrafting cells [6]. The engrafting cells included not just $CD34^+38^-$ cells but many instances of $CD34^+CD38^+$ and $CD34^-CD38^+$ cells [6]. Again, when analyzed in primary bone marrow and correlated with outcomes, this signature was found to predict worse outcomes in a cohort of patients with normal cytogenetics AML, including in multivariate analyses [6]. It should be appreciated, however, that in using an MPP surface phenotype, and including $CD34^+38^+$ and $CD34^-CD38^+$ positive cells as functionally defined LSCs, these studies were already contradicting the original model that suggested LSCs were restricted to cells with an HSC surface phenotype. Accordingly,

The global gene expression profile of LSC resembles normal MPP or GMP: As expected from the methods by which these signatures were defined, neither the 52 nor 42-gene signatures distinguish between normal HSC and normal MPP [6,21] (public data E-MTAB-2262 [22]) (Fig. 4.1). In fact, when the global gene expression of LSC was compared to global gene expression of normal hematopoietic hierarchy subsets, LSC gene expression resembled MPP or GMP, in accordance with their surface phenotype [18].

CSC signatures are compendia of MYC-regulated genes: Moreover, when chromatin immunoprecipitation was used to biochemically verify target genes of master transcription factors including MYC, the ancient coordinator of cell growth and division, it was found that the CSC signature from a number of studies was a compendia of MYC-regulated genes [23]. Notably, MYC activity is especially increased with lineage commitment; even the earliest MPPs have substantially more MYC and are substantially more proliferative than HSCs [22,24,25] (Fig. 4.1D).

Although the LSC signature is not a signature of stem cell–like cells, but a signature of MYC-enriched progenitors, it is still possible that having more of such cells has prognostic implications. Other parameters, however, are also pertinent, since several other prognosticating gene expression signatures, not based

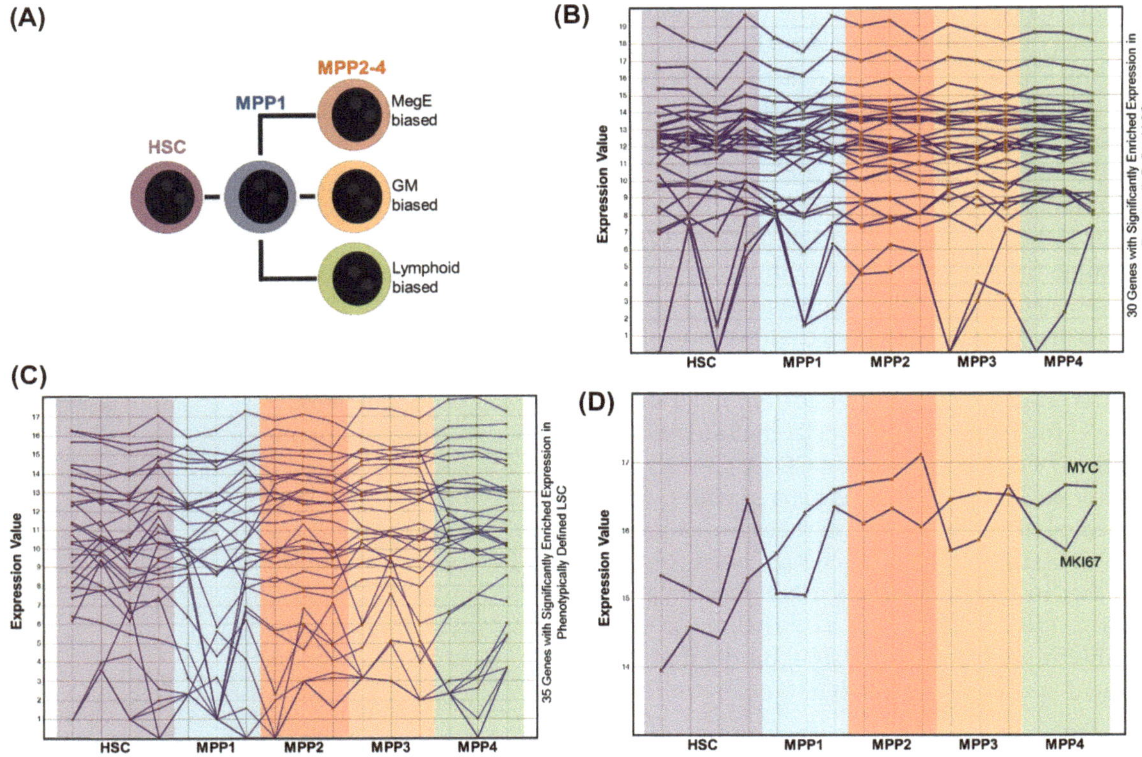

FIGURE 4.1 Gene expression signatures defined for leukemia stem cells (LSCs) do not distinguish between normal hematopoietic stem cells (HSCs) and multipotent progenitors (MPPs).

(A) Different types of MPP (gene expression by RNA-sequencing public data E-MTAB-2262 [22]); (B, C) The 42 and 52-gene signatures of functionally or phenotypically defined LSC [6,21] do not distinguish between normal HSC and normal MPP (not all the genes from the microarray signatures were available in the RNA-sequencing database). (D) Even the earliest MPPs have significantly higher MYC expression and are substantially more proliferative than HSCs (MKI67 as proliferation marker) [22,24,25].

on LSC, have been identified in AML [26–29]. More comprehensive, integrated genetic and gene expression analyses such as that of the TCGA project suggest that gene expression signatures track recurrent genetic alterations in AML [30].

OTHER MECHANISTIC EXPLANATIONS FOR RESISTANCE

There are other mechanisms of resistance that are biologically plausible, consistent with clinical experience, and based upon which practical in vivo interventions have been effected. One form of resistance relates to interactions between genetic alterations in leukemia cells and specific drugs: cytotoxic treatments, despite differing proximal actions, eg, topoisomerase inhibition (daunorubicin) or termination of DNA chain synthesis (cytarabine), are common in intending apoptosis, regulated by the master transcription factor p53 (*TP53*) (reviewed in Ref. [31]). As such, *TP53* mutation/deletion is associated with resistance to cytotoxic treatments in vitro, eg, selection by cytarabine for *TP53*-mutated leukemia subclones [32,33] and in vivo: (1) treatment selects for relapse with *TP53*-mutated AML subclones [34]; (2) *TP53*-mutated AML treated with cytotoxic chemotherapy has a substantially lower response rate than *TP53* wild-type AML—22–33% versus 76–81% [35,36]; (3) the poorest risk AML subtypes, eg, AML with complex cytogenetic abnormalities, have the highest rate of *TP53* mutations, exceeding 70% in some series [37,38]; and (4) even if *TP53* itself is not mutated, alterations in other key p53-system genes are frequent in poor risk AML, eg, gains in *MDM4*, that inactivates p53, are very common in chemorefractory AML [39,40]. As expected from this, treatments that do not rely on p53 and apoptosis produce significantly better outcomes than standard cytotoxic chemotherapy in vivo, including in randomized clinical trials, especially in high-risk myeloid malignancy containing complex chromosome alterations and *TP53* mutations [41–44].

Metabolism is intrinsically plastic, and another mechanism of resistance observed in numerous in vitro and in vivo studies is selection for leukemia cells that avoid drugs via simple metabolic shifts, not necessarily correlated with specific genetic alterations. For example, 5-azacytidine selects for leukemia cells favoring the pyrimidine salvage enzyme deoxycytidine kinase that bypasses 5-azacytidine, while decitabine selects for leukemia cells favoring the pyrimidine salvage enzyme uridine cytidine kinase 2 (UCK2) that bypasses decitabine (reviewed in Ref. [45]).

Mutations in *FLT3* (*FLT3*-ITD) are another genetic alteration consistently linked with poor outcomes, but via unknown mechanisms. The fact that the prognostic impact is restricted to AMLs treated with cytotoxic chemotherapy, however, and not to acute promyelocytic leukemia (APL) treated with retinoic acid and arsenic differentiation-based therapy and which has a *FLT3* mutation rate exceeding 30%, again suggests that resistance is related to an interaction between the genetic aberration and mechanisms/metabolism of specific drugs.

MASTER TRANSCRIPTION FACTOR EXPRESSION IN LEUKEMIA STEM CELLS AND HEMATOPOIETIC STEM CELLS

A few of the hundreds of transcription factors expressed in hematopoietic cells are masters that specify and drive cell lineage fates, demonstrated in murine knockout, lineage conversion, and enhancer studies [46–50]. The LSC model as originally framed predicts that LSC should express high levels of the master

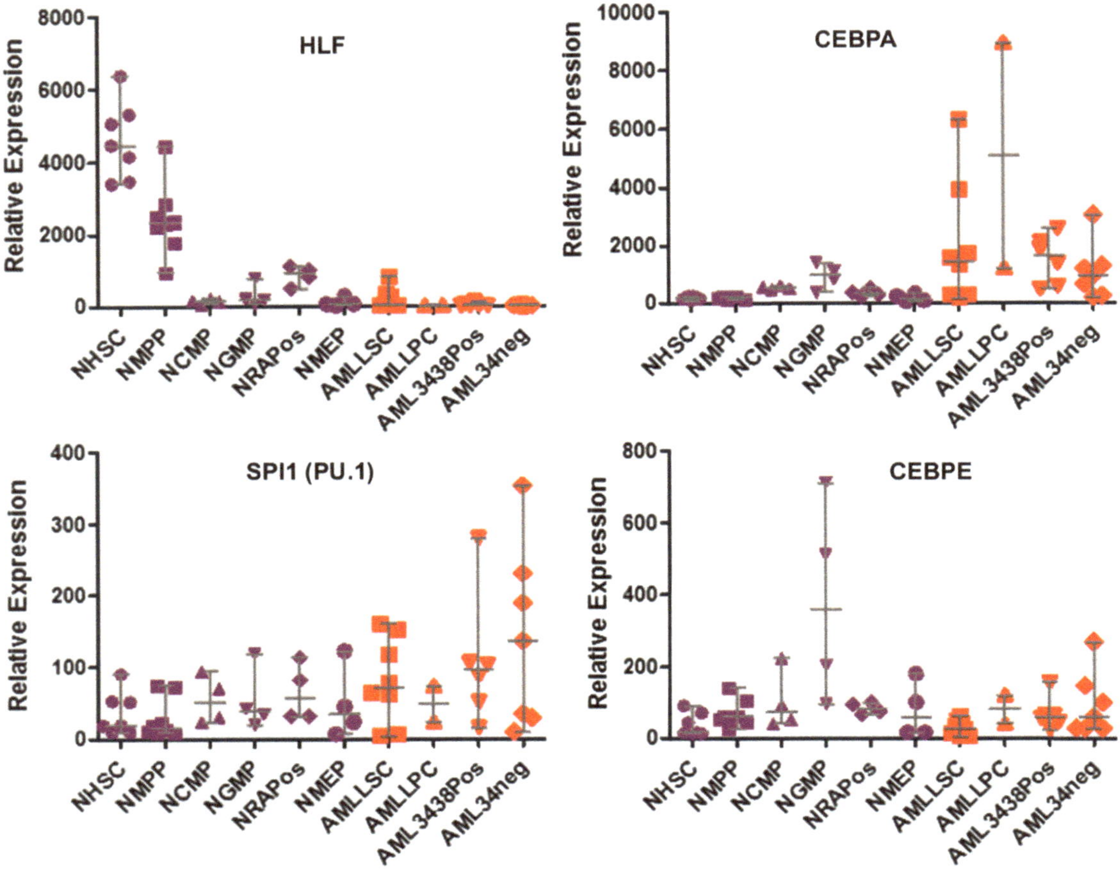

FIGURE 4.2 **Master transcription factor expression in hematopoietic stem cells versus leukemia stem cells.**
HLF drives stem cell fate, CEBPA and SPI1 (PU.1) drive myeloid lineage fate, and CEBPE drives terminal differentiation. Gene expression by microarray, public data GSE24006 [21]. Normal and AML cell populations were flow sorted by surface phenotype. AML cases were representative of the morphologic and genetic spectrum of disease. *AML LSC*, leukemia stem cells; *AML LPC*, leukemia progenitor cells; *NCMP*, normal common myeloid progenitor; *NHSC*, normal hematopoietic stem cells; *NMEP*, normal megakaryocyte erythroid progenitor; *NMPP*, normal multipotent progenitors; *NRAPos*, normal mature myeloid cells. Error bars = median ± range.

transcription factors that make HSC (eg, HLF), explaining self-renewal and consequently differentiation impairment [51]. Conspicuously, however, LSCs express very low levels of these master transcription factor relative to normal HSC (Figs. 4.2 and 4.3). Instead, across the genetic and cytogenetic spectrum of AML, there are supranormal levels of commitment/early-differentiation master transcription factors (CEBPA, PU.1) [42,52] yet repression of key terminal differentiation target genes (eg, *CEBPE* [49,71–77]) of these master transcription factors (Figs. 4.2 and 4.3).

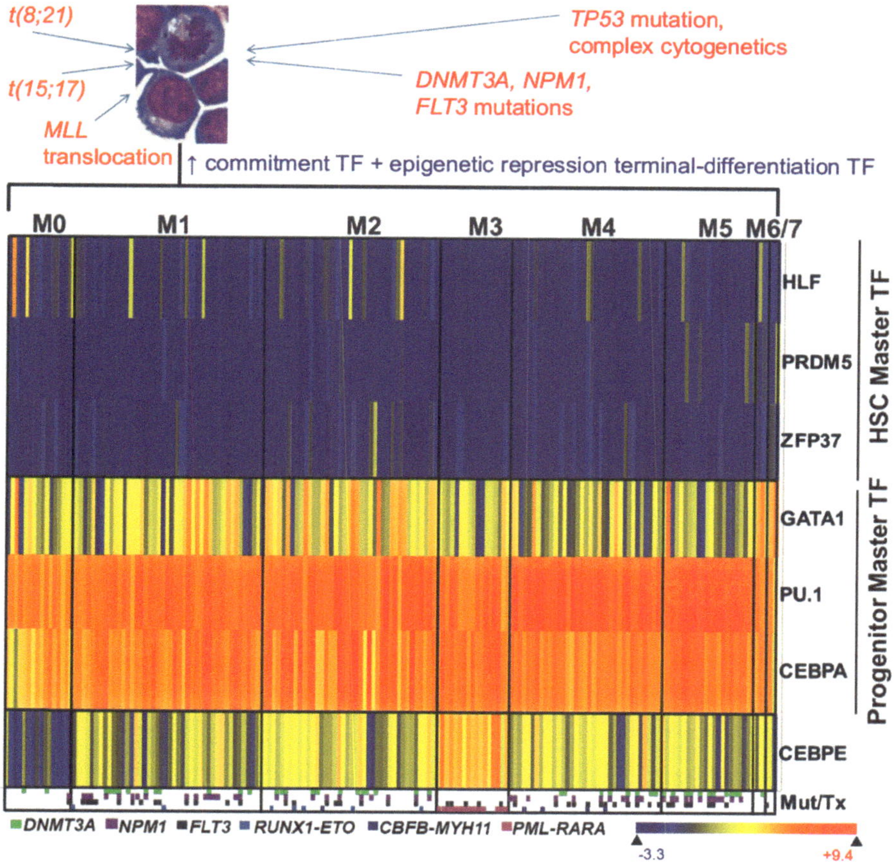

FIGURE 4.3 **Different genetics, same strategy.**
All acute myeloid leukemia (AMLs), regardless of genetics and evolutionary times, have a similar profile of high master commitment transcription factor (TF) expression but aberrant epigenetic repression of key terminal differentiation TF. AML morphologic subtypes (M0–M7) correspond to relative amounts of CEBPA, PU.1, and GATA1 that drive granulocytic, monocytic, and erythroid lineage fates, respectively. Master stem cell TF: HLF, PRDM5, ZFP37; master commitment/early-differentiation TF: CEBPA, PU.1, GATA1; master late/terminal-differentiation TF: CEBPE. Gene expression by RNA sequencing, The Cancer Genome Atlas [30]. Each column in the heat map is one AML bone marrow sample, $n=174$. Partial annotation is provided for contained genetic alterations.

THE MOST FREQUENT MUTATIONS IN ACUTE MYELOID LEUKEMIA, IN *NPM1* AND *FLT3*, ARE IN COMMITTED PROGENITORS, NOT HEMATOPOIETIC STEM CELLS

The founder mutation or *n-1* hit in a process of leukemogenesis that requires *n* or more hits, originates in germ line or HSC, demonstrated unequivocally by familial AML syndromes (eg, *RUNX1*, *CEBPA* mutations) [53]. The most highly recurrent mutations in de novo AML, targeting *FLT3* and *NPM1*, however, do not originate in HSC—these mutations are not detectable in cells with an HSC/MPP surface phenotype isolated from AML patient bone marrow [19,20,54]. Accordingly, HSCs/MPPs from AML patient bone marrow, containing some but not all the AML mutations, do not initiate leukemia in immune-deficient mice, but instead yield multilineage engraftment, as expected of functional HSCs/MPPs [19,20,54]. That is, these HSCs/MPPs are preleukemic. In contrast, cells with committed progenitor surface phenotype from the same bone marrows did contain mutations in *FLT3* and/or *NPM1*, did initiate leukemia in immune-deficient mice, did not produce multilineage engraftment, and could not be detected in T-cells [19,20,54]. Thus, leukemia originates in germ line or an HSC, but the final steps of transformation to LSC, eg, by mutations in *FLT3* and/or *NPM1*, occur in committed daughter cells (progenitors).

DO FOUNDER MUTATIONS EXPAND HEMATOPOIETIC STEM CELLS AND/OR EXPAND COMMITTED PROGENITORS?

Some observations can be construed as supporting the original model in that a founder mutation deregulates self-renewal pathways in HSC: (1) preleukemic HSCs/MPPs from the bone marrow of AML patients demonstrated a high *DNMT3A* mutation allele frequency (2–69%) [19,20,54]; (2) between weeks 8 and 16, there was a >2-fold increase in *DNMT3A* mutant allele frequency in a human graft in immune-deficient mice [20]; and (3) in ~10% of general population individuals >65 years of age, elevated variant allele frequencies (median ~8%) for mutated *DNMT3A*, *ASXL1*, and *TET2* were detected in DNA from peripheral blood mononuclear cells [55,56]. One explanation for these observations is that mutated DNMT3A (or ASXL1 or TET2) increases self-renewal of HSC, thereby generating a functional repopulation advantage.

If these *n-1* founder mutations are serving to expand HSC, however, the expectation would be that transforming hit *n* would also occur in an abnormally expanding, destabilized HSC compartment. The several lines of reproduced evidence described earlier demonstrate that this is not the case. Thus, another explanation deserves serious consideration: well-known *n-1* hits including *DNMT3A* mutation permit lineage commitment, and the MYC upregulation that accompanies this, but impede terminal differentiation. This latter explanation is entirely consistent with the observations in the previous paragraph but also a wider range of findings: (1) the overall trajectory of neoplastic evolution that culminates with transformation of committed progenitors into LSC; (2) the committed-progenitor compartment, not HSC compartment, expands in AML patient bone marrow [18]; (3) *DNMT3A* or *IDH2* founder mutation allele frequency was consistently higher by a log or more in progenitors (common myeloid progenitors, megakaryocyte erythroid progenitors, granulocyte monocyte progenitors) than in HSCs/MPPs or T-cells, suggesting DNMT3A mutation facilitated expansion of myeloid-biased progenitors rather than HSCs [20]—even when only a founder *DNMT3A* mutation could be detected during 36 months of remission in one patient, variant allele frequency went to zero in HSC/MPP while at the same time, it considerably increased in multiple committed lineages [20]; and (4) the most common cause of familial AML is loss-of-function mutations in *RUNX1*—biochemical and functional studies demonstrated that RUNX1 is more important for activation of key terminal differentiation genes (eg, *CEBPE*)—a late-differentiation master transcription factor [49,71–77] than key commitment genes (eg, *CEBPA*) [57]. Accordingly, *Runx1*-deficient mice have an expanded progenitor compartment, not HSCs [58]. Similarly, point mutations in *CEBPA* that also cause familial AML have been shown to permit commitment but impede subsequent maturation [8].

In short, a number of observations indicate that *n-1* mutations, although originating in germ line or HSC, exercise their consequences most prominently in committed daughter cells, setting the stage for transforming mutations in *FLT3* and/or *NPM1* in the latter compartment. Some founder mutations, eg, in *DNMT3A*, may also expand HSC, since this has been suggested in some, but not all, murine studies [59,60]. Although it seems intuitive that founder mutations expand HSC, similar intuition was misleading with regards to LSC—clonal expansion requires proliferation, by orders of magnitude a trait more of committed progenitors than HSCs.

THE REVISED MODEL

The original model of LSC had intuitive appeal—stem cells intrinsically and indefinitely self-renew, use error-prone nonhomologous end joining for DNA repair [61], and are long lived enough to sustain multiple mutations. Also, founder mutations clearly originate and can be detected in germ line or stem cells. Evolution requires reproduction, however, and adult stem cells are quiescent, with intervals between cell divisions measured in months [62–64] (reviewed in Ref. [65]). Committed daughter cells, in contrast, activate and stabilize the master transcription factor coordinator of cell growth and division MYC [24,25,66–68] and have intervals between divisions measured in hours [62–64]. This vigorous growth is controlled in normal committed progenitors by differentiation/maturation. From an evolutionary perspective, differentiation programs are recent metazoan overlays on substantially more ancient cell growth and division (proliferation) programs [69,70]. These young programs have many ways to fail, as the varied genetic alterations in AML, a disease pathologically defined by differentiation failure, attest (Fig. 4.3). Founder mutations are propagated from germ line or HSC to committed daughter cells where they impede maturation, thereby introducing ectopic cell divisions in an intrinsically proliferative compartment. Additional genetic alterations acquired in this destabilized compartment produce even more obvious differentiation failure culminating in malignant self-renewal (proliferation without differentiation). In short, the key to clonal advantage is not induction of self-renewal in intrinsically quiescent stem cells, but failure of differentiation in intrinsically proliferative progenitors (Fig. 4.4).

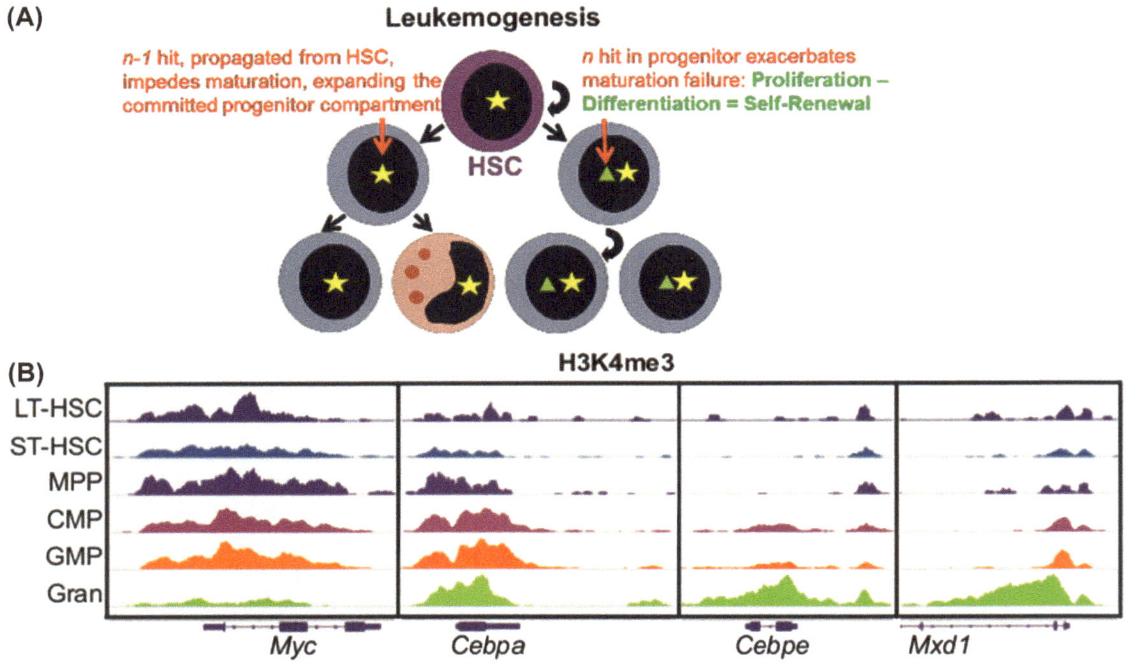

FIGURE 4.4

(A) Founding mutations originate in germ line or hematopoietic stem cells (HSCs), but transformation is of committed progenitors. The founder mutation or *n-1* hit in a process of leukemogenesis that requires *n* or more hits, originates in germ line or HSC. These founder mutations permit commitment/early differentiation, but impede activation of maturation genes, introducing ectopic cell divisions into the committed-progenitor compartment. The most highly recurrent mutations in de novo AML, targeting *FLT3* and *NPM1* (*n* hit), originate in this destabilized progenitor compartment, impeding maturation to the point of creating aberrant self-renewal: proliferation without differentiation equals to self-renewal. **(B) The greater chromatin remodeling needed at key terminal differentiation genes than at key commitment genes facilitates this mode of neoplastic evolution.** Key commitment genes (*Myc*, *Cebpa*) are already epigenetically primed for activation even in long-term HSC (LT-HSC), with little need for additional activation mark H3K4me3 in the transition to committed progenitors. Key terminal differentiation genes (*Cebpe*, *Mxd1*), by contrast, require extensive H3K4me3 augmentation for terminal granulocyte differentiation (Gran). Interestingly, the H3K4 methyltransferase domain is invariably deleted by recurrent translocations of *KMT2A* (MLL) in AML. ChIP-Seq data normal hematopoiesis [79] (GSE60103), scale identical across samples.

Biochemically, there is a common theme underlying terminal-differentiation failure: alterations to master differentiation-driving transcription factors (eg, *CEBPA*) or their cofactors (eg, *RUNX1*, *KMT2A*, splicing factors, cohesins) alter coactivator/corepressor balance at terminal-differentiation target genes (eg, *CEBPE*) [49,71–77], such that these genes are epigenetically repressed instead of activated [41,52,57,78]. Thus, single genetic alterations produce aberrant epigenetic repression of hundreds of differentiation target genes [52]. Facilitating this mode of neoplastic evolution is the more extensive chromatin remodeling needed for activation of terminal-differentiation genes than for commitment/early-differentiation genes [79]—commitment/early-differentiation genes, such as *CEBPA* and *MYC* are already poised for activation with H3K4me3 marks even in long-term HSC (Fig. 4.4). Thus, during progressive myeloid differentiation, these genes require substantially less remodeling toward activation marks than terminal-differentiation genes *CEBPE* and *MXD1* (Fig. 4.4). Perhaps therefore, it is not surprising that the H3K4 methyltransferase domain is invariably deleted by recurrent MLL (*KMT2A*) translocations in AML, and leukemogenesis selects for other genetic alterations that shift coactivator/corepressor dynamics at terminal-differentiation genes toward corepressors [78,80].

PRACTICAL, IMMEDIATE IMPLICATIONS OF THE REVISED MODEL

Molecular mechanism implications: Thus, LSCs have a phenotype of commitment but with aberrant epigenetic repression of terminal-differentiation genes. It seems likely that recurrent loss-of-function alterations to gene-activating machinery in AML (MLL family, SWI/SNF family, splicing factors, cohesins, etc.) contributes to this empirically observable LSC phenotype. That *RUNX1* and *CEBPA* founder mutations, the two most common causes of familial AML, permit commitment but impede activation of terminal-differentiation genes has already been shown [8,57]. The two most common secondary mutations in AML, in *NPM1* and *FLT3*, have also been found to directly impact transactivating function of CEBPA and PU.1 [81,82], and reversing these effects of mutated FLT3 and mutated NPM1 on CEBPA and PU.1 triggers terminal granulocytic and monocytic differentiation of AML cells, respectively [82,83]. Loss-of-function genetic alterations in polycomb repressor complex 2 (PRC2) components such as *EZH2*, *ASXL1*, *DNMT3A*, etc. may skew commitment decisions in the myeloid direction [84] and favor MYC transactivating function [85]. Do alterations in p53, the master regulator of apoptosis, contribute to differentiation failure? p53-null mice have essentially normal differentiation, and cell cycle exits by differentiation are readily induced in vitro in p53/p16-null cells (reviewed in Ref. [42]).

Thus, it seems more likely that selection for p53-system alterations during neoplastic evolution is to attenuate cell cycle exits by apoptosis, thereby facilitating cytogenetic instability, invasion, and other stress-inducing behaviors of malignant cells.

Translational implications: One explanation for clinical experience in which 1st line therapy somehow selects for leukemia cells resistant to 2nd and 3rd line treatments, is that treatments overlap in needing p53, that is frequently impaired by genetic alterations. Cytotoxic treatments meanwhile destroy HSC with intact p53, causing substantial toxicities including death [41,86–90]. Cell cycle exits by differentiation do not require p53 [41,52,91,92], and differentiation can salvage chemorefractory disease: clinical IDH2-inhibitors decrease levels of a histone methyltransferase-inhibiting oncometabolite produced by mutant-IDH2, and thereby salvage chemorefractory AML by terminal differentiation [93]. Inhibitors of FLT3 and of mutant-NPM1 nuclear export also restore terminal differentiation, observations in active clinical translation for chemorefractory AML [94,95]. APL was converted from the AML subtype with the worst survival to the best by using ATRA and arsenic to reverse differentiation failure caused by unbalanced recruitment of corepressors to the leukemia fusion protein PML-RARA [94,95]. In addition to these examples of "precision oncology," the revised model generates a therapeutic index rationale for broad-based differentiation therapy: LSCs of all genotypes require chromatin modifying enzymes (eg, DNMT1) to epigenetically repress terminal-differentiation target genes of highly expressed master commitment transcription factors [41,42,52,78,80] ("nononcogene addiction") (reviewed in Ref. [96]) (Fig. 4.3). Clinical trials in which these enzymes are drugged with minimal or no cytotoxicity have been conducted, with encouraging, broad-based clinical impact [42–44] (reviewed in Ref. [45]). The same treatments preserve self-renewal of normal HSCs which express master stem cell–transcription factors, not master differentiation-driving transcription factors (reviewed in Refs. [45,96]) (Fig. 4.2). Currently, there are very few compounds that can be positioned for this noncytotoxic, p53-independent, normal stem cell sparing, epigenetic-differentiation mode of action; expansion of this armamentarium, to match the tens of drugs available for cytotoxic intent, would be logical (reviewed in Ref. [45]).

Global gene expression profiles of LSC show that they resemble MPP or GMP [18]. Thus, LSC and HSC differ in expression of key kinases or growth factor receptors (Fig. 4.5). These differences may contribute to a favorable therapeutic index for treatment with specific kinase inhibitors or growth factor receptor agonists—eg, thrombopoietin receptor agonists (MPL agonists) have been found to favor functional hematopoiesis over leukemia in vitro and in vivo [97].

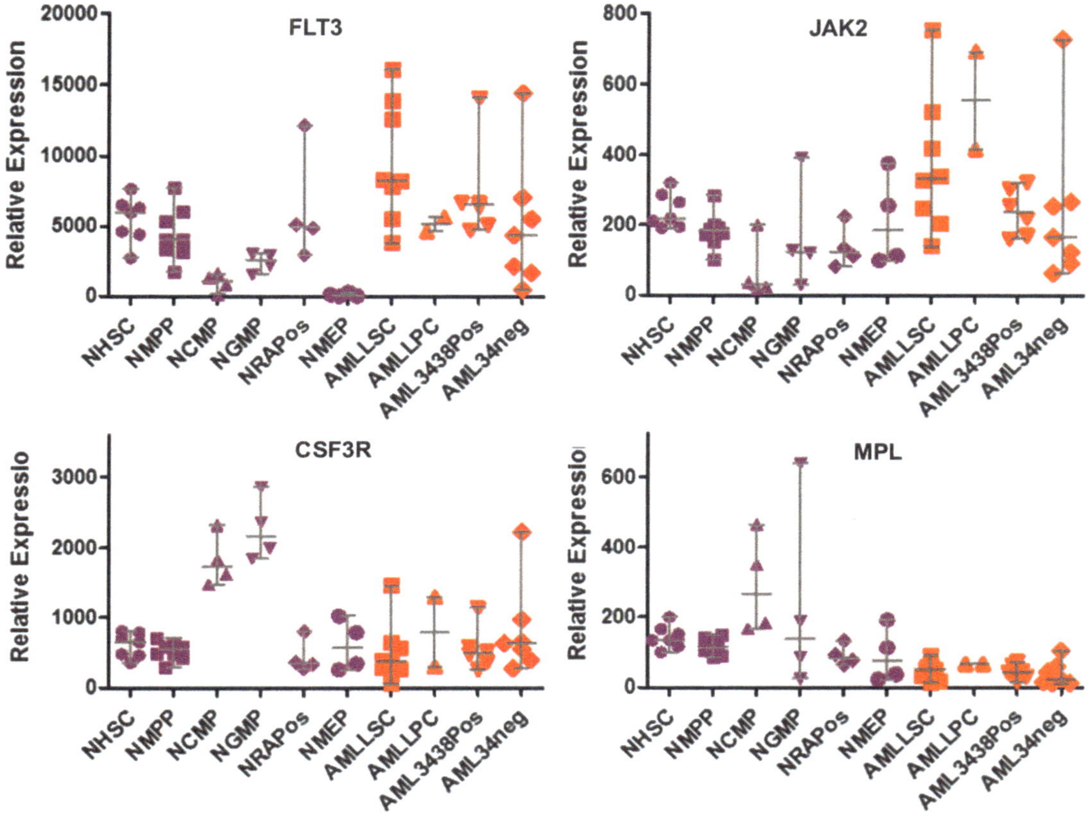

FIGURE 4.5 Expression of clinically relevant kinases and growth factor receptors in HSC versus LSC. JAK2 and FLT3 are kinases for which there are clinically available inhibitors, and CSF3R (G-CSF receptor) and MPL (thrombopoietin receptor) are growth factor receptors for which there are clinically available agonists. Gene expression by microarray, public data GSE24006 [21]. Normal and AML cell populations were flow sorted by surface phenotype. AML cases were representative of the morphologic and genetic spectrum of disease. *AML LPC*, leukemia progenitor cells; *AML LSC*, leukemia stem cells; *NCMP*, normal common myeloid progenitor; *NHSC*, normal hematopoietic stem cells; *NMEP*, normal megakaryocyte erythroid progenitor; *NMPP*, normal multipotent progenitors; *NRAPos*, normal mature myeloid cells. Error bars = median ± range.

CONCLUSION

The original model of LSC had intuitive appeal—stem cells intrinsically self-renew and are long lived enough to sustain multiple mutations. Also, founder mutations clearly originate in germ line or HSC. Evolution requires reproduction, however, and adult stem cells are quiescent, with intervals between cell divisions measured in months [62–64]. Committed daughter cells, in contrast, have intervals between divisions measured in hours [62–64]. Differentiation is the usual control on this vigorous growth. From an evolutionary perspective, differentiation programs are recent adaptations that have many ways to fail [69,70], facilitated by the longer journey to terminal differentiation than to commitment. Thus, founder mutations originate in germ line or HSC, but consequences and transformation is by terminal-differentiation failure in committed progenitors, wherein mutations are propagated and supplemented. The key to clonal expansion is not induction of self-renewal in intrinsically quiescent stem cells, but failure of differentiation in intrinsically proliferative progenitors. Until recently, this would have been considered a controversial view. The weight of reproduced evidence, however, indicates that this model is considerably closer to biological reality than the original, and that it can guide novel, rational clinical interventions, with the better therapeutic index needed for clinical progress.

List of Acronyms and Abbreviations

AML Acute myeloid leukemia
CSC Cancer stem cell
HSC Hematopoietic stem cell
GMP Granulocyte monocyte progenitor
LSC Leukemia stem cell
MPP Multipotent progenitor
TCGA The Cancer Genome Atlas

References

[1] Kaiser J. The cancer stem cell gamble. Science 2015;347(6219):226–9.

[2] Lapidot T, Sirard C, Vormoor J, Murdoch B, Hoang T, Caceres-Cortes J, et al. A cell initiating human acute myeloid leukaemia after transplantation into SCID mice. Nature 1994;367(6464):645–8.

[3] Kelly PN, Dakic A, Adams JM, Nutt SL, Strasser A. Tumor growth need not be driven by rare cancer stem cells. Science 2007;317(5836):337.

[4] Dick JE. Stem cell concepts renew cancer research. Blood 2008;112(13):4793–807.

[5] Reya T, Morrison SJ, Clarke MF, Weissman IL. Stem cells, cancer, and cancer stem cells. Nature 2001;414(6859):105–11.

[6] Eppert K, Takenaka K, Lechman ER, Waldron L, Nilsson B, van Galen P, et al. Stem cell gene expression programs influence clinical outcome in human leukemia. Nat Med 2011;17(9):1086–93.

[7] Taussig DC, Miraki-Moud F, Anjos-Afonso F, Pearce DJ, Allen K, Ridler C, et al. Anti-CD38 antibody-mediated clearance of human repopulating cells masks the heterogeneity of leukemia-initiating cells. Blood 2008;112(3):568–75.

[8] Kirstetter P, Schuster MB, Bereshchenko O, Moore S, Dvinge H, Kurz E, et al. Modeling of C/EBPalpha mutant acute myeloid leukemia reveals a common expression signature of committed myeloid leukemia-initiating cells. Cancer Cell 2008;13(4):299–310.

[9] Huntly BJ, Shigematsu H, Deguchi K, Lee BH, Mizuno S, Duclos N, et al. MOZ-TIF2, but not BCR-ABL, confers properties of leukemic stem cells to committed murine hematopoietic progenitors. Cancer Cell 2004;6(6):587–96.

[10] Somervaille TC, Cleary ML. Identification and characterization of leukemia stem cells in murine MLL-AF9 acute myeloid leukemia. Cancer Cell 2006;10(4):257–68.

[11] van Rhenen A, Moshaver B, Kelder A, Feller N, Nieuwint AW, Zweegman S, et al. Aberrant marker expression patterns on the CD34+. Leukemia 2007;21(8):1700–7.

[12] Blair A, Hogge DE, Ailles LE, Lansdorp PM, Sutherland HJ. Lack of expression of Thy-1 (CD90) on acute myeloid leukemia cells with long-term proliferative ability in vitro and in vivo. Blood 1997;89(9):3104–12.

[13] Blair A, Sutherland HJ. Primitive acute myeloid leukemia cells with long-term proliferative ability in vitro and in vivo lack surface expression of c-kit (CD117). Exp Hematol 2000;28(6):660–71.

[14] Wunderlich M, Chou FS, Link KA, Mizukawa B, Perry RL, Carroll M, et al. AML xenograft efficiency is significantly improved in NOD/SCID-IL2RG mice constitutively expressing human SCF, GM-CSF and IL-3. Leukemia 2010;24(10):1785–8.

[15] Feuring-Buske M, Gerhard B, Cashman J, Humphries RK, Eaves CJ, Hogge DE. Improved engraftment of human acute myeloid leukemia progenitor cells in beta 2-microglobulin-deficient NOD/SCID mice and in NOD/SCID mice transgenic for human growth factors. Leukemia 2003;17(4):760–3.

[16] Agliano A, Martin-Padura I, Mancuso P, Marighetti P, Rabascio C, Pruneri G, et al. Human acute leukemia cells injected in NOD/LtSz-scid/IL-2Rgamma null mice generate a faster and more efficient disease compared to other NOD/scid-related strains. Int J Cancer 2008;123(9):2222–7.

[17] Sarry JE, Murphy K, Perry R, Sanchez PV, Secreto A, Keefer C, et al. Human acute myelogenous leukemia stem cells are rare and heterogeneous when assayed in NOD/SCID/IL2Rgammac-deficient mice. J Clin Invest 2011;121(1):384–95.

[18] Goardon N, Marchi E, Atzberger A, Quek L, Schuh A, Soneji S, et al. Coexistence of LMPP-like and GMP-like leukemia stem cells in acute myeloid leukemia. Cancer Cell 2011;19(1):138–52.

[19] Corces-Zimmerman MR, Hong WJ, Weissman IL, Medeiros BC, Majeti R. Preleukemic mutations in human acute myeloid leukemia affect epigenetic regulators and persist in remission. Proc Natl Acad Sci USA 2014;111(7):2548–53.

[20] Shlush LI, Zandi S, Mitchell A, Chen WC, Brandwein JM, Gupta V, et al. Identification of pre-leukaemic haematopoietic stem cells in acute leukaemia. Nature 2014;506(7488):328–33.

[21] Gentles AJ, Plevritis SK, Majeti R, Alizadeh AA. Association of a leukemic stem cell gene expression signature with clinical outcomes in acute myeloid leukemia. JAMA 2010;304(24):2706–15.

[22] Cabezas-Wallscheid N, Klimmeck D, Hansson J, Lipka DB, Reyes A, Wang Q, et al. Identification of regulatory networks in HSCs and their immediate progeny via integrated proteome, transcriptome, and DNA methylome analysis. Cell Stem Cell 2014;15(4):507–22.

[23] Kim J, Woo AJ, Chu J, Snow JW, Fujiwara Y, Kim CG, et al. A Myc network accounts for similarities between embryonic stem and cancer cell transcription programs. Cell 2010;143(2):313–24.

[24] Wilson A, Murphy MJ, Oskarsson T, Kaloulis K, Bettess MD, Oser GM, et al. c-Myc controls the balance between hematopoietic stem cell self-renewal and differentiation. Genes Dev 2004;18(22):2747–63.

[25] Arnold I, Watt FM. c-Myc activation in transgenic mouse epidermis results in mobilization of stem cells and differentiation of their progeny. Curr Biol 2001;11(8):558–68.

[26] Li Z, Herold T, He C, Valk PJ, Chen P, Jurinovic V, et al. Identification of a 24-gene prognostic signature that improves the European LeukemiaNet risk classification of acute myeloid leukemia: an international collaborative study. J Clin Oncol 2013;31(9):1172–81.

[27] Metzeler KH, Hummel M, Bloomfield CD, Spiekermann K, Braess J, Sauerland MC, et al. An 86-probe-set gene-expression signature predicts survival in cytogenetically normal acute myeloid leukemia. Blood 2008;112(10):4193–201.

[28] Hackl H, Steinleitner K, Lind K, Hofer S, Tosic N, Pavlovic S, et al. A gene expression profile associated with relapse of cytogenetically normal acute myeloid leukemia is enriched for leukemia stem cell genes. Leuk Lymphoma 2015;56(4):1126–8.

[29] Bullinger L, Dohner K, Bair E, Frohling S, Schlenk RF, Tibshirani R, et al. Use of gene-expression profiling to identify prognostic subclasses in adult acute myeloid leukemia. N Engl J Med 2004;350(16):1605–16.

[30] Cancer Genome Atlas Research Network. Genomic and epigenomic landscapes of adult de novo acute myeloid leukemia. N Engl J Med 2013;368(22):2059–74.

[31] Kinzler KW, Vogelstein B. Cancer therapy meets p53. N Engl J Med 1994;331(1):49–50.

[32] Suarez L, Vidriales MB, Garcia-Larana J, Sanz G, Moreno MJ, Lopez A, et al. CD34+ cells from acute myeloid leukemia, myelodysplastic syndromes, and normal bone marrow display different apoptosis and drug resistance-associated phenotypes. Clin Cancer Res 2004;10(22):7599–606.

[33] Yin B, Kogan SC, Dickins RA, Lowe SW, Largaespada DA. Trp53 loss during in vitro selection contributes to acquired Ara-C resistance in acute myeloid leukemia. Exp Hematol 2006;34(5):631–41.

[34] Wong TN, Ramsingh G, Young AL, Miller CA, Touma W, Welch JS, et al. Role of TP53 mutations in the origin and evolution of therapy-related acute myeloid leukaemia. Nature 2015;518(7540):552–5.

[35] Wattel E, Preudhomme C, Hecquet B, Vanrumbeke M, Quesnel B, Dervite I, et al. p53 mutations are associated with resistance to chemotherapy and short survival in hematologic malignancies. Blood 1994;84(9):3148–57.

[36] Falk IJ, Willander K, Chaireti R, Lund J, Nahi H, Hermanson M, et al. TP53 mutations and MDM2(SNP309) identify subgroups of AML patients with impaired outcome. Eur J Haematol 2015;94(4):355–62.

[37] Haferlach C, Dicker F, Herholz H, Schnittger S, Kern W, Haferlach T. Mutations of the TP53 gene in acute myeloid leukemia are strongly associated with a complex aberrant karyotype. Leukemia 2008;22(8):1539–41.

[38] Grossmann V, Schnittger S, Kohlmann A, Eder C, Roller A, Dicker F, et al. A novel hierarchical prognostic model of AML solely based on molecular mutations. Blood 2012;120(15):2963–72.

[39] Toledo F, Wahl GM. Regulating the p53 pathway: in vitro hypotheses, in vivo veritas. Nat Rev Cancer 2006;6(12):909–23.

[40] Harutyunyan A, Klampfl T, Cazzola M, Kralovics R. p53 lesions in leukemic transformation. N Engl J Med 2011;364(5):488–90.

[41] Ng KP, Ebrahem Q, Negrotto S, Mahfouz RZ, Link KA, Hu Z, et al. p53 independent epigenetic-differentiation treatment in xenotransplant models of acute myeloid leukemia. Leukemia 2011;25(11):1739–50.

[42] Saunthararajah Y, Sekeres M, Advani A, Mahfouz R, Durkin L, Radivoyevitch T, et al. Evaluation of noncytotoxic DNMT1-depleting therapy in patients with myelodysplastic syndromes. J Clin Invest 2015;125(3):1043–55.

[43] Fenaux P, Mufti GJ, Hellstrom-Lindberg E, Santini V, Gattermann N, Germing U, et al. Azacitidine prolongs overall survival compared with conventional care regimens in elderly patients with low bone marrow blast count acute myeloid leukemia. J Clin Oncol 2010;28(4):562–9.

[44] Dombret H, Seymour JF, Butrym A, Wierzbowska A, Selleslag D, Jang JH, et al. International phase 3 study of azacitidine vs conventional care regimens in older patients with newly diagnosed AML with >30% blasts. Blood 2015;126(3):291–9.

[45] Saunthararajah Y. Key clinical observations after 5-azacytidine and decitabine treatment of myelodysplastic syndromes suggest practical solutions for better outcomes. Hematol Am Soc Hematol Educ Program 2013;2013:511–21.

[46] Vierbuchen T, Wernig M. Molecular roadblocks for cellular reprogramming. Mol Cell 2012;47(6):827–38.

[47] Riddell J, Gazit R, Garrison BS, Guo G, Saadatpour A, Mandal PK, et al. Reprogramming committed murine blood cells to induced hematopoietic stem cells with defined factors. Cell 2014;157(3):549–64.

[48] Iwasaki H, Akashi K. Myeloid lineage commitment from the hematopoietic stem cell. Immunity 2007;26(6):726–40.

[49] Yamanaka R, Barlow C, Lekstrom-Himes J, Castilla LH, Liu PP, Eckhaus M, et al. Impaired granulopoiesis, myelodysplasia, and early lethality in CCAAT/enhancer binding protein epsilon-deficient mice. Proc Natl Acad Sci USA 1997;94(24):13187–92.

[50] FANTOM Consortium, the RIKEN PMI, and CLST, Forrest AR, Kawaji H, Rehli M, Bailie JK, de Hoon MJ, et al. A promoter-level mammalian expression atlas. Nature 2014;507(7493):462–70.

[51] Jaiswal S, Ebert BL. MDS is a stem cell disorder after all. Cancer Cell 2014;25(6):713–4.

[52] Negrotto S, Ng KP, Jankowska AM, Bodo J, Gopalan B, Guinta K, et al. CpG methylation patterns and decitabine treatment response in acute myeloid leukemia cells and normal hematopoietic precursors. Leukemia 2012;26(2):244–54.

[53] Owen C, Barnett M, Fitzgibbon J. Familial myelodysplasia and acute myeloid leukaemia – a review. Br J Haematol 2008;140(2):123–32.

[54] Jan M, Snyder TM, Corces-Zimmerman MR, Vyas P, Weissman IL, Quake SR, et al. Clonal evolution of preleukemic hematopoietic stem cells precedes human acute myeloid leukemia. Sci Transl Med 2012;4(149):149ra118.

[55] Jaiswal S, Fontanillas P, Flannick J, Manning A, Grauman PV, Mar BG, et al. Age-related clonal hematopoiesis associated with adverse outcomes. N Engl J Med 2014;371(26):2488–98.

[56] Genovese G, Kahler AK, Handsaker RE, Lindberg J, Rose SA, Bakoum SF, et al. Clonal hematopoiesis and blood-cancer risk inferred from blood DNA sequence. N Engl J Med 2014;371(26):2477–87.

[57] Ng KP, Hu Z, Ebrahem Q, Negrotto S, Lausen J, Saunthararajah Y. Runx1 deficiency permits granulocyte lineage commitment but impairs subsequent maturation. Oncogenesis 2013;2:e78.

[58] Sun W, Downing JR. Haploinsufficiency of AML1 results in a decrease in the number of LTR-HSCs while simultaneously inducing an increase in more mature progenitors. Blood 2004;104(12):3565–72.

[59] Challen GA, Sun D, Jeong M, Luo M, Jelinek J, Berg JS, et al. Dnmt3a is essential for hematopoietic stem cell differentiation. Nat Genet 2012;44(1):23–31.

[60] Tadokoro Y, Ema H, Okano M, Li E, Nakauchi H. De novo DNA methyltransferase is essential for self-renewal, but not for differentiation, in hematopoietic stem cells. J Exp Med 2007;204(4):715–22.

[61] Mohrin M, Bourke E, Alexander D, Warr MR, Barry-Holson K, Le Beau MM, et al. Hematopoietic stem cell quiescence promotes error-prone DNA repair and mutagenesis. Cell Stem Cell 2010;7(2):174–85.

[62] Nygren JM, Bryder D, Jacobsen SE. Prolonged cell cycle transit is a defining and developmentally conserved hemopoietic stem cell property. J Immunol 2006;177(1):201–8.

[63] Schwartz GN, Vance BA, Levine BM, Fukazawa M, Telford WG, Cesar D, et al. Proliferation kinetics of subpopulations of human marrow cells determined by quantifying in vivo incorporation of $[^2H_2]$-glucose into DNA of S-phase cells. Blood 2003;102(6):2068–73.

[64] van der Wath RC, Wilson A, Laurenti E, Trumpp A, Lio P. Estimating dormant and active hematopoietic stem cell kinetics through extensive modeling of bromodeoxyuridine label-retaining cell dynamics. PLoS One 2009;4(9):e6972.

[65] Li J. Quiescence regulators for hematopoietic stem cell. Exp Hematol 2011;39(5):511–20.

[66] Reavie L, Della Gatta G, Crusio K, Aranda-Orgilles B, Buckley SM, Thompson B, et al. Regulation of hematopoietic stem cell differentiation by a single ubiquitin ligase-substrate complex. Nat Immunol 2010;11(3):207–15.

[67] Laurenti E, Varnum-Finney B, Wilson A, Ferrero I, Blanco-Bose WE, Ehninger A, et al. Hematopoietic stem cell function and survival depend on c-Myc and N-Myc activity. Cell Stem Cell 2008;3(6):611–24.

[68] Zhang J, Xiao Y, Guo Y, Breslin P, Zhang S, Wei W, et al. Differential requirements for c-Myc in chronic hematopoietic hyperplasia and acute hematopoietic malignancies in Pten-null mice. Leukemia 2011;25(12):1857–68.

[69] Domazet-Loso T, Tautz D. Phylostratigraphic tracking of cancer genes suggests a link to the emergence of multicellularity in metazoa. BMC Biol 2010;8:66.

[70] Srivastava M, Simakov O, Chapman J, Fahey B, Gauthier MEA, Mitros T, et al. The *Amphimedon queenslandica* genome and the evolution of animal complexity. Nature 2010;466(7307):720–6.

[71] Nakajima H, Ihle JN. Granulocyte colony-stimulating factor regulates myeloid differentiation through CCAAT/enhancer-binding protein epsilon. Blood 2001;98(4):897–905.

[72] Truong BT, Lee YJ, Lodie TA, Park DJ, Perrotti D, Wantanabe N, et al. CCAAT/Enhancer binding proteins repress the leukemic phenotype of acute myeloid leukemia. Blood 2003;101(3):1141–8.

[73] Gery S, Gombart AF, Fung YK, Koeffler HP. C/EBPepsilon interacts with retinoblastoma and E2F1 during granulopoiesis. Blood 2004;103(3):828–35.

[74] Nakajima H, Watanabe N, Shibata F, Kitamura T, Ikeda Y, Handa M. N-terminal region of CCAAT/enhancer-binding protein epsilon is critical for cell cycle arrest, apoptosis, and functional maturation during myeloid differentiation. J Biol Chem 2006;281(20):14494–502.

[75] Matsushita H, Nakajima H, Nakamura Y, Tsukamoto H, Tanaka Y, Jin G, et al. C/EBPalpha and C/EBPvarepsilon induce the monocytic differentiation of myelomonocytic cells with the MLL-chimeric fusion gene. Oncogene 2008;27(53):6749–60.

[76] Walkley CR, Purton LE, Snelling HJ, Yuan YD, Nakajima H, Chambon P, et al. Identification of the molecular requirements for an RAR alpha-mediated cell cycle arrest during granulocytic differentiation. Blood 2004;103(4):1286–95.

[77] Theilgaard-Monch K, Jacobsen LC, Borup R, Rasmussen T, Bjerregaard MD, Nielsen FC, et al. The transcriptional program of terminal granulocytic differentiation. Blood 2005;105(4):1785–96.

[78] Hu Z, Gu X, Baraoidan K, Ibanez V, Sharma A, Kadkol S, et al. RUNX1 regulates corepressor interactions of PU.1. Blood 2011;117(24):6498–508.

[79] Lara-Astiaso D, Weiner A, Lorenzo-Vivas E, Zaretsky I, Jaitin DA, David E, et al. Immunogenetics. Chromatin state dynamics during blood formation. Science 2014;345(6199):943–9.

[80] Gu X, Hu Z, Ebrahem Q, Crabb JS, Mahfouz RZ, Radivoyevitch T, et al. Runx1 regulation of Pu.1 corepressor/coactivator exchange identifies specific molecular targets for leukemia differentiation therapy. J Biol Chem 2014;289(21):14881–95.

[81] Radomska HS, Basseres DS, Zheng R, Zhang P, Dayaram T, Yamamoto Y, et al. Block of C/EBP alpha function by phosphorylation in acute myeloid leukemia with FLT3 activating mutations. J Exp Med 2006;203(2):371–81.

[82] Cu XR, Mahfouz RZ, Enane F, Hu Z, Clemente MJ, Przychodzen BP, et al. A specific mechanism by which NPM1 mutations impede myeloid differentiation also explains the link with DNMT3A mutation. Blood 2013;122(21).

[83] Sexauer A, Perl A, Yang X, Borowitz M, Gocke C, Rajkhowa T, et al. Terminal myeloid differentiation in vivo is induced by FLT3 inhibition in FLT3/ITD AML. Blood 2012;120(20):4205–14.

[84] Saunthararajah Y, Maciejewski J. Polycomb segment myeloid malignancies. Blood 2012;119(5):1097–8.

[85] Goodliffe JM, Wieschaus E, Cole MD. Polycomb mediates Myc autorepression and its transcriptional control of many loci in *Drosophila*. Genes Dev 2005;19(24):2941–6.

[86] Roberts DA, Wadleigh M, McDonnell AM, DeAngelo DJ, Stone RM, Steensma DP. Low efficacy and high mortality associated with clofarabine treatment of relapsed/refractory acute myeloid leukemia and myelodysplastic syndromes. Leuk Res 2015;39(2):204–10.

[87] Brennig S, Rattmann I, Lachmann N, Schambach A, Williams DA, Moritz T. In vivo enrichment of cytidine deaminase gene-modified hematopoietic cells by prolonged cytosine-arabinoside application. Cytotherapy 2012;14(4):451–60.

[88] Ben-Ishay Z, Barak V. Bone marrow stromal dysfunction in mice administered cytosine arabinoside. Eur J Haematol 2001;66(4):230–7.

[89] Mandelli F, Petti MC, Ardia A, Di Pietro N, Di Raimondo F, Ganzina F, et al. A randomised clinical trial comparing idarubicin and cytarabine to daunorubicin and cytarabine in the treatment of acute non-lymphoid leukaemia. A multicentric study from the Italian cooperative group GIMEMA. Eur J Cancer 1991;27(6):750–5.

[90] Schoch C, Kern W, Schnittger S, Buchner T, Hiddemann W, Haferlach T. The influence of age on prognosis of de novo acute myeloid leukemia differs according to cytogenetic subgroups. Haematologica 2004;89(9):1082–90.

[91] Negrotto S, Hu Z, Alcazar O, Ng KP, Triozzi P, Lindner D, et al. Noncytotoxic differentiation treatment of renal cell cancer. Cancer Res 2011;71(4):1431–41.

[92] Alcazar O, Achberger S, Aldrich W, Hu Z, Negrotto S, Saunthararajah Y, et al. Epigenetic regulation by decitabine of melanoma differentiation in vitro and in vivo. Int J Cancer 2012;131(1):18–29.

[93] Wang F, Travins J, DeLaBarre B, Penard-Lacronique V, Schalm S, Hansen E, et al. Targeted inhibition of mutant IDH2 in leukemia cells induces cellular differentiation. Science 2013;340(6132):622–6.

[94] Smith ML, Hills RK, Grimwade D. Independent prognostic variables in acute myeloid leukaemia. Blood Rev 2011;25(1):39–51.

[95] Lo-Coco F, Avvisati G, Vignetti M, Thiede C, Orlando SM, Iacobelli S, et al. Retinoic acid and arsenic trioxide for acute promyelocytic leukemia. N Engl J Med 2013;369(2):111–21.

[96] Saunthararajah Y, Triozzi P, Rini B, Singh A, Radivoyevitch T, Sekeres M, et al. p53-Independent, normal stem cell sparing epigenetic differentiation therapy for myeloid and other malignancies. Semin Oncol 2012;39(1):97–108.

[97] Tamari R, Schinke C, Bhagat T, Roth M, Braunschweig I, Will B, et al. Eltrombopag can overcome the anti-megakaryopoietic effects of lenalidomide without increasing proliferation of the malignant myelodysplastic syndrome/acute myelogenous leukemia clone. Leuk Lymphoma 2014;55(12):2901–6.

CHAPTER 5

Breast Cancer Stem Cells and the Move Toward High-Resolution Stem Cell Systems

B.T. Spike
University of Utah, Salt Lake City, UT, United States

INTRODUCTION

At its core the cancer stem cell (CSC) hypothesis is intended to functionally explain the cellular heterogeneity frequently observed in tumors. It proposes that a subset of tumor cells referred to as CSCs act in a manner analogous to self-renewing stem cells of a normal tissue. These CSCs are thought to propagate the cancer long term by giving rise to more CSCs while also giving rise to bulk tumor cells that lack long-term cancer maintenance properties.

From its inception, this straightforward description of cellular functional heterogeneity has been intertwined with questions of its relationship to normal stem cell biology at the tissue, cellular, and molecular levels. Accordingly, there have been three major trajectories of research that all fall under the theme of CSC research: (1) studies which seek to isolate and characterize the cells responsible for tumor propagation using specific biomarkers, (2) studies which seek to determine broader molecular profiles and mechanisms governing cancer cells and relate them to molecular profiles from normal stem cells, and (3) studies that seek to determine the normal cell of origin for tumors bearing a given molecular profile. These three types of studies are not mutually exclusive and often occur in combination in a single report with little distinction between them, though they are in fact separate ideas.

Breast cancer has been extensively studied in all three contexts. Yet, it often remains unclear how well the CSC model fits a particular instance of the disease, how closely breast cancer cells resemble normal mammary stem cells at the molecular level and whence the stem cell related feature of a tumor derive. For instance, it is generally unknown whether stem cell phenotypes are inherited by tumor cells from a transformed normal stem cell or whether they are acquired through tumor-associated reprogramming events.

CONTENTS

Cancer Stem Cells. http://dx.doi.org/10.1016/B978-0-12-803892-5.00005-X

In addition to highlighting some the most common markers, regulators, and biological processes uncovered for breast cancer stem cells (BCSCs) to date, this chapter will address several of the open questions and limitations inherent in the commonly used stem cell assays and their interpretation.

However, the chief aim of the chapter is to emphasize the need for, and propose avenues to obtain, a more comprehensive understanding of molecular and cellular mechanisms fueling stem cell phenotypes in breast and other cancers, including how these parallel or contrast with normal developmental biology. A deeper and broader understanding of both normal and tumorigenic stem cell biology as systems may have profound impact on how we identify good molecular targets and conceptualize curative cancer therapies.

ORIGINS IN DEVELOPMENT

The idea that tumors parallel or caricature normal development is not new (see Triolo 1965 [1]). From the mid-19th century tenets of cell theory advanced by Remak and Virchow describing the production of daughter cells from the division of parent cells [2,3], it logically follows that the generation of complex tissues would be the result of these daughter cells differentiating and, further, that the generation of tumor tissue that is also comprised of cells should represent some derangement of the same process. The outward similarities between rapidly dividing and plastic embryonic tissue and tumorigenic tissue (especially teratomas) led a number of 19th century researchers to propose a causative relationship whereby deviations in the embryo or embryonic remnants were responsible for tumorigenesis in the adult [1]; a concept that Cohnheim formalized as the embryonic rest hypothesis of cancer [4]. In the same vein, Durante opined more than 140 years ago that:

> Elements which have retained their [...] embryonal characteristics in the adult organism, or have regained them through some chemico-physiologic deviation, represent [...] the generative elements of every tumor variety and specifically those of a malignant nature. Such elements may remain enclosed within matured tissues for many years, giving no indication of their presence, until an irritation – a simple stimulus suffices – rekindles their vital cellular activities...
>
> **F. Durante 1874 [1].**

Consider how timely Durante's hypothesis would still be were it offered today with these "elements" representing today's CSCs.

Over the next century a number of additional studies corroborated this relationship between early development and cancer at the molecular level, identifying immune antigens shared between cancers and embryonic material [5]. However, it was not until the landmark works from John Dick's group, in the 1990s in the hematopoietic system, that the CSC hypothesis was formalized [6,7]. They reported that leukemias followed a cellular hierarchy related to that of the normal tissue where only a subset of cells were fully tumorigenic upon transplantation. Like their normal counterparts these cells were capable of giving rise to all of the heterogeneous cell types observed in the original tumor tissue and were inferred to be responsible for tumor propagation/perpetuation in situ [6,7]. Of course, this work drew inspiration from earlier work including, the seminal work of Till and McCullough demonstrating the presence of stem cell activity in bone marrow transplantation of irradiated mice [8]; work showing that leukemias could similarly be transferred from donor to host [9]; and a considerable body of work defining surface markers and differentiation hierarchies in the normal hematopoietic system and relating these to various blood cancers [10]. In this regard, it may also have been significant that expansion of recognizable blast populations in many leukemias correlate with blast populations in healthy hematopoiesis except for their number, emphasizing the role of defective differentiation in tumorigenesis over enhanced proliferation from mutated oncogenes per se. Thus, the many points of analogy between normal developmental cellular hierarchies and those observed in cancers undoubtedly played a central role in forming the concept of CSCs and in their experimental discovery.

THE MAMMARY GLAND AS A CLASSIC STEM CELL–DRIVEN SYSTEM

In solid tissues, cellular hierarchies and markers (especially surface markers) that define parent/progeny relationships are typically less well defined than they are in the blood system. Nevertheless, over the past decade, the cancer stem cell paradigm has been applied to a wide variety solid tissue cancers including brain [11], colon [12], pancreas [13,14], prostate [15,16], and mammary gland [17]. Like the blood system, the mammary gland is a natural candidate for stem cell–driven biology with repetitive cycles of tissue expansion and involution. This was substantiated experimentally in classic studies by DeOme and subsequent work by others showing reconstitution of mammary tissue upon transplantation in mice [18]. Viral tagging experiments used in combination with DeOmes transplant approach further showed that individual cells could recapitulate the heterogeneity of the mammary gland and could self-renew [19]. Despite this clear evidence for the existence of mammary stem cells, no reproducible prospective markers for identifying and isolating the cells exhibiting stem cell activity in these assays were known until 2006. In that year, back to back papers in Nature prospectively isolated subpopulations of cells from mouse mammary glands based on their expression of the surface markers CD24 and either CD49f or CD29 that could reconstitute the mammary gland upon transplantation much more efficiently than their counterparts and could occasionally even do so as single cells [20,21].

IDENTIFICATION OF BREAST CANCER STEM CELLS

However, 3 years prior to the prospective isolation of mammary stem cells in mice, the groups of Michael Clarke and Max Wicha, borrowing markers from studies on melanoma and neural stem cells and applying them to human breast cancer, isolated the first solid tumor CSCs to be identified [17]. In their study, the majority of human breast cancers examined (8 of 9) contained a subset of cells that were CD44+CD24- and exhibited an elevated capacity to form xenografted tumors in NOD/SCID mice that bore the same cellular heterogeneity as the original tumor. The relationship of these BCSCs to normal stem cells of the mammary gland could not initially be determined, however, owing to a lack normal mammary stem cell markers, and the authors cautiously and perhaps more aptly referred to these cells as the tumorigenic cells or tumor-initiating cells.

In 2008 Bob Weinberg's group showed that the CD44+CD24- phenotype in human breast cells from both normal and cancer contexts was associated with markers of epithelial-to-mesenchymal transition (EMT) and enhanced performance in nonadherent mammosphere growth and differentiation assays developed by Dontu et al. [22,23]. These assays ostensibly measure stem cell activity as it was demonstrated that the 3D structures called mammospheres could be dissociated and reseeded in multiple generations demonstrating self-renewal and that they produced multilineage outgrowths when plated in differentiation promoting conditions [23]. Therefore, a link was postulated between EMT, CD44+/CD24- marker phenotype and both normal and tumorigenic stem cells [22]. In further support of this interpretation, the Weinberg study profiled a set of EMT markers in the recently identified mouse mammary stem cell enriched fraction and showed that these cells also exhibited relative mesenchymal gene expression profiles when compared to the non-stem cell–enriched epithelial compartment. Based on these initial studies, much of the subsequent mechanistic work in the BCSC field has focused on this marker set, this experimental setting and the EMT phenotype as a stem cell hallmark in both the normal and tumorigenic setting, as well as in tumor progression, metastasis, and therapy resistance [24].

However, while the aforementioned studies were under way the Wicha group reported a second strategy for identifying BCSCs based on their aldehyde dehydrogenase (ALDH) activity [25]. They reported that ALDH+ breast cancer cells as well as ALDH+ normal cells exhibited robust stem cell activity in mammosphere culture and in vivo, though it should be noted that the clonality of multilineage outgrowths from normal cells in vivo was not firmly established in these experiments. Surprisingly, they also found that the CD44+CD24- population and ALDH+ population only partially overlapped in breast cancers and that xenografts could be generated from both CD44+CD24- cells and

CD44-CD24+ cells provided these cells were ALDH+. In addition, cells jointly exhibiting both types of BCSC markers were reported to be the most competent subpopulation in these experiments.

Thus, despite their discrepancies, these studies support well the idea that at any given point in time a tumor is likely comprised of heterogeneous cell types with differing potencies and associated risks, and therefore, that treatments targeting the most dangerous cell types in a tumor and reducing or eliminating them could perform better in the long run than treatments targeting the tumor bulk [25a]. In spite of this, there are a number of critical caveats to consider in these experiments.

CAVEATS OF THE CANCER STEM CELL ASSAY

The process of isolating cells, sorting them, and culturing them in vitro and/or transplanting them into a xenogeneic host presents a number of selective pressures on cells that may not be universally germane in the context of tumors in situ. Thus, it is quite possible that some cells will be classified as non-CSCs because they perform poorly during these ex vivo manipulations but they may nevertheless harbor significant potential to drive tumor progression or recurrence in vivo. In this regard, the EMT phenotype is particularly suspect as a stem cell hallmark. In the Weinberg study correlating EMT with stem cells, the authors demonstrated that artificially inducing EMT through molecular genetic means appeared to increase the number of stem cells in assays using nontransformed mammary epithelial cell lines [22]. Though this was interpreted to mean that EMT "generates" stemness, an alternate explanation is that mesenchymal characteristics specifically confer competence in a particular facet of these assays (eg, anoikis resistance, attachment to substrata, etc.), such that cells already harboring stem cell potential are not lost. Indeed, a number of groups have pointed out that while mammospheres can arise from individual cells, they frequently represent aggregates when plated at higher densities [26] so it is also possible that the presence of epithelial-to-mesenchymal transitioned (EMT'd) cells may facilitate the survival of non-EMT'd or reverted cells in these assays through aggregation.

The EMT phenotype may play a similar role in transplantation, by promoting critical cell–cell or cell–matrix interactions or facilitating migration toward nutrient supply or specific niches. This would be expected to favor the naturally more mesenchymal compartment, again quite independently of other core stem cell programs. Consistent with this caveat, lineage-tracing studies in the adult mouse mammary gland indicate the existence of two independent stem cell pools responsible for producing luminal or basal cells, respectively [27]. In those studies, as in previous studies only the basal compartment performed well upon transplantation as measured by mammary reconstitution and multilineage differentiation potential [20,21,27]. Thus, in both the normal and cancer situation, the requirements of the assay could overly restrict our view of the true stem cell pool.

It is also noteworthy that most studies on BCSC have proven to be *enrichment* studies not *purification* studies as tumor-initiating activity is also present (albeit much reduced) in the "non-BCSC" compartment. For example, in the original report by Al-Hajj and colleagues, the findings held for eight out of nine cancers examined and but even among the eight examples with demonstrative BCSCs there were sporadic successes in xenografting from the non-CD44+CD24- population [17], and in other studies take rates from the non-CD44+CD24- population were quite appreciable [25]. This has rather obvious implications for the development of truly curative therapies that seek to exclusively target and track a single BCSC-enriched fraction. Namely, if residual cells with a low but measurable potential to regenerate the tumor exist in the so-called non-BCSC fraction then a paradox emerges where these cells are both non-BCSC by marker expression but retain BCSC like properties over the longer clinically relevant term. It may be that more effective or comprehensive CSC markers will emerge and rectify this apparent tautology. Alternatively, it may not be possible to find a marker set with the ability to purify stem cells from breast cancers (or normal tissues) if the pool of stem cells is itself inherently heterogeneous.

Even more disruptive to the cancer stem cell model is the potential for bidirectional interconversion between different "stem cell" states (Fig. 5.1). Indeed, some studies examining breast cancer cell lines have demonstrated instances where tumors appear to be seeded by multiple cell compartments though they may exhibit somewhat different rates of interconversion [28–30]. Similarly, ALDH+ and CD44+CD24- BCSC populations were also shown to interconvert, but also to occupy largely distinct anatomical locations in the tumor in vivo suggesting an important role for the microenvironment in determining their molecular phenotype [31]. Again, even in cases where there is a relatively strong directionality to the BCSC/Non-BCSC hierarchy, given enough time retrograde conversion of the non-BCSC compartment to BCSC activity could lead to treatment failure. Such cellular plasticity may be a common mechanism for generating tumor cell heterogeneity.

Finally, perhaps the most notable deficiency of CSC models is a very general one—that we just do not yet know enough about their broader molecular and cellular underpinnings. As indicated earlier, initial studies were aimed at identifying markers for prospective CSC isolation or identification in situ and much of the CSC field remains focused on biomarker identification for clinical detection irrespective of molecular function.

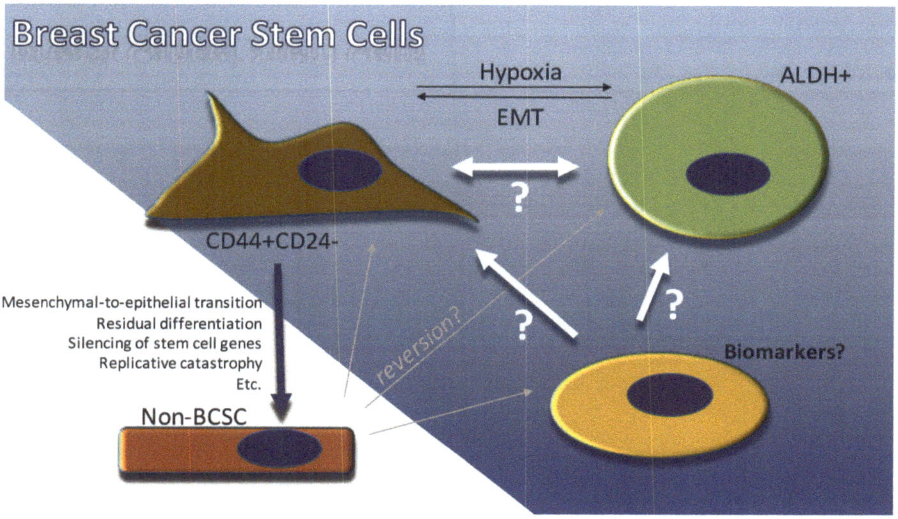

FIGURE 5.1 Breast cancer stem cell (BCSC) differentiation and interconversion.
The best characterized BCSCs, CD44+CD24- cells (top left) and ALDH+ cells (top right), have been demonstrated to comprise largely independent pools and to interconvert. Although it has been shown that factors promoting EMT increase the number of CD44+CD24- BCSCs and that hypoxia can promote an increase in ALDH+ cells, it remains to be determined if other molecular mechanism can contribute to their interconversion. Additionally, there may exist alternative stem cell states (yellow cell, bottom right) that interconvert with those identified and characterized to date, but their markers and mechanisms of plasticity naturally remain to be determined. BCSC differentiation into non-BCSCs has been shown to accompany the action of various signaling molecules, epigenetic regulators, transcription factors, and microRNAs that promote changes in cell state (eg, MET, differentiation, metabolic changes, etc.) (see Table 5.1). However, the extent to which these changes are permanent in vivo generally remains to be determined, and there may be specific microenvironmental cues or genetic lesions that promote reversion of apparent non-BCSCs to BCSCs.

ASSOCIATED CANCER STEM CELL MARKERS AND REGULATORS

Markers—These caveats aside, a growing body of work has begun to pick apart the molecular biology driving tumor propagation both globally and in select subsets of tumor cells such as the BCSC populations described earlier (see Table 5.1). Additional BCSC markers have also been proposed. PROCR and ESA were in fact proposed to be superior BCSC enrichment markers when compared to CD44+/CD24-cells in a panel of cell lines and primary tumors and the same study also found correlations between a variety of other proposed BCSC markers and progression to metastasis [32]. Among these the chemokine receptor CXCR4, has been implicated in EMT, stem cells and metastasis in breast and other cancers previously, and was specifically shown to increase mammosphere

Table 5.1 Molecular Regulators in BCSCs

BCSC Regulator	Direction of Effect	Proposed Mechanism	
BMI1	Promotes	Silencing of differentiation genes	[73]
c-JUN	Promotes	Promotes EMT and BCSC growth *in trans* via expression of SCF and CCL5	[50]
CD90	Promotes	Promotes niche interaction and Src and NFκB activation	[72]
CXCR4	Promotes	Unknown; promotes production of mammospheres	[33]
EPHA4	Promotes	Promotes niche interaction and Src and NFκB activation	[72]
HH/Smoothened	Promotes	Multiple targets including STAT-3, BCL-2, CYCLIN-D3 and matrix metaloproteinases	[48]
HIF1-α/2-α	Promotes	Multiple targets including IL-8, MDR-1, TAZ, SIAH-1 and potentialy ESC regulators and others.	[105] [106] [107] [108]
ID4	Promotes	Blocks luminal differentiation genes	[109]
IL-8/CXCR1/2	promotes	Unknown; EGFR/Her2 dependent	[110]
miR9	promotes	Blocks E-Cadherin expression	[111]
mTOR	Promotes	Promotes drug resistance and colony formation	[69] [77]
NOTCH1/4	Promotes	Differentiation blockade	[45] [46]
P63	Promotes	Activates CXCR4	[112]
SLUG	Promotes	Promotes CD44 expression; niche factor regulation	[62] [61] [113]
SNAIL	Promotes	Transcriptional activation of ZEB1	[59]
SOX10	Promotes	Cooperates with SLUG	[62]
SOX9	Promotes	Cooperates with SLUG; niche factor regulation	[62] [113]
STAT3	Promotes	Maintains ALDH+ cellular subset	[53]
TAZ	Promotes	Promotes EMT, metastasis, and chemoresistance	[114] [51]
TGF-β	Promotes	Classic promoter of EMT	[115] [49]
TWIST	Promotes	Promotes EMT; suppresses CD24	[63] [57]
WNT	Promotes	Promotes colony growth in nonadherent conditions	[43]
ZEB1	Promotes	Promotes EMT and radioresistance	[116] [117] [118]
BMP2/7	Inhibits	Blocks TGF-β	[119]
DACH1	Inhibits	Blocks NANOG and SOX2	[120]
Let-7	Inhibits	Inhibits multiple BCSC promoting genes including H-RAS and HMGA2	[121]

Table 5.1 Molecular Regulators in BCSCs—cont'd

BCSC Regulator	Direction of Effect	Proposed Mechanism	
MEL18	Inhibits	Blocks WNT and NOTCH via BMI1	[122]
miR200c	Inhibits	Reduces chemoresistance	[74]
miR27b	Inhibits	Inhibits ABCG2 mediated drug resistance by suppressing ENPP1	[123]
miR30C	Inhibits	Blocks migration and chemorsesistance through cytoskeletal effects and IL-11 downregulation	[76] [77]
miR34a	Inhibits	Downregulates NOTCH	[75]
miR99	Inhibits	Blocks mTOR	[78]
P53	Inhibits	Multiple targets including transactivation of miR200c	[124]
UTX	Inhibits	Epigenetic repression of EMT genes	[54]

forming ability of breast cancer cell lines under stimulation with ligand [33]. However, most studies have continued to utilize the CD44+/24- profile or ALDH+ to mark BCSCs. Functionally, ALDH is a detoxifying enzyme with reported roles in retinoic acid synthesis and signaling [34]. The CD44 BCSC marker is a transmembrane protein that binds components in the extracellular matrix including fibronectin and hyaluronan, among others [35]. It has been demonstrated to be functionally relevant to the CD44+/CD24- BCSC as its downregulation promotes their differentiation to the non-BCSC phenotype and inhibits their tumorigenic capacity [36].

More recently, genetically encoded marker systems have been developed that permit isolation and tracking of putative cancer stem cells based on the activity of stem cell–related gene promoters. One such system, utilizing lentiviral delivery of a green fluorescent protein (GFP) and luciferase reporter construct that responds to the Notch intracellular domain (NICD)-transcription complex, determined that breast cancer cells with active Notch signaling exhibit increased self-renewal relative to reporter-negative cells [37]. This is consistent with known roles for Notch in the regulation of normal stem and progenitor phenotypes. Critically, Notch-reporter activity also correlated with increased xenograft initiating ability in diverse breast cancer cell lines [37]. Similarly, two other recent studies used lentiviral molecular genetic systems that report the activity of the core embryonic stem (ES) cell transcription factors, Oct4 and Sox2 to isolate BCSCs [38,39]. On average, the expression of these factors and their stem cell target gene, Nanog, appears to be hundreds to thousands of times lower in breast cancer cells than it is in ES cell cultures possibly explaining their low and sporadic detection in several earlier breast cancer studies focused on bulk samples [40–42]. Nevertheless, despite these low levels, cells showing reporter construct activation exhibited enhanced self-renewal, enhanced xenograft and metastasis initiating ability, and intrinsic insensitivity to chemotherapeutic agents in vitro [38]. Reporter positive cells were highly correlated to the ALDH+ BCSC marker type but showed even greater xenograft initiating capacity than the ALDH+ population suggesting that Nanog is a higher specificity BCSC marker and leading to the identification of the tight junction protein JAM-A, as an improved specificity BCSC surface marker with therapeutic potential [39]. Of course, more studies will be needed to establish the generality of this promising new marker and its utility as a therapeutic target. It will also be interesting to determine if lentiviral delivered Notch, Sox2, and Oct4 responsive vectors show such high fidelity in reporting specific pathway signaling and identifying CSCs in other experimental settings and tumor types, and whether cancers are capable of exploiting alternative factors to achieve similar malignant cellular phenotypes.

Signaling Pathways—Beyond these markers, myriad other molecular players ranging from cell surface signaling pathways to transcription factor networks have been functionally implicated in BCSCs. For example, there are reported roles for Wnt [43,44], Notch [45,46], HH [47,48], and Tgf-β [49] signaling and at the transcriptional level critical roles for c-Jun [50], Taz [51,52], Stat3 [53], and others. In addition, chromatin modifiers are thought to add an additional layer of transcriptional control [54–56]. Potential regulators of EMT have garnered the most attention among transcription factors in BCSCs including Zeb1 [54], Twist [57,58], Snail [59,60], and Slug [61]. Weinberg's group reported cooperativity between Slug and Sox9 (or Sox10) in promoting the stem cell state in mammary cells [62], although recently they have also reported that normal stem cells and BCSCs use different molecular variations on the EMT theme, with BCSC preferring Snail to Slug [59]. Another EMT-promoting transcription factor, Twist, has also been recognized to regulate breast cancer cell EMT and metastasis and to directly repress CD24 expression in BCSC [63,57].

Microenvironment and Stroma—However, CD24 is also under the control of hypoxia inducible factor, Hif-1α [64], and broader roles for hypoxia in regulating BCSC have been reported (see Semenza [65]) and may be particularly relevant in defining the ALDH+ BCSC niche [66]. In addition to hypoxia, a number of additional reports suggest BCSCs may be sensitive to other microenvironmental cues and stresses. These stresses may specifically impinge on unique BCSC metabolism as there are several reports indicating BCSC sensitivity to Metformin [67,68]. mTOR dysregulation has also been implicated in BCSC survival [69]. Recently, Dong et al. described a mechanism coupling EMT with glucose metabolism and a reduction of mitochondrial ROS and this could play a critical role in chemotherapy insensitivity of BCSCs [60]. Autophagy was also recently reported to play a critical role in BCSC maintenance through regulation of CD24 and IL-6 [70]. BCSC may also be sensitive to their cellular microenvironment as several reports indicate immune cell modulation of BCSC function. CD8T cells were reported to actively promote the

BCSC phenotype in tumors in $FVB^{neu\text{-}tg}$ mice [71] though the precise mechanism remains to be elucidated. Monocytes and macrophages may also contribute to BCSCs maintenance through juxtacrine signaling involving CD90 and EphA4 [72].

MicroRNAs—A sizeable proportion of the mechanistic literature concerning BCSCs has focused on their regulation by microRNAs (miRNAs, miR). In 2009, Clarke's group identified a set of 37 miRNAs that were differentially expressed between BCSC and non-BCSCs [73]. Several of these miRNAs have been shown to target pathways critical to BCSC function. Among these miR200c was found to regulate Bmi1, a polycomb chromatin modifier critical in normal mammary cells and stemlike embryonal carcinoma cells where it is similarly targeted by miR200c [73]. miR200c expression was also subsequently shown to enhance chemosensitivity and reverse the metastatic potential of Claudin-low breast cancer cells [74]. The MCF7 breast cancer cell line has a high proportion of CD44+/CD24- cells, and another microRNA, miR34a, was reported to function through Notch to sensitize MCF7 cells to Doxorubicin [75]. Similarly, miR30c was shown to effect BCSC chemoresistence, though in this case it acts through regulation of the actin-binding protein TWF and the cytokine, IL-11 [76]. miR30c was also shown to regulate cellular migration and metastasis through its effects on cytoskeletal components [77]. In two other examples, among many, Let-7 repression downstream of Wnt has been shown to promote BCSCs [44], and miR99 [78] has been reported to integrate their EMT and mTor signaling.

Together these studies, a brief survey of the total literature really, have provided inroads for understanding BCSC at the molecular level. It will be important to continue expanding this mechanistic knowledge base within and outside of the BCSC compartment and to identify associated opportunities for therapeutic intervention in breast cancer, especially for those tumor types where molecularly targeted therapies are lacking and where existing approaches are frequently ineffective and bear harmful side effects.

GENOME WIDE MOLECULAR PROFILING IN BREAST CANCER

Historically, the molecular biology of cancer has been understood only in terms of a handful of markers and pathways that could be feasibly picked apart through gene-by-gene molecular genetic approaches similar to those described earlier. Though enlightening in its own right, the resulting somewhat narrow view of tumor biology is naturally prone to miss many integrated biological processes. The advent of genome wide technologies (eg, genomics, transcriptomics, proteomics, etc.) has promised to permit a more comprehensive and systems-oriented view. Through such comprehensive methods, we might expect identification of novel candidate therapeutic targets even when these are not directly or commonly mutated, or even when they are not proximal to a commonly mutated gene in the cellular signaling network. In addition to facilitating the identification of new molecular targets for the bulk of tumor cells, comprehensive knowledge of tumor cell genes and transcripts can facilitate comparisons to the molecular underpinnings of the normal stem cells state and potentially, by broad molecular similarity, point to a cell of origin for a given manifestation of the disease.

Genome wide approaches, particularly genomics and transcriptomics, have already been extensively applied in breast cancer. Large-scale sequencing and cataloging projects, most notably The Cancer Genome Atlas (TCGA) have begun compiling genetic lesion databases across many samples in many tumor types. In breast cancer these studies reveal that (1) the most prevalent driver mutations were those already uncovered by traditional molecular genetic approaches, although novel significantly mutated genes were identified and (2) these mutations alone and in combination are insufficient to wholly describe tumor type, since several of the common genetic lesions (eg, PIK3CA mutation, p53 inactivation) are distributed across multiple subtypes [79].

Therefore, these studies further support the idea that understanding tumor cell biology and ultimately BCSC biology will require not only an understanding of the genetic lesions in a tumor, but how these integrate with the tumor cells' broader molecular circuitries. Ultimately this should extend to analysis of the epigenome, transcriptome, proteome, metabolome, etc. and will need to be considered not only on the level of the bulk tumor but also in terms of its subpopulations and infiltrating nontransformed cells. Among these technologies transcriptomics (gene expression profiling) is by far the most well-developed and widely used approach to date.

For more than a decade in fact, multisample data sets comprised of expression profiles from various normal tissues, cancers and cell lines have been produced by multiple laboratories and made publically available for analysis. The resulting data sets have been used to identify candidate genes for study in experimental systems, for tumor classification, for prognostic and predictive clinical tool development, and in comparative studies; for instance, those examining the similarity of a given tumor to a normal stem or progenitor cell population. Studies of this later type in the mammary gland and breast cancer are among the most prolific.

Breast cancer has long been recognized as a diverse set of diseases, with some cases exhibiting significant cellular differentiation and others less. Clinically, it is defined by this heterogeneity and other standard histopathological features such as nuclear atypia, mitotic index, the integrity of tissue boundaries, etc. Over the past couple of decades, a handful of molecular markers have also become routine in the clinical classification of breast cancers. The hormone receptors for estrogen and progesterone (ER and PR, respectively), and HER2/neu, an epithelial growth factor receptor family member (ErbB2), serve not only as clinical designations but as targets for molecular therapy. In contrast, triple-negative breast cancers (TNBCs) lacking these markers are considered the most clinically recalcitrant and are presently treated with relatively nonspecific chemotherapy approaches in the clinical arsenal [80].

However, beginning in 2000, gene expression profiling efforts began to vastly broaden the molecular profile available in breast cancer classification and analysis [81]. Perou and colleagues identified "intrinsic" subtypes of breast cancer based on clusters of similar gene expression within large multitumor gene expression data sets. These intrinsic subtypes were named based on their differential expression of genes known to be expressed in various normal mammary cellular compartments or deregulated in particular types of breast cancer. Ultimately, although alternative clustering strategies have been proposed, six distinct clusters of tumor samples were identified [82]. These included two subtypes with expression of luminal characteristics referred to as Luminal A and Luminal B, a Her2-enriched subtype with many tumors showing elevated Her2 expression, a Basal-like subtype, and a more mesenchymal subtype called Claudin-low due to reduced expression of claudins 3, 4, 7, and other markers, as well as a cluster that was more similar to normal mammary tissue, the so-called Normal-like subtype [83].

STEM CELL PROGRAMS IN BREAST CANCER

Though the overlap between clinically defined tumor types and intrinsic subtypes is not an exact match, the Basal-like intrinsic subtype includes the vast majority of TNBCs and these were initially thought to best reflect a dedifferentiated or stem cell–like molecular phenotype [83]. However, based partly on mesenchymal features reported in the CSC studies described earlier and partly based on comparative transcriptomic studies between luminal progenitor–enriched populations and Basal-like breast cancers [84], Perou and colleagues generated a differentiation score for tumors as a prognostic indicator that equated Claudin-low tumors with the stem cell transcriptional state [28]. In this schema, Basal-like breast cancers were further along the differentiation scale toward the phenotype of the normal luminal progenitors. These studies also inferred like many others that the observed relationships were likely to reflect the cell of origin for the tumors or at least the state of differentiation arrest since a tumor could be driven by rare BCSCs with a more primitive molecular phenotype than the tumor bulk. Indeed, one study reported that the degree to which a given tumor exhibits a stem cell gene expression profile is directly correlated to the number of stem cells in that tumor [85].

However, in contrast to the differentiation score, comparative analysis from Ben-Porath et al. demonstrated significant enrichment for ES cell signatures in Basal-like breast cancers [41]; a finding that we subsequently showed was tightly correlated to the tumor's p53 status [42]. Interestingly, attenuated p53 may be a natural feature of the stem cells state [86], and several groups demonstrated that p53 attenuation facilitates cellular reprogramming [87]. In our study, we also showed that induced pluripotent stem (iPS) cell–derived signatures were enriched in Basal-like and p53 mutated cancers, an analysis meant to highlight the potential for tumor associate reprogramming as means to acquiring a stem cell profile during tumorigenesis. The stem cell signatures we used in these analyses were typically comprised of hundreds of genes including Oct4, Sox2, and Nanog, but we did not see a correlation between expression of these three factors and any group of breast cancers although Ben-Porath et al. reported that expression of their target genes correlates with the Basal-like subtype and high tumor grade. In our analysis, neither the iPS cell signature nor the ES cell signatures were well represented in the highly EMT-related Claudin-low breast cancer intrinsic subtype.

Both the ES cell and iPS cell signatures were derived from culture contexts where the percentage of stem cells is expected to be very high and where the expression profile of the bulk sample is consequently unlikely to diverge significantly from the average stem cell. This is in contrast to the putative adult stem cell signatures analyzed to date where the actual stem cells are a relatively small minority of cells. The differences in the spectrum of tumors enriched for the ES/iPS cell signatures verses the putative adult stem cell signatures could relate to the different stem cell content of their source populations. It could also be true that the ES cell expression profile is significantly different from tissue specific stem cell profiles at the cellular level, leading to different enrichment patterns in breast cancers. However, we subsequently isolated and profiled a highly enriched population of mammary stem cells from the mouse embryo that did show strong enrichment for Basal-like breast cancers similar to ES and iPS cell signatures [88]. In examining the fetal mammary stem cell profiles, we noted that these fetal mammary stem cells were clearly more epithelial in nature than the associated stroma as they express high levels of CD24 and coexpress both basal and luminal cytokeratins though they did express some mesenchymal markers (eg, vimentin). This is in contrast to previous studies on adult mammary stem cells and BCSCs that would have predicted them to be highly EMT'd. Also in contrast to the common expectation for stem cells, the fetal stem cell fraction did not grow efficiently in nonadherent mammosphere culture though the more mesenchymal stromal, non-stem cells did. It remains to be determined how similar the expression profiles of fetal mammary stem cells and highly purified adult mammary stem cells would be pending purification of the later, and it is also unknown how much of a role microenvironment would play in these differences.

AN INDUCIBLE STEMNESS

Normal stem cells in other tissue systems are well known to require highly controlled culture conditions in vitro and to depend upon highly specialized niches in vivo to self-renew, differentiate, or possibly dedifferentiate. One notable example of the later is that of tuft cells in the intestine, which do not fit the definition of tissue stem cells "until an irritation—a simple stimulus suffices" to induce them to behave as such; in this case chronic injury and inflammation [89]. Though the mammary stem cell niche and the cues that can evoke latent stem cell capacity remain largely undefined, several groups have demonstrated that mammary stem cells are under similar contextual regulation. Transplantation analysis is significantly affected by the genetics, hormone levels and innate immune system of the host [90]. Furthermore, transplantation efficiencies are dramatically augmented by supplementation with reconstituted extracellular matrix [90]. The performance of normal mammary stem cells in culture is similarly sensitive to matrix and growth factor control, with some factors promoting self-renewal and others promoting differentiation [91]. Furthermore, fetal mammary stem cells can be dramatically and reversibly shifted from an epithelial state to a mesenchymal state by modulating levels of Sox10 potentially in coordination with specific fibroblast growth factor (FGF) signaling [92]. Such flexibility may be essential to normal organogenesis and wound healing. In addition, recent studies show that oncogene activation can reactivate bilineage differentiation potential in adult basal and luminal cells in vivo prior to significant tumor progression [93], similar to the bilineage potential we observed for fetal mammary stem cells (fMaSC) in the rapidly developing and remodeling mammary rudiment [88].

This context dependence for both molecular profile and stem cell fitness (ie, performance in certain assays or tumor propagation in vivo) is likely at play in cancer as well. For instance, the distinct anatomical niche of ALDH+ cells appears to be in the more hypoxic zones of the tumor relative to the CD44+CD24- pool. This raises the hypothesis that hypoxia may specifically evoke the ALDH+ phenotype and/or that the ALDH+ cells are particularly fit for survival and growth in hypoxia [31]. Studies showing that ALDH+ BCSC increase in tumors after antiangiogenic treatment with Avastin advance this idea [66] and fits well with hypotheses we and others have also raised that nutrient deprivation and hypoxic stresses could specifically promote stem cell phenotypes [91]. Thus, it may be a general principal in complex tissues that many cells possess latent stem cell potential needing only the right cue to "rekindle their vital cellular activities...."

TUMOR-ASSOCIATED REPROGRAMMING

The ability of particular stimuli or stresses to restore primitive, stem cell–like behaviors in cancer cells, akin to in vitro cellular reprogramming, would no doubt also be colored by oncogene activation and loss of tumor suppressor activity (eg, p53). Given these added factors, it seems unlikely that such a process would achieve complete and faithful recapitulation of normal stem cell gene expression programs. It may be relevant in this regard that the stem cell signature enrichment we have observed in cancers are virtually always modular [88,94]. That is, when we derive stem cell expression profiles and compare them to human tumor data, we typically find large coordinately expressed groups of genes as well as some genes and groups of genes that are not expressed. Critically, we do not find that this undermines the value of comparisons between developmental programs and cancer, and recently, we reported that modules from both luminal progenitors and fetal mammary stem cells were useful in predicting breast cancer response to neoadjuvant chemotherapy [94]. However, it is not always clear why tumors would selectively reactivate a given portion of a stem cell gene expression program but we can speculate that cancers are accessing combinatorial biological programs not only from the repertoire of what *does* happen in normal development but also what *can* happen in all possible functional combinations. Thus, stem cell activity in tumors may be achieved through "unnatural" stem cell states, crafted by co-opting and corrupting whichever developmental programs can respond to the stress at hand and there could be many such states (Fig. 5.2). Such a model could explain

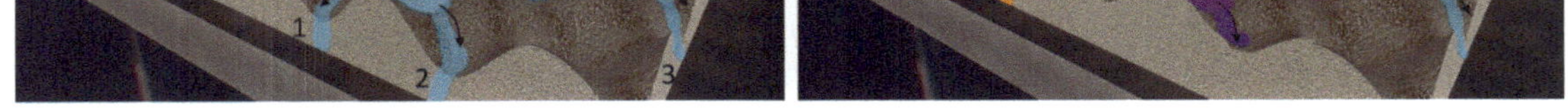

FIGURE 5.2 Model of normal and cancer stem cell (CSC) hierarchies.
In this take on a Waddington diagram, (A) represents a normal differentiation hierarchy (cyan) where the stem cell (SC) pool cascades along several paths of lineage commitment to produce three distinct differentiated cell types within the tissue. A different tissue (yellow) is shown in the background with a similar hierarchy. (B) Panel B represents a model cancer hierarchy. In this model, the bulk of the tumor (purple) is derived from a CSC pool resembling the normal stem cell (red, CSC1) but is potentially fed or interconverting with a second CSC pool (CSC2) representing a more primitive stem cell state or alternative niche not usually utilized in the adult. A third CSC pool (CSC3) may be able to yield cells atypical for the tissue that are functional in some other capacity such as providing blood flow (orange) but these are also not normally represented in the adult. There may be many possible CSC states potentially available to the growing cancer represented by the additional valleys in the diagram.

Weinberg's recent observation that different EMT programs are used by BCSCs than are used by normal mammary stem cells [59]. This model could also explain reports of transdifferentiation and non-tissue of origin mimicry and could help to explain a significant amount of intertumor heterogeneity even within an intrinsic subtype as there could be multiple ways to achieve the same overall tumor phenotype. While the model is perhaps not as simple as the original CSC model, it would predict that along with the significant similarities between normal stem cells and CSCs there are fundamental differences that could be exploited to target tumor cells without damaging the host. However, under this model it is unclear how many stem cell states might be achievable by a tumor or how many of them would need to be targeted to cure the disease.

TACKLING HETEROGENEITY AT SINGLE-CELL RESOLUTION

Ultimately, it will be clinically essential to know how many functional/dangerous cell states are accessed by a tumor. Here, "dangerous" means those cells capable of operating as CSCs in the broadest sense of propagating or disseminating tumor growth and perpetuating the disease in active or dormant states. Single-cell sequencing has begun to provide inroads into this daunting task. At the genomic level, a study by Navin et al. analyzed 100 single TNBC cells for copy number variations identifying punctuated disequilibrium in cancer evolution and rare subclones with massive KRAS amplification, two novel observations that were unlikely to have emerged from traditional analysis [95]. A subsequent work identified an even broader potential heterogeneity of genetic subclones and speculated that these could represent differing fitness for specific environments in the tumor and in the tumor's future including the stresses we have highlighted earlier such as nutrient deprivation and hypoxia [96].

At the gene expression level, single-cell studies are just now beginning to emerge in numbers. While there are some quite early studies that amplified single-cell cDNA libraries for analysis on arrays. Many studies have turned to a "many cells × many genes" RT-PCR approach, sometimes with the aid of rapidly advancing microfluidic technologies. These studies are proving informative in terms of identifying and validating the expression of various candidate stem cell regulator genes, but also revealing general principles of cell biology overlooked by bulk gene expression analysis. For instance, single-cell profiling on fMaSC revealed the expression of key candidates stem cell regulators in subsets of cells (eg, lgr5, Sox10), validated the coexpression of luminal and basal characteristics within individual cells and uncovered a striking amount of diversity of gene expression patterns even within a cluster of significantly related cells [88]. A very recent study used a similar approach to identify stem cell programs in metastatic breast cancer [97]. However, a significant drawback to these approaches is their limitation on the discovery of novel-involved genes and splice variants since a limited set of genes must be chosen in the assay.

Recent, advances in single-cell transcriptional profiling using RNA sequencing (RNA-Seq), which lacks this limitation, have raised the unprecedented possibility of deconstructing complex tissues at single-cell resolution and reintegrating this information for a systems-level understanding of cancer and cancer stem cells. For instance, single-cell RNA-Seq may aid in establishing how many distinct cancer stem cell pools there are in a particular tumor as well as how internally heterogeneous these pools are. One recent study profiled 430 single cells from glioblastoma and determined that cells of the tumor exhibit a gradient of relationships to stem cell profiles rather than having a distinct and clearly identifiable subset of stemlike cells [98]. Recently, Lee et al. profiled individual MDA-MB-231 cells before and after treatment with Paclitaxel to study drug resistance dynamics using RNA-Seq and identified microtubule stabilization effects correlated with drug tolerance [99]. In addition, a recent methodological report suggested that cell types that are extremely rare and even those represented by an individual cell in a given experiment can be identified and characterized [100]. Though this procedure still requires some vetting in practice, it raises the very exciting prospect of characterizing early neoplasias, residual disease following therapy, dormant cells, circulating tumor cells, micrometastases, and many other rare, previously inaccessible or unculturable cell types.

The extensive cellular heterogeneity indicated by existing single-cell profiling data in both normal and neoplastic contexts represents a daunting challenge but also a tremendous opportunity to uncover new principals of tissue organization and new therapeutic targets.

In this regard it may be worth remembering that several clinical tests for existing molecularly targeted therapies (eg, Her2) are based not only on the identification of receptor expression in the tumor but on the number of cells expressing the target. Despite this heterogeneity, molecularly targeted therapies can sometimes be effective although response is often significantly lower in patients with heterogeneous versus homogenous HER2 positivity [101]. There are several possible reasons for this. For instance, it may be that the Her2 positive fraction of cells represents the sole BCSC for this disease. Alternatively, all tumorigenic cells may need to traffic through a receptor positive status even if the individual cell would have been diagnostically negative at the time of the test, or there may be bystander effects where targeting of the positive cells leads to destruction of nearby negative cells. In some tumors there may be obligate heterogeneity with some cells playing niche roles and others playing stem cell roles, and either stem cell or niche may provide valid therapeutic target, as would targeting their physical interaction. Some tumors may recruit a normal niche cell from the stroma. Such a recruited niche might have less potential for developing therapy resistance. Still other tumors may differentiate to provide their own niche cells, while still others may be relatively cell autonomous for niche signaling. All of these possibilities may color the hierarchical structure of a given tumor and influence how CSCs are best defined in each case. Understanding these scenarios could easily prove impenetrable without single-cell resolution studies and potentially the integration of this information across genomics, epigenomics, proteomics, metabolomics analyses as new technologies enable these to function at high resolution.

CONCLUSION

Since the identification of BCSCs in 2003, the cancer stem cell literature has grown dramatically. Indeed, it would be somewhat impractical to recount here in any appreciable detail the myriad molecules and pathways that have been reported to play critical, stem cell–related roles in breast cancer. Furthermore, as several dozen reviews covering many specialized molecular aspects of BCSC biology have been published even within the last year, this chapter has focused instead on challenges and critical next steps. The hope is that future work along these and related trajectories will lead to clinically relevant improvements in our understanding of the problem that the cancer stem cell hypothesis at its core has sought to address; namely that of cellular heterogeneity and plasticity in breast and other cancers [102–104].

Though some pieces of the puzzle have been assembled, no consensus regulatory circuitry has been defined for the mammary gland stem cells or BCSCs. We can anticipate that BCSC circuitries would borrow extensively from developmental paradigms, that they will involve multigene networks and systems-level processes, and that these could point to new strategies for molecularly targeted therapies. However, we do not yet know the catalog of molecular regulators, microenvironmental/niche cues and selective pressures that diversify upon developmental themes in breast cancer or BCSCs. We do know that multiple stem cell states exist as they have been reported in both the normal mammary gland and in tumors. However, we know little about their underlying mechanisms or mechanisms facilitating their functional plasticity. Indeed, it may be necessary to piece together many individual stem cell profiles to understand this continuum.

Recent advances in molecular profiling especially as these progress to higher resolution and even single cells hold great promise for obtaining the systems-level view of stem cell function in normal mammary tissue and in breast cancer. They promise to help answer a number of critical questions: How similar are the mammary stem cell and BCSC states at the molecular level? How many BCSC states are there? How do they interconvert? Do molecular profiles identify cells with latent stem cell potential? What cues can "rekindle" them? Answering these questions will in turn allow us to make better sense of the outward similarities between stem cells, development and cancer and to better choose cellular and molecular therapeutic targets at all stages of the disease.

List of Acronyms and Abbreviations

ALDH Aldehyde dehydrogenase
Bmi1 B lymphoma Moloney murine leukemia virus insertion region 1 homolog
BCSC Breast cancer stem cell
c-Jun Cellular Jun protooncogene
CD24,49f,44,90 Cluster of differentiation
CSC Cancer stem cell
CXCR4 c-x-c chemokine receptor type 4
EMT Epithelial-to-mesenchymal transition
EphA4 Ephrin type-4 receptor 4
ER Estrogen receptor
ErbB2 Erythroblastic leukemia B2 protooncogene
ES Embryonic stem (cell)
ESA Epithelial surface antigen
FGF Fibroblast growth factor
fMaSC Fetal mammary stem cell
FVBneu-tg Friend Virus B-Type mouse strain with neu transgene
GFP Green fluorescent protein
HER2 Human epidermal growth factor receptor 2
HH Hedge hog
Hif1-a Hypoxia inducible factor
IL-11 Interleukin 11
iPS Induced pluripotent stem (cell)
JAM-A Junctional adhesion molecule -A
Let-7 Lethal 7 gene
Lgr5 Leucine-rich repeat-containing G-protein receptor 5
MCF7 Michigan cancer foundation 7 brest cancer cell line
MDAMB231 MD Anderson metastatic breast cancer 231
miR Micro-ribonucleic acid
miRNA Micro-ribonucleic acid
mTOR Mechanistic target of rapamycin
neu Neuronal protooncogene
NICD Notch intracellular domain
NOD/SCID Nonobese diabetic/severe combined immunodeficiency mouse strain
Oct4 Octamer binding transcription factor 4
p53 Protein product of transformation-related protein 53 gene
PIK3CA Phosphatidylinositol-4,5-bisphosphate 3-kinase catalytic subunit A
PR Progesterone receptor

PROCR Protein c receptor
RNA-Seq Ribonucleic acid sequencing
Sox2, 9, 10 (Sex determining region y)-box region 2, 9, 10
Stat3 Signal transducer and activator of transcription 3
TAZ Tafazzin
TCGA The cancer genome atlas
Tgf-β Transforming growth factor-β
TNBC Triple-negative breast cancer
Wnt Wingless/integration
Zeb1 Zinc finger e-box homeobox 1

References

[1] Triolo VA. Nineteenth century foundations of cancer research advances in tumor pathology, nomenclature, and theories of oncogenesis. Cancer Res 1965;25:75–106.

[2] Virchow R. Cellular pathologie. Arch fuer Pathol Anat fuer Klin Med 1855;8:3–39.

[3] Silver GA. Virchow, the heroic model in medicine: health policy by accolade. Am J Public Health 1987;77:82–8.

[4] Cohnheim, J. Ein Handbuch fuer Aertzte un Studirende. 1882;1:622–91 (Hirschwald, 1877–1880).

[5] Brewer BG, Mitchell RA, Harandi A, Eaton JW. Embryonic vaccines against cancer: an early history. Exp Mol Pathol 2009;86:192–7. http://dx.doi.org/10.1016/j.yexmp.2008.12.002.

[6] Lapidot T, Sirard C, Vormoor J, Murdoch B, Hoang T, Caceres-Cortes J, et al. A cell initiating human acute myeloid leukaemia after transplantation into SCID mice. Nature 1994;367:645–8. http://dx.doi.org/10.1038/367645a0.

[7] Bonnet D, Dick JE. Human acute myeloid leukemia is organized as a hierarchy that originates from a primitive hematopoietic cell. Nat Med 1997;3:730–7.

[8] McCulloch EA, Till JE. The radiation sensitivity of normal mouse bone marrow cells, determined by quantitative marrow transplantation into irradiated mice. Radiat Res 1960;13:115–25.

[9] Furth J, Kahn MC. The transmission of leukemia of mice with a single cell. Am J Cancer 1937;31:276–82.

[10] Chao MP, Seita J, Weissman IL. Establishment of a normal hematopoietic and leukemia stem cell hierarchy. Cold Spring Harb Symp Quant Biol 2008;73:439–49. http://dx.doi.org/10.1101/sqb.2008.73.031.

[11] Singh SK, Hawkins C, Clarke ID, Squire JA, Bayani J, Hide T, et al. Identification of human brain tumour initiating cells. Nature 2004;432:396–401. http://dx.doi.org/10.1038/nature03128.

[12] O'Brien CA, Pollett A, Gallinger S, Dick JE. A human colon cancer cell capable of initiating tumour growth in immunodeficient mice. Nature 2007;445:106–10. http://dx.doi.org/10.1038/nature05372.

[13] Li C, Heidt DG, Dalerba P, Burant CF, Zhang L, Adsay V, et al. Identification of pancreatic cancer stem cells. Cancer Res 2007;67:1030–7. http://dx.doi.org/10.1158/0008-5472.CAN-06-2030.

[14] Hermann PC, Huber SL, Herrler T, Aicher A, Ellwart JW, Guba M, et al. Distinct populations of cancer stem cells determine tumor growth and metastatic activity in human pancreatic cancer. Cell Stem Cell 2007;1:313–23. http://dx.doi.org/10.1016/j.stem.2007.06.002.

[15] Collins AT, Berry PA, Hyde C, Stower MJ, Maitland NJ. Prospective identification of tumorigenic prostate cancer stem cells. Cancer Res 2005;65:10946–51. http://dx.doi.org/10.1158/0008-5472.CAN-05-2018.

[16] Patrawala L, Calhoun T, Schneider-Broussard R, Li H, Bhatia B, Tang S, et al. Highly purified CD44+ prostate cancer cells from xenograft human tumors are enriched in tumorigenic and metastatic progenitor cells. Oncogene 2006;25:1696–708. http://dx.doi.org/10.1038/sj.onc.1209327.

[17] Al-Hajj M, Wicha MS, Benito-Hernandez A, Morrison SJ, Clarke MF. Prospective identification of tumorigenic breast cancer cells. Proc Natl Acad Sci USA 2003;100:3983–8. http://dx.doi.org/10.1073/pnas.0530291100.

[18] Deome KB, Faulkin Jr LJ, Bern HA, Blair PB. Development of mammary tumors from hyperplastic alveolar nodules transplanted into gland-free mammary fat pads of female C3H mice. Cancer Res 1959;19:515–20.

[19] Kordon EC, Smith GH. An entire functional mammary gland may comprise the progeny from a single cell. Development 1998;125:1921–30.

[20] Stingl J, Eirew P, Ricketson I, Shackleton M, Vaillant F, Choi D, et al. Purification and unique properties of mammary epithelial stem cells. Nature 2006;439:993–7. http://dx.doi.org/10.1038/nature04496.

[21] Shackleton M, Vaillant F, Simpson KJ, Stingl J, Smyth GK, Asselin-Labat ML, et al. Generation of a functional mammary gland from a single stem cell. Nature 2006;439:84–8. http://dx.doi.org/10.1038/nature04372.

[22] Mani SA, Guo W, Liao MJ, Eaton EN, Ayyanan A, Zhou AY, et al. The epithelial-mesenchymal transition generates cells with properties of stem cells. Cell 2008;133:704–15. http://dx.doi.org/10.1016/j.cell.2008.03.027.

[23] Dontu G, Al-Hajj M, Abdallah WM, Clarke MF, Wicha MS. Stem cells in normal breast development and breast cancer. Cell Prolif 2003;36(Suppl. 1):59–72.

[24] Singh A, Settleman J. EMT, cancer stem cells and drug resistance: an emerging axis of evil in the war on cancer. Oncogene 2010;29:4741–51. http://dx.doi.org/10.1038/onc.2010.215.

[25] Ginestier C, Hur MH, Charafe-Jauffret E, Monville F, Dutcher J, Brown M, et al. ALDH1 is a marker of normal and malignant human mammary stem cells and a predictor of poor clinical outcome. Cell Stem Cell 2007;1:555–67. http://dx.doi.org/10.1016/j.stem.2007.08.014.

[25a] Clarke MF, Dick JE, Dirks PB, Eaves CJ, Jamieson CHM, Jones DL, et al. Cancer stem cells—perspectives on current status and future directions: AACR workshop on cancer stem cells. Cancer Research. http://dx.doi.org/10.1158/0008-5472.CAN-06-3126. Published October 1, 2006.

[26] Liao MJ, Zhang CC, Zhou B, Zimonjic DB, Mani SA, Kaba M, et al. Enrichment of a population of mammary gland cells that form mammospheres and have in vivo repopulating activity. Cancer Res 2007;67:8131–8. http://dx.doi.org/10.1158/0008-5472.CAN-06-4493.

[27] Van Keymeulen A, Rocha AS, Ousset M, Beck B, Bouvencourt G, Rock J, et al. Distinct stem cells contribute to mammary gland development and maintenance. Nature 2011;479:189–93. http://dx.doi.org/10.1038/nature10573.

[28] Prat A, Karginova O, Parker JS, Fan C, He X, Bixby L, et al. Characterization of cell lines derived from breast cancers and normal mammary tissues for the study of the intrinsic molecular subtypes. Breast Cancer Res Treat 2013;142:237–55. http://dx.doi.org/10.1007/s10549-013-2743-3.

[29] Gupta PB, Fillmore CM, Jiang G, Shapira SD, Tao K, Kuperwasser C, et al. Stochastic state transitions give rise to phenotypic equilibrium in populations of cancer cells. Cell 2011;146:633–44. http://dx.doi.org/10.1016/j.cell.2011.07.026.

[30] Chaffer CL, Brueckmann I, Scheel C, Kaestli AJ, Wiggins PA, Rodrigues LO, et al. Normal and neoplastic nonstem cells can spontaneously convert to a stem-like state. Proc Natl Acad Sci USA 2011;108:7950–5. http://dx.doi.org/10.1073/pnas.1102454108.

[31] Liu S, Cong Y, Wang D, Sun Y, Deng L, Liu Y, et al. Breast cancer stem cells transition between epithelial and mesenchymal states reflective of their normal counterparts. Stem Cell Rep 2014;2:78–91. http://dx.doi.org/10.1016/j.stemcr.2013.11.009.

[32] Hwang-Verslues WW, Kuo WH, Chang PH, Pan CC, Wang HH, Tsai ST, et al. Multiple lineages of human breast cancer stem/progenitor cells identified by profiling with stem cell markers. PLoS One 2009;4:e8377. http://dx.doi.org/10.1371/journal.pone.0008377.

[33] Ablett MP, O'Brien CS, Sims AH, Farnie G, Clarke RB. A differential role for CXCR4 in the regulation of normal versus malignant breast stem cell activity. Oncotarget 2014;5:599–612.

[34] Vasiliou V, Pappa A, Estey T. Role of human aldehyde dehydrogenases in endobiotic and xenobiotic metabolism. Drug Metab Rev 2004;36:279–99. http://dx.doi.org/10.1081/DMR-120034001.

[35] Chanmee T, Ontong P, Kimata K, Itano N. Key roles of hyaluronan and its CD44 receptor in the stemness and survival of cancer stem cells. Front Oncol 2015;5:180. http://dx.doi.org/10.3389/fonc.2015.00180.

[36] Pham PV, Phan NL, Nguyen NT, Truong NH, Duong TT, Le DV, et al. Differentiation of breast cancer stem cells by knockdown of CD44: promising differentiation therapy. J Transl Med 2011;9:209. http://dx.doi.org/10.1186/1479-5876-9-209.

[37] D'Angelo RC, Ouzounova M, Davis A, Choi D, Tchuenkam SM, Kim G, et al. Notch reporter activity in breast cancer cell lines identifies a subset of cells with stem cell activity. Mol Cancer Ther 2015;14:779–87. http://dx.doi.org/10.1158/1535-7163.MCT-14-0228.

[38] Tang B, Raviv A, Esposito D, Flanders KC, Daniel C, Nghiem BT, et al. A flexible reporter system for direct observation and isolation of cancer stem cells. Stem Cell Rep 2015;4:155–69. http://dx.doi.org/10.1016/j.stemcr.2014.11.002.

[39] Thiagarajan PS, Hitomi M, Hale JS, Alvarado AG, Otvos B, Sinyuk M, et al. Development of a fluorescent reporter system to delineate cancer stem cells in triple-negative breast Cancer. Stem Cells 2015;33:2114–25. http://dx.doi.org/10.1002/stem.2021.

[40] Shafee N, Smith CR, Wei S, Kim Y, Mills GB, Hortobagyi GN, et al. cancer stem cells contribute to cisplatin resistance in Brca1/p53-mediated mouse mammary tumors. Cancer Res 2008;68:3243–50. http://dx.doi.org/10.1158/0008-5472.CAN-07-5480.

[41] Ben-Porath I, Thomson MW, Carey VJ, Ge R, Bell GW, Regev A, et al. An embryonic stem cell-like gene expression signature in poorly differentiated aggressive human tumors. Nat Genet 2008;40:499–507. http://dx.doi.org/10.1038/ng.127.

[42] Mizuno H, Spike BT, Wahl GM, Levine AJ. Inactivation of p53 in breast cancers correlates with stem cell transcriptional signatures. Proc Natl Acad Sci USA 2010;107:22745–50. http://dx.doi.org/10.1073/pnas.1017001108.

[43] Lamb R, Ablett MP, Spence K, Landberg G, Sims AH, Clarke RB. Wnt pathway activity in breast cancer sub-types and stem-like cells. PLoS One 2013;8:e67811. http://dx.doi.org/10.1371/journal.pone.0067811.

[44] Cai WY, Wei TZ, Luo QC, Wu QW, Liu QF, Yang M, et al. The Wnt-beta-catenin pathway represses let-7 microRNA expression through transactivation of Lin28 to augment breast cancer stem cell expansion. J Cell Sci 2013;126:2877–89. http://dx.doi.org/10.1242/jcs.123810.

[45] Harrison H, Farnie G, Howell SJ, Rock RE, Stilianou S, Brennan KR, et al. Regulation of breast cancer stem cell activity by signaling through the Notch4 receptor. Cancer Res 2010;70:709–18. http://dx.doi.org/10.1158/0008-5472.CAN-09-1681.

[46] Lombardo Y, Filipovic A, Molyneux G, Periyasamy M, Giamas G, Hu Y, et al. Nicastrin regulates breast cancer stem cell properties and tumor growth in vitro and in vivo. Proc Natl Acad Sci USA 2012;109:16558–63. http://dx.doi.org/10.1073/pnas.1206268109.

[47] Sims-Mourtada J, Opdenaker LM, Davis J, Arnold KM, Flynn D. Taxane-induced hedgehog signaling is linked to expansion of breast cancer stem-like populations after chemotherapy. Mol Carcinog 2015;54:1480–93. http://dx.doi.org/10.1002/mc.22225.

[48] Wang L, Duan W, Kang L, Mao J, Yu X, Fan S, et al. Smoothened activates breast cancer stem-like cell and promotes tumorigenesis and metastasis of breast cancer. Biomed Pharmacother 2014;68:1099–104. http://dx.doi.org/10.1016/j.biopha.2014.09.012.

[49] Chen ZY, Wang PW, Shieh DB, Chiu KY, Liou YM. Involvement of gelsolin in TGF-beta 1 induced epithelial to mesenchymal transition in breast cancer cells. J Biomed Sci 2015;22:90. http://dx.doi.org/10.1186/s12929-015-0197-0.

[50] Jiao X, Katiyar S, Willmarth NE, Liu M, Ma X, Flomenberg N, et al. c-Jun induces mammary epithelial cellular invasion and breast cancer stem cell expansion. J Biol Chem 2010;285:8218–26. http://dx.doi.org/10.1074/jbc.M110.100792.

[51] Bartucci M, Dattilo R, Moriconi C, Pagliuca A, Mottolese M, Federici G, et al. TAZ is required for metastatic activity and chemoresistance of breast cancer stem cells. Oncogene 2015;34:681–90. http://dx.doi.org/10.1038/onc.2014.5.

[52] Xiang L, Gilkes DM, Hu H, Takano N, Luo W, Lu H, et al. Hypoxia-inducible factor 1 mediates TAZ expression and nuclear localization to induce the breast cancer stem cell phenotype. Oncotarget 2014;5:12509–27.

[53] Lin L, Hutzen B, Lee HF, Peng Z, Wang W, Zhao C, et al. Evaluation of STAT3 signaling in ALDH+ and ALDH+/CD44+/CD24- subpopulations of breast cancer cells. PLoS One 2013;8:e82821. http://dx.doi.org/10.1371/journal.pone.0082821.

[54] Choi HJ, Park JH, Park M, Won HY, Joo HS, Lee CH, et al. UTX inhibits EMT-induced breast CSC properties by epigenetic repression of EMT genes in cooperation with LSD1 and HDAC1. EMBO Rep 2015;16:1288–98. http://dx.doi.org/10.15252/embr.201540244.

[55] Lu Z, Jia J, Di L, Song G, Yuan Y, Ma B, et al. DNA methyltransferase inhibitor CDA-2 synergizes with high-dose thiotepa and paclitaxel in killing breast cancer stem cells. Front Biosci (Elite Ed) 2011;3:240–9.

[56] Lu S, Labhasetwar V. Drug resistant breast cancer cell line displays cancer stem cell phenotype and responds sensitively to epigenetic drug SAHA. Drug Deliv Transl Res 2013;3:183–94. http://dx.doi.org/10.1007/s13346-012-0113-z.

[57] Vesuna F, Lisok A, Kimble B, Raman V. Twist modulates breast cancer stem cells by transcriptional regulation of CD24 expression. Neoplasia 2009;11:1318–28.

[58] Li J, Zhou BP. Activation of beta-catenin and Akt pathways by Twist are critical for the maintenance of EMT associated cancer stem cell-like characters. BMC Cancer 2011;11:49. http://dx.doi.org/10.1186/1471-2407-11-49.

[59] Ye X, Tam WL, Shibue T, Kaygusuz Y, Reinhardt F, Ng Eaton E, et al. Distinct EMT programs control normal mammary stem cells and tumour-initiating cells. Nature 2015;525:256–60. http://dx.doi.org/10.1038/nature14897.

[60] Dong C, Yuan T, Wu Y, Wang Y, Fan TW, Miriyala S, et al. Loss of FBP1 by snail-mediated repression provides metabolic advantages in basal-like breast cancer. Cancer Cell 2013;23:316–31. http://dx.doi.org/10.1016/j.ccr.2013.01.022.

[61] Bhat-Nakshatri P, Appaiah H, Ballas C, Pick-Franke P, Goulet Jr R, Badve S, et al. SLUG/SNAI2 and tumor necrosis factor generate breast cells with CD44+/CD24- phenotype. BMC Cancer 2010;10:411. http://dx.doi.org/10.1186/1471-2407-10-411.

[62] Guo W, Keckesova Z, Donaher JL, Shibue T, Tischler V, Reinhardt F, et al. Slug and Sox9 cooperatively determine the mammary stem cell state. Cell 2012;148:1015–28. http://dx.doi.org/10.1016/j.cell.2012.02.008.

[63] Yang J, Mani SA, Donaher JL, Ramaswamy S, Itzykson RA, Come C, et al. Twist, a master regulator of morphogenesis, plays an essential role in tumor metastasis. Cell 2004;117:927–39. http://dx.doi.org/10.1016/j.cell.2004.06.006.

[64] Overdevest JB, Knubel KH, Duex JE, Thomas S, Nitz MD, Harding MA, et al. CD24 expression is important in male urothelial tumorigenesis and metastasis in mice and is androgen regulated. Proc Natl Acad Sci USA 2012;109:E3588–96. http://dx.doi.org/10.1073/pnas.1113960109.

[65] Semenza GL. Regulation of the breast cancer stem cell phenotype by hypoxia-inducible factors. Clin Sci (Lond) 2015;129:1037–45. http://dx.doi.org/10.1042/CS20150451.

[66] Conley SJ, Gheordunescu E, Kakarala P, Newman B, Korkaya H, Heath AN, et al. Antiangiogenic agents increase breast cancer stem cells via the generation of tumor hypoxia. Proc Natl Acad Sci USA 2012;109:2784–9. http://dx.doi.org/10.1073/pnas.1018866109.

[67] Vazquez-Martin A, Oliveras-Ferraros C, Cufi S, Del Barco S, Martin-Castillo B, Menendez JA. Metformin regulates breast cancer stem cell ontogeny by transcriptional regulation of the epithelial-mesenchymal transition (EMT) status. Cell Cycle 2010;9:3807–14.

[68] Jung JW, Park SB, Lee SJ, Seo MS, Trosko JE, Kang KS. Metformin represses self-renewal of the human breast carcinoma stem cells via inhibition of estrogen receptor-mediated OCT4 expression. PLoS One 2011;6:e28068. http://dx.doi.org/10.1371/journal.pone.0028068.

[69] Zhou J, Wulfkuhle J, Zhang H, Gu P, Yang Y, Deng J, et al. Activation of the PTEN/mTOR/STAT3 pathway in breast cancer stem-like cells is required for viability and maintenance. Proc Natl Acad Sci USA 2007;104:16158–63. http://dx.doi.org/10.1073/pnas.0702596104.

[70] Maycotte P, Jones KL, Goodall ML, Thorburn J, Thorburn A. Autophagy supports breast cancer stem cell maintenance by regulating IL6 secretion. Mol Cancer Res 2015;13:651–8. http://dx.doi.org/10.1158/1541-7786.MCR-14-0487.

[71] Santisteban M, Reiman JM, Asiedu MK, Behrens MD, Nassar A, Kalli KR, et al. Immune-induced epithelial to mesenchymal transition in vivo generates breast cancer stem cells. Cancer Res 2009;69:2887–95. http://dx.doi.org/10.1158/0008-5472.CAN-08-3343.

[72] Lu H, Clauser KR, Tam WL, Frose J, Ye X, Eaton EN, et al. A breast cancer stem cell niche supported by juxtacrine signalling from monocytes and macrophages. Nat Cell Biol 2014;16:1105–17. http://dx.doi.org/10.1038/ncb3041.

[73] Shimono Y, Zabala M, Cho RW, Lobo N, Dalerba P, Qian D, et al. Downregulation of miRNA-200c links breast cancer stem cells with normal stem cells. Cell 2009;138:592–603. http://dx.doi.org/10.1016/j.cell.2009.07.011.

[74] Knezevic J, Pfefferle AD, Petrovic I, Greene SB, Perou CM, Rosen JM. Expression of miR-200c in claudin-low breast cancer alters stem cell functionality, enhances chemosensitivity and reduces metastatic potential. Oncogene 2015. http://dx.doi.org/10.1038/onc.2015.48.

[75] Kang L, Mao J, Tao Y, Song B, Ma W, Lu Y, et al. MicroRNA-34a suppresses the breast cancer stem cell-like characteristics by downregulating Notch1 pathway. Cancer Sci 2015;106:700–8. http://dx.doi.org/10.1111/cas.12656.

[76] Bockhorn J, Dalton R, Nwachukwu C, Huang S, Prat A, Yee K, et al. MicroRNA-30c inhibits human breast tumour chemotherapy resistance by regulating TWF1 and IL-11. Nat Commun 2013;4:1393. http://dx.doi.org/10.1038/ncomms2393.

[77] Chang WW, et al. The expression and significance of insulin-like growth factor-1 receptor and its pathway on breast cancer stem/progenitors. Breast Cancer Res 2013;15(3):R39.

[78] Yang Z, Han Y, Cheng K, Zhang G, Wang X. miR-99a directly targets the mTOR signalling pathway in breast cancer side population cells. Cell Prolif 2014;47:587–95. http://dx.doi.org/10.1111/cpr.12146.

[79] Cancer Genome Atlas N. Comprehensive molecular portraits of human breast tumours. Nature 2012;490:61–70. http://dx.doi.org/10.1038/nature11412.

[80] Coates AS, Winer EP, Goldhirsch A, Gelber RD, Gnant M, Piccart-Gebhart M, et al. Tailoring therapies-improving the management of early breast cancer: St Gallen International Expert consensus on the primary therapy of early breast cancer 2015. Ann Oncol 2015;26:1533–46. http://dx.doi.org/10.1093/annonc/mdv221.

[81] Perou CM, Sorlie T, Eisen MB, van de Rijn M, Jeffrey SS, Rees CA, et al. Molecular portraits of human breast tumours. Nature 2000;406:747–52. http://dx.doi.org/10.1038/35021093.

[82] Herschkowitz JI, Simin K, Weigman VJ, Mikaelian I, Usary J, Hu Z, et al. Identification of conserved gene expression features between murine mammary carcinoma models and human breast tumors. Genome Biol 2007;8:R76. http://dx.doi.org/10.1186/gb-2007-8-5-r76.

[83] Prat A, Perou CM. Deconstructing the molecular portraits of breast cancer. Mol Oncol 2011;5:5–23. http://dx.doi.org/10.1016/j.molonc.2010.11.003.

[84] Lim E, Wu D, Pal B, Bouras T, Asselin-Labat ML, Vaillant F, et al. Transcriptome analyses of mouse and human mammary cell subpopulations reveal multiple conserved genes and pathways. Breast Cancer Res 2010;12:R21. http://dx.doi.org/10.1186/bcr2560.

[85] Pece S, Tosoni D, Confalonieri S, Mazzarol G, Vecchi M, Ronzoni S, et al. Biological and molecular heterogeneity of breast cancers correlates with their cancer stem cell content. Cell 2010;140:62–73. http://dx.doi.org/10.1016/j.cell.2009.12.007.

[86] Aladjem MI, Spike BT, Rodewald LW, Hope TJ, Klemm M, Jaenisch R, et al. ES cells do not activate p53-dependent stress responses and undergo p53-independent apoptosis in response to DNA damage. Curr Biol 1998;8:145–55.

[87] Krizhanovsky V, Lowe SW. Stem cells: the promises and perils of p53. Nature 2009;460:1085–6. http://dx.doi.org/10.1038/4601085a.

[88] Spike BT, Engle DD, Lin JC, Cheung SK, La J, Wahl GM. A mammary stem cell population identified and characterized in late embryogenesis reveals similarities to human breast cancer. Cell Stem Cell 2012;10:183–97. http://dx.doi.org/10.1016/j.stem.2011.12.018.

[89] Westphalen CB, Asfaha S, Hayakawa Y, Takemoto Y, Lukin DJ, Nuber AH, et al. Long-lived intestinal tuft cells serve as colon cancer-initiating cells. J Clin Invest 2014;124:1283–95. http://dx.doi.org/10.1172/JCI73434.

[90] Makarem M, Spike BT, Dravis C, Kannan N, Wahl GM, Eaves CJ. Stem cells and the developing mammary gland. J Mammary Gland Biol Neoplasia 2013;18:209–19. http://dx.doi.org/10.1007/s10911-013-9284-6.

[91] Spike BT, Kelber JA, Booker E, Kalathur M, Rodewald R, Lipianskaya J, et al. CRIPTO/GRP78 signaling maintains fetal and adult mammary stem cells ex vivo. Stem Cell Rep 2014;2:427–39. http://dx.doi.org/10.1016/j.stemcr.2014.02.010.

[92] Dravis C, Spike BT, Harrell JC, Johns C, Trejo CL, Southard-Smith EM, et al. Sox10 regulates stem/progenitor and mesenchymal cell states in mammary epithelial cells. Cell Rep 2015;12:2035–48. http://dx.doi.org/10.1016/j.celrep.2015.08.040.

[93] Van Keymeulen A, Lee MY, Ousset M, Brohee S, Rorive S, Giraddi RR, et al. Reactivation of multipotency by oncogenic PIK3CA induces breast tumour heterogeneity. Nature 2015;525:119–23. http://dx.doi.org/10.1038/nature14665.

[94] Pfefferle AD, Spike BT, Wahl GM, Perou CM. Luminal progenitor and fetal mammary stem cell expression features predict breast tumor response to neoadjuvant chemotherapy. Breast Cancer Res Treat 2015;149:425–37. http://dx.doi.org/10.1007/s10549-014-3262-6.

[95] Navin N, Kendall J, Troge J, Andrews P, Rodgers L, McIndoo J, et al. Tumour evolution inferred by single-cell sequencing. Nature 2011;472:90–4. http://dx.doi.org/10.1038/nature09807.

[96] Cai X, Evrony GD, Lehmann HS, Elhosary PC, Mehta BK, Poduri A, et al. Single-cell, genome-wide sequencing identifies clonal somatic copy-number variation in the human brain. Cell Rep 2014;8:1280–9. http://dx.doi.org/10.1016/j.celrep.2014.07.043.

[97] Lawson DA, Bhakta NR, Kessenbrock K, Prummel KD, Yu Y, Takai K, et al. Single-cell analysis reveals a stem-cell program in human metastatic breast cancer cells. Nature 2015;526:131–5. http://dx.doi.org/10.1038/nature15260.

[98] Patel AP, Tirosh I, Trombetta JJ, Shalek AK, Gillespie SM, Wakimoto H, et al. Single-cell RNA-seq highlights intratumoral heterogeneity in primary glioblastoma. Science 2014;344: 1396–401. http://dx.doi.org/10.1126/science.1254257.

[99] Lee MC, Lopez-Diaz FJ, Kahn SY, Tarig MA, Dayn Y, Vaske CJ, et al. Single-cell analyses of transcriptional heterogeneity during drug tolerance transition in cancer cells by RNA sequencing. Proc Natl Acad Sci USA 2014;111:E4726–35. http://dx.doi.org/10.1073/pnas.1404656111.

[100] Grun D, Lyubimova A, Kester L, Wiebrands K, Basak O, Sasaki N, et al. Single-cell messenger RNA sequencing reveals rare intestinal cell types. Nature 2015;525:251–5. http://dx.doi.org/10.1038/nature14966.

[101] Lee HJ, Seo AN, Kim EJ, Jang MH, Suh KJ, Ryu HS, et al. HER2 heterogeneity affects trastuzumab responses and survival in patients with HER2-positive metastatic breast cancer. Am J Clin Pathol 2014;142:755–66. http://dx.doi.org/10.1309/AJCPIRL4GUVGK3YX.

[102] Spike BT, Wahl GM. p53, stem cells, and reprogramming: tumor suppression beyond guarding the genome. Genes Cancer 2011;2:404–19. http://dx.doi.org/10.1177/1947601911410224.

[103] Meacham CE, Morrison SJ. Tumour heterogeneity and cancer cell plasticity. Nature 2013;501:328–37. http://dx.doi.org/10.1038/nature12624.

[104] Brooks MD, Burness ML, Wicha MS. Therapeutic implications of cellular heterogeneity and plasticity in breast cancer. Cell Stem Cell 2015;17:260–71. http://dx.doi.org/10.1016/j.stem.2015.08.014.

[105] Samanta D, et al. Hypoxia-inducible factors are required for chemotherapy resistance of breast cancer stem cells. Proc Natl Acad Sci U S A 2014;111(50):E5429–38.

[106] Lu H, et al. Chemotherapy triggers HIF-1-dependent glutathione synthesis and copper chelation that induces the breast cancer stem cell phenotype. Proc Natl Acad Sci U S A 2015;112(33):E4600–9.

[107] Xiang L, et al. HIF-1alpha and TAZ serve as reciprocal co-activators in human breast cancer cells. Oncotarget 2015;6(14):11768–78.

[108] Mathieu J, et al. HIF induces human embryonic stem cell markers in cancer cells. Cancer Res 2011;71(13):4640–52.

[109] Junankar S, et al. ID4 controls mammary stem cells and marks breast cancers with a stem cell-like phenotype. Nat Commun 2015;6:6548.

[110] Singh JK, et al. Targeting CXCR1/2 significantly reduces breast cancer stem cell activity and increases the efficacy of inhibiting HER2 via HER2-dependent and -independent mechanisms. Clin Cancer Res 2013;19(3):643–56.

[111] Ma L, et al. miR-9, a MYC/MYCN-activated microRNA, regulates E-cadherin and cancer metastasis. Nat Cell Biol 2010;12(3):247–56.

[112] DeCastro AJ, et al. DeltaNP63alpha transcriptionally activates chemokine receptor 4 (CXCR4) expression to regulate breast cancer stem cell activity and chemotaxis. Mol Cancer Ther 2015;14(1):225–35.

[113] Fazilaty H, et al. SLUG and SOX9 cooperatively regulate tumor initiating niche factors in breast cancer. Cancer Microenviron 2016;9(1):71–4.

[114] Cordenonsi M, et al. The Hippo transducer TAZ confers cancer stem cell-related traits on breast cancer cells. Cell 2011;147(4):759–72.

[115] Shipitsin M, et al. Molecular definition of breast tumor heterogeneity. Cancer Cell 2007;11(3):259–73.

[116] Yang J, Weinberg RA. Epithelial-mesenchymal transition: at the crossroads of development and tumor metastasis. Dev Cell 2008;14(6):818–29.

[117] Jang MH, et al. Expression of epithelial-mesenchymal transition-related markers in triple-negative breast cancer: ZEB1 as a potential biomarker for poor clinical outcome. Hum Pathol 2015;46(9):1267–74.

[118] Zhang P, et al. ATM-mediated stabilization of ZEB1 promotes DNA damage response and radioresistance through CHK1. Nat Cell Biol 2014;16(9):864–75.

[119] Buijs JT, et al. The BMP2/7 heterodimer inhibits the human breast cancer stem cell subpopulation and bone metastases formation. Oncogene 2012;31(17):2164–74.

[120] Wu K, et al. Cell fate determination factor Dachshund reprograms breast cancer stem cell function. J Biol Chem 2011;286(3):2132–42.

[121] Yu F, et al. let-7 regulates self renewal and tumorigenicity of breast cancer cells. Cell 2007;131(6):1109–23.

[122] Won HY, et al. Loss of Mel-18 enhances breast cancer stem cell activity and tumorigenicity through activating Notch signaling mediated by the Wnt/TCF pathway. FASEB J 2012;26(12):5002–13.

[123] Takahashi K, et al. High-fat diet increases vulnerability to atrial arrhythmia by conduction disturbance via miR-27b. J Mol Cell Cardiol 2016;90:38–46.

[124] Chang CJ, et al. p53 regulates epithelial-mesenchymal transition and stem cell properties through modulating miRNAs. Nat Cell Biol 2011;13(3):317–23.

CHAPTER 6

Lung Cancer Stem Cells: Drivers of a Genetically and Histologically Complex Disease

C. Fillmore Brainson[1,2,3], C.F. Kim[1,2,3]

[1]Stem Cell Program, Children's Hospital Boston, Boston, MA, United States; [2]Harvard Stem Cell Institute, Cambridge, MA, United States; [3]Harvard Medical School, Boston, MA, United States

NORMAL LUNG STEM CELLS

To understand the hierarchies being explored in lung cancer, one must first be familiar with the extensive research characterizing the epithelial cell types in the murine and human lung. Fluorescence-activated cell sorting (FACS) has revolutionized the stem cell field by allowing for enrichment of living cell populations from dissociated tissue. Typically, the lung tissue is dissociated with a cocktail of enzymes, including elastase, dispase, pronase, and collagenase type 1. Each research group that has published isolation of lung stem cells has its own preferred method to dissociate lungs. In general, sorting strategies begin with gating for viable cells based on a live/dead fluorescent dye such as 7-amino-actinomycin D (7AAD) or 4′,6-diamidino-2-phenylindole (DAPI), followed by exclusion of the $CD31^+$ endothelial lineage and the $CD45^+$ hematopoietic lineage. Many groups then gate on the epithelial marker epithelial cell adhesion molecule (EpCAM, also known as TACSTD1 and ESA) [1–4]. The cell surface markers used to sort different populations of epithelial cells include integrin α6 (CD49f), integrin β4 (CD104), CD166 (ALCAM), CD44, nerve growth factor receptor (NGFR, CD271), Sca1 (Ly6A), aldehyde dehydrogenase (ALDH1), and CD133 (prominin) in varied combinations [1–3,5–8]. Through FACS-isolation followed by 3D culture, analysis of lung injury repair and lineage tracing in vivo, many stem/progenitor cell populations within the lung have been shown to have region- and injury-specific functions. We summarize the major findings here (Fig. 6.1), moving from the more proximal airway, or trachea, to the distal lung spaces where gas exchange takes place. The take-home message is that the lung epithelium consists of a complex array of cell types that each can serve as stem and progenitor cells in certain contexts.

In the proximal respiratory system, including the trachea and main bronchi, the basal cells are the cells with the most repair and differential potential (Fig. 6.1(A)).

CONTENTS

Cancer Stem Cells. http://dx.doi.org/10.1016/B978-0-12-803892-5.00006-1

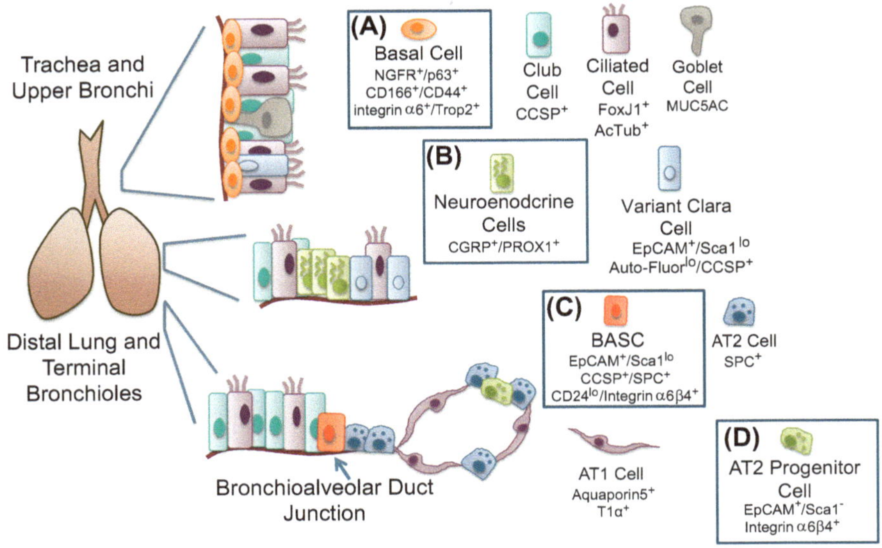

FIGURE 6.1 **Stem and progenitor cells in the lung.**

Diverse lung stem/progenitor cell populations are thought to have ability to drive lung oncogenesis in different contexts. (A) The proximal trachea and bronchi are lined by a layer of "basal cells" that can give rise to club, goblet, and ciliated cells. (B) In the bronchi of the central lung there is a columnar epithelium which has neuroendocrine bodies containing both neuroendocrine cells and variant club cells. (C) At the bronchioalveolar duct junction, the "BASC" can give rise to both club cells and AT2 cells.
(D) In the alveolar space, "AT2 progenitor cells" can give rise to both AT2 and AT1 cells. Markers used to fluorescence-activated cell sorting purify and distinguish these cell types are indicated (integrin α6 also known as CD49f, integrin β4 also known as CD104).

This pseudostratified epithelium also contains secretory club cell secretory protein (CCSP)-expressing club cells (previously known as Clara cells), mucus-producing goblet cells, and host-defending multiciliated cells. At the base of this epithelial layer reside the tracheal basal cells [5]. Basal cells can be isolated from murine tracheas based on their expression of the cell surface markers NGFR, CD44, and podoplanin (PDPN, also known as T1α) and are known to express cytokeratin 5 (KRT5) and the transcription factor p63. In addition, basal cells, and some but not all of the columnar cells throughout the airway, express the transcription factor SOX2. In mouse lineage-tracing studies using a *Krt5*-CreER transgene mouse model, basal cells give rise to ciliated cells in the proximal airways during both homeostasis and after sulfur dioxide inhalation injury [5,9]. A perhaps distinct pool of murine ALDH1$^+$ basal cells in the tracheal submucosal gland were shown to regenerate submucosal gland tubules, duct cells, and airway surface epithelium after severe hypoxic-ischemic injury [10,11]. Basal cells have also been isolated from human proximal airway and express NGFR, CD166, ALDH1, CD44, and integrin α6 (CD49f) [5,12]. Recently, a transient population of p63$^+$/KRT5$^+$ cells was shown to appear in the distal lung and proliferate after H1N1 viral infection in vivo [13]. These cells were shown to expand locally into "pods" with an organized sphere shape with a lumen and then suggested to gain expression of alveolar-specific proteins. Given that these cells express p63 and KRT5, it is possible that they share cellular identity with basal cells. However, other studies have suggested that lineage negative cells in the distal lung are the precursors for these p63^{+}/KRT5^{+} cells [14]. Because squamous lung cancers also express p63 and KRT5, it is often postulated that basal cells are the cell of origin of this tumor type [15,16].

Many studies have focused on understanding the microenvironmental and cell-intrinsic cues that control basal cell self-renewal and differentiation, and these pathways may prove important in lung squamous cell carcinoma (SCC). FACS-isolated basal cells (p63^{+}, NGFR^{+}, KRT5^{+}) can be cultured in two ways: on a layer of Matrigel in what is termed an air–liquid interface culture [17], or embedded in Matrigel to form clonal spherical structures positive for markers of both ciliated and club cells [5]. Both assays include use of growth medium containing epidermal growth factor (EGF), fibroblast growth factor (FGF), insulin, cholera toxin, retinoic acid, and bovine pituitary extract for colony growth. Growth of the cells without these factors is limited. Notch signaling has also been identified as a key regulator of basal cell self-renewal and differentiation. Basal cells express high levels of the NOTCH1 protein, as well as high levels of the Notch ligands JAGGED1, JAGGED2, and DELTA-LIKE1 and use Notch signaling to control both basal cell self-renewal and differentiation [18]. Furthermore, reactive oxygen species (ROS) levels have been shown to be low in a resting basal cell, but increase when basal cells are needed to proliferate [19]. The ROS regulator NRF2 is then upregulated to scavenge the ROS, and through a change in Notch signaling the basal cells self-renew. Together these data indicate that the niche of the basal cell (ie, growth factors, Notch signaling) and cell-intrinsic factors (ROS) are both critical for the decisions of self-renewal and differentiation.

The plasticity of the tracheal epithelial cells has been appreciated for decades [20] and importantly, studies have demonstrated that under certain circumstances, club cells can "dedifferentiate" and adopt a basal cell–like role [21,22]. Using an inducible diphtheria toxin under the *Krt5* promoter, basal cells were depleted from the adult murine trachea by inhalation of the chemical that induced the diphtheria toxin fragment A (DTA) expression. Using a lineage-tracing approach with the *Ccsp*-CreER, it was demonstrated that CCSP-expressing cells, when not in contact with any basal cells, were capable of "dedifferentiation" into cells indistinguishable from endogenous basal cells by functional and expression assays [21]. This finding underscores the importance of understanding what the barriers to reprogramming are, both in the normal setting and in the genetically mutated lung cancer state. If cells are able to transition between stem and less-stem states even when genetically normal, we must assume that they are capable of this transition when there are genetic alterations in pathways key to cell proliferation and fate. This idea is critical to the interpretation of therapies that would fit the CSC hypothesis—it will not be sufficient to target the CSCs and expect that the tumor will die, such as a weed without its roots. Instead, combining therapies that target both the CSC and non-CSC compartments are likely to have the greatest impact on survival.

In the central lung, there are at least two distinct stem and progenitor cell populations, the variant club cells (previously known as variant Clara cells) and the neuroendocrine cells (Fig. 6.1(B)). The variant club cells, also known as bronchiolar progenitor cells, can give rise to ciliated and club cell lineages after injuries such as naphthalene exposure [2]. These progenitors are known to reside specifically in the neuroendocrine bodies and bronchioalveolar duct junctions (BADJs) [23,24]. Also within the neuroendocrine bodies are calcitonin gene-related peptide (CGRP)-expressing progenitor cells that can differentiate into club and ciliated cells after injury [25]. Neuroendocrine cells also express the marker prospero homeobox 1 (PROX1) and like basal cells, rely on Notch signaling for proper differentiation [18]. These populations are the likely cells of origin of small cell lung cancer (SCLC) [26,27].

At the BADJ, where the columnar epithelium ends and the airspace begins, a cell population termed the bronchioalveolar stem cells (BASC) has been identified that coexpresses both CCSP and surfactant protein C (SPC) (Fig. 6.1(C)). BASCs proliferate after brochiolar and alveolar injury in vivo, and can give rise to both CCSP-expressing club cells and SPC-expressing alveolar epithelial type 2 cells (AT2) cell lineages in vitro when cultured with lung-derived endothelial cells [1]. The endothelial cells in these cocultures are stimulated by bone morphogenic protein 4 to provide factors necessary for the alveolar differentiation of BASCs, namely thrombospondin 1. BASCs can be FACS-isolated from murine lung by a $CD31^-/CD45^-/EpCAM^+/CD24^{low}/Sca1^{low}$ phenotype [1,28,29]. In a separate study, matrix metalloproteinase 14 produced by endothelial cells was shown to be required to unmask cryptic EGFR ligands that would in turn drive alveolarization [30]. Both studies found that endothelial–epithelial cell interactions were required for proper repair of distal lung in response to varied lung injuries and that the vascular niche is therefore critical for the proper function of distal lung stem and progenitor cells. A population similar to the BASC, in that it is bipotent for the club and AT2 lineages, can be also isolated by FACS for $CD31^-/CD45^-/EpCAM^+/CD24^{low}/Int\beta4^+$ and grown in culture with lung-derived mesenchymal stem cells, isolated by $CD31^-/CD45^-/EpCAM^-/Sca1^+$ [7]. These cultures unveiled the roles of FGF10 and TGFβ in supporting and preventing lung epithelial colony growth, respectively. Several groups have demonstrated that a CCSP-expressing cell lineage can contribute to AT2 cell repair in vivo by using the *Ccsp*-CreER, LSL:YFP lineage-tracing analysis coupled with bleomycin injury [31,32]. Like the basal cell studies, these examinations of the micro-environmental signaling that control BASC and other distal lung progenitor cells could prove useful when considering molecular pathways required in distal lung cancers such as adenocarcinomas (ADCs).

Lastly, alveolar progenitor cells have been identified that can give rise to AT2 and AT1 cells after injuries such as by bleomycin (Fig. 6.1(D)) [8]. The distal lung space is composed of AT2 cells, which highly express the transcription factor thyroid transcription factor 1 (TTF1, also known as NKX2.1) and surfactant proteins, and the AT1 cells, which express PDPN also known as T1α and perform gas exchange. Many groups have demonstrated that AT2 cells have progenitor cell capacity and can repair alveolar damage; however, it has not been demonstrated that AT2 cells can contribute to club cell regeneration and therefore the AT2 lineage may contain both fully differentiated cells, as well as cells that can act as more restricted, facultative, progenitors [4,8,33]. One population of these AT2 cells that were isolated as $\alpha6\beta4^+$ cells could form organoids, that when transplanted under the renal capsule, could form highly vascularized alveolar structures [8]. Further, when $CD31^-/CD45^-/EpCAM^+/Sca1^-$ cells are sorted and grown in 3D culture with endothelial cells, they are capable of clonogenic growth of SPC-expressing organoids [1]. Given the intrinsic proliferation capacities of both BASCs and AT2 progenitor cells, both populations could be suitable cell of origin pools for lung ADCs.

THE CANCER STEM CELL MODEL IN LUNG CANCER

It is well known that not all cancer cells within a tumor of a given tissue are equal in their ability to initiate or propagate disease, and lung cancer is no exception. Since the development and characterization of many key lung cancer cell lines at the National Cancer Institute, it has been appreciated that only certain lung cancer cells are "clonogenic," or able to propagate disease in laboratory cultures. In the clinical sense, disease propagation may refer to increase of primary tumor mass, recurrence following therapy, or establishment of tumors at distant sites, also known as metastasis. Generally stated, the cancer stem cell hypothesis predicts that a tumor contains a population of cells that is able to self-renew and give rise to the phenotypically distinct cells found within the heterogeneous tumor mass [34,35]. This hypothesis, coupled with evidence that CSCs are relatively chemo-resistant, has led to the idea that identification of CSCs in a tumor would enable development of treatments to specifically target the cells responsible for recurrence and metastasis [34,35]. If these CSC-specific therapies were combined with tumor debulking strategies, which may be represented by surgery, chemotherapy, targeted therapy, or radiation therapy, then, in theory, treatments could be much more complete and leave behind fewer cells capable of driving disease recurrence or metastasis.

The gold standard of orthotopic serial transplantation remains the system of choice to identify putative lung cancer stem cells, or functionally defined, tumor-propagating cells. Several methods to introduce tumor cells into the murine lung are used, including injection into the vasculature via the tail vein, instillation of the cells into the airspace by intratracheal catheter, and injection of the cells into the airspace by intrathoracic injection (Fig. 6.2(A)). Each method has advantages and disadvantages. Tail vein injection is very efficient at producing multifocal tumors, but only when the cell type injected is competent to transverse the endothelial barrier, an attribute associated with metastasis. Therefore, some researchers argue that tail vein injection is an assay of metastatic potential, even though the resulting tumor is growing in the orthotopic site of the lung. Intratracheal instillation is less efficient and may not work well with human cells, even when they are delivered to highly

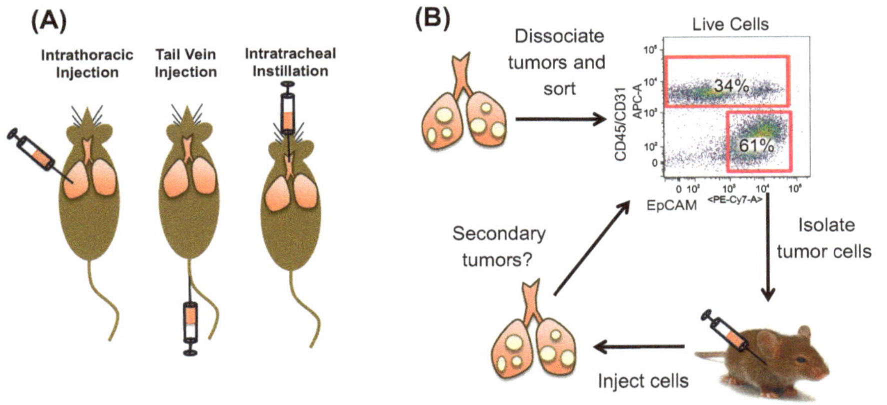

FIGURE 6.2 Types of orthotopic injection for lung cancer.
(A) Though many researchers have relied upon subcutaneous injection of lung cells to test tumor initiation capacity, orthotopic transplantation into the lung of animals will ultimately be a better test of lung cancer cells' growth capacity in their native microenvironment. Intrathoracic injection mimics subcutaneous injection in that the cells can be injected with the solidifying matrix Matrigel and the cells will initially be concentrated at the injection site. Typically up to one million cells embedded in a 1:1 vol:vol Matrigel and media mixture will be injected. Another way to achieve delivery of cells to the lung is to inject the cells into the tail vein. Because the blood in the tail vein goes to the lung for the reoxygenization, the cells travel with it and become trapped in the small capillaries of the lung. Typically 10,000–100,000 cells are resuspended in 200 μL saline for injection. The third method is to instill the cells in liquid suspension into the lungs through a catheter inserted into the trachea, termed intratracheal instillation. A small volume of only 50 μL is used to deliver cell numbers from several hundred to hundreds of thousands. (B) No matter which method is used to introduce the tumor cells into a new host, the cancer stem cell (CSC) hypothesis posits that tumors can be dissociated, and that more and less tumorigenic fractions can be isolated by flow cytometry or other enrichment methods. These single tumor cells, when injected into new mouse recipients, will be described as having CSC activity if they can propagate disease in a serial fashion.

immunocompromised mice. Intrathoracic injection is a more invasive procedure, but has advantages that the cells can be injected with the solidifying matrix Matrigel, and one tumor, rather than multifocal tumor formation, is likely to be formed, which mimics more closely the human disease. No matter which method is used to introduce the tumor cells into a new host, the central tenant of the CSC hypothesis postulates that a CSC-enriched cell population should be able to initiate disease that histologically mimics the parental disease, and that these secondary tumors could be dissociated to be injected a third, fourth, and fifth time (Fig. 6.2(B)). A population with true CSC potential would be fit to propagate the tumor several times in this serial passaging approach, while a non-CSC population may be able to initiate secondary disease, but will fail upon tertiary and quaternary transplants.

More recently, lineage-tracing technologies developed in murine systems hold promise to help understand the cellular dynamics within growing, metastatic, and treatment-resistant tumors. One well-designed way to control lineage tagging of cells is to use a knock-in Cre recombinase allele that is responsive to administration of Tamoxifen. These CreER-T2 alleles are knocked into endogenous loci of genes restricted to a specific cell lineage(s). The *Cre* gene is fused to a modified estrogen receptor that is specifically sensitive to the estrogen mimic Tamoxifen. When Tamoxifen is administered to the system, Cre protein, which is sequestered in the cytoplasm by HSP90, is allowed to translocate to the nucleus. Many fluorescent reporter alleles have been generated such that when Cre is nuclear, a stop cassette before the reporter will be excised and the cell will irreversibly be labeled. The fluorescent reporter alleles can also be used to visualize cells in which virally introduced Cre, such as Adeno-viral Cre, was once active. Lastly, integration of lentiviruses and retroviruses can allow for tracing of cellular contributions based on the unique sequence surrounding the integration site. These Cre alleles or viruses, coupled with oncogene and tumor-suppressor alleles that will be modified upon Cre expression (floxed alleles), will ultimately allow researchers to understand clonal tumor dynamics during disease progression, metastasis, and response to therapies.

Central to the cancer stem cell hypothesis is the idea that each tissue contains a cellular hierarchy, at the top of which are cells with more potential to divide and differentiate, and at the bottom of which are the differentiated and terminally differentiated (ie, nondividing) cells. As discussed in the previous section, the mammalian lung contains at least 40 different cell types with different functions, anatomical locations, and potentials to contribute to injury repair. There is no cell to date that can give rise to the whole epithelial tree of the lung, which may be due to inadequate transplantation systems to test this concept, or due to the inexistence of this type of multipotent lung stem cell. The vast complexity of the lung epithelium has made it difficult to establish an overarching lung cell hierarchy with which to query the existence of similar hierarchical differentiation structures within lung carcinomas.

Adding further complexity to the field, lung carcinomas can be subdivided into many distinct pathological subtypes with unique mutational signatures, gene expression, and microenvironments. Lung cancer is subdivided into two major groups: SCLCs and non-small cell lung cancers (NSCLCs) [36]. SCLC accounts for about ~20% of lung malignancies and is characterized by neuroendocrine cell morphology and gene expression. NSCLCs account for the remaining ~80% of lung malignances. Within the NSCLC group, at least two distinct histological subtypes exist: ADC and SCC [37]. ADC is the predominant subtype, accounting for approximately ~60% of NSCLCs. ADCs have the histology and biomarker expression consistent with distal lung, including expression of surfactant proteins [36,38,39]. SCC is the next most prevalent subtype, with ~33% of NSCLC patients receiving this diagnosis. SCC in the lung are characterized by squamous differentiation, which is most reminiscent of normal basal and secretory cell subtypes found in the more proximal lung and trachea [36,39–41]. Some tumors have areas of histology that represent several subtypes, and often if the subtypes are commixed the pathologist will term the lesion "adenosquamous" [36]. While smoking is associated with all types of lung tumors, it is most highly correlated with small cell and SCCs [42,43]. For all three of these tumor types, the mutational load and spectrum is much higher than that of other tumor types, eclipsed only by that of melanoma [44]. For example, lung ADC has an average of 225 nonsynonymous mutations per cancer genome, which is equivalent to about 5.1 mutations per megabase of DNA, SCC has an average of 8.1 mutations per megabase and small cell has 8.6 mutations per megabase [40,45,46]. In comparison, breast cancer only has an average of only one mutation per megabase of DNA. Given these complexities, the research to uncover the predominant cellular states within each genetic class of lung cancer is vast.

Despite advances in characterization of the heterogeneity of the disease, lung cancers are often treated with identical treatment regiments regardless of phenotype. For any patient, tumors deemed Stage I–IIIA are typically resected, and the patient will receive combination chemotherapy that differs according to subtype [36,47]. The therapies were often chosen initially because they were the drugs available, and some success has solidified them as the choice for each histologic subtype and stage. However, while chemotherapy regiments can help to consolidate disease and prolong life, in general it is unclear why some patients respond much better than others to the current chemotherapies. For tumors Stage IIIB–IV, which cannot be surgically resected, and in Stage IV have metastasized to distant organs, chemotherapy and radiation treatment is often ineffective. The molecular characterization of tumors and the use of genetic mouse models of lung tumors has led to identification and preclinical testing of novel drug targets in lung cancer. These advances are paving the way toward more personalized lung cancer treatment regiments, which have already improved survival rates for groups of patients [39,48,49]. Mounting genomic data demonstrate that distinct gene expression programs and driver mutations distinguish SCLC, ADC, and SCC. However, whether these three tumor types could arise from a common cell of origin remains unclear (Fig. 6.3). Also unclear are the predominant cellular hierarchies in each of these lung cancer subtypes, and how these cellular states pertain to treatment strategies. In the following sections, the research to understand cellular states with each of these tumor subtypes, with the goal to design rational therapeutic strategies that target specific vulnerabilities of each cancer type, will be summarized.

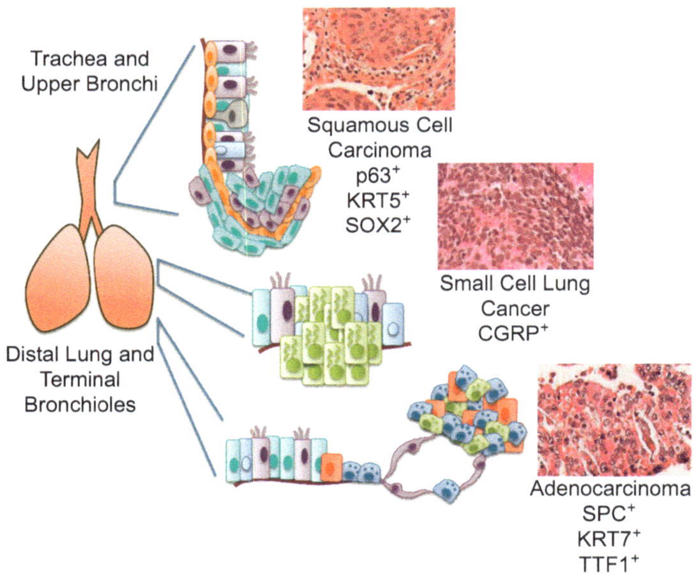

FIGURE 6.3 Three distinct subtypes of lung cancer.
In the proximal lung, the most common malignancies are termed squamous cell carcinoma (SCC), which display squamous differentiation patterns and express the markers p63, SOX2, and keratin 5 (KRT5). Squamous tumors are modeled in mice by *KrasG12D* expression and *Lkb1* knockout (20% of lesions are squamous), or knocking out both *Lkb1* and *Pten* (100% of lesions are squamous). Squamous lung tumors are found in the distal lung as well (30–50% of SCC); however, more predominant in the distal lung is the subtype termed adenocarcinoma (ADC). These tumors often retain characteristic of distal airways such as SPC, KRT7, and TTF1. Lung ADC is modeled in mice by *KrasG12D* expression and p53 knockout, or *EGFR-T790M/L858R* expression, among other models and are thought to arise from distal airway cells. Lastly, small cell lung cancers (SCLC) are classified as tumors with neuroendocrine cell features, such as expression of the neuroendocrine cell marker calcitonin gene-related peptide. SCLC is modeled in mice by deletion of both *p53* and *Rb*.

Lung Adenocarcinoma

Lung ADC is the most prevalent subtype of lung cancer, and generally speaking, resembles normal distal lung epithelial cells. Many lung ADCs express genes associated with AT2 cells, such as the transcription factor thyroid transcription factor 1 TTF1 (NKX2.1) and surfactants, such as SPC [50]. Genetically, lung ADCs can be subdivided into many distinct classes. Inactivating mutations in p53 are the most common genetic alteration, with 30% having loss of the gene, and an additional 45% showing expression of a mutated form of p53 such as p53R175H [45]. Activating mutations in KRAS and EGFR are both common, but mutually exclusive [45]. Within early-stage tumors with KRAS mutation, there are several groups, including the KRAS with LKB1 (*STK11*) loss, KRAS with *CDKN2A* loss, and KRAS with *p53* loss [51]. Later in tumor development, when the tumor is higher grade and more genetically complex, the triple mutation of KRAS, *LKB1*, and *p53* is observed. Another gene often lost in ADC is *SMARCA4*, also known as BRG1, a chromatin remodeling factor and its inactivation is also mutually exclusive with EGFR activation mutations [52,53]. Other driver, or activating, mutations, include mutations in the TGFβ receptor ALK and in the MAPK signaling pathway component BRAF. TTF1 positive ADCs often have EGFR and BRAF activating mutations, while TTF1 negative tumors often have *KEAP1* loss [54].

A breakthrough in lung ADC treatment is the use of specific tyrosine kinase inhibitors (TKIs) to treat EGFR, ALK and BRAF mutant tumors [39,48,49]. In addition to being sensitive to EGFR TKIs, EGFR-driven lung tumors may be particularly amenable to blockade of immune-suppression pathways [55]. Tumors can evade immune surveillance by expressing molecules that maintain tolerance to normal peripheral tissues, including the interaction of the tumor-associated pd-ligand-1 (PD-L1) with the immune receptor programmed cell death-1 (PDCD1 or PD1). Recently, the use of antibodies targeting the PD1/PD-L1 checkpoint has resulted in some dramatic responses in early-stage clinical trials for a large panel of therapy-refractory cancer subtypes including advanced melanoma, NSCLC and renal-cell cancer, with a fraction of responding patients showing persistent long-term benefits [56,57]. The PD1/PD-L1 interaction inhibits $CD8^+$ cytotoxic T lymphocyte proliferation, survival, and effector function and can induce apoptosis of tumor-infiltrating T cells. Despite the promising clinical benefits, there is currently no defined subset of lung cancer patients that are particularly sensitive to PD1/PD-L1 blockade.

Studies of human lung cancers include large-scale sequencing studies, expression studies, and correlations of expression and markers with clinical parameters [40,41,45,58]. To functionally define CSCs, a large amount of homogenous tumor material is needed, and the limited amount of fresh patient tissue has prevented large-scale analysis of CSCs. Patient-derived xenografts (PDXs), where a fresh or cryopreserved piece of resected lung tumor is embedded onto the side of an immunocompromised mouse, help to expand the numbers of cells available from a single tumor for analysis. One study sorted for different putative CSC markers (EpCAM, CD44, CD133, and CD166) from several ADC and SCC PDX samples. The authors showed that cell surface marker CD166 and the gene Lin28B mark cells enriched 100-fold for CSC activity when the cells were transplanted subcutaneously into very immunocompromised mice (NOD/SCID/IL2Rγ-null) with Matrigel [59]. In contrast, the $CD166^-$ population was devoid of tumor propagation ability, which precluded the ability of the authors to assess serial passaging of this population. The $CD166^+$ cells also had increased levels of glycine decarboxylase leading to changes in pyrimidine biosynthesis upon which the cells relied to survive and proliferate. This study was one of the first to support the theory that there is an intrinsic reliance of CSCs on altered metabolic pathways.

Adherent cell lines have been established from over 200 primary lung cancers [60], although whether cellular hierarchies are maintained in cancer cell lines is an ongoing debate. One study demonstrated that aldehyde dehydrogenase (ALDH) activity can enrich for lung cancer cells with increased tumorigenic potential when cells purified from two ADC cell lines H125 and H358 were injected subcutaneously into Nude mice [61]. These $ALDH1^+$ cells appeared to copurify with $CD133^+$ fractions. In a separate study, $ALDH1^+$ cells from ADC cell lines H1299 and H358 were also shown to have increased tumorigenic capacity when injected subcutaneously into NOD/SCID mice. These cells again copurified with $CD133^+$ and also copurified with Notch signaling-positive cells [62]. Knockdown of *Notch3* reduced clonogenic growth of cell lines in vivo. The first study on CD133 as a human lung cancer CSC marker showed that 7 of 19 primary patient tumors (2 were SCLC) could be cultured as nonadherent spheres, which were significantly enriched for the marker CD133, and that these sphere cultures were more tumorigenic in vivo when injected in SCID mice subcutaneously than cultures of the same cells switched to adherent plates and serum-containing medium [63]. Later studies supported these findings, and one found that $CD133^+/EpCAM^+$ cells isolated from PDX samples could serially transplant disease [64–66]. However, other studies have challenged the usefulness of CD133 as a marker for every tumor and exemplify the importance of assessing CSC activity in each new tumor type [59,67]. It has also become clear through deep sequencing efforts, that the panel of cell lines available does not cover the totality of the mutational spectrum of lung cancer [60]. Modified culture methods, including 3D Matrigel techniques, may allow for more patient tissue to be successfully grown ex vivo and culture techniques described earlier for lung stem cell propagation are currently being translated to use with primary tumor specimens.

Though lung tumors can be induced in rodent models through chemical administration, the ability to produce transgenic mice has transformed the cancer research field by allowing for precise control of the genetic drivers of model tumors. The first genetically engineered mouse model (GEMM) of lung cancer, in which oncogenic *Kras-G12D* can be specifically induced in the murine lung, was described in 2001 [68,69]. This model allows for activation of oncogenic KRAS through Cre recombinase–mediated excision of the Lox-Stop-Lox cassette preceding the *Kras-G12D* oncogene. Lung specificity is obtained through administration of Cre-expressing viral vector by inhalation, targeting lung cells along the proximal to distal tract. The *KrasG12D* mouse has been bred to many conditional knockout allele backgrounds (also known as floxed), including the tumor-suppressor *p53*, yielding *KrasG12D/p53*-null tumors [70]. These *Kras-G12D/p53*-null tumors are more advanced in stage than tumors from the *Kras-G12D* mice. In similar conditional systems, lung-specific promoters such as CCSP can be used to drive expression of transgenes in a doxycycline-inducible manner. This system has been particularly useful for expression of specific point mutants of EGFR that were identified in patient samples [71,72].

Experimental data from GEMMs suggest that stem and progenitor cells are primary targets for transformation in the lung and distinct lung cancer subtypes arise from different lung progenitor populations [73]. In models of lung ADC, the SPC- and CCSP-expressing BASCs are the first cells to proliferate in response to oncogenic *Kras* activation induced by Adeno-Cre administration [28]. In a recent lineage-tracing study, two different tamoxifen-responsive Cre transgenic strains, *Ccsp*:CreER and *Spc*:CreER, were bred to *Kras-G12D, p53*$^{flox/+}$ mice. Upon injection of tamoxifen into cohorts of mice, Cre was activated in only CCSP-expressing (bronchiolar cells and BASCs) or SPC-expressing (AT2 cells and BASCs) cells. In these studies, SPC^+ cells were most fit to give rise to tumors in the alveolar space, while $CCSP^+/SPC^-$ or $CCSP^+/SPC^+$ bronchiolar cells appear limited to driving papillary hyperplasia [74]. Tumors that did arise from *Ccsp*:CreER appeared to be in the alveolar space, again suggesting that a population of $CCSP^+/SPC^+$ cells distinct from those at the BADJ may exist in the alveolar region. The conclusion that mature AT2 cells were the most predominant cells of origin for Kras-driven tumors is supported by an inducible *Sca1*:Cre targeting some club cells and BASCs driving bronchiolar hyperplasia, while a mature AT2 lineage-restricted *LysM*:Cre gave rise to frank ADC when combined with the same *Kras-G12D* allele [33,75]. However, studies in the *Kras/p53* mice, it was observed that both *Ccsp*:CreER and *Spc*:CreER could drive frank ADC formation [76], and that these tumors often arose from the BADJ.

Several studies suggest that the genetic perturbations and microenvironment can both dramatically alter which cells serve as cells of origin in lung cancer models. For example, when *Sox2* was overexpressed during *Ccsp*:CreER *Kras* induced tumorigenesis, it was demonstrated that SOX2 inhibited NOTCH1 and NOTCH2 and inhibited alveolar ADC formation while promoting papillary carcinomas at the BADJ and squamous hyperplasia in the alveolar region [77]. Furthermore, when naphthalene, which kills club cells and activates repair in variant club cells and BASCs, was administered to *Kras* mice prior to Adeno-Cre instillation, the resulting tumor numbers and sizes were dramatically bigger than in naphthalene-naïve cohorts [28]. Injury or inflammation probably more closely mimics the scenario of tumor initiation in humans, where environmental influences and ongoing injury occur in contrast to the relatively sterile mouse colony. These questions are unexplored in other models of ADC that use distinct oncogenes. Thus, it remains possible that club cells, AT2 cells and BASCs are each possible initiators of lung ADC in different contexts. Future development of more precise lineage-specific Cre drivers combined with approaches to study tumorigenesis in the context of injury and inflammation, situations more likely mimicking carcinogenesis in humans, will be needed to better examine the comparable ADC-initiating activity of these populations.

One outstanding question in the field is whether CSCs will retain cell of origin phenotypes as the cancer evolves. In addition to shedding light on the cell of origin of different genetic subtypes of lung cancer, GEMMs have allowed for systematic studies demonstrating that tumor genetics must be considered when searching for CSC phenotypes. In the *Kras/p53* model of lung ADC, negative selection for CD31 and CD45 followed sorting for the BASC marker Sca1 enriched for tumor-propagating potential [78]. However, in lung tumor models that rely upon only *Kras* activation, both $Sca1^{+}$ and $Sca1^{-}$ cells could serially transplant disease via intratracheal instillation in Nude (*Foxn1*$^{nu/nu}$) immunocompromised mice. These results are significant as they mirror cell of origin studies; the studies described earlier suggest that *Kras* tumors may arise from BASCs ($Sca1^{+}$) or AT2 ($Sca1^{-}$) cell populations, while *Kras/p53* tumors may preferentially be derived from BASCs. Using the same *Kras/p53* mouse model, another group identified the $NOTCH3^{+}/CD24^{+}/INT\beta4^{+}$ cells as the CSCs within these primary murine tumors [79]. This study used an LSL:YFP transgene to mark cells that had been infected by Cre, sorting for $YFP^{+}/CD31^{-}/CD45^{-}$ cells, which were predominately $EpCAM^{+}$, and subsequently gated for $CD24^{+}$, $INT\beta4^{+}$, and $NOTCH^{+}$ cells in succession. Like the $Sca1^{+}$ cells, this cell population could serially transplant disease, though if this population overlaps with the $Sca1^{+}$ cells remains unknown. In EGFR-T790M; L858R driven tumors, CSC activity appeared to be enriched in the $Sca1^{-}$ cell fraction when compared to the $Sca1^{+}$ fraction. This research exemplifies the importance of oncogenotype in determining the phenotype of CSC populations. One hypothesis derived from this data is that tumor genotype directly influences the cell of origin, and whether CSCs retain cell of origin marker expression as the cancer evolves. It also exemplifies the importance of functionally validating the tumor initiation capacity of sorted populations in each new tumor type under study. Therefore, it is essential going forward that studies of human disease characterize the major genetic alterations of the tumors when assessing for CSC presence and phenotype.

Another useful phenotype of the *Kras/p53* mouse model is that the lung tumors will spontaneously metastasize to distant sites such as liver, adrenal gland, and lymph nodes [80] and this allows the examination of metastatic CSCs. Tracing of the tumors that give rise to metastases in this model was accomplished by using lentiviral-Cre, which integrates into the genome of the cells, to initiate the tumors. Cre integration sites in the metastatic lesions were then matched back to the tumors within the lung [80]. Importantly, the tumors capable of metastatic dissemination were significantly different in expression profiles from the nonmetastatic tumors. While nonmetastatic tumors had high levels of TTF1 and lower levels of HMGA2, metastatic tumors had an opposite phenotype of $TTF1^-/HMGA2^+$, which was maintained in the metastatic lesions. The follow-up study to using Sca1 as a CSC marker in the *Kras/p53* model also observed a subset of cells that were more efficient at distant metastatic colonization, and these cells could be enriched by the $CD31^-/CD45^-/CD24^+/Sca1^+$ phenotype [81]. The $CD24^+/Sca1^+$ tumor cell transcriptional profile was also enriched for targets of the YAP and TAZ transcription factors, and these profiles appeared to also predict metastatic potential in patient ADC samples. Together these studies indicate that metastatic potential may be a characteristic of a subset of lung CSCs that have cell-intrinsic adaptations such as TTF1 downregulation and YAP/TAZ signaling that allow them to survive the metastatic process. Learning which pathways are essential to these cells could help us design therapies to specifically prevent metastatic spread of lung cancer, which would significantly improve the outcome of the disease.

Lung Squamous Cell Carcinoma

Lung SCC is a tumor characterized by squamous epithelial differentiation. This is curious, as there is no normal squamous tissue in the lung. The tissue that most closely resembles a squamous pattern is the pseudostratified epithelium of the trachea and upper bronchi. In response to specific types of epithelial injury, the pseudostratified epithelium can undergo a transient squamous stage [9,37]. Predominant mutations in lung SCC include PI3KCA, NRF2, FGFR2, and SOX2 activation, and PTEN, NOTCH1, and TGFβR2 loss [40,50]. Often SCCs are found in the more proximal airway, but peripheral SCCs also exist [82,83].

Although many researchers were actively trying to model lung SCC in mice, the GEMMs made between the years of 2001 and 2007 all produced predominantly the ADC subtype of the disease. One potential reason for the skew toward ADC was the frequent use of oncogenic *Kras*, or chemicals that induce Ras mutations, to drive tumors—oncogenic *Kras* is associated almost exclusively with ADC rather than SCC in patient tissues [40,45]. Basal cells of the trachea are proposed as the cells of origin for lung SCC, not only because lung SCCs arise more frequently in the proximal lung [37], but also because they often express KRT5, SOX2, and p63, markers of the normal basal cell population [9,15,84]. Targeted deletion of squamous tumor suppressors *Pten* and *TgfbrII* in more proximal cells by targeting keratin (*Krt5*:CrePR and *Krt14*:CrePR) have produced only partial SCC differentiation, and only with *Kras-G12D* alone or the *Pten*$^{+/-}$, *TgfrbrII*$^{-/-}$ combination of alleles [85]. Expression of the squamous oncogene *Sox2* directed by an inducible *Ccsp*:CreER produces lung tumors that express the marker p63, but histologically these tumors are ADCs [84]. This intriguing finding suggests that the *Ccsp*-Cre allele targets a subset of cells, possibly variant club cells and BASCs, that are unable to produce a fully squamous phenotype despite expression of a SCC transcription factor. A transgenic mouse that expresses a kinase-dead IKKα transgene in all tissues was shown to develop spontaneous lung SCCs [86], and while this study showed the importance of NFκB signaling in squamous tumor development, it did not pinpoint whether NFκB signaling was important in the tumor cells themselves, the microenvironmental cells, or both. The use of a transgenic mouse also precludes the types of lineage-specific studies described earlier for ADC.

Finally, in 2007, a genetic mouse model was produced that had partial squamous differentiation—the *Kras-G12D/Lkb1*-null mouse model [87]. Not only could the *Kras/Lkb1* mouse model produce tumors that were fully squamous, *Lkb1* loss also conferred metastatic ability on the normally nonmetastatic *Kras* mutant cells. In-depth study of the development of the *Kras/Lkb1* tumors suggests that first, predominantely ADCs form, and these same tumors can later transdifferentiate into SCC tumors [88]. These findings mirror observations in the clinic that suggest that human ADCs can transdifferentiate to SCC during the course of therapy. Importantly, these studies also implicate a common cell of origin for both the ADC and SCCs in this model. YAP signaling was shown to inhibit the change from ADC to SCC, in part through direct transcriptional repression of *p63* by YAP [88]. Just as ROS has a role in regulating the normal basal cell, ROS is also implicated in the transition of *Kras/Lkb1* tumors from the ADC to SCC state [89]. *Lkb1* loss is known to cause oxidative stress, both in human and mouse tumors, but appears to be more pronounced when the cells queried are in an ADC differentiation state. To overcome this stress, the cells are driven to transdifferentiate from ADC to SCC by upregulating, among other genes, *Nrf2* and *p63*. Although the exact mechanism of this cellular reprogramming is still under investigation, it does suggest that therapies targeting vulnerabilities of tumors such as high ROS will be overcome by tumor cell transdifferentiation. How the CSC population changes during such a transdifferentiation will be an important aspect of this phenomenon to study.

Despite these interesting advances, and the unmistakable utility of the *Kras/Lkb1* GEMM to model the transition between ADC and SCC, the model was still not fully squamous. Again it was postulated that oncogenic Kras was predisposing to a more ADC phenotype. However, PTEN loss and PI3K pathway activation were both significantly associated with human SCCs [40], and therefore it was reasoned that combining *Lkb1* and *Pten* floxed alleles could possibly produce a more complete SCC phenotype. Indeed, upon administration of intranasal Adeno-Cre virus to *Lkb1*$^{fl/fl}$, *Pten*$^{fl/fl}$ mice, fully penetrant SCCs arose at a median onset of 50 weeks [16]. ADC histology was never observed in this model. Very soon after, a second publication demonstrated that combination of *Lkb1*$^{fl/fl}$ alleles with a lentivirus expressing SOX2 and Cre could also drive purely squamous lung disease after intranasal viral administration [90]. These studies demonstrate that with only genetic alterations, and no additional environmental stressors added, a purely squamous lung cancer can be produced. While *Lkb1* is not commonly mutated in human SCC, it appears that its loss is required for pure SCC formation in the murine lung. Regardless of this difference, both the *Lkb1/Pten* and *Lkb1/Sox2* models are important tool for investigating the molecular mechanisms of SCC carcinogenesis. The *Lkb1/Pten* tumors were compared by gene expression profiling to human SCC tumors and had striking similarities despite the differences in genetic mutations. These models also provide platforms to study new therapeutic opportunities for lung SCC through the targeting of a CSC-specific interaction with the immune cell microenvironment.

Based on the markers for basal cells of the trachea, NGFR was chosen as a prospective marker for CSCs within the *Lkb1/Pten* mouse model [16]. When tumors were dissociated, $CD31^-/CD45^-/EpCAM^+$ cells were gated and subsequently sorted into four fractions based on Sca1 and NGFR expression (Fig. 6.4(A)), 3D Matrigel growth was enriched in the $Sca1^+/NGFR^+$ population (Fig. 6.4(B)). Next, the three major cell populations were sorted and transplanted into immunocompromised mouse lungs. $Sca1^+/NGFR^+$ fractions were enriched for CSCs that could serially transplant the disease in orthotopic assays (Fig. 6.4(C)). These CSCs in the LP model, and $NGFR^+$ cells in human SCCs, both highly expressed PD-L1, suggesting a mechanism of immune escape for CSCs. The next important step will be to use lineage-restricted Cre alleles, such as the *Krt5*:CreER, to dissect which lung cells null for *Lkb1* and *Pten* are able to drive squamous disease. Thus far, limited analysis of primary human lung SCCs that rely upon sphere culture have shown some heterogeneity within clonal tumor populations [91]. In the future, examination of the normal basal cell markers NGFR, CD44, and ALDH1 in human SCC with special consideration of genotype and cellular heterogeneity of the tumors will be necessary to translate the promise of exploiting CSC biology for therapeutic benefit in lung SCC.

Small Cell Lung Cancer

SCLC is the most aggressive of the three subtypes, with less than a 5% 5-year survival rate, and is composed of cells with neuroendocrine features. Recent sequencing efforts have demonstrated that SCLC have loss or epigenetic

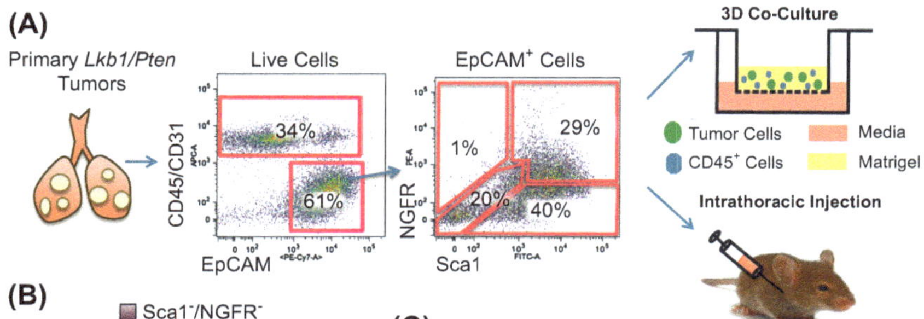

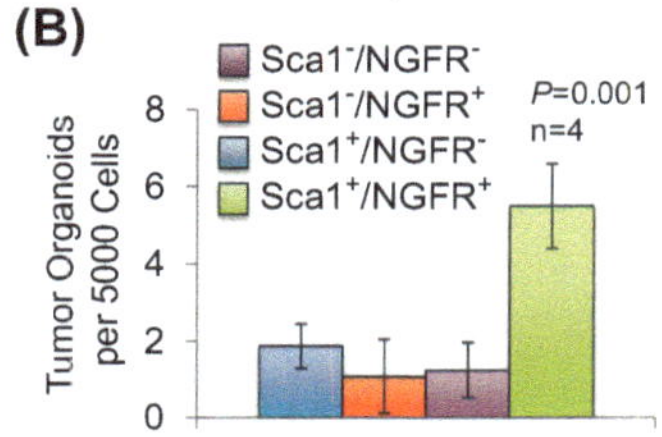

(C)

Phenotype	Secondary Tumors	Tertiary Tumors
SCA1⁻NGFR⁻	0/4	0/3
SCA1⁺NGFR⁻	0/5	0/5
SCA1⁺NGFR⁺	*4/4	*4/4

FIGURE 6.4 Testing the cancer stem cell (CSC) hypothesis in a lung squamous cell carcinoma mouse model.

(A) The sorting scheme used for *Lkb1/Pten* tumors shows first depletion of endothelial and hematopoietic cells by negative selection of CD31 and CD45, followed by enrichment for tumor cells by epithelial cell adhesion molecule (EpCAM). Nerve growth factor receptor (NGFR) and Sca1 fractions were then sorted. (B) Results from 3D organoid assay clearly show Sca1+/NGFR+ cells harbor the most growth capacity. (C) Results from serial intrathoracic injections show that Sca1+/NGFR+ cells are bona fide CSCs in this model (Xu and Fillmore et al., Cancer Cell 2014).

suppression of the tumor-suppressor *retinoblastoma* (Rb) and almost all have loss of the tumor-suppressor p53 [46,92]. Just as in lung ADC and SCC, Notch signaling has also emerged as a very important pathway both in normal neuroendocrine cells and SCLC. In addition, MYC pathway activation appears critical for SCLC pathogenesis. Almost all SCLCs are treated with combination chemotherapy of cisplatinum and etoposide, and targeted therapies are desperately needed for this disease subtype.

Several GEMMs recapitulate the characteristics of human SCLC, including lung-specific inactivation of both *p53* and *Rb* (*p53*-null/*Rb*-null), which exhibits highly penetrant SCLC after Adeno-Cre administration [93]. These tumors show high levels of genomic instability, which often leads to inactivation of *Pten*. Additional of *Pten* conditional knockout alleles accelerates tumorigenesis [94]. Because SCLC has neuroendocrine features, it was long proposed that these tumors arise from neuroendocrine bodies within the lung, which are thought to contain their own stem cell population. In support of this, the majority of early lesions in the *p53*-null/*Rb*-null model infected with Adeno-Cre comprised proliferating neuroendocrine cells [26]. Likewise, when a variety of cell type–specific Cre viruses were used, the virus driving Cre expression in CGRP+ cells was most efficient at driving SCLC formation [26,27]. In contrast, *Ccsp*-restricted Cre virus or *Ccsp*:CreER both led to very minimal disease, while *Spc*-restricted Cre viruses could drive some NSCLCs, but only after more than 400 days post-Cre did any tumors show neuroendocrine features. Together these data suggest that CGRP⁺ neuroendocrine progenitor cells originate tumors.

Despite this clear evidence that SCLC is preferentially derived from the neuroendocrine cells, a fascinating clinical observation further suggests that neuroendocrine tumors and EGFR-driven ADCs may share a common cell of origin: 14% of patients with EGFR-driven ADCs show development of SCLC after long-term treatment with EGFR TKIs gefitinib or erlotinib [95]. These results suggest that there is a previously unappreciated lineage relationship between ADC and SCLC, which could be supported by the fact that club cells can differentiate from the neuroendocrine lineage [25]. Using the *Cgrp*-restricted Cre virus to drive *EGFR* mutations would be an interesting experiment—it is possible that the *EGFR* oncogene favors differentiation of the NE progenitor cell down a CCSP⁺ ADC lineage, while loss of *Rb* drives a neuroendocrine fate. Given that existing *EGFR* alleles are in a TetO system, this experiment will have to await the development of different mouse strains.

Although no CSC population has been identified by cell sorting in SCLC, there are several treatments that may influence the stemlike transcriptional programs in these tumors. One study showed that inhibition of CDK7 with a novel small molecule inhibitor lead to reduction of tumor volume in the *p53/Rb* GEMM [96]. CDK7 could also kill human SCLC cell line in vitro. The authors demonstrated that super-enhancers, which are defined as tandem active enhancer elements converging upstream of important regulatory genes, where most affected by CDK7 inhibition. Gene transcription of these super-enhancer containing genes was significantly reduced by CDK7 inhibition, consistent with the role of CDK7 in promoting transcriptional initiation and elongation. Of these super-enhancer regulated genes, downregulation of *Myc* family member and the neuroendocrine-related genes *Ascl1* and *NeuroD1* were most pronounced and suggest that the cells rely upon the neuroendocrine differentiation state to survive. Another study examined the role of the histone H3 lysine 4 demethylase called LSD1 in treatment of SCLC [97]. Again LSD1 inhibition was demonstrated to decrease in vivo tumor growth, this time with xenografts of human SCLC cell lines. Inhibition of LSD1 resulted in accumulation of the histone H3 lysine 4 methylation mark, which indicates transcriptional activation in a subset of transcriptional start sites. Although there was no apparent overlap of which genes were upregulated in response to LSD1 between patient samples, one unifying theme of cell lines that responded strongly to LSD1 inhibition was a global DNA hypomethylation signature. This signature could predict sensitivity to the LSD1 inhibitor and could be used in the future after cisplatinum and etoposide treatment to prevent residual tumor cells from proliferation.

TARGETING LUNG CANCER STEM CELLS THERAPEUTICALLY

There are many potential therapies that could be explored as a result of the data described earlier, all with the hope that targeting CSCs could allow for more complete treatment response.

For metastatic *Kras/p53* ADCs, targeting the YAP/TAZ transcription factors could prove useful [81]. However, the technology to target transcription factors is still lacking. Furthermore, it is not clear whether YAP, or the closely homologous but not identical transcription factor TAZ are the most important in lung ADC metastasis. These mechanistic details, along with the likely gene targets of YAP and TAZ in metastatic CSCs may lead to identification of a targetable downstream regulator of metastasis. However, it is important to remember that YAP and TAZ were shown to be required to maintain an ADC phenotype in the *Kras/Lkb1* tumors [88]. Therefore, it is possible that inhibition of YAP and TAZ may lead to derepression of p63 and drive the transdifferentiation of ADC to SCC. This may serve as an "escape" mechanism for YAP/TAZ targeted tumor cells and instead of treating disease, may simply lead to tumor evolution from ADC to SCC.

Notch signaling, which appears critical to both normal lung cell self-renewal and differentiation and lung tumorigenesis, is certainly a potential targetable pathway for lung cancer. However, given that there are four distinct Notch receptors, which each have unique effects on downstream pathways, strategies to target just one Notch receptor need to be developed. For example in the *Kras/p53* lung ADC GEMM, inhibition of *Notch3* by small hairpin was demonstrated to decrease CSC-derived sphere growth, but hairpins targeting *Notch1*, *Notch2*, or *Notch4* actually led to an increase in sphere growth, suggesting that only NOTCH3 should be targeted in this genetic subtype of lung cancer [79]. Likewise, while global Notch inhibition prevented *Kras*-induced lung ADCs, it did potentiate squamous metaplasia [77]. Given that *NOTCH1* and *NOTCH2* are often lost in lung SCC, and *NOTCH1*, *NOTCH2*, and *NOTCH3* are mutated in SCLC, it will be important to find ways to inhibit NOTCH3 without targeting NOTCH1 and NOTCH2, and it will be important to be wary of driving a transdifferentiation to SCLC with NOTCH3 inhibition. Likewise, a NOTCH1 or NOTCH2 agonist approach could allow for forced differentiation of SCC and SCLC tumor cells. In the case of SCC, NOTCH1 activation could lead to activation of differentiation pathways and ameliorate disease. In the case of SCLC, genetic expression of NOTCH1 intracellular domain was sufficient to significantly reduce tumor burden [46]. Together these data suggest that finding chemical methods to specifically activate or repress Notch signaling could have great therapeutic potential for many types of lung cancer.

Immunotherapy is another exciting potential treatment, especially for CSCs within lung SCC tumors. Anti-PD1 immunotherapy was just demonstrated to have a significant treatment advantage in late stage squamous tumors when compared to the standard of care docetaxel regiment [98]. One important point of this study is that PD-L1 expression did not appear predictive of efficacy of anti-PD1 immunotherapy. This result may be due to the heterogeneity of the tumor—a single biopsy may not capture the PD-L1 expressing CSC compartment, if it is indeed a small fraction of the total tumor cells. Other environmental influences, such as immune cell infiltration, may also be important for efficacy of this treatment. Combining immunotherapy with other targeted therapies may help to increase the percentage of patients that benefit from this exciting new treatment modality.

Lastly, epigenetic pathways may allow for targeting the CSC pools by targeting the chromatin machinery they require to maintain their stemlike cell states. Proof of principle studies include the inhibition of CDK7 and LSD1 in SCLC, which effects global transcriptional programs and also leads to reduction of tumor proliferation. Combining epigenetic therapies with common chemotherapy strategies could allow for more complete treatment response, as has been observed with inhibition of the epigenetic enzyme EZH2 combined with the common chemotherapy etoposide in lung ADC and SCC [53]. While further studies of which cells within tumor cell hierarchies are specifically targeted by each of these therapies, data suggest that altering the global transcriptional identity of tumor cells is a useful way to prevent propagation of tumors.

Understanding the cellular origins and molecular determinants of all subtypes of lung cancer will allow for more specific and effective treatment options. While a universal CSC phenotype may not exist for human lung cancers, there certainly are cells within tumors that behave as CSCs at various stages of disease, including during tumor recurrence and metastatic spread of disease. Understanding what transcriptional programs and microenvironmental signals these cells rely upon when they are in the CSC state has the possibility of transforming our treatment strategies for this deadly disease, and for allowing a better understanding of how cancer begins and evolves.

Glossary

Adenocarcinoma This is the most common type of lung cancer, typified by expression of normal distal lung markers TTF1, SPC, and KRT7.

Alveolar type 2 cells In the distal lung, AT2 cells serve as facultative progenitor cells to repair lung injuries and maintain alveoli during homeostasis. They express the transcription factor TTF1 and surfactants. They are another probable cell of origin pool for several types of lung adenocarcinomas.

Basal cells In the trachea and upper bronchi, these cells, which are defined in part by their basal anatomic location, can both self-renew and give rise the multiple lineages of the tracheal epithelium. They are a putative cell of origin of squamous lung tumors given that they express SOX2, p63, and KRT5.

Bronchioalveolar stem cells These cells, often found at the bronchioalveolar duct junction, express both CCSP—a marker of the airway club cells, and SPC—a marker of the alveolar type II cells. BASCs can self-renew and give rise to both airway and alveolar lineages. They are a putative cell of origin for lung adenocarcinomas.

Genetically engineered mouse models The species *Mus musculus* has been engineered to express certain oncogenes, such as KRASG12D, and to have inducible deletion of certain tumor suppressor, such as *p53*. The resulting GEMMs are used extensively to explore many aspects of lung tumorigenesis, including initiation, heterogeneity, and immune evasion.

Lung cancer Typically carcinoma, or of epithelial origin, lung cancer is defined as a malignant growth of cells of lung origin within the lung. Lung cancer can metastasize to distant organs and interfere with normal lung function. It remains the leading cause of cancer-related deaths both in the United States and worldwide.

Small cell lung cancer This is the third most prevalent type of lung cancer and arguably the deadliest. It is characterized by neuroendocrine features and expression of the markers CGRP and PROX1.

Squamous cell carcinoma This is the second most common type of lung cancer, typified by a squamous differentiation pattern and expression of the squamous markers SOX2, p63, and KRT5.

List of Acronyms and Abbreviations

7AAD 7-Amino-actinomycin D
ADC Adenocarcinoma
ALDH1 Aldehyde dehydrogenase
AT2 Alveolar epithelial type 2 cells
BADJ Bronchioalveolar duct junction
BASC Bronchioalveolar stem cells
CCSP Club cell secretory protein
CD104 Integrin β4
CD49f Integrin α6
CGRP Calcitonin gene-related peptide
CSC Cancer stem cell
DAPI 4′,6-Diamidino-2-phenylindole
EGR Epidermal growth factor
EpCAM, also known as TACSTD1 and ESA Epithelial cell adhesion molecule
FGF Fibroblast growth factor
GEMM Genetically engineered mouse model
KRT5 Cytokeratin 5
NGFR, CD271 Nerve growth factor receptor
NSCLC Non-small cell lung cancers
PDCD1 or PD1 Programmed cell death-1
PD-L1 PD-ligand-1
PDPN, also known as T1α Podoplanin
PROM1, CD133 Prominin
PROX1 Prospero homeobox 1
ROS Reactive oxygen species

Sca1, Ly6A Stem cell antigen 1
SCC Lung squamous cell carcinoma
SCLC Small cell lung cancer
SPC Surfactant protein C
TTF1, NKX2.1 Thyroid transcription factor 1

References

[1] Lee J-H, Bhang DH, Beede A, Huang TL, Stripp BR, Bloch KD, et al. Lung stem cell differentiation in mice directed by endothelial cells via a BMP4-NFATc1-thrombospondin-1 Axis. Cell 2014;156(3):440–55.

[2] Teisanu RM, Chen H, Matsumoto K, McQualter JL, Potts E, Foster WM, et al. Functional analysis of two distinct bronchiolar progenitors during lung injury and repair. Am J Respir Cell Mol Biol 2011;44(6):794–803.

[3] McQualter JL, Yuen K, Williams B, Bertoncello I. Evidence of an epithelial stem/progenitor cell hierarchy in the adult mouse lung. Proc Natl Acad Sci 2010;107(4):1414–9.

[4] Barkauskas CE, Cronce MJ, Rackley CR, Bowie EJ, Keene DR, Stripp BR, et al. Type 2 alveolar cells are stem cells in adult lung. J Clin Investig 2013;123(7):3025–36.

[5] Rock J, Onaitis M, Rawlins E, Lu Y, Clark C, Xue Y, et al. Basal cells as stem cells of the mouse trachea and human airway epithelium. Proc Natl Acad Sci USA 2009;106(31):12771–5.

[6] Teisanu RM, Lagasse E, Whitesides JF, Stripp BR. Prospective isolation of bronchiolar stem cells based upon immunophenotypic and autofluorescence characteristics. Stem Cells 2009;27(3):612–22.

[7] McQualter J, Bertoncello I. Clonal culture of adult mouse lung epithelial stem/progenitor cells. In: Rich IN, editor. Rich IN, editor. Stem cell protocols. Methods in molecular biology, ;1235. New York: Springer; 2015. p. 231–41.

[8] Chapman HA, Li X, Alexander JP, Brumwell A, Lorizio W, Tan K, et al. Integrin α6β4 identifies an adult distal lung epithelial population with regenerative potential in mice. J Clin Investig 2011;121(7):2855–62.

[9] Rock JR, Randell SH, Hogan BLM. Airway basal stem cells: a perspective on their roles in epithelial homeostasis and remodeling. Dis Models Mech 2010;3(9–10):545–56.

[10] Hegab A, Ha V, Gilbert J, Zhang K, Malkoski S, Chon A, et al. Novel stem/progenitor cell population from murine tracheal submucosal gland ducts with multipotent regenerative potential. Stem Cells 2011;29(8):1283–93.

[11] Hegab AE, Nickerson DW, Ha VL, Darmawan DO, Gomperts BN. Repair and regeneration of tracheal surface epithelium and submucosal glands in a mouse model of hypoxic-ischemic injury. Respirology 2012;17(7):1101–13.

[12] Hegab AE, Ha VL, Darmawan DO, Gilbert JL, Ooi AT, Attiga YS, et al. Isolation and in vitro characterization of basal and submucosal gland duct stem/progenitor cells from human proximal airways. Stem Cells Transl Med 2012;1(10):719–24.

[13] Kumar PA, Hu Y, Yamamoto Y, Hoe NB, Wei TS, Mu D, et al. Distal airway stem cells yield alveoli in vitro and during lung regeneration following H1N1 influenza infection. Cell 2011;147(3):525–38.

[14] Vaughan AE, Brumwell AN, Xi Y, Gotts JE, Brownfield DG, Treutlein B, et al. Lineage-negative progenitors mobilize to regenerate lung epithelium after major injury. Nature 2015;517(7536):621–5.

[15] Giangreco A, Lu L, Vickers C, Teixeira VH, Groot KR, Butler CR, et al. β-Catenin determines upper airway progenitor cell fate and preinvasive squamous lung cancer progression by modulating epithelial–mesenchymal transition. J Pathol 2012;226(4):575–87.

[16] Xu C, Fillmore Christine M, Koyama S, Wu H, Zhao Y, Chen Z, et al. Loss of Lkb1 and Pten leads to lung squamous cell carcinoma with elevated pd-L1 expression. Cancer Cell 2014;25(5):590–604.

[17] You Y, Richer EJ, Huang T, Brody SL. Growth and differentiation of mouse tracheal epithelial cells: selection of a proliferative population. Am J Physiol Lung Cell Mol Physiol 2002;283(6):L1315–21.

[18] Rock Jason R, Gao X, Xue Y, Randell Scott H, Kong Y-Y, Hogan Brigid LM. Notch-Dependent differentiation of adult airway basal stem cells. Cell Stem Cell 2011;8(6):639–48.

[19] Paul Manash K, Bisht B, Darmawan Daphne O, Chiou R, Ha Vi L, Wallace William D, et al. Dynamic changes in intracellular ROS levels regulate airway basal stem cell homeostasis through NRF2-dependent Notch signaling. Cell Stem Cell 2014;15(2):199–214.

[20] Randell SH, Comment CE, Ramaekers FCS, Nettesheim P. Properties of rat tracheal epithelial cells separated based on expression of cell surface α-galactosyl end groups. Am J Respir Cell Mol Biol 1991;4(6):544–54.

[21] Tata PR, Mou H, Pardo-Saganta A, Zhao R, Prabhu M, Law BM, et al. Dedifferentiation of committed epithelial cells into stem cells in vivo. Nature 2013;503(7475):218–23.

[22] Lui JY, Nettesheim P, Randell SH. Growth and differentiation of tracheal epithelial progenitor cells. Am J Physiol 1994:L296–307.

[23] Reynolds SD, Giangreco A, Power JHT, Stripp BR. Neuroepithelial bodies of pulmonary airways serve as a reservoir of progenitor cells capable of epithelial regeneration. Am J Pathol 2000;156(1):269–78.

[24] Giangreco A, Reynolds SD, Stripp BR. Terminal bronchioles harbor a unique airway stem cell population that localizes to the bronchoalveolar duct junction. Am J Pathol 2002;161(1):173–82.

[25] Song H, Yao E, Lin C, Gacayan R, Chen M-H, Chuang P-T. Functional characterization of pulmonary neuroendocrine cells in lung development, injury, and tumorigenesis. Proc Natl Acad Sci 2012;109(43):17531–6.

[26] Park K-S, Liang M-C, Raiser DM, Zamponi R, Roach RR, Curtis SJ, et al. Characterization of the cell of origin for small cell lung cancer. Cell Cycle 2011;10(16):2806–15.

[27] Sutherland KD, Proost N, Brouns I, Adriaensen D, Song J-Y, Berns A. Cell of origin of small cell lung cancer: inactivation of Trp53 and Rb1 in distinct cell types of adult mouse lung. Cancer Cell 2011;19(6):754–64.

[28] Kim CF, Jackson EL, Woolfenden AE, Lawrence S, Babar I, Vogel S, et al. Identification of bronchioalveolar stem cells in normal lung and lung cancer. Cell 2005;121(6):823–35.

[29] Zacharek SJ, Fillmore CM, Lau AN, Gludish DW, Chou A, Ho JWK, et al. Lung stem cell self-renewal relies on BMI1-dependent control of expression at imprinted loci. Cell Stem Cell 2010;9(3):272–81.

[30] Ding B-S, Nolan Daniel J, Guo P, Babazadeh Alexander O, Cao Z, Rosenwaks Z, et al. Endothelial-derived angiocrine signals induce and sustain regenerative lung alveolarization. Cell 2011;147(3):539–53.

[31] Tropea KA, Leder E, Aslam M, Lau AN, Raiser DM, Lee J-H, et al. Bronchioalveolar stem cells increase after mesenchymal stromal cell treatment in a mouse model of bronchopulmonary dysplasia. Am J Physiol Lung Cell Mol Physiol 2012;302(9):L829–37.

[32] Rock JR, Barkauskas CE, Cronce MJ, Xue Y, Harris JR, Liang J, et al. Multiple stromal populations contribute to pulmonary fibrosis without evidence for epithelial to mesenchymal transition. Proc Natl Acad Sci 2011;108(52):E1475–83.

[33] Desai TJ, Brownfield DG, Krasnow MA. Alveolar progenitor and stem cells in lung development, renewal and cancer. Nature 2014. [Advance online publication].

[34] Beck B, Blanpain C. Unravelling cancer stem cell potential. Nat Rev Cancer 2013;13(10):727–38.

[35] Visvader JE, Lindeman GJ. Cancer stem cells: current status and evolving complexities. Cell Stem Cell 2012;10(6):717–28.

[36] Travis WD, Brambilla E, Riely GJ. New pathologic classification of lung cancer: relevance for clinical practice and clinical trials. J Clin Oncol 2013;31(8):992–1001.

[37] Davidson MR, Gazdar AF, Clarke BE. The pivotal role of pathology in the management of lung cancer. J Thorac Dis 2013:S463–78.

[38] Imielinski M, Berger AH, Hammerman PS, Hernandez B, Pugh TJ, Hodis E, et al. Mapping the hallmarks of lung adenocarcinoma with massively parallel sequencing. Cell 2012;150(6):1107–20.

[39] Herbst RS, Heymach JV, Lippman SM. Lung cancer. N Engl J Med 2008;359(13):1367–80.

[40] TCGA. Comprehensive genomic characterization of squamous cell lung cancers. Nature 2012;489(7417):519–25.

[41] Wilkerson MD, Yin X, Hoadley KA, Liu Y, Hayward MC, Cabanski CR, et al. Lung squamous cell carcinoma mRNA expression subtypes are reproducible, clinically important, and correspond to normal cell types. Clin Cancer Res 2010;16(19):4864–75.

[42] Sun S, Schiller JH, Gazdar AF. Lung cancer in never smokers – a different disease. Nat Rev Cancer 2007;7(10):778–90.

[43] Govindan R, Ding L, Griffith M, Subramanian J, Dees ND, Kanchi KL, et al. Genomic landscape of non-small cell lung cancer in smokers and never-smokers. Cell 2012;150(6):1121–34.

[44] Lawrence MS, Stojanov P, Mermel CH, Robinson JT, Garraway LA, Golub TR, et al. Discovery and saturation analysis of cancer genes across 21 tumour types. Nature 2014;505(7484):495–501.

[45] TCGA. Comprehensive molecular profiling of lung adenocarcinoma. Nature 2014;511:543–50.

[46] George J, Lim JS, Jang SJ, Cun Y, Ozretic L, Kong G, et al. Comprehensive genomic profiles of small cell lung cancer. Nature 2015;524(7563):47–53.

[47] Ettinger DS, Akerley W, Borghaei H, Chang AC, Cheney RT, Chirieac LR, et al. Non-small cell lung cancer, version 2.2013. J Natl Compr Cancer Netw 2013;11(6):645–53.

[48] Lynch TJ, Bell DW, Sordella R, Gurubhagavatula S, Okimoto RA, Brannigan BW, et al. Activating mutations in the epidermal growth factor receptor underlying responsiveness of non-small-cell lung cancer to gefitinib. N Engl J Med 2004;350(21):2129–39.

[49] Pao W, Kris MG, Iafrate AJ, Ladanyi M, Jänne PA, Wistuba II, et al. Integration of molecular profiling into the lung cancer clinic. Clin Cancer Res 2009;15(17):5317–22.

[50] Rekhtman N, Ang DC, Sima CS, Travis WD, Moreira AL. Immunohistochemical algorithm for differentiation of lung adenocarcinoma and squamous cell carcinoma based on large series of whole-tissue sections with validation in small specimens. Mod Pathol 2011;24(10):1348–59.

[51] Skoulidis F, Byers LA, Diao L, Papadimitrakopoulou VA, Tong P, Izzo J, et al. Co-occurring genomic alterations define major subsets of KRAS-mutant lung adenocarcinoma with distinct biology, immune profiles, and therapeutic vulnerabilities. Cancer Discov 2015;5(8):860–77.

[52] Oike T, Ogiwara H, Tominaga Y, Ito K, Ando O, Tsuta K, et al. A synthetic lethality–based strategy to treat cancers harboring a genetic deficiency in the chromatin remodeling factor BRG1. Cancer Res 2013;73(17):5508–18.

[53] Fillmore CM, Xu C, Desai PT, Berry JM, Rowbotham SP, Lin Y-J, et al. EZH2 inhibition sensitizes BRG1 and EGFR mutant lung tumours to TopoII inhibitors. Nature 2015;520(7546):239–42.

[54] Cardnell RJG, Behrens C, Diao L, Fan Y, Tang X, Tong P, et al. An integrated molecular analysis of lung adenocarcinomas identifies potential therapeutic targets among TTF1-negative tumors, including DNA repair proteins and NRF2. Clin Cancer Res 2015;21(15):3480–91.

[55] Akbay EA, Koyama S, Carretero J, Altabef A, Tchaicha JH, Christensen CL, et al. Activation of the PD-1 pathway contributes to immune escape in EGFR-driven lung tumors. Cancer Discov 2013;3(12):1355–63.

[56] Brahmer JR, Tykodi SS, Chow LQ, Hwu WJ, Topalian SL, Hwu P, et al. Safety and activity of anti-PD-L1 antibody in patients with advanced cancer. N Engl J Med 2012;366(26):2455–65.

[57] Topalian SL, Hodi FS, Brahmer JR, Gettinger SN, Smith DC, McDermott DF, et al. Safety, activity, and immune correlates of anti-PD-1 antibody in cancer. N Engl J Med 2012;366(26):2443–54.

[58] Shedden K, Taylor JMG, Enkemann SA, Tsao M-S, Yeatman TJ, Gerald WL, et al. Gene expression-based survival prediction in lung adenocarcinoma: a multi-site, blinded validation study. Nat Med 2008;14(8):822–7.

[59] Zhang WC, Shyh-Chang N, Yang H, Rai A, Umashankar S, Ma S, et al. Glycine decarboxylase activity drives non-small cell lung cancer tumor-initiating cells and tumorigenesis. Cell 2012;148(1–2):259–72.

[60] Barretina J, Caponigro G, Stransky N, Venkatesan K, Margolin AA, Kim S, et al. The cancer cell line encyclopedia enables predictive modelling of anticancer drug sensitivity. Nature 2012;483(7391):603–7.

[61] Jiang F, Qiu Q, Khanna A, Todd NW, Deepak J, Xing L, et al. Aldehyde dehydrogenase 1 is a tumor stem cell-associated marker in lung cancer. Mol Cancer Res 2009;7(3):330–8.

[62] Sullivan JP, Spinola M, Dodge M, Raso MG, Behrens C, Gao B, et al. Aldehyde dehydrogenase activity selects for lung adenocarcinoma stem cells dependent on Notch signaling. Cancer Res 2010;70(23):9937–48.

[63] Eramo A, Lotti F, Sette G, Pilozzi E, Biffoni M, Di Virgilio A, et al. Identification and expansion of the tumorigenic lung cancer stem cell population. Cell Death Differ 2008;15(3):504–14.

[64] Chen Y-C, Hsu H-S, Chen Y-W, Tsai T-H, How C-K, Wang C-Y, et al. Oct-4 expression maintained cancer stem-like properties in lung cancer-derived CD133-positive cells. PLoS ONE 2008;3(7):e2637.

[65] Tirino V, Camerlingo R, Franco R, Malanga D, La Rocca A, Viglietto G, et al. The role of CD133 in the identification and characterisation of tumour-initiating cells in non-small-cell lung cancer. Eur J Cardio Thoracic Surg 2009;36(3):446–53.

[66] Bertolini G, Roz L, Perego P, Tortoreto M, Fontanella E, Gatti L, et al. Highly tumorigenic lung cancer $CD133^+$ cells display stem-like features and are spared by cisplatin treatment. Proc Natl Acad Sci 2009;106(38):16281–6.

[67] Meng X, Li M, Wang X, Wang Y, Ma D. Both $CD133^+$ and CD133– subpopulations of A549 and H446 cells contain cancer-initiating cells. Cancer Sci 2009;100(6):1040–6.

[68] Johnson L, Mercer K, Greenbaum D, Bronson RT, Crowley D, Tuveson DA, et al. Somatic activation of the K-ras oncogene causes early onset lung cancer in mice. Nature 2001;410(6832):1111–6.

[69] Jackson EL, Willis N, Mercer K, Bronson RT, Crowley D, Montoya R, et al. Analysis of lung tumor initiation and progression using conditional expression of oncogenic K-ras. Genes Dev 2001;15(24):3243–8.

[70] Jackson EL, Olive KP, Tuveson DA, Bronson R, Crowley D, Brown M, et al. The differential effects of mutant p53 alleles on advanced murine lung cancer. Cancer Res 2005;65(22):10280–8.

[71] Ji H, Li D, Chen L, Shimamura T, Kobayashi S, McNamara K, et al. The impact of human EGFR kinase domain mutations on lung tumorigenesis and in vivo sensitivity to EGFR-targeted therapies. Cancer Cell 2006;9(6):485–95.

[72] Politi K, Zakowski MF, Fan P-D, Schonfeld EA, Pao W, Varmus HE. Lung adenocarcinomas induced in mice by mutant EGF receptors found in human lung cancers respond to a tyrosine kinase inhibitor or to down-regulation of the receptors. Genes Dev 2006;20(11):1496–510.

[73] Kwon M, Berns A. Mouse models for lung cancer. Mol Oncol 2013;7(2):165–77.

[74] Xu X, Rock JR, Lu Y, Futtner C, Schwab B, Guinney J, et al. Evidence for type II cells as cells of origin of K-ras-induced distal lung adenocarcinoma. Proc Natl Acad Sci 2012;109(13):4910–5.

[75] Mainardi S, Mijimolle N, Francoz S, Vicente-Dueñas C, Sánchez-García I, Barbacid M. Identification of cancer initiating cells in K-ras driven lung adenocarcinoma. Proc Natl Acad Sci 2014;111(1):255–60.

[76] Sutherland KD, Song J-Y, Kwon MC, Proost N, Zevenhoven J, Berns A. Multiple cells-of-origin of mutant K-ras-induced mouse lung adenocarcinoma. Proc Natl Acad Sci 2014.

[77] Xu X, Huang L, Futtner C, Schwab B, Rampersad RR, Lu Y, et al. The cell of origin and subtype of K-ras-induced lung tumors are modified by Notch and Sox2. Genes Dev 2014;28(17):1929–39.

[78] Curtis SJ, Sinkevicius KW, Li D, Lau AN, Roach RR, Zamponi R, et al. Primary tumor genotype is an important determinant in identification of lung cancer propagating cells. Cell Stem Cell 2010;7(1):127–33.

[79] Zheng Y, de la Cruz CC, Sayles LC, Alleyne-Chin C, Vaka D, Knaak TD, et al. A rare population of $CD24^+ITGB4^+$ Notchhi cells drives tumor propagation in NSCLC and requires Notch3 for self-renewal. Cancer Cell 2013;24(1):59–74.

[80] Winslow MM, Dayton TL, Verhaak RGW, Kim-Kiselak C, Snyder EL, Feldser DM, et al. Suppression of lung adenocarcinoma progression by Nkx2-1. Nature 2011;473(7345):101–4.

[81] Lau AN, Curtis SJ, Fillmore CM, Rowbotham SP, Mohseni M, Wagner DE, et al. Tumor-propagating cells and Yap/Taz activity contribute to lung tumor progression and metastasis. EMBO J 2014.

[82] Funai K, Yokose T, G-i I, Araki K, Yoshida J, Nishimura M, et al. Clinicopathologic characteristics of peripheral squamous cell carcinoma of the lung. Am J Surg Pathol 2003;27(7):978–84.

[83] Saijo T, Ishii G, Nagai K, Funai K, Nitadori J, Tsuta K, et al. Differences in clinicopathological and biological features between central-type and peripheral-type squamous cell carcinoma of the lung. Lung Cancer 2006;52(1):37–45.

[84] Lu Y, Futtner C, Rock JR, Xu X, Whitworth W, Hogan BLM, et al. Evidence that Sox2 overexpression is oncogenic in the lung. PLoS One 2010;5(6).

[85] Malkoski SP, Cleaver TG, Thompson JJ, Sutton WP, Haeger SM, Rodriguez KJ, et al. Role of PTEN in basal cell derived lung carcinogenesis. Mol Carcinog 2013.

[86] Xiao Z, Jiang Q, Willette-Brown J, Xi S, Zhu F, Burkett S, et al. The pivotal role of IKKα in the development of spontaneous lung squamous cell carcinomas. Cancer Cell 2013;23(4):527–40.

[87] Ji H, Ramsey MR, Hayes DN, Fan C, McNamara K, Kozlowski P, et al. LKB1 modulates lung cancer differentiation and metastasis. Nature 2007;448(7155):807–10.

[88] Gao Y, Zhang W, Han X, Li F, Wang X, Wang R, et al. YAP inhibits squamous transdifferentiation of Lkb1-deficient lung adenocarcinoma through ZEB2-dependent DNp63 repression. Nat Commun 2014:5.

[89] Li F, Han X, Li F, Wang R, Wang H, Gao Y, et al. LKB1 inactivation elicits a redox imbalance to modulate non-small cell lung cancer plasticity and therapeutic response. Cancer Cell 2015;27(5):698–711.

[90] Mukhopadhyay A, Berrett Kristofer C, Kc U, Clair Phillip M, Pop Stelian M, Carr Shamus R, et al. Sox2 cooperates with Lkb1 loss in a mouse model of squamous cell lung cancer. Cell Rep 2014;8(1):40–9.

[91] Justilien V, Walsh Michael P, Ali Syed A, Thompson EA, Murray Nicole R, Fields Alan P. The PRKCI and SOX2 oncogenes are coamplified and cooperate to activate hedgehog signaling in lung squamous cell carcinoma. Cancer Cell 2014;25(2):139–51.

[92] Peifer M, Fernandez-Cuesta L, Sos ML, George J, Seidel D, Kasper LH, et al. Integrative genome analyses identify key somatic driver mutations of small-cell lung cancer. Nat Genet 2012;44(10):1104–10.

[93] Gazdar AF, Savage TK, Johnson JE, Berns A, Sage J, Linnoila RI, et al. The comparative pathology of genetically engineered mouse models for neuroendocrine carcinomas of the lung. J Thorac Oncol 2015;10(4).

[94] McFadden David G, Papagiannakopoulos T, Taylor-Weiner A, Stewart C, Carter Scott L, Cibulskis K, et al. Genetic and clonal dissection of murine small cell lung carcinoma progression by genome sequencing. Cell 2014;156(6):1298–311.

[95] Sequist LV, Waltman BA, Dias-Santagata D, Digumarthy S, Turke AB, Fidias P, et al. Genotypic and histological evolution of lung cancers acquiring resistance to EGFR inhibitors. Sci Transl Med 2011;3(75):75ra26.

[96] Christensen Camilla L, Kwiatkowski N, Abraham Brian J, Carretero J, Al-Shahrour F, Zhang T, et al. Targeting transcriptional addictions in small cell lung cancer with a covalent CDK7 inhibitor. Cancer Cell 2014;26(6):909–22.

[97] Mohammad Helai P, Smitheman Kimberly N, Kamat Chandrashekhar D, Soong D, Federowicz Kelly E, Van Aller Glenn S, et al. A DNA hypomethylation signature predicts antitumor activity of LSD1 inhibitors in SCLC. Cancer Cell 2015;28(1):57–69.

[98] Brahmer J, Reckamp KL, Baas P, Crinò L, Eberhardt WEE, Poddubskaya E, et al. Nivolumab versus docetaxel in advanced squamous-cell non-small-cell lung cancer. N Engl J Med 2015;373(2):123–35.

CHAPTER 7

Colorectal Cancer Stem Cells

S. Rehman[1,2], J. Haynes[1,2], E. Lima-Fernandes[3], A. Puri[1,2], A. Haller[1,2], C. Leung[1,2], L. Agro[1,2], Y. Wang[1,2], C.A. O'Brien[1,2,4]

[1]University Health Network, Toronto, ON, Canada; [2]University of Toronto, Toronto, ON, Canada; [3]Structural Genomics Consortium, Toronto, ON, Canada; [4]Toronto General Hospital, Toronto, ON, Canada

CONTENTS

INTRODUCTION

Colorectal Cancer Stem Cell Markers

Cancer stem cells (CSCs) are defined first and foremost by their capacity to self-renew and the ability to differentiate into the heterogeneous populations of cancer cells that comprise a tumor. Self-renewal refers to the ability to form new stem cells with an intact potential for proliferation, expansion, and differentiation, thus maintaining the CSC pool [1–3]. The term CSC refers to their potential to self-renew; however, it does not imply that the cell of origin for the tumor is a normal stem cell [2]. CSC markers in colorectal cancer (CRC) can be divided into two groups: phenotypic and functional [2,4–6] (Table 7.1). Numerous phenotypic markers have been employed to isolate the CSC in CRC, including a wide range of cell surface antigens: CD133, CD44, CD166, CD24, CDCP1, DCAMKL-1, CD26, CD49f, EphB2, and LGR5 [11,12,14–25]. Functional markers utilized to enrich the CSC fraction in CRC include: ALDEFLOUR assay, side population (SP), and the Wnt reporter. Similar to the phenotypic markers, the functional markers are also wrought with controversy and conflicting reports regarding their ability to enrich for the CSC fraction.

The ALDEFLOUR assay measures aldehyde dehydrogenase (ALDH) activity as a marker of CSC. It is well established that ALDH activity is important for multiple biological activities including drug resistance, cell differentiation, and oxidative stress response [7,8]. The use of ALDH remains controversial as a standard CSC marker in CRC [9]. The SP assay exploits the high levels of expression of cell membrane ATP-binding cassette (ABC) transporter proteins, which provide a cell with the ability to efflux chemotherapeutic drugs and certain dyes, such as Hoechst 33342. A recent publication by Feng et al.

Cancer Stem Cells. http://dx.doi.org/10.1016/B978-0-12-803892-5.00007-3

Table 7.1 List of Stem Cell Markers in Colorectal Cancer (CRC)

Stem Cell Marker		Description	References
Functional	ALDH	Aldehyde dehydrogenase activity	[7–9]
Functional	Side population	Expression of cell membrane ATP-binding cassette transporter proteins	[10]
Functional	Wnt reporter	Marker of nuclear β-catenin activity	[11–13]
Phenotypic	CD133 (Prominin 1)	Self-renewal, tumor angiogenesis	[14]
Phenotypic	CD44 (CDW44)	Cell adhesion molecule, hyaluronic acid receptor	[15,16]
Phenotypic	CD166 (ALCAM)	Cell adhesion molecule	[16]
Phenotypic	CD24 (HSA)	Cell adhesion molecule	[11]
Phenotypic	CDCP1	Loss of cell adhesion	[17]
Phenotypic	CD49f	Vascularization	[18,19]
Phenotypic	CD26	Cell adhesion molecule	[20]
Phenotypic	EphB2 (ephrin-B2)	Motility, division, and differentiation	[21]
Phenotypic	Lgr5 (GPR49)	Wnt signaling pathway	[22,23]
Phenotypic	DCAMKL-1	Encodes a protein kinase	[24,25]

demonstrated enrichment of the CSC population using the SP assay [10]. Vermeulen et al. were the first to publish the use of the Wnt reporter (TOP-GFP), a marker of nuclear β-catenin activity, as a functional tool to enrich for the CSC fraction in CRC. Both the Wnt^{hi} and Wnt^{lo} fractions were able to produce tumors; however, the initiating capacity was much greater in the Wnt^{hi} fraction. Subsequent publications have reported that the majority of CRCs have equal tumor-initiating frequencies between the Wnt^{hi} and Wnt^{lo} fractions [12,26,27].

There continues to be considerable controversy in the field with respect to the most robust phenotypic or functional CSC marker [4–6]. Currently there is no universally accepted CSC marker for CRC. It is also important to appreciate that existing CSC markers enrich for, but do not purify, the CSC population in CRC [5]. Therefore, the vast majority of cells that express these markers are not CSCs. These findings reinforce the point that all CSC markers need to be functionally validated in the context of each CRC case tested by carrying out the gold standard assay of serial passage in vivo limiting dilution assay (LDA) [28]. One possible reason for the conflicted findings with CSC markers maybe related to the existence of cell state switching or plasticity, where non-CSCs are able to convert to CSCs [29,30]. To date this data does not exist for CRC; however, a convincing case for CSC plasticity has been made in breast cancer [31]. Chaffer et al. showed that plastic non-CSCs maintain the *ZEB1* promoter in a bivalent chromatin configuration, thereby allowing non-CSCs to convert to a CSC state [32]. It is possible that a similar mechanism is at play in CRC, which may explain some of the current controversies.

Intestinal Stem Cells and Colorectal Cancer Cell of Origin

Knowledge of the mechanisms that regulate normal intestinal epithelial homeostasis is of central importance to our understanding of the biology of CRC. Intestinal tissue is highly regenerative, and has a unique architecture suited for its specialized physiological functions of nutrient absorption and transport of intestinal contents along the lumen. The epithelial cells lining the large intestine, or colon, form a flat surface with regular invaginations known as crypts, which maintain a gradient of cellular differentiation from bottom to top. Undifferentiated stem cells at the crypt base give rise to highly proliferative transit-amplifying (TA) progenitor cells that migrate up the crypt as they divide and differentiate into the following mature cells types: absorptive colonocytes, mucous-secreting goblet cells, hormone-producing enteroendocrine cells, and less-well characterized tuft cells [33]. Colonic stem cells are thought to be maintained in an undifferentiated state at the crypt base by niche factors provided by a subset of crypt base goblet cells and from as yet unidentified sources, which likely include stromal cells found in the underlying tissue [34].

The identification of intestinal stem cells (ISCs) has been greatly facilitated by genetic mouse models for cell lineage tracing (fate mapping) alone or combined with conditional cell ablation [35]. In vivo genetic lineage tracing first identified proliferative crypt base columnar (CBC) cells expressing leucine-rich-repeat-containing G-protein-coupled receptor 5 (Lgr5) as stem cells of both the small intestine and colon, capable of generating all epithelial cell types [36]. Further establishing their role as bona fide stem cells, single Lgr5+ cells purified from mouse intestine by fluorescence-activated cell sorting (FACS) were subsequently shown to initiate growth of cryptlike organoids with long-term proliferation and multilineage differentiation capacity under in vitro culture conditions [37]. Moreover, epithelial organoids derived from single Lgr5+ colonic stem cells can be cultured and expanded in vitro and subsequently transplanted into recipient mice to repair damaged colonic epithelium in vivo [38]. Lineage-tracing experiments also identified a separate pool of ISCs at the colonic crypt base different from Lgr5+ stem cells that express leucine-rich repeats and immunoglobulin-like domains 1 (Lrig1) and are predominantly quiescent, but proliferate to repopulate crypts post damage [39]. In addition, a recent study used lineage tracing to reveal a distinct stem cell population located above the crypt base that expresses the intermediate filament keratin-19 (Krt19) and generates new Lgr5+ CBCs in the colon and small intestine after Lgr5+ cell-specific ablation [40]. These studies also demonstrate the existence of fate plasticity for ISC populations, which is further supported by work showing that progenitor cells have the potential to dedifferentiate to regain ISC activity in certain contexts, such as during tissue regeneration following injury [41].

Like ISCs, colon CSCs have properties of both self-renewal and multilineage differentiation potential. Consistent with their functional similarities, gene expression profiles of CRCs resemble that of ISCs. Consistent with most CRCs having Wnt pathway-activating mutations in genes such as adenomatous polyposis coli (*APC*), an intestinal Wnt/T-cell factor transcription factor signature gene set is expressed in CRC [13]. Furthermore, a gene signature specific for adult ISCs, as identified and isolated by high surface expression of ephrin type-B receptor 2 (EphB2), is enriched in aggressive CRCs and in CSCs [21]. Validating its use as an ISC marker, Ephb2-high colonic stem cells can be isolated from human colonic epithelium and expanded in vitro as organoids that are enriched in stem cell markers and retain multilineage differentiation capacity in long-term cultures [42].

Due to the rapid turnover of the intestinal epithelium, it is generally thought that stem cells are the most likely cell of origin of CRC, as they are the only population sufficiently long-lived to accumulate the mutations required for cancer development [43]. In support of this, deletion of *Apc* in murine intestinal Lgr5+ stem cells causes unrestricted growth and rapidly produces macroscopic adenomas in the small intestine and colon [44]. However, *Apc* deletion in short-lived TA cells predominantly yields only small lesions in the intestinal epithelium that rarely progress to adenoma. Other studies evaluating the contribution of additional stem cell pools to the origin of CRC have shown that both Lrig1+ and Krt19+ stem cells also give rise to colonic tumors in mice after targeted deletion of *Apc* in these cell populations [39]. Stem cells as the cell of origin of CRC could account for much of the phenotypic heterogeneity frequently observed within tumors (intratumoral heterogeneity).

Although transformation of stem cells likely represents an important path leading to tumor development, recent studies have shown that non-stem cells can also act as the cell of origin for CRC. Conditional loss of *Apc* in a subpopulation of long-lived intestinal tuft cells was shown to drive the formation of poorly differentiated colonic tumors in a dextran sulfate sodium (DSS)-induced colitis mouse model [45]. Furthermore, nuclear factor (NF)-kB-dependent inflammatory signaling can enhance Wnt signaling and induce dedifferentiation of Lgr5− intestinal epithelial cells to acquire cancer-initiating capacity [46]. A recent study investigating the contribution of the microenvironment to CRC initiation showed that aberrant epithelial expression of the gene-encoding mesenchymal bone morphogenetic protein antagonist, GREM1, leads to the development of colonic tumors from Lgr5− cells outside the stem cell niche [47]. Taken together, these studies suggest that non-stem cell populations may require additional mutations, activation of inflammatory pathways, or other microenvironmental cues for tumor initiation and progression. Furthermore, it is possible that similar genetic mutations in distinct cells of origin may contribute to the extensive heterogeneity seen among CRC patients (intertumoral heterogeneity).

MOLECULAR MECHANISMS AND PATHWAYS REGULATING COLORECTAL CANCER STEM CELLS

It is becoming increasingly accepted that cancers are organized in a hierarchical manner which is maintained by CSCs, a rare population believed to exist in the tumor [48]. Others believe that cells within a tumor can acquire a stem cell state enabling the sustenance of the tumor [49]. Experimentally, the ability of cells to establish tumors upon serial transplantation is a well-established hallmark of CSCs [50]. Emerging evidence indicates that CSCs are clinically relevant and are capable of surviving conventional chemotherapy [51–53]. These CSCs were found to exhibit characteristics similar to the rare cells within a tumor that are capable of invasion and metastasis implying the possibility that CSCs may be the cells primarily responsible for establishing metastases [54].

CRC is one of the best understood cancers from a genetic perspective due in large part to the work of Fearon and Vogelstein who were the first to propose a genetic model of CRC [55]. Inactivation of the tumor-suppressor gene *APC* on chromosome 5q21 was noted as the earliest genetic event in colorectal carcinogenesis, followed by mutations in various genes (*KRAS, BMI-1, TGF-β, IDs*) and microRNAs (miRNAs) [55,56]. There is increasing evidence demonstrating that CRC has significant genetic heterogeneity courtesy of three major molecular subtypes: microsatellite instability (MSI), chromosomal instability, and the CpG island methylator phenotype (CIMP) [57]. The resulting genome instability and mutations, a hallmark of cancer, serve as the initiating factors for CRC. Importantly, a number of these altered pathways are known to upregulate self-renewal capacity, thereby supporting the initiation and maintenance of a CSC subset [58]. Understanding the molecular pathways driving CSCs in CRC will be the key to developing novel strategies to target this subpopulation of cells (Fig. 7.1).

Adenomatous Polyposis Coli and the Wnt Pathway

Wnt pathway activation via disruption of APC is one of the earliest genetic occurrences in CRCs. Evidently, in sporadic CRC, *APC* mutations are present in ~50% of small adenomas (<0.5 cm) and advanced adenomas and carcinomas, indicating that inactivating *APC* mutations are an early event in tumorigenesis [59–61]. Alternatively, *APC* gene inactivation may occur as a result of APC promoter hypermethylation, which has been reported in 18% of colorectal adenomas and carcinomas [62]. In the normal colonic mucosa, APC works with β-catenin signaling as part of the Wnt pathway to regulate self-renewal, differentiation, and other cellular processes [63]. In the absence of Wnt ligand signaling, APC binds to Axin protein and promotes β-catenin phosphorylation by casein kinase1 (CK1) and glycogen synthase kinase-3β (GSK3β), marking β-catenin for ubiquitination and proteasomal degradation. In CRC, lack of

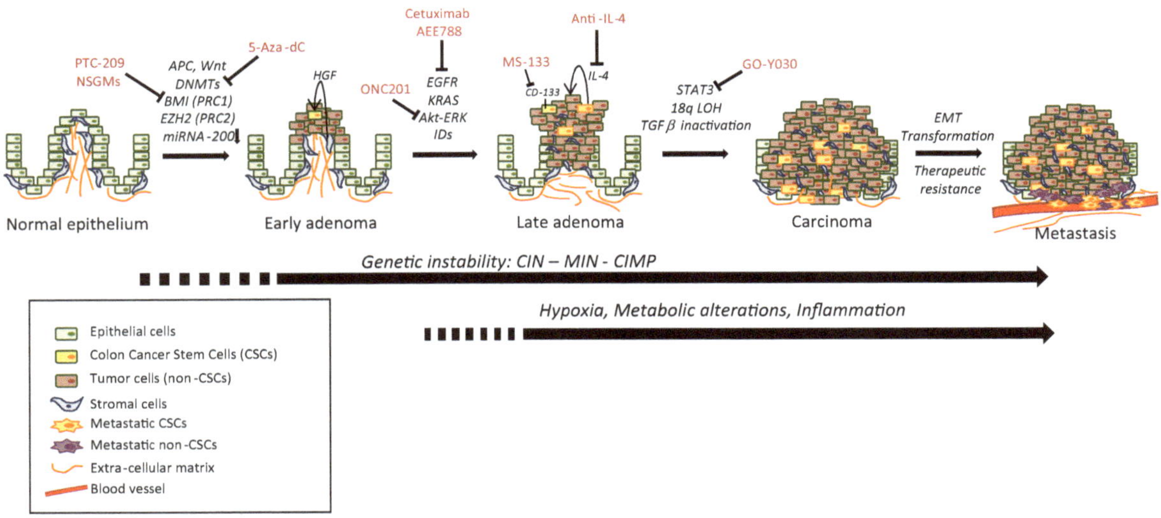

FIGURE 7.1

Pathways in colorectal cancer stem cells (CSCs) that regulate tumorigenesis. Genetic instability–driven mutations or pathway alterations initiate tumorigenesis, resulting in small benign tumors (adenomas) with colorectal CSCs. Alterations in other pathway genes enable CSC self-renewal and accumulate as the tumors progress, driven by factors such as hypoxia, metabolism, or inflammatory cytokines. Epithelial-to-mesenchymal (EMT) transformation and/or persistence through therapy followed with EMT may result in CSC metastasis. Therapies to known targets are shown in red.

APC protein due to mutations enables cytoplasmic accumulation of β-catenin, which binds to the T-cell factor/lymphoid enhancer family (TCF/LEF) of proteins and translocates to the nucleus. Here, it regulates transcription of multiple genes with TCF/LEF DNA-binding sites involved in tumorigenesis and self-renewal [64]. Other mutations in *β-CATENIN*, such as gain-of-function mutations that affect its phosphorylation and/or ubiquitination motifs have been identified in about 50% of CRC with wild-type APC, underscoring the importance of the Wnt pathway in CRC [65]. While our understanding of Wnt activation in CSCs is still building, preliminary work indicates that the Wnt pathway plays a critical role in the initiation and maintenance of CSCs [66–68].

KRAS

The RAS family of small G-proteins (K-RAS4A, K-RAS4B, H-RAS, and N-RAS) signal downstream of multiple pathways, including growth factor receptors such as epidermal growth factor receptor (EGFR) [69]. While less than 5% of CRCs have EGFR alterations (gene amplification or mutations), ~40% of CRCs exhibit KRAS oncogene mutations which sets the enzyme in the guanosine triphosphate bound (GTP) constitutively activated state. In a mouse model for sporadic CRC, the activating *KRASG12D* mutation accelerated tumor growth [70–72]. More recently, it was shown that mutant KRAS imposes an embryonic stem cell (ESC)-like gene expression signature during human colon cancer initiation from colon adenoma to stage I carcinoma implicating a role for *KRAS* in CSCs [73]. The most well-known KRAS effector is the Raf-mitogen-activated protein kinase (MEK)-extracellular signal-regulated kinase (ERK) pathway. The Raf serine–threonine kinase family (A-RAF, B-RAF, and C-RAF) phosphorylate MEK1 and MEK2, which then activate ERK1 and ERK2. ERK subsequently phosphorylates transcription factors such as JUN and ELK1 that regulate genes involved in cell cycle control (such as *CYCLIN D1*) [74].

BMI-1

BMI-1 was an essential component of the polycomb repressor complex 1 (PRC1) which has a fundamental role in chromatin structure and organization and subsequently regulates expression of various genes in stem cell regulation [75,76]. BMI-1 was one of the initial regulators found to be associated with self-renewal and implicated in stem cell maintenance and its expression is regulated by multiple transduction pathways, including Wnt pathway upregulation [77–80]. Recent work has shown that *Bmi-1*-deficient mice have intestinal defects that closely resemble a Notch inhibition phenotype, and that both Notch and β-catenin are simultaneously required to activate transcription of the *Bmi-1* gene [81]. Thus, BMI-1 regulates ISC proliferation and self-renewal downstream of both Wnt and Notch pathways to maintain intestinal homeostasis in vivo.

The oncogene *BMI-1* is found to be overexpressed in various cancers, including CRC, with the degree of upregulation correlating with the extent of disease progression and is predictive of poor patient survival [82–87]. BMI-1 was shown to be required for tumorigenesis [88] and shown to mediate chemotherapy and radiation resistance [89]. Few studies have shown the role of BMI-1 in the regulation of CSC and therapeutic resistance [90,91]. BMI-1 was found to be essential for CSC function in CRC and targeting BMI-1 was effective in reducing tumor burden in primary human CRC tumor xenograft models [92].

Inhibitor of Differentiation

The inhibitor of differentiation (ID) family of proteins (ID1, ID2, ID3, and ID4) are basic helix-loop-helix (bHLH) proteins that heterodimerize with other bHLH transcription factors to regulate cell fate and differentiation in stem and progenitor cells. *ID* genes are frequently dysregulated in many cancers which endow the cancer cells with biological characteristics similar to that of normal stem cells [93,94]. More recently, the ID proteins have been found to be markers of stem cells in various organs [95–97]. In the normal colonic crypt, ID1 is expressed in the cycling columnar cells, the +4 position cells, and transit-amplifying cells and is a functional marker for intestinal stem and progenitor cells [96]. Similarly, in CRC, ID1 and ID3 were shown to operate together to govern CSC self-renewal through the cell cycle inhibitor p21, preventing the accumulation of excess DNA damage and functional exhaustion of CSCs in the tumor [98]. This study supports the notion that CSCs play a supportive role in tumorigenesis and, furthermore, a protective role when the CSCs are hit with DNA-damaging agents (chemotherapy), thereby ensuring their survival. Indeed, upon ID1 and ID3 inhibition, CSCs exhibited increased sensitivity to a chemotherapeutic agent Oxaliplatin [98].

TGF-β and SMAD4

Mutations in the TGF-β type II receptor are found in >55% of CRCs, typically in CRCs with MSI. Upon binding TGF-β ligand, TGF-β-mediated receptor phosphorylation regulates the function of the SMAD2/SMAD3 proteins which bind SMAD4 [99]. This complex translocates to the nucleus to transcriptionally regulate target genes. Allelic loss at chromosome 18q, which carries the *SMAD4* gene, or mutations in *SMAD4* have been identified in >70% of CRCs [100–102]. SMAD4 was found to inhibit β-catenin transcription, and loss of SMAD4 signaling resulted in upregulation of β-catenin activity and Wnt signaling pathway in progression of CRC, suggesting this regulation may occur in the CSCs [103].

Epigenetics

Epigenetics refers to heritable changes in gene expression that are due to covalent modifications of DNA and the proteins that package DNA, such as histones, and through noncoding RNAs. Epigenetics regulate gene expression in a dynamic and reversible manner, to dictate cell fate through proper balance of pluripotent (normal and CSCs), and stem cell–like or differentiated states by fine-tuning the accessibility of DNA [49,104,105]. Recent advances in the epigenome of CRC have revealed that the majority of CRCs display aberrantly methylated genes and altered gene expression. It was originally proposed that in CRC, tumor formation is a stepwise process from normal to adenoma progression starting with mutations in *APC* as the principal early step leading to cancer. Several studies have now shown that colorectal tumor cells undergo extensive epigenetic reprogramming during CRC initiation that may even precede the known driver mutations of CRC [106–108]. These epigenetic alterations may lead not only to genomic instability, but also to reprogramming cells into a stem-like state. A seminal study showed that individual CSCs sharing the same genomic background displayed variation in growth dynamics and response to chemotherapy [49]. This suggests that the functional heterogeneity of cells and CSC in a given tumor might be driven by epigenetics. Unlike other cancer types, the epigenetic changes driving and regulating colorectal CSCs remain to be elucidated.

Epigenetics may contribute to the CSC state by modulating key pathways such as WNT signaling, where high activity has been shown to designate the CSC population and also shown to dedifferentiate Lgr5− cells into tumor-initiating cells [26,46]. Lysine-specific demethylase 1 (LSD1), an enzyme catalyzing specific demethylation H3K4me1/2 and H3K9, is highly expressed in CRC and correlates with TNM stage (TNM staging—tumor, node, metastasis) and metastatic disease [109]. LSD1 also downregulates the Wnt pathway antagonist DKK1, thus it may be contributing to CSC self-renewal [110]. BMI-1, a subunit of the PRC1 complex, which binds to H3K27me3, has been well documented for its role in adult stem cell maintenance and tumorigenicity in different cancer types. In CSC, inhibition of BMI-1 reduces CSC self-renewal in parallel to decreasing Wnt pathway activation through reduction of β-catenin levels [92]. EZH2, the catalytic subunit of the polycomb repressor complex 2 (PRC2), plays a role in self-renewal in ESCs and is reported to be overexpressed in many cancer types including CRC [111–113]. EZH2 has been shown to downregulate Wnt antagonists in other cancer types [114]. In CRC, it has been shown that EZH2, through its binding with PAF, contributes to hyperactivation of the Wnt pathway and induces tumorigenesis [115]. PAF is also overexpressed in CRC and ISCs and is a known self-renewal factor in hematopoietic stem cells [115,116]. Wnt pathway antagonists such as the DKKs are also modulated through DNA methylation in CRC. Reexpression of TET1, the DNA demethylation-promoting enzyme, which is downregulated in CRC, reduces tumor growth in vivo [117].

The interferon response pathway has been recently shown to be induced in response to a DNA-demethylating agent, which reduces CSC self-renewal. Indeed, low-dose exposure to 5-aza-deoxycytidine (5-aza-dC) significantly decreases CSC self-renewal in vitro and in vivo [118]. A low dose of 5-aza-dC triggers a viral mimicry state that induces the expression of double-stranded RNA, which in turn activates the interferon response through MDA5 receptor, and the downstream activation of the MAVS/IRF7 pathway [118].

There is preliminary evidence suggesting that epigenetics might play a key role in the establishment of the cell of origin in the CRC, a heretofore-unanswered question in the field. Loss of imprinting (LOI), defined by the disruption of the parent-of-origin-specific inheritable epigenetic marks, can lead to aberrant activation of a normally silenced allele and is considered an abundant and early alteration in cancer [119]. In an LOI/APC mouse model, it has been shown that high levels of IGF2 increases tumor initiation, as well as the length of intestinal crypts stained positive for progenitor cell marker Musashi1, through a shift from differentiated to undifferentiated cells [120]. Future work is required to determine whether LOI represents the driving force in the creation of the CRC cell of origin [121].

It has been proposed that cancers arise from CSCs that share epigenetic properties with ESCs. Bivalent chromatin, a key feature of ESCs, is characteristic of genes that are repressed but poised for expression by bearing both the active H3K4me3 and the repressive H3K27me3 marks. In ESCs, poised genes are thought to maintain self-renewal and control lineage commitment [122,123]. These developmental genes, bivalently marked in normal ESC, are found to be more permanently repressed through DNA methylation in cancer, including CRC [124–126]. This includes differentiation genes such as *HAND1*, a tumor suppressor downregulated in CRC [126]. Restoration of HAND1 expression in CRC lines reduces colony formation in vitro and tumor growth in vivo [126]. In human CRC, the de novo DNA methyltransferase (DNMT) 3a is predominantly expressed in Lgr5-positive tumor stem/progenitor cells [127]. Accordingly, deletion of *Dnmt3a* in $APC^{(min/+)}$ mice reduced colon tumor formation [127]. These observations support a stem cell origin for CRC, in which the permanent silencing of ESC differentiation genes through replacement of H3K27me3 by DNA methylation, locks cells in a state of self-renewal [125].

The key aspect of epigenetic signaling is its dynamic and reversible nature. Thus, altering the epigenome that dictates cell fate specification by targeting epigenetic modifiers represents a novel therapeutic strategy to reprogram CSCs and render them sensitive to existing therapies that target the bulk tumor cells.

MicroRNAs

miRNAs are a class of small, noncoding RNA involved in posttranscriptional regulation of gene expression and as such, they are epigenetic modulators. miRNAs function to maintain stem cell properties of both normal and CSCs [128]. A subset of miRNAs, which are overexpressed in tumor cells, appear to function through inhibition of tumor-suppressor genes. High levels of miR-21, an onco-miRNA overexpressed in multiple cancer types, have been detected in FOLFOX-resistant CRC lines that are enriched for CSCs [129,130]. Highly expressed in CRC, miR-146a has been shown to play a key role in controlling asymmetric cell division, a key feature of stem cells to maintain self-renewal [131,132]. The epithelial-to-mesenchymal (EMT) inducer Snail increases expression of miR-146a, which stabilizes β-catenin through the targeting of Numb, and contributes to the expansion of the stem cell pool [132].

On the contrary, miRNAs downregulated in cancer might control the expression of oncogenes in normal tissue. Disruption of miRNA expression, one of the hallmarks in many cancer types, can result from altered promoter DNA methylation or defects in miRNA-processing machinery. Impaired function of DICER1, a protein involved in the miRNA biogenesis pathway, increases the levels of stem cell markers Sox9, Sox2, and Lgr5 and enhances tumor-initiating capacity and metastasis formation in vivo [133]. This increase in CSC was associated with the downregulation of miR-34a, miR-126, and the miR-200 family [133]. In another study, the levels of the tumor-suppressor miR-34a regulate the balance between self-renewal and differentiation [134]. Low miR-34a promotes CSC self-renewal, whereas high levels of miR-34a are found in the differentiated progeny, where it suppresses Notch signaling to promote differentiation [134]. Two studies using miRNA profiling have identified the downregulation of miR-451 and miR-328 in a subpopulation CRC cells that demonstrated drug resistance, higher tumorigenicity and were capable of self-renewal [135,136]. MiR-328 directly reduces the levels of the transporter ABCG2 and the matrix metalloproteinase MMP16, suggesting that reexpression of miR-328 in CSC increases drug sensitivity and decreases invasiveness [135]. Overexpression of miR-451 reduces Wnt signaling and the expression of the drug transporter ABCB1, resulting in reduced sphere formation and increased chemosensitivity [136]. The miR-200 family (miR-200a, -200b, -200c, -141, -429) has been shown to have critical suppressive effects on cell transformation, tumor growth, and self-renewal in multiple cancer types [32,137,138]. In CRC cells, the EMT-activator Zeb1 inhibits expression of miR-200 family members to promote the tumor-initiating capacity of CSCs(138). Conversely, reexpression of miR-200c inhibits BMI1, a key self-renewal regulator in CSCs [92,137]. Another BMI1-targeting miRNA, miR-215, has also been shown to be downregulated in CSCs [139]. Dysregulated miRNAs that have been associated with development and progression of CRC are now emerging as major regulators of CSCs. More recently, studies are exploring the use of circulating miRNAs in the serum as novel biomarkers for diagnosis and prognosis of CRC and may be exploited for therapeutics targeting in CRC [140,141].

COLORECTAL CANCER STEM CELLS AND MICROENVIRONMENT

An increasing body of work in the solid tumor CSC field has pointed to the microenvironment as playing a central role in the regulation of stemness and therapy resistance. Vermeulen et al. were one of the first groups to observe interplay between the microenvironment and CSCs in CRC. Using the Wnt reporter as a marker of CSCs, they showed that Wnt^{lo} (non-CSCs) could upregulate Wnt activity when situated in close proximity to a hepatocyte growth factor (HGF) secreting stromal cell [26]. This work demonstrated for the first time that non-CSCs could convert into CSCs based on extrinsic microenvironmental signals in CRC. This can also be observed pathologically where tumor cells located next to or within myofibroblast-rich regions have a much higher incidence of nuclear-localized β-catenin [142]. Colak and Medema expanded on the initial work demonstrating that differentiated cells can reacquire stemness through additional factors secreted from myofibroblasts including: stromal derived factor-1α (SDF-1α) and osteopontin (OPN). The induced cell state also coincided with the reacquisition of resistance to chemotherapy, driven by the antiapoptotic molecule BCLXL. Furthermore, the inhibition of BCLXL activity resensitized the cancer cells to chemotherapy [143].

Tumor hypoxia represents another microenvironmental factor that can influence stemness. Hypoxia is a general term for when the oxygen demand of a cell outpaces the apparent supply of oxygen. This phenomenon is of particular importance in cancer where cells often outgrow the existing vasculature. In areas of hypoxia, hypoxia-inducible factors (HIFs) are stabilized, which results in the activation of gene programs that promote cellular survival by decreasing oxygen consumption and inhibiting apoptosis [144–146]. HIFs directly bind β-catenin in the nucleus therefore linking β-catenin to the hypoxic response [147]. These programs have also been associated with the activation of stemlike gene programs, and the induction of stemlike genes through Wnt signaling [148]. Observations in a number of solid tumors, including CRC, indicate hypoxia can drive a CSC phenotype [149]. In CRC, culturing cells under hypoxic conditions (2% O_2) inhibits differentiation and increases clonogenicity, primarily by increasingly nuclear β-catenin activity through effects of HIF-1α [150]. Interestingly, HIF-1α expression is found mainly in the CD133+ chemoresistant population, where CD133+ cells appear to activate HSP27 and p38MAPK-MAPKAPK2 escape hypoxia-induced apoptosis [151,152]. Furthermore, hypoxia induces expression of stem cell markers in vivo and increases the number and size of liver metastases after hepatic xenotransplantation of human CRC cells into immunodeficient mice [153].

Inflammation has also been shown to contribute to the formation, maintenance, and expansion of CSCs in CRC [154]. Wang et al. demonstrated that levels of the inflammatory mediator prostaglandin E2 (PGE2) correlated with a number of CSC markers (including CD133, CD44, LRG5, and SOX2) in human CRC samples. Inhibition or genetic knockdown of PGE2 receptor 4 (EP4) reduced the ability of PGE2 to induce CSC expansion in orthotopic CRC tumors, as well as decreasing the formation of liver metastases. Multiple PGE downstream signaling molecules were tested to determine the mechanism driving CSC expansion including phosphoinositide 3-kinase (PI3K) p85α, ERK1, and NF-κB. Knockdown experiments showed that NF-κB was required for PGE induction of CSCs and metastasis in mice [155].

COLORECTAL CANCER STEM CELLS AND METASTASES

The literature on the role of CSCs in CRC metastases remains limited [156,157]. Whether the CSC markers that have been identified also play a role in initiating metastatic deposits remains to be determined. However, understanding whether the cell surface phenotypes commonly used to identify CSCs are mechanistically linked to the cells that drive metastasis formation remains an important question in the field. Pang et al. were the first to identify a CSC subpopulation with metastatic potential, characterized by the expression of CD26 [20]. The expression of CD26 was uniformly present in both the primary and metastatic tumors in colorectal patients with liver metastases. Interestingly, in patients without metastases at the time of presentation, the expression of CD26 in their primary tumors predicted distant metastasis on follow-up. Only the $CD26^+$ subpopulation possessed the capacity to form liver metastases upon orthotopic cecal injection into immunocompromised mice. Treatment with standard of care (SOC) chemotherapy, 5-fluoruracil (5-FU) or Oxaliplatin, resulted in decreased cell viability and enhanced apoptosis, but led to enrichment of the $CD133^+CD26^+$ CSC population [20]. These results imply that chemotherapeutic treatments that fail to eliminate CSC may enrich for a CSC subpopulation that is both chemoresistant and metastatic.

More recently, Todaro et al. identified CD44v6 as a marker of CSCs with metastatic potential [158]. CD44v6 expression was typically low in the primary tumors; however, cytokines secreted from cancer-associated fibroblasts (including HGF, OPN, and SDF-1α) were able to upregulate CD44v6 expression in CSCs by activating nuclear β-catenin signaling, which in turn promoted migration and metastasis. At baseline $CD44v6^-$ CRC cells did not give rise to metastases, but after exposure to the aforementioned cytokines the $CD44v6^-$ cells acquired expression and metastatic capacity. In this model, the inhibition of PI3K selectively targeted the $CD44v6^-$expressing CSC subset and reduced metastatic capacity [158]. This represents an example of the tumor microenvironment playing an important role in defining whether or not a CSC has metastatic potential. Further work is required to better understand the metastatic CSC subset and factors driving their formation.

CANCER STEM CELL AND CHEMORESISTANCE

In the majority of CRC patients the ability of conventional chemotherapy regimens to cure metastatic or recurrent disease is impaired by intrinsic and/or acquired tumor resistance to cytotoxic agents [159]. According to the CSC model, self-renewing CSCs that drive tumor growth and recurrence are difficult to target with conventional chemotherapy [4]. This paradigm suggests that conventional treatment targets the highly proliferative daughter cells in the tumor but fails to target the CSC population [3,5]. This notion has been supported by publications demonstrating an increase in phenotypic CSC markers postchemotherapy, including CD133, CD44, and ALDH1 [160,161]. Albeit many of these publications rely heavily on changes in phenotypic markers, there is also evidence in CRC that conventional chemotherapies (5-FU and Oxaliplatin) enrich for CSCs based on functional in vitro and in vivo LDAs [53,98]. It is well established in both CRC, as well as other solid tumors, that some CSCs exist in a quiescent state [49]. Since chemotherapeutic drugs typically target rapidly proliferating cells, it is rational to conclude that the reversible entry into quiescence is one of the mechanisms regulating chemoresistance in CSCs [162]. Apart from exiting the cell cycle there are numerous other protective mechanisms that characterize both CSC biology and chemoresistance including enhanced DNA damage repair, altered cell cycle checkpoint control, inhibition of apoptosis pathways, and increased expression of drug transporters and efflux pumps [4]. Todaro et al. published the initial paper showing that colon CSCs are resistant to 5-FU and Oxaliplatin [53]. Expression of ID1 and ID3 has also been linked to CSC chemoresistance. ID1 and ID3 were shown to drive CSC self-renewal in CRC through preventing the accumulation of excess DNA damage and subsequent functional exhaustion of CSCs. Simultaneous silencing of ID1 and ID3 decreased self-renewal capacity and increased sensitivity to Oxaliplatin both in vitro and in vivo [98]. These are two examples, however, there are numerous in the literature connecting the pathways that drive self-renewal with chemoresistance [63,69–72,77–79,103].

Another probable driver of chemoresistance in CSCs is related to intratumoral oxygen levels, where hypoxia has been linked to increased CSC activity in a number of solid tumors including CRC [149,150,163]. Furthermore, studies suggest that CSCs are preferentially found in hypoxic tumor niches. Hypoxia exerts a powerful pressure on tumor cells, resulting in increased genomic instability and worse overall prognosis [164,165]. It is typically characterized by increased HIF-1α expression, which in turn drives a chemoresistant program characterized by cell cycle arrest, inhibition of apoptosis, and altered DNA damage repair. Tumors treated with traditional chemotherapies, 5-FU and Oxaliplatin, show enrichment for a subpopulation of cancer cells coexpressing HIF-1α and CD133, giving further evidence that hypoxic CSCs are chemoresistant [151,166].

An emerging field in CSC research is connecting CSC metabolism as a potential driver of chemoresistance. One of the first publications to identify metabolic changes in therapy-resistant cell populations demonstrated significant overexpression of a metabolic regulator SIRT-1 in 5-FU-resistant CRC cells. Furthermore, inhibition of SIRT-1 expression in these CRC cells rendered them sensitive to 5-FU [167]. More recently, Denise et al. identified that 5-FU-resistant colon cancer cells are addicted to oxidative phosphorylation [168]. The resistant cells were characterized by enhanced stem cell traits including increased spherogenic potential. Combined treatment with 5-FU and a pharmacological inhibitor of oxidative phosphorylation increased chemosensitivity, abolished spherogenic potential, and diminished the expression of CD133+ CRC cells [168]. This work is supported by findings in pancreatic cancer where Viale et al. demonstrated that a population of dormant tumors cells that survive oncogene ablation and cause tumor relapse has features of CSCs and relies on oxidative phosphorylation for survival [169]. This represents an exciting avenue of emerging research in understanding the drivers of CSC chemoresistance.

THERAPEUTICALLY TARGETING COLORECTAL CANCER STEM CELLS

As has been reviewed in this chapter so far, CSCs employ numerous mechanisms (extrinsic and intrinsic) for tumor survival and progression. Thus, developing methods to directly target CSCs represents an attractive approach to obtain improved CRC disease control. Ideal CSC-targeted therapies would effectively inhibit CSC survival pathways and/or target the CSC microenvironment [159,170,171]. Numerous strategies are being tested to target this CSC subpopulation in tumors and are reviewed here (Fig. 7.1) (Table 7.2).

Table 7.2 List of Cancer Stem Cell (CSC)-Targeted Therapies

Drug	Class	Target	References
Celecoxib	Nonsteroidal antiinflammatory drug	Cyclooxygenase-2 (COX-2)	[172]
AEE788	Dual tyrosine kinase inhibitor	Epidermal growth factor receptor/ vascular endothelial growth factor receptor	[172]
Inhibitors of Wnt production	Small molecule inhibitor	Porcupine (PORCN)	[173]
Anti-hDLL4 21M18	Antibody	Delta-like 4 ligand	[174]
OMP-18R5	Small molecule inhibitor	Frizzled (FZD)	[175]
FJ9	Small molecule inhibitor	PDZ domain of dishevelled	[176]
NSC668036	Small molecule inhibitor	PDZ domain of dishevelled	[177]
3289-8625	Small molecule inhibitor	PDZ domain of dishevelled	[178]
JW55	Small molecule inhibitor	Tankyrases	[179]
JW67, JW74	Small molecule inhibitor	Tankyrases	[180]
PFK115-584, CGP049090	Small molecule inhibitor	β-Catenin/T-cell factor complex	[181]
ICG-001	Small molecule inhibitor	cAMP-response element-binding protein	[182]
PTC-209	Small molecule inhibitor	BMI-1	[92]
G0-Y030	Small molecule inhibitor	STAT3	[183]
5-Aza-deoxycytidine (5-aza-dC)	Small molecule inhibitor	DNA methyltransferases	[184]
MS133	Bispecific antibody (BiAb)	CD133 and CD3	[185]
Zoledronate	Aminobisphosphonates (N-BP), small molecule compound	Vγ9Vδ2 T-cell activator	[186]
IL-4Rα antagonist/IL-4 neutralizing antibody	Antibody	IL-4Rα/IL-4	[14]

Monoclonal Antibodies

One of the predominant strategies for targeting the CSC subset has been to target cell surface CSC markers using monoclonal antibodies against known CSC markers including CD133, CD44, LGR5, and ALDH1 [6,187]. This approach is challenged by the fact that many of the CSC markers are also expressed on normal intestinal epithelial stem cells, thereby potentially resulting in a narrow or nonexistent therapeutic window. One study showed that CD133+ cells derived from a primary CRC tumor were significantly more resistant to both Oxaliplatin and 5-FU compared to the CD133- population of cells derived from the same tumor [53]. The CD133+ CRC cells used interleukin-4 (IL-4) in an autocrine manner which resulted in their chemoresistance and treatment with an IL-4 neutralizing antibody sensitized the CD133+ population to both 5-FU and Oxaliplatin [14,53]. This study demonstrated an important role of IL-4 in CSC resistance to SOC chemotherapy and that IL-4 inhibition could prevent the enrichment of CSCs following treatment.

Monoclonal antibodies have also been used to target the Notch signaling pathway, which is overexpressed in colorectal CSCs, and has been shown to play a role in CSC viability and tumorigenicity. Hoey et al. specifically targeted the CSC subpopulation by inhibiting a key regulator in the Notch signaling pathway, delta-like 4 ligand (DLL4), with a monoclonal antibody [174]. Treatment with the DLL4-monoclonal antibody in combination with the SOC chemotherapeutic agent Irinotecan resulted in reduced tumor growth and stem cell frequency in a colorectal xenograft model [174].

Using multiple inhibitors in combination may increase their efficacy and decrease colorectal CSC self-renewal. EGFR and vascular endothelial growth factor receptor (VEGFR) inhibitors alone have limited success in treating CRC, however, when taken together the antitumoral effects are enhanced [172]. Past studies have suggested combining antiangiogenic tyrosine kinase inhibitors with Celecoxib, a selective inhibitor of cyclooxygenase-2 (COX-2), which increases its effectiveness against late stage CRC [172]. COX-2 has been shown to promote CRC growth via Wnt/β-catenin signaling, which is a main driver of colorectal CSC cell self-renewal and function [172]. Valverde et al. used a dual tyrosine kinase inhibitor (AEE788) that causes a blockade in both EGFR/VEGFR signaling pathways in combination with Celecoxib to target the CSC fraction [172]. The combination of AEE788 with Celecoxib prevented β-catenin translocation to the nucleus and resulted in the downregulation of numerous stem cell signature genes including FOXM1, Oct3/4, Nanog, Sox-2, and Snail [172]. In CRC cells, AEE788/Celecoxib treatment was associated with the inhibition of cell proliferation, induction of apoptosis, and reduction in cell migration in the CSC subset [172].

Targeting Self-Renewal

CSCs are functionally defined by their capacity to self-renew, and understanding the pathways driving self-renewal has led to a number of strategies designed to target the self-renewal machinery in an effort to eradicate the CSC fraction. Wnt pathway signaling is frequently increased in CRCs and mutations in the pathway often lead to cancer. It is also noted as a primary driver of colorectal CSC proliferation and maintenance [188]. See Section "Adenomatous Polyposis Coli and the Wnt Pathway". Targeting the Wnt pathway, a major driver of colorectal CSC proliferation and maintenance, is a promising colorectal CSC therapeutic strategy [188].

Generic Wnt inhibitors such as Celecoxib (NSAIDs), vitamin A, and vitamin D have been reported to repress Wnt/β-catenin signaling via a decrease in Wnt/β-catenin transcription, decrease in β-catenin protein levels, or competition for TCF binding, respectively [188,189]. Another way to target the Wnt pathway is through the inhibition of the Wnt protein. Porcupine (PORCN) adds the palmitoyl group to Wnt proteins which is critical for Wnt ligand secretion [173]. Chen et al. identified a class of small molecules that inhibit Wnt production through inhibition of PORCN [173].

A secondary method to target Wnt signaling is to disrupt Wnt pathway receptor/ligand interfaces with therapeutic monoclonal antibodies to Wnt-1 has been effective in driving apoptosis in CRC [188]. Inhibiting Wnt binding to frizzled (FZD), a receptor of Wnt required for pathway activation, with a monoclonal antibody OMP-18R5, resulted in inhibition of tumor growth in patient-derived xenograft models [175]. In addition to disrupting Wnt/FZD interactions, inhibiting downstream interactions of FZD would prevent Wnt signaling. FZD binds to the scaffold protein Dishevelled (DVL) which transduces Wnt signaling from receptors to downstream effectors [188,189]. Inhibiting the FZD/DVL interaction with small molecules NSC668036, FJ9, and 3289-8625 resulted in inhibition of Wnt signaling and suppression of tumor growth in CRC [176–178]. Other small molecules are used to inhibit tankyrases (JW55, JW67, JW74), which mediate AXIN degradation, part of the β-catenin destruction complex [179,180]. Stabilizing AXIN proteins and inhibiting Wnt signaling using these small molecule inhibitors is especially effective in *APC* mutant CRCs; however, since these molecules nonspecifically inhibit Wnt signaling, there are some safety concerns for clinical use.

A further strategy to inhibit Wnt signaling is to target downstream effectors, specifically the β-catenin/TCF complex. While there are some inhibitors are available such as PFK115-584 and CGP049090, they lack specificity and may prove toxic to normal tissues [181]. There have been more promising results targeting other downstream effectors such as the transcriptional coactivator cAMP-response element-binding protein (CBP), which binds to β-catenin to form a transcriptionally active complex upon Wnt signaling. The small molecule ICG-001 binds specifically to CBP, resulting in a disruption of the CBP/β-catenin interaction and a decrease in transcription of Wnt signaling genes and decrease in vitro CRC growth without causing toxicity to normal colorectal cells [182,190].

Another canonical regulator of self-renewal is the protooncogene polycomb ring finger protein BMI-1. It is required for the maintenance of adult stem cells in numerous tissues. It functions to repress differentiation and senescence programs through its role in chromatin modification. See Section "BMI-1". BMI-1 was shown to be overexpressed in numerous cancers, including CRCs, where it has been implicated driving CSC self-renewal capacity [92]. Genetic knockdown of BMI-1 or treatment with a small molecule inhibitor that specifically targets BMI-1 activity (PTC-209) resulted in decreased CRC tumor volume in primary xenograft models [92]. Most importantly, serial passage LDAs carried out with CRC cells from mice previously treated with PTC-209 showed a significant and sustained decrease in the CSC frequency [92].

The signal transducer and activator of transcription (STAT) family of transcription factors are frequently constitutively active in CRC [183]. In CRC it is well established that STAT3 has numerous important roles in tumorigenesis, including the ability to confer resistance to chemotherapy-induced apoptosis [183]. In the context of CSC function, STAT3 has been shown to play a central role in the maintenance of colorectal CSC self-renewal capacity [183]. Using a small molecule inhibitor of STAT3 phosphorylation (G0-Y030), CSC self-renewal was inhibited as demonstrated by the decrease in both tumorsphere formation and stem cell frequency. STAT3 inhibition induced apoptosis in the CSC subset in vitro and inhibited tumor growth in vivo [183].

Therapeutically targeting canonical regulators of self-renewal represents a novel approach to treating cancer and overcoming chemoresistance. In the context of preclinical xenograft models, targeting self-renewal has demonstrated great promise. However, the effectiveness of this approach will ultimately be determined by patient response in the context of human clinical trials, which are currently underway for a number of small molecule inhibitors targeting self-renewal.

Induction of Differentiation

Another approach to targeting CSCs is the idea of inducing CSC differentiation, thereby resulting in a loss of self-renewal capacity and the inhibition of tumorigenesis. Induction of differentiation in colon CSCs has been achieved by exposing the cells to bone morphogenic protein 4 (BMP4), which initiates a differentiation program resulting in decreased tumorigenic capacity. In addition, treatment with BMP4 sensitized CRC cells to the effects of chemotherapy

(5-FU or Oxaliplatin) in vivo, resulting in complete and long-term regression of CRC xenografts [191]. Although further work is needed in this area the current results imply that unlike CSCs, cells forced to undergo differentiation are rendered responsive to SOC chemotherapies in CRC.

Inhibition of DNA methylation is another strategy aimed at inducing CSC self-renewal. Seminal work showed that a mouse model with loss of both DNMT1 and DNMT3b was incapable of DNA methylation and exhibited a loss of tumorsphere-forming ability compared to control mice [192]. Building on this initial observation, colorectal CSCs were found to be sensitive to treatment with a DNA methyltransferase inhibitor, 5-aza-deoxycytidine (5-aza-dC), indicating that DNA methylation may play a central role in maintaining CSCs in the undifferentiated multipotential state [92,184,192]. 5-aza-dC is known to strongly induce cell differentiation in other systems, therefore it is hypothesized that in the context of colorectal CSCs, 5-Aza-dC causes CSCs to be forced into cell cycle and undergoes differentiation which results in the subsequent loss of tumorigenic potential [192]. Published work in breast cancer has shown that epigenetic modulators can drive non-CSCs into a CSC state [31,32]. Therefore, the idea of manipulating epigenetic pathways to drive cells from a CSC to non-CSC state represents an area of research that holds significant promise.

Immunotherapy

Immunotherapy is an emerging field for the treatment of malignant tumors. Clinical studies have shown that a combination of chemotherapy and immune therapy can increase survival in CRC patients. The development of bispecific antibodies (BiAbs) to target tumor-specific antigens and T lymphocytes may enable effective targeting of colorectal CSCs [185]. BiAbs link T lymphocytes to tumor cells by binding both a T-cell receptor and a molecule specific to the tumor cell, thus allowing more effective targeting. This therapeutic method is a novel way to enhance targeting of colorectal CSCs based on their unique cell surface properties. Zhao et al. developed a BiAb (MS133) that targets both CD133 and CD3, causing the activation of T cells resulting in increased cytotoxicity toward CD133^{high} CRC cells in vitro and inhibited tumor growth in vivo [185]. Antibodies that target both colorectal CSC markers and tumor antigens could increase the specificity of colorectal CSC-targeted therapy [4].

Another way to use immunotherapy is to activate T cells to induce apoptotic cell death in colorectal CSCs. Todaro et al. treated colorectal CSCs with Zoledronate, a Vγ9Vδ2 T-cell activator, which caused an increase in $\gamma\delta$ T lymphocyte proliferation and activation, cytokine production, and increased colorectal CSC susceptibility to perforin release verses untreated cells [186]. Since $\gamma\delta$ T cells are abundant in the intestinal mucosa, activating them may improve CRC therapy when combined with chemotherapy [186].

CONCLUSION

While some progress has been made in targeting CSCs, it is a difficult endeavor due to the complex signaling pathways involved and critical role these pathways play in normal physiology. To clinically exploit CSCs, there is a need to better understand CSC biology, including finding better CSC markers, determining the molecular mechanisms driving both self-renewal and chemoresistance, and discerning the differences between normal and CSCs that allow for targeted agents tolerable to normal tissues. Utilization of CSC markers in prediction and prognosis will be pertinent in aiding disease management by identifying patients who will have a poor response to current SOC regimens. These patients then may be suitable for novel targeted agents in the effort to outline effective treatment strategies, with the primary goal of attaining prolonged remission and improved patient survival. Since CSCs play a major role in chemoresistance and relapse, targeting them in combination with traditional chemotherapeutics may result in eradication of disease and would result in major advances in cancer treatment. The ultimate test of CSCs in CRC will come from clinical trials designed to target this subset of cells.

Glossary

Cancer Stem Cells (CSCs) Cancer cells with the capacity to undergo asymmetric division into a daughter stem cell (self-renewal) and a differentiated cell, both of which retain cancerous properties. These cells are found within the tumor mass and have properties similar to normal stem cells (embryonic or adult stem cells).

Chemotherapy Treatment of cancer disease using chemical substances/drugs, typically cytotoxic.

Differentiation Cellular differentiation is the process of a cell changing from one cell type to another, typically from a less specialized type (stem cell) to a more specialized type (organ/tissue specific cell, eg, colonocyte).

Epigenetic regulation Refers to the study of mechanisms of temporal and spatial control in gene expression due to covalent modifications of DNA and histones (the proteins that package DNA) via methylation and/or acetylation. These dynamic alterations in the transcriptional potential of a cell may or may not be heritable.

Hypoxia The state of the microenvironment of a tissue as a result of diminished availability of oxygen.

Immunotherapy Therapeutically using the patient's own immune system to target the tumor, typically achieved by activating immune cells to target the tumor and induce cell death.

Ligand The signal (protein/molecule) that binds to its corresponding receptor on a cell and elicits a cellular response.

Metastasis The development of secondary tumors at a distance (in the same organ or in another organ) from the primary tumor.

Microenvironment (of tumor) The immediate or small-scale environment surrounding the tumor mass.

Quiescence The state of a cell when it is dormant or in a period of inactivity.

Receptor A protein molecule that receives signals from the surrounding microenvironment of the cell. When signals (ligands) bind to the receptor, they result in some form of cellular/tissue response (eg, cell growth, quiescence).

Self-renewal The ability of a stem cell to form new stem cells with an intact potential for proliferation, expansion, and differentiation.

Stem cells Cells with the capacity to undergo asymmetric division into a daughter stem cell (self-renewal) and a differentiated cell.

Transformation (malignant) Malignant transformation defines the processes (genetic or epigenetic) by which cells acquire cancerous properties (uncontrolled proliferation, evasion of apoptosis).

Tumorigenesis The formation of a cancer where normal cells of an organ/tissue transform into cancer cells. This is defined by a series of changes at the genetic and epigenetic level that ultimately transforms a cell to undergo uncontrolled proliferation, forming a malignant mass.

Xenograft The tissue graft or organ transplant from a donor of a species that is different from the recipient. For example, tumors that are grown in mice using tissue/cells obtained from human patient tumors.

List of Acronyms and Abbreviations

5-Aza-Cd 5-Aza-deoxycytidine
ABC ATP-binding cassette
ALDH Aldehyde dehydrogenase
APC Adenomatous polyposis coli
bHLH Basic helix-loop-helix
BiAb Bispecific antibodies
BMI1 B cell–specific Moloney murine leukemia virus Integration site 1
BMP4 Bone morphogenic protein 4
CBC Crypt base columnar
CBP cAMP-response element-binding protein
CIMP CpG island methylator phenotype
CIN Chromosomal instability
CK1 Casein kinase1
COX-2 Cyclooxygenase-2
CRC Colorectal cancer
CSC Cancer stem cells
DLL4 Delta-like 4 ligand
DNMT DNA methyltransferase
DSS Dextran sulfate sodium
DVL Dishevelled
EGFR Epidermal growth factor receptor
EphB2 Ephrin type-B receptor 2
ERK Extracellular signal-regulated kinase
ESCs Embryonic stem cells
FACS Fluorescence-activated cell sorting
FZD Frizzled
GSK3β Glycogen synthase kinase-3β
GTP Guanosine triphosphate bound
HGF Hepatocyte growth factor
HIFs Hypoxia-inducible factors
ID Inhibitor of differentiation

IL-4 Interleukin-4
ISCs Intestinal stem cells
Krt19 Keratin-19
LDA Limiting dilution assay
Lgr5 Leucine-rich-repeat-containing G-protein-coupled receptor 5
LOI Loss of imprinting
Lrig1 Leucine-rich repeats and immunoglobulin-like domains 1
LSD1 Lysine-specific demethylase 1
MEK Mitogen-activated protein kinase
miRNA MicroRNA
MSI Microsatellite instability
NF-κβ Nuclear factor-κβ
PGE2 Prostaglandin E2
PI3K Phosphoinositide 3-kinase
PORCN Porcupine
PRC1 Polycomb repressor complex 1
PRC2 Polycomb repressor complex 2
SOC Standard of care
SP Side population
TA Transit amplifying
TCF T-cell factor
TCF/LEF T-cell factor/lymphoid enhancer family
TNM TNM staging—tumor, node, metastasis
VEGFR Vascular endothelial growth factor receptor

References

[1] Visvader JE, Lindeman GJ. Cancer stem cells: current status and evolving complexities. Cell Stem Cell June 14, 2012;10(6):717–28. PMID:22704512.

[2] O'Brien CA, Kreso A, Jamieson CH. Cancer stem cells and self-renewal. Clin Cancer Res June 15, 2010;16(12):3113–20. PMID:20530701.

[3] O'Brien CA, Kreso A, Dick JE. Cancer stem cells in solid tumors: an overview. Semin Radiat Oncol April 2009;19(2):71–7. PMID:19249644.

[4] Garza-Trevino EN, Said-Fernandez SL, Martinez-Rodriguez HG. Understanding the colon cancer stem cells and perspectives on treatment. Cancer Cell Int 2015;15(1):2. PMID:25685060, PMCID:4328053.

[5] Puglisi MA, Tesori V, Lattanzi W, Gasbarrini GB, Gasbarrini A. Colon cancer stem cells: controversies and perspectives. World J Gastroenterol May 28, 2013;19(20):2997–3006. PMID:23716979, PMCID:3662939.

[6] Kozovska Z, Gabrisova V, Kucerova L. Colon cancer: cancer stem cells markers, drug resistance and treatment. Biomed Pharmacother October 2014;68(8):911–6. PMID: 25458789.

[7] Huang EH, Hynes MJ, Zhang T, Ginestier C, Dontu G, Appelman H, et al. Aldehyde dehydrogenase 1 is a marker for normal and malignant human colonic stem cells (SC) and tracks SC overpopulation during colon tumorigenesis. Cancer Res April 15, 2009;69(8):3382–9. PMID:19336570, PMCID:2789401.

[8] Moreb JS. Aldehyde dehydrogenase as a marker for stem cells. Curr Stem Cell Res Ther December 2008;3(4):237–46. PMID:19075754.

[9] Khorrami S, Zavaran Hosseini A, Mowla SJ, Malekzadeh R. Verification of ALDH activity as a biomarker in colon cancer stem cells-derived HT-29 cell line. Iran J Cancer Prev October 2015;8(5):e3446. PMID:26634106, PMCID:4667234.

[10] Feng L, Wu JB, Yi FM. Isolation and phenotypic characterization of cancer stem-like side population cells in colon cancer. Mol Med Rep September 2015;12(3):3531–6. PMID:25997915.

[11] Vermeulen L, Todaro M, de Sousa Mello F, Sprick MR, Kemper K, Perez Alea M, et al. Single-cell cloning of colon cancer stem cells reveals a multi-lineage differentiation capacity. Proc Natl Acad Sci USA September 09, 2008;105(36):13427–32. PMID:18765800, PMCID:2533206.

[12] Shenoy AK, Fisher RC, Butterworth EA, Pi L, Chang LJ, Appelman HD, et al. Transition from colitis to cancer: high Wnt activity sustains the tumor-initiating potential of colon cancer stem cell precursors. Cancer Res October 01, 2012;72(19):5091–100. PMID:22902411, PMCID:3463774.

[13] Van der Flier LG, Sabates-Bellver J, Oving I, Haegebarth A, De Palo M, Anti M, et al. The intestinal Wnt/TCF signature. Gastroenterology February 2007;132(2):628–32. PMID:17320548.

[14] Todaro M, Alea MP, Di Stefano AB, Cammareri P, Vermeulen L, Iovino F, et al. Colon cancer stem cells dictate tumor growth and resist cell death by production of interleukin-4. Cell Stem Cell October 11, 2007;1(4):389–402. PMID:18371377.

[15] Haraguchi N, Ohkuma M, Sakashita H, Matsuzaki S, Tanaka F, Mimori K, et al. CD133+CD44+ population efficiently enriches colon cancer initiating cells. Ann Surg Oncol October 2008;15(10):2927–33. PMID:18663533.

[16] Dalerba P, Dylla SJ, Park IK, Liu R, Wang X, Cho RW, et al. Phenotypic characterization of human colorectal cancer stem cells. Proc Natl Acad Sci USA June 12, 2007;104(24): 10158–63. PMID:17548814, PMCID:1891215.

[17] Gao W, Chen L, Ma Z, Du Z, Zhao Z, Hu Z, et al. Isolation and phenotypic characterization of colorectal cancer stem cells with organ-specific metastatic potential. Gastroenterology September 2013;145(3):636–46.e5. PMID:23747337.

[18] Sukhdeo K, Paramban RI, Vidal JG, Elia J, Martin J, Rivera M, et al. Multiplex flow cytometry barcoding and antibody arrays identify surface antigen profiles of primary and metastatic colon cancer cell lines. PLoS One 2013;8(1):e53015. PMID:23308131, PMCID:3538639.

[19] Haraguchi N, Ishii H, Mimori K, Ohta K, Uemura M, Nishimura J, et al. CD49f-positive cell population efficiently enriches colon cancer-initiating cells. Int J Oncol August 2013;43(2):425–30. PMID:23708747.

[20] Pang R, Law WL, Chu AC, Poon JT, Lam CS, Chow AK, et al. A subpopulation of CD26+ cancer stem cells with metastatic capacity in human colorectal cancer. Cell Stem Cell June 04, 2010;6(6):603–15. PMID:20569697.

[21] Merlos-Suarez A, Barriga FM, Jung P, Iglesias M, Cespedes MV, Rossell D, et al. The intestinal stem cell signature identifies colorectal cancer stem cells and predicts disease relapse. Cell Stem Cell May 06, 2011;8(5):511–24. PMID:21419747.

[22] Takahashi H, Ishii H, Nishida N, Takemasa I, Mizushima T, Ikeda M, et al. Significance of Lgr5(+ve) cancer stem cells in the colon and rectum. Ann Surg Oncol April 2011;18(4):1166–74. PMID:21125339.

[23] Kobayashi S, Yamada-Okabe H, Suzuki M, Natori O, Kato A, Matsubara K, et al. LGR5-positive colon cancer stem cells interconvert with drug-resistant LGR5-negative cells and are capable of tumor reconstitution. Stem Cells December 2012;30(12):2631–44. PMID:23081779.

[24] Nakanishi Y, Seno H, Fukuoka A, Ueo T, Yamaga Y, Maruno T, et al. Dclk1 distinguishes between tumor and normal stem cells in the intestine. Nat Genet January 2013;45(1):98–103. PMID:23202126.

[25] Vega KJ, May R, Sureban SM, Lightfoot SA, Qu D, Reed A, et al. Identification of the putative intestinal stem cell marker doublecortin and CaM kinase-like-1 in Barrett's esophagus and esophageal adenocarcinoma. J Gastroenterol Hepatol April 2012;27(4):773–80. PMID:21916995, PMCID:3289765.

[26] Vermeulen L, De Sousa EMF, van der Heijden M, Cameron K, de Jong JH, Borovski T, et al. Wnt activity defines colon cancer stem cells and is regulated by the microenvironment. Nat Cell Biol May 2010;12(5):468–76. PMID:20418870, Epub 2010/04/27. eng.

[27] Horst D, Chen J, Morikawa T, Ogino S, Kirchner T, Shivdasani RA. Differential WNT activity in colorectal cancer confers limited tumorigenic potential and is regulated by MAPK signaling. Cancer Res March 15, 2012;72(6):1547–56. PMID:22318865, PMCID:3571091.

[28] Fazekas de St G. The evaluation of limiting dilution assays. J Immunol Methods March 12, 1982;49(2):R11–23. PMID:7040548.

[29] Marjanovic ND, Weinberg RA, Chaffer CL. Cell plasticity and heterogeneity in cancer. Clin Chem January 2013;59(1):168–79. PMID:23220226.

[30] Marjanovic ND, Weinberg RA, Chaffer CL. Poised with purpose: cell plasticity enhances tumorigenicity. Cell Cycle September 01, 2013;12(17):2713–4. PMID:23966153, PMCID:3899180.

[31] Chaffer CL, Brueckmann I, Scheel C, Kaestli AJ, Wiggins PA, Rodrigues LO, et al. Normal and neoplastic nonstem cells can spontaneously convert to a stem-like state. Proc Natl Acad Sci USA May 10, 2011;108(19):7950–5. PMID:21498687, PMCID:3093533.

[32] Chaffer CL, Marjanovic ND, Lee T, Bell G, Kleer CG, Reinhardt F, et al. Poised chromatin at the ZEB1 promoter enables breast cancer cell plasticity and enhances tumorigenicity. Cell July 03, 2013;154(1):61–74. PMID:23827675, PMCID:4015106.

[33] Barker N. Adult intestinal stem cells: critical drivers of epithelial homeostasis and regeneration. Nat Rev Mol Cell Biol January 2014;15(1):19–33. PMID:24326621.

[34] Rothenberg ME, Nusse Y, Kalisky T, Lee JJ, Dalerba P, Scheeren F, et al. Identification of a cKit(+) colonic crypt base secretory cell that supports Lgr5(+) stem cells in mice. Gastroenterology May 2012;142(5):1195–205.e6. PMID:22333952, PMCID:3911891.

[35] Barker N, van Oudenaarden A, Clevers H. Identifying the stem cell of the intestinal crypt: strategies and pitfalls. Cell Stem Cell October 05, 2012;11(4):452–60. PMID:23040474.

[36] Barker N, van Es JH, Kuipers J, Kujala P, van den Born M, Cozijnsen M, et al. Identification of stem cells in small intestine and colon by marker gene Lgr5. Nature October 25, 2007;449(7165):1003–7. PMID:17934449.

[37] Sato T, Vries RG, Snippert HJ, van de Wetering M, Barker N, Stange DE, et al. Single Lgr5 stem cells build crypt-villus structures in vitro without a mesenchymal niche. Nature May 14, 2009;459(7244):262–5. PMID:19329995.

[38] Sato T, Stange DE, Ferrante M, Vries RG, Van Es JH, Van den Brink S, et al. Long-term expansion of epithelial organoids from human colon, adenoma, adenocarcinoma, and Barrett's epithelium. Gastroenterology November 2011;141(5):1762–72. PMID:21889923.

[39] Powell AE, Wang Y, Li Y, Poulin EJ, Means AL, Washington MK, et al. The pan-ErbB negative regulator Lrig1 is an intestinal stem cell marker that functions as a tumor suppressor. Cell March 30, 2012;149(1):146–58. PMID:22464327, PMCID:3563328.

[40] Asfaha S, Hayakawa Y, Muley A, Stokes S, Graham TA, Ericksen RE, et al. Krt19(+)/Lgr5(-) cells are radioresistant cancer-initiating stem cells in the colon and intestine. Cell Stem Cell June 04, 2015;16(6):627–38. PMID:26046762, PMCID:4457942.

[41] Vermeulen L, Snippert HJ. Stem cell dynamics in homeostasis and cancer of the intestine. Nat Rev Cancer July 2014;14(7):468–80. PMID:24920463.

[42] Jung P, Sato T, Merlos-Suarez A, Barriga FM, Iglesias M, Rossell D, et al. Isolation and in vitro expansion of human colonic stem cells. Nat Med October 2011;17(10):1225–7. PMID:21892181.

[43] Huels DJ, Sansom OJ. Stem vs non-stem cell origin of colorectal cancer. Br J Cancer June 30, 2015;113(1):1–5. PMID:26110974, PMCID:4647531.

[44] Barker N, Ridgway RA, van Es JH, van de Wetering M, Begthel H, van den Born M, et al. Crypt stem cells as the cells-of-origin of intestinal cancer. Nature January 29, 2009;457(7229):608–11. PMID:19092804.

[45] Westphalen CB, Asfaha S, Hayakawa Y, Takemoto Y, Lukin DJ, Nuber AH, et al. Long-lived intestinal tuft cells serve as colon cancer-initiating cells. J Clin Invest March 2014;124(3):1283–95. PMID:24487592, PMCID:3934168.

[46] Schwitalla S, Fingerle AA, Cammareri P, Nebelsiek T, Goktuna SI, Ziegler PK, et al. Intestinal tumorigenesis initiated by dedifferentiation and acquisition of stem-cell-like properties. Cell January 17, 2013;152(1–2):25–38. PMID:23273993.

[47] Davis H, Irshad S, Bansal M, Rafferty H, Boitsova T, Bardella C, et al. Aberrant epithelial GREM1 expression initiates colonic tumorigenesis from cells outside the stem cell niche. Nat Med January 2015;21(1):62–70. PMID:25419707, PMCID:4594755.

[48] Meacham CE, Morrison SJ. Tumour heterogeneity and cancer cell plasticity. Nature September 19, 2013;501(7467):328–37. PMID:24048065, PMCID:4521623.

[49] Kreso A, Dick JE. Evolution of the cancer stem cell model. Cell Stem Cell March 06, 2014;14(3):275–91. PMID:24607403.

[50] Clarke MF, Dick JE, Dirks PB, Eaves CJ, Jamieson CH, Jones DL, et al. Cancer stem cells–perspectives on current status and future directions: AACR Workshop on cancer stem cells. Cancer Res October 01, 2006;66(19):9339–44. PMID:16990346.

[51] Dylla SJ, Beviglia L, Park IK, Chartier C, Raval J, Ngan L, et al. Colorectal cancer stem cells are enriched in xenogeneic tumors following chemotherapy. PLoS One 2008;3(6):e2428. PMID:18560594, PMCID:2413402.

[52] Horst D, Kriegl L, Engel J, Kirchner T, Jung A. CD133 expression is an independent prognostic marker for low survival in colorectal cancer. Br J Cancer October 21, 2008;99(8):1285–9. PMID:18781171, PMCID:2570510.

[53] Todaro M, Perez Alea M, Scopelliti A, Medema JP, Stassi G. IL-4-mediated drug resistance in colon cancer stem cells. Cell Cycle February 01, 2008;7(3):309–13. PMID:18235245.

[54] Dieter SM, Ball CR, Hoffmann CM, Nowrouzi A, Herbst F, Zavidij O, et al. Distinct types of tumor-initiating cells form human colon cancer tumors and metastases. Cell Stem Cell October 04, 2011;9(4):357–65. PMID:21982235.

[55] Fearon ER, Vogelstein B. A genetic model for colorectal tumorigenesis. Cell June 01, 1990;61(5):759–67. PMID:2188735.

[56] Mamoori A, Gopalan V, Anthony Smith R, King-Yin Lam A. Modulatory roles of microRNAs in the regulation of different signalling pathways in large bowel cancer stem cells. Biol Cell March 2016;108(3):51–64. PMID:26712035.

[57] Samowitz WS. Genetic and epigenetic changes in colon cancer. Exp Mol Pathol August 2008;85(1):64–7. PMID:18482722.

[58] Hanahan D, Weinberg RA. Hallmarks of cancer: the next generation. Cell March 04, 2011;144(5):646–74. PMID:21376230.

[59] Powell SM, Zilz N, Beazer-Barclay Y, Bryan TM, Hamilton SR, Thibodeau SN, et al. APC mutations occur early during colorectal tumorigenesis. Nature September 17, 1992;359(6392):235–7. PMID:1528264.

[60] Miyaki M, Konishi M, Kikuchi-Yanoshita R, Enomoto M, Igari T, Tanaka K, et al. Characteristics of somatic mutation of the adenomatous polyposis coli gene in colorectal tumors. Cancer Res June 01, 1994;54(11):3011–20. PMID:8187091.

[61] Cottrell S, Bicknell D, Kaklamanis L, Bodmer WF. Molecular analysis of APC mutations in familial adenomatous polyposis and sporadic colon carcinomas. Lancet September 12, 1992;340(8820):626–30. PMID:1355210.

[62] Esteller M, Sparks A, Toyota M, Sanchez-Cespedes M, Capella G, Peinado MA, et al. Analysis of adenomatous polyposis coli promoter hypermethylation in human cancer. Cancer Res August 15, 2000;60(16):4366–71. PMID:10969779.

[63] Aoki K, Taketo MM. Adenomatous polyposis coli (APC): a multi-functional tumor suppressor gene. J Cell Sci October 01, 2007;120(Pt 19):3327–35. PMID:17881494.

[64] Mann B, Gelos M, Siedow A, Hanski ML, Gratchev A, Ilyas M, et al. Target genes of beta-catenin-T cell-factor/lymphoid-enhancer-factor signaling in human colorectal carcinomas. Proc Natl Acad Sci USA February 16, 1999;96(4):1603–8. PMID:9990071, PMCID:15532.

[65] Sparks AB, Morin PJ, Vogelstein B, Kinzler KW. Mutational analysis of the APC/beta-catenin/Tcf pathway in colorectal cancer. Cancer Res March 15, 1998;58(6):1130–4. PMID:9515795.

[66] Korkaya H, Paulson A, Charafe-Jauffret E, Ginestier C, Brown M, Dutcher J, et al. Regulation of mammary stem/progenitor cells by PTEN/Akt/beta-catenin signaling. PLoS Biol June 02, 2009;7(6):e1000121. PMID:19492080, PMCID:2683567.

[67] Abrahamsson AE, Geron I, Gotlib J, Dao KH, Barroga CF, Newton IG, et al. Glycogen synthase kinase 3beta missplicing contributes to leukemia stem cell generation. Proc Natl Acad Sci USA March 10, 2009;106(10):3925–9. PMID:19237556, PMCID:2646624.

[68] Takahashi-Yanaga F, Kahn M. Targeting Wnt signaling: can we safely eradicate cancer stem cells?. Clin Cancer Res June 15, 2010;16(12):3153–62. PMID:20530697.

[69] Malumbres M, Barbacid M. RAS oncogenes: the first 30 years. Nat Rev Cancer June 2003;3(6):459–65. PMID:12778136.

[70] Shirasawa S, Furuse M, Yokoyama N, Sasazuki T. Altered growth of human colon cancer cell lines disrupted at activated Ki-ras. Science April 02, 1993;260(5104):85–8. PMID:8465203.

[71] Hung KE, Maricevich MA, Richard LG, Chen WY, Richardson MP, Kunin A, et al. Development of a mouse model for sporadic and metastatic colon tumors and its use in assessing drug treatment. Proc Natl Acad Sci USA January 26, 2010;107(4):1565–70. PMID:20080688, PMCID:2824379.

[72] Liu X, Jakubowski M, Hunt JL. KRAS gene mutation in colorectal cancer is correlated with increased proliferation and spontaneous apoptosis. Am J Clin Pathol February 2011;135(2):245–52. PMID:21228365.

[73] Le Rolle AF, Chiu TK, Zeng Z, Shia J, Weiser MR, Paty PB, et al. Oncogenic KRAS activates an embryonic stem cell-like program in human colon cancer initiation. Oncotarget January 19, 2016;7(3):2159–74. PMID:26744320.

[74] Pruitt K, Der CJ. Ras and Rho regulation of the cell cycle and oncogenesis. Cancer Lett September 28, 2001;171(1):1–10. PMID:11485822.

[75] Sparmann A, van Lohuizen M. Polycomb silencers control cell fate, development and cancer. Nat Rev Cancer November 2006;6(11):846–56. PMID:17060944.

[76] Sauvageau M, Sauvageau G. Polycomb group proteins: multi-faceted regulators of somatic stem cells and cancer. Cell Stem Cell September 03, 2010;7(3):299–313. PMID:20804967.

[77] Park IK, Qian D, Kiel M, Becker MW, Pihalja M, Weissman IL, et al. Bmi-1 is required for maintenance of adult self-renewing haematopoietic stem cells. Nature May 15, 2003;423(6937):302–5. PMID:12714971.

[78] Lessard J, Sauvageau G. Bmi-1 determines the proliferative capacity of normal and leukaemic stem cells. Nature May 15, 2003;423(6937):255–60. PMID:12714970.

[79] Molofsky AV, Pardal R, Iwashita T, Park IK, Clarke MF, Morrison SJ. Bmi-1 dependence distinguishes neural stem cell self-renewal from progenitor proliferation. Nature October 30, 2003;425(6961):962–7. PMID:14574365, PMCID:2614897.

[80] Yu T, Chen X, Zhang W, Colon D, Shi J, Napier D, et al. Regulation of the potential marker for intestinal cells, Bmi1, by beta-catenin and the zinc finger protein KLF4: implications for colon cancer. J Biol Chem February 03, 2012;287(6):3760–8. PMID:22170051, PMCID:3281718.

[81] Lopez-Arribillaga E, Rodilla V, Pellegrinet L, Guiu J, Iglesias M, Roman AC, et al. Bmi1 regulates murine intestinal stem cell proliferation and self-renewal downstream of Notch. Development January 1, 2015;142(1):41–50. PMID:25480918.

[82] Bruggeman SW, Hulsman D, Tanger E, Buckle T, Blom M, Zevenhoven J, et al. Bmi1 controls tumor development in an Ink4a/Arf-independent manner in a mouse model for glioma. Cancer Cell October 2007;12(4):328–41. PMID:17936558.

[83] Leung C, Lingbeek M, Shakhova O, Liu J, Tanger E, Saremaslani P, et al. Bmi1 is essential for cerebellar development and is overexpressed in human medulloblastomas. Nature March 18, 2004;428(6980):337–41. PMID:15029199.

[84] Lukacs RU, Memarzadeh S, Wu H, Witte ON. Bmi-1 is a crucial regulator of prostate stem cell self-renewal and malignant transformation. Cell Stem Cell December 03, 2010;7(6):682–93. PMID:21112563, PMCID:3019762.

[85] Glinsky GV, Berezovska O, Glinskii AB. Microarray analysis identifies a death-from-cancer signature predicting therapy failure in patients with multiple types of cancer. J Clin Invest June 2005;115(6):1503–21. PMID:15931389, PMCID:1136989.

[86] Tateishi K, Ohta M, Kanai F, Guleng B, Tanaka Y, Asaoka Y, et al. Dysregulated expression of stem cell factor Bmi1 in precancerous lesions of the gastrointestinal tract. Clin Cancer Res December 01, 2006;12(23):6960–6. PMID:17145814.

[87] Li DW, Tang HM, Fan JW, Yan DW, Zhou CZ, Li SX, et al. Expression level of Bmi-1 oncoprotein is associated with progression and prognosis in colon cancer. J Cancer Res Clin Oncol July 2010;136(7):997–1006. PMID:20024662.

[88] Maynard MA, Ferretti R, Hilgendorf KI, Perret C, Whyte P, Lees JA. Bmi1 is required for tumorigenesis in a mouse model of intestinal cancer. Oncogene July 10, 2014;33(28):3742–7. PMID:23955081, PMCID:3931743.

[89] Siddique HR, Parray A, Tarapore RS, Wang L, Mukhtar H, Karnes RJ, et al. BMI1 polycomb group protein acts as a master switch for growth and death of tumor cells: regulates TCF4-transcriptional factor-induced BCL2 signaling. PLoS One 2013;8(5):e60664. PMID:23671559, PMCID:3645992.

[90] Chiba T, Miyagi S, Saraya A, Aoki R, Seki A, Morita Y, et al. The polycomb gene product BMI1 contributes to the maintenance of tumor-initiating side population cells in hepatocellular carcinoma. Cancer Res October 01, 2008;68(19):7742–9. PMID:18829528.

[91] Abdouh M, Facchino S, Chatoo W, Balasingam V, Ferreira J, Bernier G. BMI1 sustains human glioblastoma multiforme stem cell renewal. J Neurosci July 15, 2009;29(28):8884–96. PMID:19605626.

[92] Kreso A, van Galen P, Pedley NM, Lima-Fernandes E, Frelin C, Davis T, et al. Self-renewal as a therapeutic target in human colorectal cancer. Nat Med January 2014;20(1):29–36. PMID:24292392, Epub 2013/12/03. eng.

[93] Lasorella A, Benezra R, Iavarone A. The ID proteins: master regulators of cancer stem cells and tumour aggressiveness. Nat Rev Cancer February 2014;14(2):77–91. PMID:24442143.

[94] Manrique I, Nguewa P, Bleau AM, Nistal-Villan E, Lopez I, Villalba M, et al. The inhibitor of differentiation isoform Id1b, generated by alternative splicing, maintains cell quiescence and confers self-renewal and cancer stem cell-like properties. Cancer Lett January 28, 2015;356(2 Pt B):899–909. PMID:25449776.

[95] Sun F, Xu Q, Zhao D, Chen CD. Id4 marks spermatogonial stem cells in the mouse testis. Sci Rep 2015;5:17594. PMID:26621350, PMCID:4665196.

[96] Zhang N, Yantiss RK, Nam HS, Chin Y, Zhou XK, Scherl EJ, et al. ID1 is a functional marker for intestinal stem and progenitor cells required for normal response to injury. Stem Cell Rep November 11, 2014;3(5):716–24. PMID:25418719, PMCID:4235234.

[97] Ono Y, Masuda S, Nam HS, Benezra R, Miyagoe-Suzuki Y, Takeda S. Slow-dividing satellite cells retain long-term self-renewal ability in adult muscle. J Cell Sci March 01, 2012;125(Pt 5):1309–17. PMID:22349695.

[98] O'Brien CA, Kreso A, Ryan P, Hermans KG, Gibson L, Wang Y, et al. ID1 and ID3 regulate the self-renewal capacity of human colon cancer-initiating cells through p21. Cancer Cell June 12, 2012;21(6):777–92. PMID:22698403.

[99] Grady WM, Myeroff LL, Swinler SE, Rajput A, Thiagalingam S, Lutterbaugh JD, et al. Mutational inactivation of transforming growth factor beta receptor type II in microsatellite stable colon cancers. Cancer Res January 15, 1999;59(2):320–4. PMID:9927040.

[100] Tanaka T, Watanabe T, Kitayama J, Kanazawa T, Kazama Y, Tanaka J, et al. Chromosome 18q deletion as a novel molecular predictor for colorectal cancer with simultaneous hepatic metastasis. Diagn Mol Pathol December 2009;18(4):219–25. PMID:19861895.

[101] Maru D, Wu TT, Canada A, Houlihan PS, Hamilton SR, Rashid A. Loss of chromosome 18q and DPC4 (Smad4) mutations in appendiceal adenocarcinomas. Oncogene January 22, 2004;23(3):859–64. PMID:14647445.

[102] Kodach LL, Bleuming SA, Musler AR, Pesssppelenbosch MP, Hommes DW, van den Brink GR, et al. The bone morphogenetic protein pathway is active in human colon adenomas and inactivated in colorectal cancer. Cancer January 15, 2008;112(2):300–6. PMID:18008360.

[103] Freeman TJ, Smith JJ, Chen X, Washington MK, Roland JT, Means AL, et al. Smad4-mediated signaling inhibits intestinal neoplasia by inhibiting expression of beta-catenin. Gastroenterology March 2012;142(3):562–71.e2. PMID:22115830, PMCID:3343368.

[104] van Engeland M, Derks S, Smits KM, Meijer GA, Herman JG. Colorectal cancer epigenetics: complex simplicity. J Clin Oncol April 01, 2011;29(10):1382–91. PMID:21220596, Epub 2011/01/12. eng.

[105] Easwaran H, Tsai HC, Baylin SB. Cancer epigenetics: tumor heterogeneity, plasticity of stem-like states, and drug resistance. Mol Cell June 05, 2014;54(5):716–27. PMID:24905005, Epub 2014/06/07. eng.

[106] Akhtar-Zaidi B, Cowper-Sal-lari R, Corradin O, Saiakhova A, Bartels CF, Balasubramanian D, et al. Epigenomic enhancer profiling defines a signature of colon cancer. Science May 11, 2012;336(6082):736–9. PMID:22499810, PMCID:3711120, Epub 2012/04/14. eng.

[107] Karpf AR, Matsui S. Genetic disruption of cytosine DNA methyltransferase enzymes induces chromosomal instability in human cancer cells. Cancer Res October 01, 2005;65(19):8635–9. PMID:16204030, Epub 2005/10/06. eng.

[108] Rodriguez J, Frigola J, Vendrell E, Risques RA, Fraga MF, Morales C, et al. Chromosomal instability correlates with genome-wide DNA demethylation in human primary colorectal cancers. Cancer Res September 01, 2006;66(17):8462–9468. PMID:16951157, Epub 2006/09/05. eng.

[109] Jie D, Zhongmin Z, Guoqing L, Sheng L, Yi Z, Jing W, et al. Positive expression of LSD1 and negative expression of E-cadherin correlate with metastasis and poor prognosis of colon cancer. Dig Dis Sci June 2013;58(6):1581–9. PMID:23314859, Epub 2013/01/15. eng.

[110] Huang Z, Li S, Song W, Li X, Li Q, Zhang Z, et al. Lysine-specific demethylase 1 (LSD1/KDM1A) contributes to colorectal tumorigenesis via activation of the Wnt/beta-catenin pathway by down-regulating Dickkopf-1 (DKK1) [corrected]. PLoS One 2013;8(7):e70077. PMID:23922913, PMCID:3724785, Epub 2013/08/08. eng.

[111] Liu YL, Gao X, Jiang Y, Zhang G, Sun ZC, Cui BB, et al. Expression and clinicopathological significance of EED, SUZ12 and EZH2 mRNA in colorectal cancer. J Cancer Res Clin Oncol April 2015;141(4):661–9. PMID:25326896, Epub 2014/10/20. eng.

[112] Vire E, Brenner C, Deplus R, Blanchon L, Fraga M, Didelot C, et al. The Polycomb group protein EZH2 directly controls DNA methylation. Nature February 16, 2006;439(7078):871–4. PMID:16357870, Epub 2005/12/17. eng.

[113] Fluge O, Gravdal K, Carlsen E, Vonen B, Kjellevold K, Refsum S, et al. Expression of EZH2 and Ki-67 in colorectal cancer and associations with treatment response and prognosis. Br J Cancer October 20, 2009;101(8):1282–9. PMID:19773751, PMCID:2768450, Epub 2009/09/24. eng.

[114] Cheng AS, Lau SS, Chen Y, Kondo Y, Li MS, Feng H, et al. EZH2-mediated concordant repression of Wnt antagonists promotes beta-catenin-dependent hepatocarcinogenesis. Cancer Res June 01, 2011;71(11):4028–39. PMID:21512140, Epub 2011/04/23. eng.

[115] Jung HY, Jun S, Lee M, Kim HC, Wang X, Ji H, et al. PAF and EZH2 induce Wnt/beta-catenin signaling hyperactivation. Mol Cell October 24, 2013;52(2):193–205. PMID:24055345, PMCID:4040269, Epub 2013/09/24. eng.

[116] Amrani YM, Gill J, Matevossian A, Alonzo ES, Yang C, Shieh JH, et al. The Paf oncogene is essential for hematopoietic stem cell function and development. J Exp Med August 29, 2011;208(9):1757–65. PMID:21844206, PMCID:3171089, Epub 2011/08/17. eng.

[117] Neri F, Dettori D, Incarnato D, Krepelova A, Rapelli S, Maldotti M, et al. TET1 is a tumour suppressor that inhibits colon cancer growth by derepressing inhibitors of the WNT pathway. Oncogene August 06, 2015;34(32):4168–76. PMID:25362856, Epub 2014/11/05. eng.

[118] Roulois D, Yau HL, Singhania R, Wang Y, Amavaz D, Shen SY, et al. DNA-demethylating agents target colorectal cancer cells by inducing viral mimicry by endogenous transcripts. Cell August 27, 2015;162(5):961–73.

[119] Jelinic P, Shaw P. Loss of imprinting and cancer. J Pathol February 2007;211(3):261–8. PMID:17177177, Epub 2006/12/21. eng.

[120] Sakatani T, Kaneda A, Iacobuzio-Donahue CA, Carter MG, de Boom Witzel S, Okano H, et al. Loss of imprinting of Igf2 alters intestinal maturation and tumorigenesis in mice. Science March 25, 2005;307(5717):1976–8. PMID:15731405, Epub 2005/02/26. eng.

[121] Sharma SV, Lee DY, Li B, Quinlan MP, Takahashi F, Maheswaran S, et al. A chromatin-mediated reversible drug-tolerant state in cancer cell subpopulations. Cell April 02, 2010;141(1):69–80. PMID:20371346, PMCID:2851638, Epub 2010/04/08. eng.

[122] Mikkelsen TS, Ku M, Jaffe DB, Issac B, Lieberman E, Giannoukos G, et al. Genome-wide maps of chromatin state in pluripotent and lineage-committed cells. Nature August 02, 2007;448(7153):553–60. PMID:17603471, PMCID:2921165, Epub 2007/07/03. eng.

[123] Ohm JE, Mali P, Van Neste L, Berman DM, Liang L, Pandiyan K, et al. Cancer-related epigenome changes associated with reprogramming to induced pluripotent stem cells. Cancer Res October 01, 2010;70(19):7662–73. PMID:20841480, PMCID:2980296, Epub 2010/09/16. eng.

[124] Easwaran H, Johnstone SE, Van Neste L, Ohm J, Mosbruger T, Wang Q, et al. A DNA hypermethylation module for the stem/progenitor cell signature of cancer. Genome Res May 2012;22(5):837–49. PMID:22391556, PMCID:3337430, Epub 2012/03/07. eng.

[125] Widschwendter M, Fiegl H, Egle D, Mueller-Holzner E, Spizzo G, Marth C, et al. Epigenetic stem cell signature in cancer. Nat Genet February 2007;39(2):157–8. PMID:17200673, Epub 2007/01/04. eng.

[126] Tan J, Yang X, Jiang X, Zhou J, Li Z, Lee PL, et al. Integrative epigenome analysis identifies a Polycomb-targeted differentiation program as a tumor-suppressor event epigenetically inactivated in colorectal cancer. Cell Death Dis 2014;5:e1324. PMID:25032847, PMCID:4123077, Epub 2014/07/18. eng.

[127] Weis B, Schmidt J, Maamar H, Raj A, Lin H, Toth C, et al. Inhibition of intestinal tumor formation by deletion of the DNA methyltransferase 3a. Oncogene April 02, 2015;34(14):1822–30. PMID:24837369, Epub 2014/05/20. eng.

[128] Shukla S, Meeran SM. Epigenetics of cancer stem cells: pathways and therapeutics. Biochim Biophys Acta December 2014;1840(12):3494–502. PMID:25240776, Epub 2014/09/23. eng.

[129] Yu Y, Kanwar SS, Patel BB, Oh PS, Nautiyal J, Sarkar FH, et al. MicroRNA-21 induces stemness by downregulating transforming growth factor beta receptor 2 (TGFbetaR2) in colon cancer cells. Carcinogenesis January 2012;33(1):68–76. PMID:22072622, PMCID:3276336, Epub 2011/11/11. eng.

[130] Volinia S, Calin GA, Liu CG, Ambs S, Cimmino A, Petrocca F, et al. A microRNA expression signature of human solid tumors defines cancer gene targets. Proc Natl Acad Sci USA February 14, 2006;103(7):2257–61. PMID:16461460, PMCID:1413718. Epub 2006/02/08. eng.

[131] Omrane I, Kourda N, Stambouli N, Privat M, Medimegh I, Arfaoui A, et al. MicroRNAs 146a and 147b biomarkers for colorectal tumor's localization. Biomed Res Int 2014;2014:584852. PMID:24800242, PMCID:3985145, Epub 2014/05/07. eng.

[132] Hwang WL, Jiang JK, Yang SH, Huang TS, Lan HY, Teng HW, et al. MicroRNA-146a directs the symmetric division of Snail-dominant colorectal cancer stem cells. Nat Cell Biol March 2014;16(3):268–80. PMID:24561623, Epub 2014/02/25. eng.

[133] Iliou MS, da Silva-Diz V, Carmona FJ, Ramalho-Carvalho J, Heyn H, Villanueva A, et al. Impaired DICER1 function promotes stemness and metastasis in colon cancer. Oncogene July 24, 2014;33(30):4003–15. PMID:24096488, PMCID:4114136, Epub 2013/10/08. eng.

[134] Bu P, Chen KY, Chen JH, Wang L, Walters J, Shin YJ, et al. A microRNA miR-34a-regulated bimodal switch targets Notch in colon cancer stem cells. Cell Stem Cell May 02, 2013;12(5):602–15. PMID:23642368, PMCID:3646336, Epub 2013/05/07. eng.

[135] Xu XT, Xu Q, Tong JL, Zhu MM, Nie F, Chen X, et al. MicroRNA expression profiling identifies miR-328 regulates cancer stem cell-like SP cells in colorectal cancer. Br J Cancer March 27, 2012;106(7):1320–30. PMID:22453125, PMCID:3314795, Epub 2012/03/29. eng.

[136] Bitarte N, Bandres E, Boni V, Zarate R, Rodriguez J, Gonzalez-Huarriz M, et al. MicroRNA-451 is involved in the self-renewal, tumorigenicity, and chemoresistance of colorectal cancer stem cells. Stem Cells November 2011;29(11):1661–71. PMID:21948564, Epub 2011/09/29. eng.

[137] Wellner U, Schubert J, Burk UC, Schmalhofer O, Zhu F, Sonntag A, et al. The EMT-activator ZEB1 promotes tumorigenicity by repressing stemness-inhibiting microRNAs. Nat Cell Biol December 2009;11(12):1487–95. PMID:19935649, Epub 2009/11/26. eng.

[138] Brabletz S, Bajdak K, Meidhof S, Burk U, Niedermann G, Firat E, et al. The ZEB1/miR-200 feedback loop controls Notch signalling in cancer cells. EMBO J February 16, 2011;30(4):770–82. PMID:21224848, PMCID:3041948, Epub 2011/01/13. eng.

[139] Jones MF, Hara T, Francis P, Li XL, Bilke S, Zhu Y, et al. The CDX1-microRNA-215 axis regulates colorectal cancer stem cell differentiation. Proc Natl Acad Sci USA March 31, 2015;112(13):E1550–8. PMID:25775580, PMCID:4386393, Epub 2015/03/17. eng.

[140] Rokkas T, Kothonas F, Rokka A, Koukoulis G, Symvoulakis E. The role of circulating microRNAs as novel biomarkers in diagnosing colorectal cancer: a meta-analysis. Eur J Gastroenterol Hepatol July 2015;27(7):819–25. PMID:25856691.

[141] Zeng W, Tu Y, Zhu Y, Wang Z, Li C, Lao L, et al. Predictive power of circulating miRNAs in detecting colorectal cancer. Tumour Biol April 2015;36(4):2559–67. PMID:25527153.

[142] Fodde R, Brabletz T. Wnt/beta-catenin signaling in cancer stemness and malignant behavior. Curr Opin Cell Biol April 2007;19(2):150–8. PMID:17306971.

[143] Colak S, Zimberlin CD, Fessler E, Hogdal L, Prasetyanti PR, Grandela CM, et al. Decreased mitochondrial priming determines chemoresistance of colon cancer stem cells. Cell Death Differ July 2014;21(7):1170–7. PMID:24682005, PMCID:4207483.

[144] Koukourakis MI, Giatromanolaki A, Harris AL, Sivridis E. Comparison of metabolic pathways between cancer cells and stromal cells in colorectal carcinomas: a metabolic survival role for tumor-associated stroma. Cancer Res January 15, 2006;66(2):632–7. PMID:16423989.

[145] Papandreou I, Cairns RA, Fontana L, Lim AL, Denko NC. HIF-1 mediates adaptation to hypoxia by actively downregulating mitochondrial oxygen consumption. Cell Metab March 2006;3(3):187–97. PMID:16517406.

[146] Zeng M, Kikuchi H, Pino MS, Chung DC. Hypoxia activates the K-ras proto-oncogene to stimulate angiogenesis and inhibit apoptosis in colon cancer cells. PLoS One 2010;5(6):e10966. PMID:20532039, PMCID:2881039.

[147] Kaidi A, Williams AC, Paraskeva C. Interaction between beta-catenin and HIF-1 promotes cellular adaptation to hypoxia. Nat Cell Biol February 2007;9(2):210–7. PMID:17220880.

[148] Mazumdar J, O'Brien WT, Johnson RS, LaManna JC, Chavez JC, Klein PS, et al. O_2 regulates stem cells through Wnt/beta-catenin signalling. Nat Cell Biol October 2010;12(10):1007–13. PMID:20852629, PMCID:3144143.

[149] Li Z, Bao S, Wu Q, Wang H, Eyler C, Sathornsumetee S, et al. Hypoxia-inducible factors regulate tumorigenic capacity of glioma stem cells. Cancer Cell June 02, 2009;15(6):501–13. PMID:19477429, PMCID:2693960.

[150] Yeung TM, Gandhi SC, Bodmer WF. Hypoxia and lineage specification of cell line-derived colorectal cancer stem cells. Proc Natl Acad Sci USA March 15, 2011;108(11):4382–7. PMID:21368208, PMCID:3060223.

[151] Mao Q, Zhang Y, Fu X, Xue J, Guo W, Meng M, et al. A tumor hypoxic niche protects human colon cancer stem cells from chemotherapy. J Cancer Res Clin Oncol February 2013;139(2):211–22. PMID:23052691.

[152] Lin SP, Lee YT, Wang JY, Miller SA, Chiou SH, Hung MC, et al. Survival of cancer stem cells under hypoxia and serum depletion via decrease in PP2A activity and activation of p38-MAPKAPK2-Hsp27. PLoS One 2012;7(11):e49605. PMID:23185379, PMCID:3502468.

[153] Govaert KM, Emmink BL, Nijkamp MW, Cheung ZJ, Steller EJ, Fatrai S, et al. Hypoxia after liver surgery imposes an aggressive cancer stem cell phenotype on residual tumor cells. Ann Surg April 2014;259(4):750–9. PMID:24253142.

[154] Zhou C, Liu J, Tang Y, Liang X. Inflammation linking EMT and cancer stem cells. Oral Oncol November 2012;48(11):1068–75. PMID:22766510.

[155] Wang D, Fu L, Sun H, Guo L, DuBois RN. Prostaglandin E2 promotes colorectal cancer stem cell expansion and metastasis in mice. Gastroenterology December 2015;149(7):1884–95. e4. PMID:26261008.

[156] Bleau AM, Agliano A, Larzabal L, de Aberasturi AL, Calvo A. Metastatic dormancy: a complex network between cancer stem cells and their microenvironment. Histol Histopathol December 2014;29(12):1499–510. PMID:24887025.

[157] Borovski T, De Sousa EMF, Vermeulen L, Medema JP. Cancer stem cell niche: the place to be. Cancer Res February 01, 2011;71(3):634–9. PMID:21266356.

[158] Todaro M, Gaggianesi M, Catalano V, Benfante A, Iovino F, Biffoni M, et al. CD44v6 is a marker of constitutive and reprogrammed cancer stem cells driving colon cancer metastasis. Cell Stem Cell March 06, 2014;14(3):342–56. PMID:24607406.

[159] Paldino E, Tesori V, Casalbore P, Gasbarrini A, Puglisi MA. Tumor initiating cells and chemoresistance: which is the best strategy to target colon cancer stem cells?. BioMed Res Int 2014;2014:859871. PMID:24527460, PMCID:3914574.

[160] Nautiyal J, Kanwar SS, Yu Y, Majumdar AP. Combination of dasatinib and curcumin eliminates chemo-resistant colon cancer cells. J Mol Signal 2011;6:7. PMID:21774804, PMCID:3162943.

[161] Ong CW, Kim LG, Kong HH, Low LY, Iacopetta B, Soong R, et al. CD133 expression predicts for non-response to chemotherapy in colorectal cancer. Mod Pathol March 2010;23(3):450–7. PMID:20081809.

[162] Touil Y, Igoudjil W, Corvaisier M, Dessein AF, Vandomme J, Monte D, et al. Colon cancer cells escape 5FU chemotherapy-induced cell death by entering stemness and quiescence associated with the c-Yes/YAP axis. Clin Cancer Res February 15, 2014;20(4):837–46. PMID:24323901, PMCID:4387277.

[163] Xing F, Okuda H, Watabe M, Kobayashi A, Pai SK, Liu W, et al. Hypoxia-induced Jagged2 promotes breast cancer metastasis and self-renewal of cancer stem-like cells. Oncogene September 29, 2011;30(39):4075–86. PMID:21499308, PMCID:3145824.

[164] Koshiji M, To KK, Hammer S, Kumamoto K, Harris AL, Modrich P, et al. HIF-1alpha induces genetic instability by transcriptionally downregulating MutSalpha expression. Mol Cell March 18, 2005;17(6):793–803. PMID:15780936.

[165] Rajaganeshan R, Prasad R, Guillou PJ, Poston G, Scott N, Jayne DG. The role of hypoxia in recurrence following resection of Dukes' B colorectal cancer. Int J Colorectal Dis November 2008;23(11):1049–55. PMID:18594846.

[166] Roberts DL, Williams KJ, Cowen RL, Barathova M, Eustace AJ, Brittain-Dissont S, et al. Contribution of HIF-1 and drug penetrance to oxaliplatin resistance in hypoxic colorectal cancer cells. Br J Cancer October 20, 2009;101(8):1290–7. PMID:19755992, PMCID:2768443.

[167] Chen X, Sun K, Jiao S, Cai N, Zhao X, Zou H, et al. High levels of SIRT1 expression enhance tumorigenesis and associate with a poor prognosis of colorectal carcinoma patients. Sci Rep 2014;4:7481. PMID:25500546, PMCID:4265776.

[168] Denise C, Paoli P, Calvani M, Taddei ML, Giannoni E, Kopetz S, et al. 5-Fluorouracil resistant colon cancer cells are addicted to OXPHOS to survive and enhance stem-like traits. Oncotarget December 08, 2015;6(39):41706–21. PMID:26527315.

[169] Viale A, Pettazzoni P, Lyssiotis CA, Ying H, Sanchez N, Marchesini M, et al. Oncogene ablation-resistant pancreatic cancer cells depend on mitochondrial function. Nature October 30, 2014;514(7524):628–32. PMID:25119024, PMCID:4376130.

[170] Ricci-Vitiani L, Fabrizi E, Palio E, De Maria R. Colon cancer stem cells. J Mol Med November 2009;87(11):1097–104. PMID:19727638.

[171] Todaro M, Francipane MG, Medema JP, Stassi G. Colon cancer stem cells: promise of targeted therapy. Gastroenterology June 2010;138(6):2151–62. PMID:20420952.

[172] Valverde A, Penarando J, Canas A, Lopez-Sanchez LM, Conde F, Hernandez V, et al. Simultaneous inhibition of EGFR/VEGFR and cyclooxygenase-2 targets stemness-related pathways in colorectal cancer cells. PLoS One 2015;10(6):e0131363. PMID:26107817, PMCID:4479446.

[173] Chen B, Dodge ME, Tang W, Lu J, Ma Z, Fan CW, et al. Small molecule-mediated disruption of Wnt-dependent signaling in tissue regeneration and cancer. Nat Chem Biol February 2009;5(2):100–7. PMID:19125156, PMCID:2628455.

[174] Hoey T, Yen WC, Axelrod F, Basi J, Donigian L, Dylla S, et al. DLL4 blockade inhibits tumor growth and reduces tumor-initiating cell frequency. Cell Stem Cell August 07, 2009;5(2):168–77. PMID:19664991.

[175] Gurney A, Axelrod F, Bond CJ, Cain J, Chartier C, Donigan L, et al. Wnt pathway inhibition via the targeting of Frizzled receptors results in decreased growth and tumorigenicity of human tumors. Proc Natl Acad Sci USA July 17, 2012;109(29):11717–22. PMID:22753465, PMCID:3406803.

[176] Fujii N, You L, Xu Z, Uematsu K, Shan J, He B, et al. An antagonist of dishevelled protein-protein interaction suppresses beta-catenin-dependent tumor cell growth. Cancer Res January 15, 2007;67(2):573–9. PMID:17234765.

[177] Shan J, Shi DL, Wang J, Zheng J. Identification of a specific inhibitor of the dishevelled PDZ domain. Biochemistry November 29, 2005;44(47):15495–503. PMID:16300398.

[178] Grandy D, Shan J, Zhang X, Rao S, Akunuru S, Li H, et al. Discovery and characterization of a small molecule inhibitor of the PDZ domain of dishevelled. J Biol Chem June 12, 2009;284(24):16256–63. PMID:19383605, PMCID:2713547.

[179] Waaler J, Machon O, Tumova L, Dinh H, Korinek V, Wilson SR, et al. A novel tankyrase inhibitor decreases canonical Wnt signaling in colon carcinoma cells and reduces tumor growth in conditional APC mutant mice. Cancer Res June 01, 2012;72(11):2822–32. PMID:22440753.

[180] Waaler J, Machon O, von Kries JP, Wilson SR, Lundenes E, Wedlich D, et al. Novel synthetic antagonists of canonical Wnt signaling inhibit colorectal cancer cell growth. Cancer Res January 01, 2011;71(1):197–205. PMID:21199802.

[181] Lepourcelet M, Chen YN, France DS, Wang H, Crews P, Petersen F, et al. Small-molecule antagonists of the oncogenic Tcf/beta-catenin protein complex. Cancer Cell January 2004;5(1):91–102. PMID:14749129.

[182] Emami KH, Nguyen C, Ma H, Kim DH, Jeong KW, Eguchi M, et al. A small molecule inhibitor of beta-catenin/CREB-binding protein transcription [corrected]. Proc Natl Acad Sci USA August 24, 2004;101(34):12682–7. PMID:15314234, PMCID:515116.

[183] Lin L, Liu Y, Li H, Li PK, Fuchs J, Shibata H, et al. Targeting colon cancer stem cells using a new curcumin analogue, GO-Y030. Br J Cancer July 12, 2011;105(2):212–20. PMID:21694723, PMCID:3142799.

[184] Jones PA, Taylor SM. Cellular differentiation, cytidine analogs and DNA methylation. Cell May 1980;20(1):85–93. PMID:6156004.

[185] Zhao L, Yang Y, Zhou P, Ma H, Zhao X, He X, et al. Targeting CD133high colorectal Cancer cells in vitro and in vivo with an asymmetric bispecific antibody. J Immunother July–August 2015;38(6):217–28. PMID:26049545.

[186] Todaro M, D'Asaro M, Caccamo N, Iovino F, Francipane MG, Meraviglia S, et al. Efficient killing of human colon cancer stem cells by gammadelta T lymphocytes. J Immunol June 01, 2009;182(11):7287–96. PMID:19454726.

[187] Junttila MR, Mao W, Wang X, Wang BE, Pham T, Flygare J, et al. Targeting LGR5+ cells with an antibody-drug conjugate for the treatment of colon cancer. Sci Transl Med November 18, 2015;7(314):314ra186. PMID:26582901.

[188] Novellasdemunt L, Antas P, Li VS. Targeting Wnt signaling in colorectal cancer. A review in the theme: cell signaling: proteins, pathways and mechanisms. Am J Physiol Cell Physiol October 15, 2015;309(8):C511–21. PMID:26289750, PMCID:4609654.

[189] Song L, Li Y, He B, Gong Y. Development of small molecules targeting the Wnt signaling pathway in cancer stem cells for the treatment of colorectal cancer. Clin Colorectal Cancer September 2015;14(3):133–45. PMID:25799881.

[190] Lazarova DL, Chiaro C, Wong T, Drago E, Rainey A, O'Malley S, et al. Cbp activity mediates effects of the histone deacetylase inhibitor butyrate on WNT activity and apoptosis in colon cancer cells. J Cancer 2013;4(6):481–90. PMID:23901348, PMCID:3726710.

[191] Lombardo Y, Scopelliti A, Cammareri P, Todaro M, Iovino F, Ricci-Vitiani L, et al. Bone morphogenetic protein 4 induces differentiation of colorectal cancer stem cells and increases their response to chemotherapy in mice. Gastroenterology January 2011;140(1):297–309. PMID:20951698.

[192] Wang S. Enrichment and selective targeting of cancer stem cells in colorectal cancer cell lines. Hum Genet Embryol 2013;01(S2).

CHAPTER 8

Targeting Bladder Cancer Stem Cells: One Step Closer to the Clinic?

A.V. Kurtova, K.S. Chan
Baylor College of Medicine, Houston, TX, United States

CONTENTS

INTRODUCTION

Bladder cancer is the second most common urological malignancy after prostate cancer, accounting for 16,000 deaths in 2015 in the United States. More than 90% of bladder cancers are pathologically urothelial carcinomas (UCs) that arise from the urothelium, containing functional adult stem cells that give rise to downstream differentiated cells. Cancer arising from such hierarchically organized tissue architecture would likely retain part of cellular differentiation. Indeed, decades of pathological analyses revealed the existence of bladder cancers representing various differentiation states. For instance, poorly differentiated (or high-grade) bladder UCs are either invasive or have the propensity to become invasive [1]. On the contrary, well-differentiated (or low-grade) bladder UCs generally have better clinical outcome [1]. To date, pathological grading remains one of the most important independent factors for clinical prognosis of bladder cancers [1]. Pathological grading defines cancer differentiation on the basis of histology, thus supporting the notion that cellular differentiation becomes aberrant within bladder cancers (to be discussed in Section Bladder Cancer Stem Cell Signaling).

The cancer stem cell (CSC) hypothesis builds upon similar pathological observations and understanding from normal developmental biology and provides a biological and molecular explanation for intratumoral heterogeneity of bladder UCs [2]. However, it should be noted that the term "cancer stem cell" still most accurately describes a tumor subpopulation that can enrich for tumorigenic cells, with the self-renewal and differentiation capacity to generate cellular heterogeneity within a tumor [3]. These CSCs need not necessarily arise from normal stem cells and could also derive from differentiated progenies by acquiring the functional properties of CSCs via genetic or epigenetic

Cancer Stem Cells. http://dx.doi.org/10.1016/B978-0-12-803892-5.00008-5

alterations. Prospective isolation of the first CSCs from human acute myeloid leukemia and breast cancer was extensively reviewed elsewhere [3,4] and will not be discussed here. In this book chapter, we will focus on reviewing recent literature on the isolation and characterization of normal urothelial and bladder CSCs, and how current understanding of relevant signaling could translate into therapeutic targeting of bladder CSCs.

Bladder Organ-Specific Stem Cells

The adult urinary bladder is composed of the *urothelium* (epithelium of the urinary bladder or the mucosa), which is supported by the *lamina propria* (submucosa) containing sheets of extracellular matrix, and a detrusor smooth muscle layer of *muscularis propria* arranged as inner longitudinal, middle circular, and outer longitudinal muscle bundles, all encapsulated by an outer layer of perivesical adipose or soft tissues. The human adult urothelium is histologically classified as a transitional epithelium, which comprises of at least three to six layers of basal, intermediate, and superficial/umbrella cells [5,6]. Basal cells are small and polygonal in shape (~10 μm in diameter) and usually form a single layer in direct contact with the basement membrane. Intermediate cells are pyriform in shape (~10–40 μm in diameter) and form one to several cell layers. Umbrella cells are in direct contact with the urinary space, and they form a single layer of cells that are large bi- or multinucleated cuboidal cells (~70–100 μm in diameter) [5,7]. A unique feature of transitional urothelium is that it can accommodate bladder filling and emptying through distension and subsequent retraction (Fig. 8.1A).

The adult human urothelium is estimated to renew itself every 3–6 months during normal homeostasis [6]. However, upon pathological damage (for example, bacterial induced) or chemically induced injury (for example, following exposure to cyclophosphamide) [8], basal cells of the murine urothelium rapidly proliferate, and the urothelium is completely regenerated within 72 hours [9]. Similarly, when rat urothelium is exposed to protamine sulfate, which induces restricted abrasion of umbrella cells while sparing the intermediate cell layer, the intermediate cells possess the regenerative capacity to form a functionally intact urothelium by day 10 [10].

The urothelium displays properties of plasticity, where it can undergo metaplastic transdifferentiation. For example, squamous metaplasia of the trigonal region of the bladder is commonly seen [11] as is metaplastic transformation in cystitis glandularis [12]. The normal characteristics of a slow-cycling urothelium with the capacity for rapid regeneration in response to insult and possessing a robust capacity to transdifferentiate into other cell types as pathological response supports the existence of normal adult stem cells within the urothelium (Fig. 8.1B).

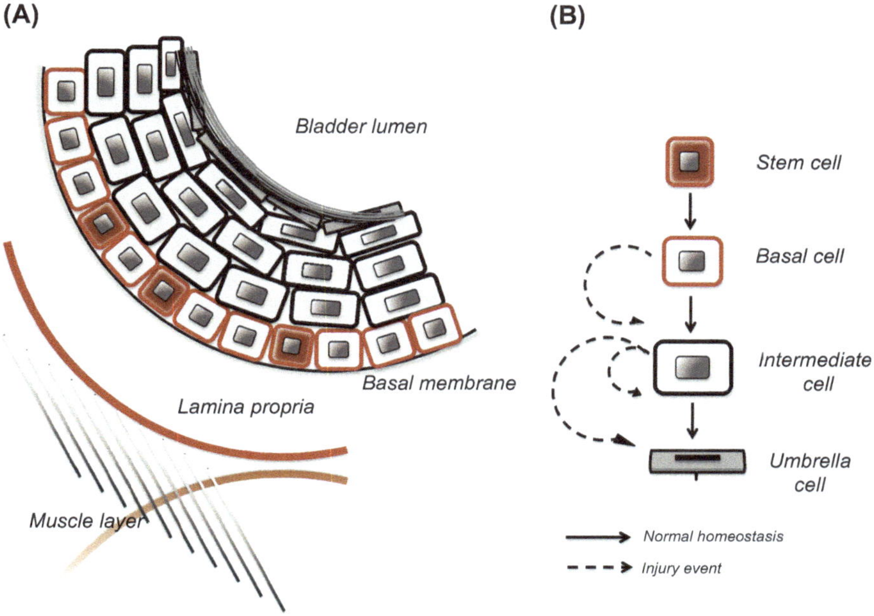

FIGURE 8.1 **Cellular organization of normal bladder urothelium.**
(A) Spatial localization of basal, intermediate, and umbrella cells. Basal cells are located in direct contact with basal membrane and contain a subpopulation of stem cells. Intermediate cells usually form several layers apical to basal cells. Umbrella cells form a single layer of cells in direct contact with the urinary space. (B) Schematic representation of renewal patterns in bladder urothelium under normal homeostasis (*solid lines*) and injury conditions (*dashed lines* represent hypothetical alternative pathways employed in tissue regeneration).

Slow-Cycling Characteristics of Urothelial Stem Cells

Epithelial stem cells have been demonstrated to possess a unique label-retention property, resulting from either asymmetric segregation of DNA strand and/or slow-cycling characteristics of stem cells [13]. In label-retaining assays, cells are exposed to labeled or synthetic nucleosides, such as ^{3}H-thymidine and 5-bromo-2′-deoxyuridine (BrdU), respectively. Replicating cells incorporate these nucleosides into nascent DNA strands during S phase of cell cycle. Repeated incorporation of these modified nucleosides ensures labeling of all cellular populations, including stem cells. Upon successive cell divisions and replications, the presence of the labeled nucleoside becomes diluted and, therefore, diminishes. Consequently, rapidly proliferating cells dilute their nucleoside labels while stem cells retain the label due to slow-cycling characteristics or asymmetric retention of the labeled DNA strand. Therefore, stem cells identified by such method are referred to as label-retaining stem cells (LRCs).

Using this label-retaining approach, adult stem cells have been localized in the crypts of intestine [14], the bulge region of the hair follicle [15], the corneal limbus [16], the endometrium [17], the prostatic ducts [18], and the bladder [19,20]. While the adult urothelium is known to be slow cycling, LRCs are found to localize restrictedly to the basal layer and are estimated to comprise ~9% of all basal cells [20]. These LRCs also express β4 integrin and are morphologically small with low granularity–characteristics of basal cells. They are found to be highly clonogenic in an in vitro stem cell assay [20]. However, another study using another synthetic nucleoside 5-ethynyl-2-deoxyuridine (EdU) to locate LRC in rat bladder resulted in a different outcome [21]. Results from this study demonstrated that the distribution of LRCs is random within the layers of urothelial cells, with no evidence of preferential labeling to basal cells [21]. Additionally, subsequent studies in mice were not able to specifically localize LRCs to a particular cell layer within the urothelium [22,23]; however, one of these studies implicated their preferential localization to a region called the trigone [22]. These seemingly contrasting findings appear to result from different scheduling and timing of label incorporation and the chase periods, and we speculate that the overall slow-cycling property of the homeostatic urothelium in contrast to other rapid proliferating epithelial tissues such as intestine and skin epidermis may play some roles to these discrepancies. Additionally, an alternative technical limitation may be the inability for these thymidine analogs to label all G0 and postmitotic differentiation cells.

Controversies on Urothelial Stem Cell Pools and Location

Lineage tracing experiments in mice using the Sonic Hedgehog (Shh) $Shh^{CreERT2}$ promoter revealed that Shh-expressing basal cells can give rise to all layers of mouse urothelium upon injury-induced tissue repair [24]. This suggests the existence of an adult stem cell population within the basal compartment capable of differentiating into all layers of the urothelium. In human urothelial specimens, one study utilized naturally occurring mitochondrial DNA mutations as a marker of clonal expansion. Their results demonstrated that patches of clonally related urothelial cells from the basal layer always connected to patches of intermediate and umbrella cells [25], supporting a basal cell origin of urothelial stem cell. However, another lineage tracing study using cytokeratin 5 $K5^{CreERT2}$ promoter-driven mice for tracking urothelial basal cells revealed that these $K5^+$ basal cells are incapable of giving rise to either intermediate or umbrella cells during normal homeostasis or injury-induced by cyclophosphamide [26]. Using the $Foxa2^{CreERT2}$ promoter-driven mice, these authors demonstrated a cell type with embryonic origin that could be traced to the intermediate cells and can further give rise to the umbrella cells of the urothelium. Other studies using $p63^{-/-}$ mice demonstrated that the urothelium of $p63^{-/-}$ embryonic and adult mouse bladder are defective and only contain umbrella cells [27,28], suggesting functional p63 in urothelial basal cells is not required for the generation of umbrella cells. By complementing $p63^{-/-}$ mouse blastocysts with wild-type ES cells, the authors revealed that in these chimeric mouse bladders, both $p63^-$ and $p63^+$ cells can form the umbrella cells, while $p63^+$ cells only contributed to 0–15% of umbrella cells [27]. Collectively, these findings suggested the possible existence of two stem cell pools, in contrast to studies implicating the sole existence of a basal origin of stem cells, which remained to be reconciled.

BLADDER CANCER STEM CELLS

Cell Surface Markers

Even back in the 1980s, researchers had generated monoclonal antibodies that preferentially bind to the basal layer (MoAb 21.48) of normal urothelium and superficial cells (MoAb 5.48) [29]. They subsequently analyzed the staining patterns of these cell surface proteins in UCs. They found a high degree of MoAb 21.48 (basal) staining in high-grade UC specimens, whereas MoAb 5.48, an antibody with higher specificity toward superficial cells, showed more diffuse staining in well-differentiated tumors [29]. These early studies are likely the first to examine the parallel between bladder cancer development and normal urothelial developmental biology. These results support the notion that cellular heterogeneity and hierarchical tissue organization as seen in normal urothelium are partially conserved in tumors.

Cell surface markers have remained a routine method for the isolation of viable unadulterated CSCs. Since the immortalization process of cancer cells from patients and subsequent long-term culture in vitro could induce altered behavior and antigenic expression—and therefore promote deviation from characteristics of the source tumor—primary or early in vivo passage tumor cells from patients are the most preferable source for CSC isolation and characterization [2,30]. To this end, our laboratory screened a panel of cell surface markers to analyze cell suspensions created from primary human bladder cancer specimens [2,30]. In these studies, CD44 stood out as a marker consistently expressed by ~50% of tumors [2,30]. Using fluorescence-activated cell sorting (FACS), infiltrating hematopoietic and endothelial cells were first excluded, and tumor cells were then subfractionated into those that were $CD44^+$ and those that were $CD44^-$. The relative tumorigenic potential of these two tumor cell subpopulations was interrogated on the basis of their ability to form tumors in immunocompromised $RAG2^-/\gamma c^-$ mice. $CD44^+$ tumor cells were at least 10- to 200-fold enriched for tumorigenic potential when compared to $CD44^-$ tumor cells. The $CD44^+$ tumor cells could be serially passaged and recapitulate the heterogeneity of the original tumor [30]. Therefore, these $CD44^+$ tumor cells fulfilled all the functional criteria of a CSC. Using a colony formation assay in vitro to assess for tumor cells' stemness, another group independently used the CD44 spliced variant (CD44v6), and identified CD44v6$^+$ EMA$^-$ tumor cells as bladder CSCs [31]. In another study, xenografts from the established bladder cancer cell line SW780 and from a once-passaged xenograft line, XBL8, derived from an invasive urothelial carcinoma, were used to identify tumorigenic cells. 67L-kDa laminin receptor (67LR), a marker expressed at the tumor–stroma interface [32] and another marker CEACAM6 (CD66C) were used to isolate tumorigenic cells. 67LR bright cells and CEACAM6 negative cells were the tumorigenic tumor cells, and these markers are normally expressed in this configuration in basal cells [33]. Collectively, these early and recent studies all point to a distinct characteristic of CSCs expressing markers representing normal urothelial basal cells. However, bladder CSCs could not be isolated by a one-marker-fits-all type of approach, since >50% of tumors do not express the basal cell marker CD44. Combining a computational approach to predict markers that correspond with bladder cancer differentiation via exploiting existing gene expression databases, we further demonstrated that marker combinations including CD90, CD44, and CD49f could be used to isolate CSCs from different subtype of bladder cancers (Fig. 8.2) [34].

Cytokeratins

Cytokeratins are intermediate filaments expressed in cytoskeleton of epithelial tissues and are differentially expressed in different cell types, ie, basal,

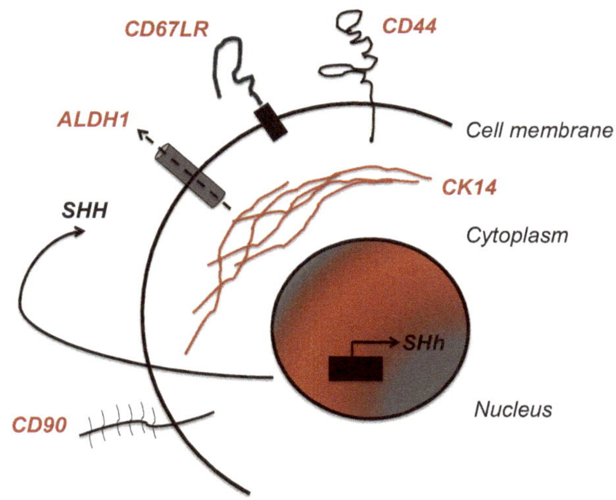

FIGURE 8.2 Phenotypic markers associated with bladder cancer stem cells (CSCs). Depicted markers summarize common phenotypic signature associated with "CSC" state. Importantly, not all of the markers listed may be expressed simultaneously and loss of certain markers may be observed in different subtypes of bladder cancers. *67LR*, 67L-kDa laminin receptor; *CD*, cluster of differentiation; *CK14*, cytokeratin 14; *SHH*, sonic hedgehog protein.

intermediate, and umbrella cells as a result of cellular differentiation. Although cytokeratins are intracellular proteins and cannot presently be used to isolate viable tumor cells by FACS, this family of protein is very useful to provide additional supporting evidence to indicate the differentiation status of tumor cells.

Intracellular cytokeratin 5 (CK5) was shown to be expressed in normal basal cells that can differentiate into CK20-expressing umbrella cells. Immunofluorescence analysis of xenografts from $CD44^+$ tumors cells revealed colocalization of CD44 and CK5, whereas $CD44^-$ cells were often CK20 expressing [30]. In some patients, $CD44^+$ tumor cells are morphologically small and homogeneous in size, with a high nuclear to cytoplasmic ratio, known characteristics of basal cells [30]. Conversely, $CD44^-$ tumor cells were heterogeneously moderate to large in size, a characteristic expected of differentiated cells [30]. In a different study, cytokeratin 17 (CK17)—another urothelial basal cell marker, was shown to colocalize with 67LR tumor cells in SW780 xenografts [33]. Further study demonstrated that cytokeratin 14 (CK14) is a more primitive stem cell marker preceding the basal markers CK5 and CK17, respectively [34]. Importantly, Kaplan–Meier analyses of formalin-fixed-paraffin-embedded (FFPE) samples from two independent cohorts of patients revealed that patients with $CK14^+$ bladder UCs demonstrated significantly reduced overall survival compared to patients with $CK14^-$ tumors. *KRT14* gene expression is a strong predictor of poor survival—independent of tumor stage and grade [34].

Side Populations and Aldehyde Dehydrogenase

In addition to cell surface and cytokeratin markers, distinct biological properties of CSCs have been exploited for their prospective isolation. Stem cells are known to express a high level of ATP-binding cassette (ABC) transporters, with a distinct ability to effectively efflux drugs [35]. This subpopulation of stem cells with the ability to efflux the vital dye Hoechst 33342 is commonly referred as "side population." This unique property of stem cells has also been used to isolate CSCs. Two independent groups used established bladder cancer cell lines SW780 and T24, respectively, and isolated their side population cells by flow cytometry. These cells had characteristics of colony formation ability, self-renewal, and differentiation, as seen in CSCs, which were distinct from nonside population cells [36,37].

Another promising marker for bladder CSCs is aldehyde dehydrogenase 1 A1 (ALDH1A1), which demonstrates increased expression in bladder CSCs [38]. Using in vitro sphere-forming assay and in vivo xenograft assays, one study demonstrated greater clonogenicity and tumorigenicity of $ALDH1A1^+$ bladder cancer cells. The xenografts from $ALDH1A1^+$ cells retained the phenotypic heterogeneity of the source tumor with every serial passage [38]. Functional disruption of the ALDH1A1 gene with an shRNA construct led to the reduction of these cells' clonogenic and tumorigenic potential. Interestingly, $ALDH1A1^+$ cells appeared to be a subpopulation of $CD44^+$ cells and may represent more primitive CSCs.

Collectively, current literature supports the existence of a functional CSC subpopulation within primary bladder cancer specimens, xenografts, and immortalized cancer cell lines. While bladder CSC isolation has been approached with different antigenic markers and functional assays, the degree of overlap has only been examined for some of these markers (eg, CD44 and ALDH1A1), while overlap of other markers have not been extensively explored. As novel markers and approaches to isolate CSCs continue to emerge, there will be a need to elucidate the relationship and reconciliation of these different CSC subpopulations to each other [2].

Lineage Tracing Studies Suggest Different Cells-of-Origin for Different Bladder Cancers

Lineage tracing experiments have been performed using various promoter-driven Cre mouse lines to track urothelial cells from different compartments of the urothelium, followed by induction of early lesions and bladder cancers by the procarcinogen N-Butyl-N-(4-hydroxybutyl)nitrosamine (BBN), a similar class of chemical found in tobacco products). Rosa26^{mTmG} mice harbor red (mT^{+}) cells that can be turned into green (mG^{+}) cells in response to specific promoter-driven Cre in target tissues. Using Shh.Cre^{ERT2} mice crossed with Rosa26^{mTmG} mice [39], it was found that both BBN carcinogen-induced carcinoma in situ and invasive UCs contain primarily green (mG^{+}) cells, indicating they arise from Shh^{+} cells and thus probably from basal cells where the Shh promoter is active. This concept was independently validated by another group using the K5.Cre^{ERT2}-driven mice that were crossed with Rosa26^{LacZ} mice to mark urothelial basal cells; upon BBN carcinogen induction, they found that K5-expressing basal cells can contribute to carcinoma in situ, muscle-invasive UCs, and squamous cell carcinomas [40]. These findings independently supported basal cells as the cell-of-origin for BBN-induced invasive lesions. On the other hand, another group utilized uroplakin 2 (Upk2) promoter–driven Cre mice, a gene more restrictedly expressed in differentiated urothelial cells, to perform lineage tracing. When Upk2.Cre^{ERT2} was used to track urothelial intermediate and differentiated cells upon BBN-induction, 9/13 noninvasive papillary lesions arose from Upk2 LacZ^{+} intermediate cells, suggestive of a predominantly differentiated cell-of-origin; while none of the carcinoma in situ or invasive UCs that arose appeared to derive from Upk2 LacZ^{+} cells, indirectly supporting basal cells as the origin of those tumors [40]. Finally, by crossing Shh.Cre^{ERT2} mice to Rosa26^{DTA} mice, activation of Cre will induce diphtheria toxin to ablate Shh^{+} basal cells, and these mice were significantly depleted of carcinoma in situ or invasive urothelial carcinoma formation [39]. Collectively, these findings supported the notion that invasive lesions likely arose from more primitive basal cells, while noninvasive papillary lesion likely arose from the intermediate cells. Although the methodologies are different, these findings support studies using human patient–derived xenograft specimens and showing that different pools of CSCs exist in different cancer subtypes [34], as well as studies using gene expression profiling from laser-captured human basal and differentiated urothelial cells to implicate different cell-of-origin for muscle-invasive and noninvasive bladder cancer [41].

BLADDER CANCER STEM CELL SIGNALING

Understanding the signaling pathways regulating bladder CSCs and their functional interactions with other cellular compartments (eg, the stroma) will provide a means for possible therapeutic intervention and would be the focus on the following section.

Peroxisome Proliferator–Activated Receptor Signaling

Peroxisome proliferator–activated receptor (PPAR) is a family of nuclear receptors and transcription factors that are known to promote cellular differentiation in a variety of tissue types, including the urothelium. All PPARs heterodimerize with the retinoid X receptor (RXR) and bind to the peroxisome proliferator–hormone response elements for mediating transcriptional activation. There are three major members of the PPAR family—PPARα, PPARβ/δ, and PPARγ. The PPARγ agonists, such as troglitazone, have been shown to repress urothelial basal cell proliferation by inducing cell cycle arrest at G0/G1, via upregulation of cyclin-dependent kinase inhibitor p21 [42]. Further, activation of PPARγ by troglitazone has been demonstrated to promote urothelial differentiation in normal human urothelial cells in vitro, as demonstrated by upregulation of late/terminal differentiation markers such as CK13, UPKII, UPKIb, and CK20 [43–45]. Pretreatment of human urothelial cell culture with the PPARγ antagonist GW9662 attenuated the troglitazone-induced differentiation response [43]. Interestingly, tissue expression of PPARγ in human bladder UC patients has an inverse association with stage, tumor grade, and the expression of proliferation marker Ki-67 [44]. Furthermore, exome sequencing results from The Cancer Genome Atlas (TCGA) project confirmed that PPARγ gene amplification is more prevalent in the clusters I/II bladder tumors (or differentiated subtype of bladder tumors), which highly express cytokeratins and other markers indicative of terminal differentiation [46]. These results are consistent with the concept that PPARγ signaling is involved in both normal urothelial and neoplastic differentiation.

Epidermal Growth Factor Receptor Signaling

Early in vitro data revealed that epidermal growth factor (EGF) ligands and their receptors (EGFR) are activated transiently early on during wound response in both rodent [47,48] and human [49,50] urothelial cells. This implicated their early involvement during urothelial cell renewal. The ligands for EGFR implicated in wound response include transforming growth factor alpha, EGF, keratinocyte growth factor, heparin-binding EGF-like growth factor (HB-EGF), and amphiregulin. A neutralizing antibody toward amphiregulin attenuated wound repair in vitro [50], implicating functional involvement of EGFR signaling during wound-induced urothelial cell renewal. Recently, nuclear active SOX9 was implicated as a downstream mediator to the EGFR signaling. SOX9 was found to be transiently induced following urothelial repair, in conjunction with increased EGFR, Her2, and Her3 mRNA expression [51]. SOX9 expression was induced by exogenous treatment of EGF and HB-EGF in immortalized human urothelial cells [51]. Its induction by EGF ligands was diminished by MAPK kinase inhibitor, which suggests a role of EGFR-MAPK-SOX-9 signaling during wound-induced urothelial cell renewal [51]. In bladder UCs, EGFR overexpression correlates with poor prognosis parameters including histological grade, tumor stage, and recurrence [52]. Recent RNA sequencing results confirmed that EGFR signaling is enriched in the basal subtype of bladder cancer, which highly expresses basal cytokeratins including KRT14/5/17 [53]. These results support that functional EGFR signaling is active during urothelial regeneration and neoplastic transformation.

Epithelial–Stromal Interaction: Sonic Hedgehog and Paracrine Stromal Wnt/BMP Signaling

Signaling regulating urothelial renewal and cellular differentiation is not only restricted to the urothelial cell compartment. Recent findings reveal an interplay between the epithelial and stromal compartment in modulating regeneration of injured urothelium from basal stem cells [24]. Using genetically modified mice marked by enhanced green fluorescent protein, it was demonstrated that the sonic hedgehog ligand SHH is expressed in urothelial basal cells during steady state conditions, while its active effector transcription factor Gli1 is localized in neighboring stromal cells in the lamina propria [24]. After bacterial-induced injury of the urothelium, lineage tracing revealed that SHH-expressing urothelial basal cells can give rise to all layers of the urothelium, while regeneration of urothelium from $SHH^{+}CK5^{+}$ urothelial basal cells depends on feedback paracrine signals from the neighboring stromal cells. SHH secretion from epithelial basal cells signals the activation of Gli1 signaling in the stromal compartment, which in turn secrete paracrine stromal Wnt signals to induce the expansion of urothelial basal cells to replenish the injured urothelium [24]. This finding during urothelial regeneration mirrors the emerging roles of cancer-associated stromal cells acting in a paracrine fashion on epithelial tumor cells during the transforming process of cancer development. In fact, it has been shown in another study during BBN carcinogen-induced bladder cancer development, that urothelial basal SHH ligand could induce stromal expression of BMP, which in turn limits bladder cancer progression by stimulating urothelial differentiation [54]. However, SHH ligand was progressively lost during temporal development of bladder cancer progression, therefore limiting stromal secretion of the differentiation-inducing factor bone morphogenetic protein (BMP) and leading to an undifferentiated urothelial stem cell phenotype within muscle-invasive bladder cancers [54]. Therefore, interplay of paracrine signals between the urothelial and stromal compartments during tissue regeneration and cancer progression can modulate final stem cell or CSC phenotypes within the urothelial compartment and may be exploited for therapeutic purposes (Fig. 8.3A).

Molecular Heterogeneity of Self-Renewal Pathways

Several other signaling pathways involved in the self-renewal of adult and embryonic stem cells are often activated during tumorigenesis [30]. β-Catenin is a downstream activator of the Wnt ligand, which has been shown to

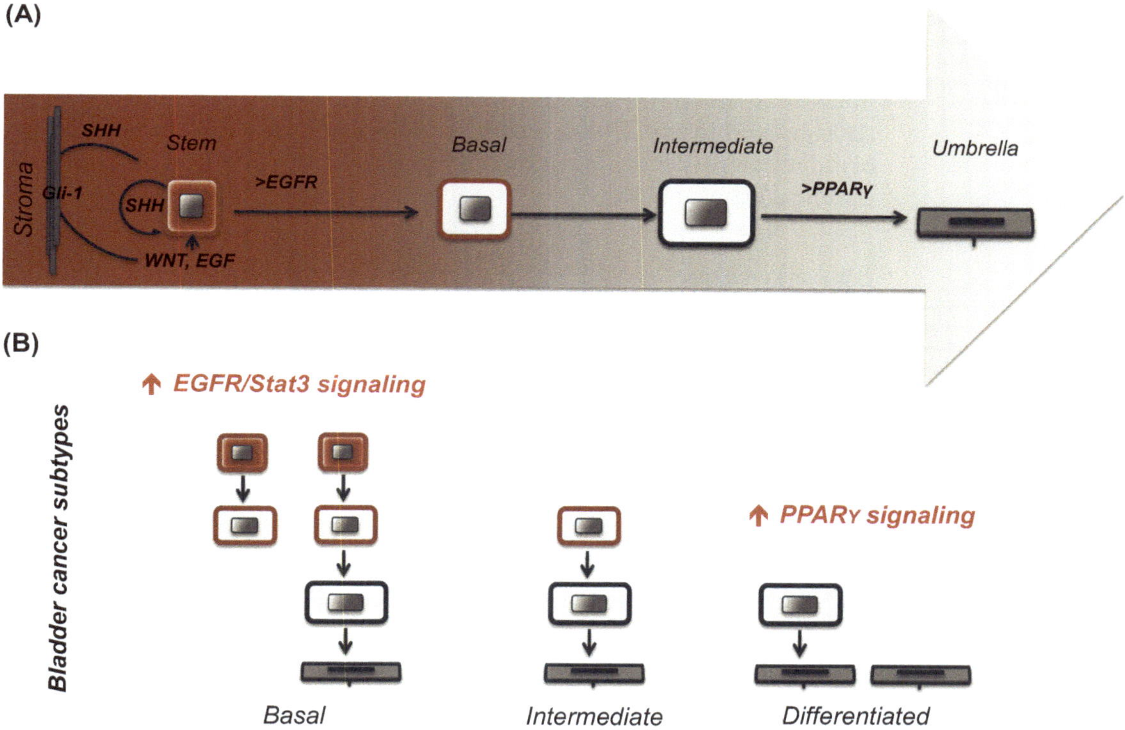

FIGURE 8.3 Molecular pathways associated with regulation of bladder stem cells maintenance. (A) Summary of signaling pathways important for normal urothelial differentiation upon transition from one cell type to another (*solid black lines*) and support of normal stem cells renewal. (B) Prevalence of signaling pathways associated with pathogenesis of different bladder cancer subtypes, notably parallel to those observed in corresponding differentiation stage of normal urothelial cells. *EGF*, epidermal growth factor; *EGFR*, epidermal growth factor receptor; *GLI1*, zinc finger protein GLI1; *MAPK*, mitogen-activated protein kinase; *PPARγ*, peroxisome proliferator–activated receptor γ; *SHH*, sonic hedgehog protein; *SOX9*, transcription factor SOX-9; *STAT3*, signal transducer and activator of transcription 3 (acute-phase response factor); *Wnt*, protein Wnt.

play a role in both self-renewal and oncogenesis as seen in hematopoietic and chronic myelogenous leukemic stem cells [55]. Another gene, Bmi-1, essential for the self-renewal of adult hematopoietic stem cells and neural stem cells [56,57], is also implicated in the tumorigenesis of many malignancies. Signal transducer and activator of transcription 3, Stat3, is a protein essential for maintaining self-renewal of human embryonic stem cells [58] and that is constitutively active in a number of epithelial cancers [59,60]. Oct-4 and Nanog are key transcription factors to maintain self-renewal of embryonic stem cells [61] and are expressed in some somatic and germ cell tumors. In epithelial tissues, Oct-4 expression leads to proliferation of progenitor cells, dysplastic growth, and block of differentiation [61,62]. Comprehensive analysis of these oncoproteins in human bladder UCs revealed a tremendous heterogeneity in their activation status among patients. Nuclear active β-catenin was seen in only 5% of samples, nuclear localization of Bmi-1 was seen in 20%, nuclear active Stat3 was seen in 40% of bladder UCs, and 80% of bladder UCs expressed Gli1 mRNA [30]. The expressions of Nanog and Oct-4 were not seen in any of the samples. This molecular heterogeneity in the signaling pathway within CSCs likely reflects a diversity in the molecular etiology of bladder UCs, which has a strong link to cigarette smoking and other industrial chemical exposure [30].

PROGNOSTIC ROLE OF CANCER STEM CELLS

Cancer Stem Cell Status Subclassifies Bladder Cancer Subtypes

We have previously reported the subclassification of bladder cancers into basal and differentiated subtypes on the basis of their tumor differentiation status [7,34]. A KRT14/Thy-1/CD44-positive self-renewing stem cell population was proposed to give rise to a partially differentiated KRT5/KRT17/CD44-positive progeny that is thought, in turn, to acquire KRT8/18 expression and to terminally differentiate into luminal cells expressing uroplakins and KRT20. Knowledge of the normal urothelium differentiation program was thus used to subclassify tumors [7]. The basal subtype (KRT14$^+$ group) was characterized as having poor prognosis in multiple patient cohorts [34], and these KRT14$^+$ basal tumors were resistant to neoadjuvant cisplatin-based chemotherapy in a small patient cohort [63]. In contrast, the differentiated subtype of bladder cancer highly expressing KRT18/20 correlated with a more favorable clinical outcome [7,34].

Such basal and differentiated (or luminal) subtypes of bladder cancer are independently reported by the TCGA [46] and by other groups [53,64,65]. The Cluster III/IV tumors reported by TCGA correspond well with the basal subtype, which highly expresses basal cytokeratins 14/5/17, and the Cluster I/II tumors by TCGA correspond closely with the differentiated (or luminal) subtype of bladder cancer, which highly expresses markers associated with terminal differentiation, eg, cytokeratin 18/20 and uroplakins. Interestingly, signaling pathways that modulate cellular differentiation (described in Section Bladder Cancer Stem Cell Signaling) seemed to drive these different cancer subtypes. For instance, the basal cancers were reported to exhibit strong EGFR and Stat3 signaling, which are important for maintaining stem cell self-renewal [53,66], while gene amplification of other signaling known to drive cellular differentiation, eg, PPARγ, was closely associated with the differentiated subtype of bladder cancer [46,65]. Prospective validation of cytokeratin markers of bladder cancer differentiation state may facilitate their translation into the clinic as useful prognostic markers, and therapeutic targeting of the associated cellular differentiation pathways will likely provide new opportunities for molecular targeting of different subtypes of bladder cancers (Fig. 8.3B).

TherapeuticTargeting of Bladder Cancer Stem Cells

Conventional Cytotoxic Chemotherapy

CSCs possess intrinsic unique biological properties that allow them to survive and repopulate residual tumors following cytotoxic treatment [67]. Studies in other organ systems have shown that cancer stem cells are more resistant to the conventional cytotoxic treatment approaches such as radiation and chemotherapy than non-CSCs [67]. Recently, CSCs were reported to be isolated in the 5637 bladder cancer cell line on the basis of CD44 expression [68]. The authors demonstrated that $CD44^+$ cells had greater survival benefit after exposure to cisplatin than $CD44^-$ cells (50% inhibitory concentration (IC_{50}) of 0.43 μg/ml and 0.25 μg/ml, respectively). $CD44^+$ cells also demonstrated a higher transforming ability when exposed to cisplatin in a clonogenic assay [68]. In another study, the ALDEFLUOR assay was used to isolate ALDH(High) tumor cells from T24 and 5637 bladder cancer cell lines. Therein the authors demonstrated that the abundance of ALDH(High) tumor cells correlated with resistance of cisplatin in vitro, and the ALDH(High) tumor cells demonstrated high colony formation property and other aggressive phenotypes of CSCs [69]. Using an alternative approach, a recent study generated a derivative of T24 bladder cancer cell line (DR-T24) by continuous treatment of cisplatin in vitro. They found that these DR-T24 derivatives contain a larger subpopulation of side population (SP) cells and greater clonogenic capacity in vitro, in comparison to the T24 parental line [70]. A study using the SW780 bladder cancer cells further demonstrated that the histone deacetylase inhibitor AR-42 preferentially targets $CD44^+CD49f^+$ CSCs in synergy with cisplatin [71]. Collectively, these findings support the notion that a bladder CSC population exists, that they differentially respond to cisplatin-based chemotherapy, and that they could potentially be targeted to enhance clinical therapeutic response.

Cancer Stem Cell Repopulation

In the clinics, chemotherapeutics are administered in multiple cycles of fractionated doses that are spaced out to allow normal cells and tissue stem cells to recover or repopulate between treatment cycles (reviewed in Refs. [72,73]). However, residual surviving cancer cells can also repopulate tumors during the gap periods between chemotherapy cycles—a major cause of treatment failure that is often overlooked.

During normal tissue homeostasis, quiescent tissue stem cells generally serve as a reserve until being challenged by emergency circumstances such as tissue injury or stress. Our group employed immortalized cancer cell–derived and patient-derived xenografts from bladder UCs to demonstrate the existence of a quiescent tumor subpopulation that could be enriched by CK14 promoter-driven fluorescence [63]. Combination chemotherapeutic treatment of the cytidine nucleoside analog gemcitabine and the DNA alkylating agent cisplatin was effective in diminishing proliferating UC cells, while quiescent $CK14^+$ CSCs were spared. Many proinflammatory factors including prostaglandin E2 (PGE_2) were released from urothelial carcinoma cells during chemotherapy-induced cell death, which drove these chemotherapy-treated tumors into a "wound response" state as demonstrated by global RNA sequencing analyses following by gene-set enrichment analyses confirming an enrichment of a "wound response" gene-signature, when compared with vehicle-treated xenograft tumors [63]. The proinflammatory phospholipid PGE_2 contributed to enhanced ability to generate sphere-forming stem cells in vitro and induced recruitment of quiescent CSCs into cell division for tumor repopulation [63]. Since chemotherapy was administered in multiple cycles as clinical regimen, successive rounds of "wound-induced" tumor repopulation driven by PGE_2 and other proinflammatory factors ultimately led to marked expansion of CSCs and treatment failure following multiple chemotherapy cycles [63].

Relevant to clinical translation, blocking PGE_2 release by a neutralizing antibody or using the FDA-approved drug celecoxib was effective in diminishing paracrine effects of neighboring dying cells to promote expansion of CSCs induced by chemotherapy treatment [63]. Further, celecoxib drug treatment was sufficient to block chemotherapy-treated xenografts from entering into a "wound response" state, thereby preventing subsequent wound-induced tumor repopulation and eventual chemoresistance, as measurable by significant reduced volume of xenograft tumors and significant reduction in the frequency and size of distal metastatic foci in lung [63]. In light of the fact that celecoxib is efficacious in improving chemotherapeutic response in a primary xenograft model derived from an original patient who failed chemotherapy, this provided convincing proof-of-concept evidence for testing such adjuvant therapeutic approaches with comparable concepts in other preclinical models and subsequent human clinical trials.

Hence, in additional to the previously proposed intrinsic resistance of CSCs to chemotherapeutic drugs [74–80], the earlier intriguing observations collectively confirmed an active response and "reawakening" of quiescent CSCs into cell division, contributing to repopulation of residual tumors following cytotoxic chemotherapy.

Targeted Therapy

An emerging and highly promising approach for anticancer therapy is the targeting of monoclonal antibodies to cancer-specific cell antigens. CD47, or integrin-associated protein, is one of the proteins significantly downregulated during bladder cancer cell differentiation. Immunofluorescence and FACS analysis confirmed this finding, showing widespread expression of CD47 in bulk bladder UC cells, with preferentially higher level in $CD44^+$ CSCs in comparison to $CD44^-$ tumor cell subpopulation. CD47 is a "don't eat me" signal, which interacts with signal regulatory protein alpha (SIRPα), a plasma membrane protein expressed mainly in myeloid cells—macrophages, neutrophils, basophils, and dendritic cells. Ligation of CD47 to SIRP leads to the suppression of immune cell activity, and, ultimately blocks phagocytosis by macrophages [61]. We previously showed that in vitro incubation of human bladder UC cells with macrophages in the presence of a blocking anti-CD47 antibody led to phagocytosis of tumor cells, whereas incubation with IgG1 isotype negative control did not lead to tumor cell engulfment [30]. Recently, assessment of the same blocking anti-CD47 antibody in human bladder UC xenograft in vivo revealed a significant reduction in metastatic lymph node size and micrometastasis to lung, in comparison to IgG isotype treated control [81]. These findings suggest that CD47, and potentially other proteins overexpressed by CSCs, may become promising targets for cancer therapy. Unlike conventional chemotherapy and radiation that play a palliative role in most malignancies, targeted therapy toward CSCs whether it be through immune modulation with CSC-targeted antibodies, CSC-targeted radiosensitizers, CSC-targeted chemosensitizers, or small molecule inhibitors to stem cell pathway, hopefully, will prove to be of curative benefit.

CONCLUSIONS

Recent progress in the molecular and functional characterization of CSCs in human UCs portends significant clinical relevance to bladder carcinogenesis. Several drug targets that were revealed to modulate bladder CSCs and demonstrated great success in preclinical studies should bring us one step closer to CSC directed molecular therapies in the clinic. Additionally, subtyping of bladder cancers on the basis of their CSCs status (or tumor differentiation status) revealed promise as prognostic markers to stratify high-risk bladder UCs. Future initiation of large-scale prospective clinical trials will be important to validate the ultimate clinical utility of these CSC markers in prognostics.

List of Acronyms and Abbreviations

ALDH1A1 Aldehyde dehydrogenase 1 family member A1
CD Cluster of differentiation
CEACAM6 Carcinoembryonic antigen-related cell adhesion molecule 6
CK Cytokeratin
CPP Cyclophosphamide
CSCs Cancer stem cells
EdU 5-Ethynyl-2-deoxyuridine
EGF Epidermal growth factor
ES cells Embryonic stem cells
FOXA2 Forkhead Box A2
LR67 Laminin receptor 67
LRCs Label-retaining cells
p63 Tumor protein p63
PPARs Peroxisome proliferator–activated receptor
RAG2 Recombination activating gene 2
Shh Sonic Hedgehog
SOX Sex determining region Y-box
STAT Signal transducer and activator of transcription
UCs Urothelial carcinomas

References

[1] Prasad SM, Decastro GJ, Steinberg GD. Urothelial carcinoma of the bladder: definition, treatment and future efforts. Nat Rev Urol 2011;8(11):631–42.

[2] Chan KS, Volkmer JP, Weissman I. Cancer stem cells in bladder cancer: a revisited and evolving concept. Curr Opin Urol 2010;20(5):393–7.

[3] Clarke MF, et al. Cancer stem cells—perspectives on current status and future directions: AACR Workshop on cancer stem cells. Cancer Res 2006;66(19):9339–44.

[4] Clevers H. The cancer stem cell: premises, promises and challenges. Nat Med 2011;17(3):313–9.

[5] Apodaca G. The uroepithelium: not just a passive barrier. Traffic 2004;5(3):117–28.

[6] Khandelwal P, Abraham SN, Apodaca G. Cell biology and physiology of the uroepithelium. Am J Physiol Renal Physiol 2009;297(6):F1477–501.

[7] Ho PL, Kurtova A, Chan KS. Normal and neoplastic urothelial stem cells: getting to the root of the problem. Nat Rev Urol 2012;9(10):583–94.

[8] Romih R, Jezernik K. Reorganisation of the urothelial luminal plasma membrane in the cyclophosphamide treated rats. Pflugers Arch 1996;431(6 Suppl. 2):R241–2.

[9] Kreft ME, et al. Urothelial injuries and the early wound healing response: tight junctions and urothelial cytodifferentiation. Histochem Cell Biol 2005;123(4–5):529–39.

[10] Lavelle J, et al. Bladder permeability barrier: recovery from selective injury of surface epithelial cells. Am J Physiol Renal Physiol 2002;283(2):F242–53.

[11] Kvist E, et al. Squamous cell metaplasia of the bladder urothelium. A retrospective study of 36 patients. APMIS 1992;100(7):650–4.

[12] Susmano D, et al. Cystitis glandularis and adenocarcinoma of the bladder. J Urol 1971;105(5):671–4.

[13] Potten CS, Morris RJ. Epithelial stem cells in vivo. J Cell Sci Suppl 1988;10:45–62.

[14] Bromberg JF, et al. Stat3 as an oncogene. Cell 1999;98(3):295–303.

[15] Blair K, Wray J, Smith A. The liberation of embryonic stem cells. PLoS Genet 2011;7(4):e1002019.

[16] Herzenberg LA, Sweet RG. Fluorescence-activated cell sorting. Sci Am 1976;234(3):108–17.

[17] Watt FM, Lo Celso C, Silva-Vargas V. Epidermal stem cells: an update. Curr Opin Genet Dev 2006;16(5):518–24.

[18] Spangrude GJ, Heimfeld S, Weissman IL. Purification and characterization of mouse hematopoietic stem cells. Science 1988;241(4861):58–62.

[19] Langkilde NC, Wolf H, Orntoft TF. DNA replication in experimental rat bladder tumours: immunohistochemical detection of bromodeoxyuridine labelled nuclei. Scand J Urol Nephrol Suppl 1989;125:133–9.

[20] Kurzrock EA, et al. Label-retaining cells of the bladder: candidate urothelial stem cells. Am J Physiol Renal Physiol 2008;294(6):F1415–21.

[21] Zhang H, et al. Label retaining and stem cell marker expression in the developing rat urinary bladder. Urology 2012;79(3):746.e1–6.

[22] Sun W, Wilhelmina Aalders T, Oosterwijk E. Identification of potential bladder progenitor cells in the trigone. Dev Biol 2014;393(1):84–92.

[23] Colopy SA, et al. A population of progenitor cells in the basal and intermediate layers of the murine bladder urothelium contributes to urothelial development and regeneration. Dev Dyn 2014;243(8):988–98.

[24] Shin K, et al. Hedgehog/Wnt feedback supports regenerative proliferation of epithelial stem cells in bladder. Nature 2011;472(7341):110–4.

[25] Gaisa NT, et al. The human urothelium consists of multiple clonal units, each maintained by a stem cell. J Pathol 2011;225(2):163–71.

[26] Gandhi D, et al. Retinoid signaling in progenitors controls specification and regeneration of the urothelium. Dev Cell 2013;26(5):469–82.

[27] Signoretti S, et al. p63 regulates commitment to the prostate cell lineage. Proc Natl Acad Sci USA 2005;102(32):11355–60.

[28] Karni-Schmidt O, et al. Distinct expression profiles of p63 variants during urothelial development and bladder cancer progression. Am J Pathol 2011;178(3):1350–60.

[29] Dotsikas G, et al. Cellular heterogeneity in normal and neoplastic human urothelium: a study using murine monoclonal antibodies. Br J Cancer 1987;56(4):439–44.

[30] Chan KS, et al. Identification, molecular characterization, clinical prognosis, and therapeutic targeting of human bladder tumor-initiating cells. Proc Natl Acad Sci USA 2009;106(33):14016–21.

[31] Yang YM, Chang JW. Bladder cancer initiating cells (BCICs) are among $EMA^{-}CD44v6^{+}$ subset: novel methods for isolating undetermined cancer stem (initiating) cells. Cancer Invest 2008;26(7):725–33.

[32] Edris B, et al. Antibody therapy targeting the CD47 protein is effective in a model of aggressive metastatic leiomyosarcoma. Proc Natl Acad Sci USA 2012.

[33] He X, et al. Differentiation of a highly tumorigenic basal cell compartment in urothelial carcinoma. Stem Cells 2009;27(7):1487–95.

[34] Volkmer JP, et al. Three differentiation states risk-stratify bladder cancer into distinct subtypes. Proc Natl Acad Sci USA 2012;109(6):2078–83.

[35] Goodell MA, et al. Isolation and functional properties of murine hematopoietic stem cells that are replicating in vivo. J Exp Med 1996;183(4):1797–806.

[36] She JJ, et al. Identification of side population cells from bladder cancer cells by DyeCycle Violet staining. Cancer Biol Ther 2008;7(10):1663–8.

[37] Ning ZF, et al. Subpopulations of stem-like cells in side population cells from the human bladder transitional cell cancer cell line T24. J Int Med Res 2009;37(3):621–30.

[38] Carthon BC, et al. Preoperative CTLA-4 blockade: tolerability and immune monitoring in the setting of a presurgical clinical trial. Clin Cancer Res 2010;16(10):2861–71.

[39] Shin K, et al. Cellular origin of bladder neoplasia and tissue dynamics of its progression to invasive carcinoma. Nat Cell Biol 2014;16(5):469–78.

[40] Van Batavia J, et al. Bladder cancers arise from distinct urothelial sub-populations. Nat Cell Biol 2014;16(10):982–91. 1–5.

[41] Dancik GM, et al. A cell of origin gene signature indicates human bladder cancer has distinct cellular progenitors. Stem Cells 2014;32(4):974–82.

[42] Kawakami S, et al. PPARgamma ligands suppress proliferation of human urothelial basal cells in vitro. J Cell Physiol 2002;191(3):310–9.

[43] Varley CL, et al. Role of PPARgamma and EGFR signalling in the urothelial terminal differentiation programme. J Cell Sci 2004;117(Pt 10):2029–36.

[44] Mylona E, et al. Peroxisome proliferator-activated receptor gamma expression in urothelial carcinomas of the bladder: association with differentiation, proliferation and clinical outcome. Eur J Surg Oncol 2009;35(2):197–201.

[45] Varley CL, Southgate J. Effects of PPAR agonists on proliferation and differentiation in human urothelium. Exp Toxicol Pathol 2008;60(6):435–41.

[46] Cancer Genome Atlas Research N. Comprehensive molecular characterization of urothelial bladder carcinoma. Nature 2014;507(7492):315–22.

[47] de Boer WI, et al. Characterization of distinct functions for growth factors in murine transitional epithelial cells in primary organotypic culture. Exp Cell Res 1994;214(2):510–8.

[48] Baskin LS, et al. Growth factors in bladder wound healing. J Urol 1997;157(6):2388–95.

[49] Daher A, et al. Growth, differentiation and senescence of normal human urothelium in an organ-like culture. Eur Urol 2004;45(6):799–805.

[50] Varley C, et al. Autocrine regulation of human urothelial cell proliferation and migration during regenerative responses in vitro. Exp Cell Res 2005;306(1):216–29.

[51] Ling S, et al. An EGFR-ERK-SOX9 signaling cascade links urothelial development and regeneration to cancer. Cancer Res 2011;71(11):3812–21.

[52] Mellon K, et al. Long-term outcome related to epidermal growth factor receptor status in bladder cancer. J Urol 1995;153(3 Pt 2):919–25.

[53] Rebouissou S, et al. EGFR as a potential therapeutic target for a subset of muscle-invasive bladder cancers presenting a basal-like phenotype. Sci Transl Med 2014;6(244):244ra91.

[54] Shin K, et al. Hedgehog signaling restrains bladder cancer progression by eliciting stromal production of urothelial differentiation factors. Cancer Cell 2014;26(4):521–33.

[55] Reya T, et al. A role for Wnt signalling in self-renewal of haematopoietic stem cells. Nature 2003;423(6938):409–14.

[56] Park IK, et al. Bmi-1 is required for maintenance of adult self-renewing haematopoietic stem cells. Nature 2003;423(6937):302–5.

[57] Prince ME, et al. Identification of a subpopulation of cells with cancer stem cell properties in head and neck squamous cell carcinoma. Proc Natl Acad Sci USA 2007;104(3):973–8.

[58] Niwa H, et al. Self-renewal of pluripotent embryonic stem cells is mediated via activation of STAT3. Genes Dev 1998;12(13):2048–60.

[59] Bromberg J. Stat proteins and oncogenesis. J Clin Invest 2002;109(9):1139–42.

[60] Chan KS, et al. Disruption of Stat3 reveals a critical role in both the initiation and the promotion stages of epithelial carcinogenesis. J Clin Invest 2004;114(5):720–8.

[61] Chambers I, et al. Functional expression cloning of Nanog, a pluripotency sustaining factor in embryonic stem cells. Cell 2003;113(5):643–55.

[62] Hochedlinger K, et al. Ectopic expression of Oct-4 blocks progenitor-cell differentiation and causes dysplasia in epithelial tissues. Cell 2005;121(3):465–77.

[63] Kurtova AV, et al. Blocking PGE2-induced tumour repopulation abrogates bladder cancer chemoresistance. Nature 2015;517(7533):209–13.

[64] Damrauer JS, et al. Intrinsic subtypes of high-grade bladder cancer reflect the hallmarks of breast cancer biology. Proc Natl Acad Sci USA 2014;111(8):3110–5.

[65] Biton A, et al. Independent component analysis uncovers the landscape of the bladder tumor transcriptome and reveals insights into luminal and basal subtypes. Cell Rep 2014;9(4):1235–45.

[66] Ho PL, et al. Stat3 activation in urothelial stem cells leads to direct progression to invasive bladder cancer. Cancer Res 2012;72(13):3135–42.

[67] Frank NY, Schatton T, Frank MH. The therapeutic promise of the cancer stem cell concept. J Clin Invest 2010;120(1):41–50.

[68] Tatokoro M, et al. Potential role of Hsp90 inhibitors in overcoming cisplatin resistance of bladder cancer-initiating cells. Int J Cancer 2011;131(4):987–96.

[69] Falso MJ, Buchholz BA, White RW. Stem-like cells in bladder cancer cell lines with differential sensitivity to cisplatin. Anticancer Res 2012;32(3):733–8.

[70] Zhang Y, et al. Cancer stem-like cells contribute to cisplatin resistance and progression in bladder cancer. Cancer Lett 2012;322(1):70–7.

[71] Li DR, et al. Synergy of histone-deacetylase inhibitor AR-42 with cisplatin in bladder cancer. J Urol 2015;194(2):547–55.

[72] Davis AJ, Tannock JF. Repopulation of tumour cells between cycles of chemotherapy: a neglected factor. Lancet Oncol 2000;1:86–93.

[73] Kim JJ, Tannock IF. Repopulation of cancer cells during therapy: an important cause of treatment failure. Nat Rev Cancer 2005;5(7):516–25.

[74] Frank NY, et al. ABCB5-mediated doxorubicin transport and chemoresistance in human malignant melanoma. Cancer Res 2005;65(10):4320–33.

[75] Todaro M, et al. Colon cancer stem cells dictate tumor growth and resist cell death by production of interleukin-4. Cell Stem Cell 2007;1(4):389–402.

[76] Li X, et al. Intrinsic resistance of tumorigenic breast cancer cells to chemotherapy. J Natl Cancer Inst 2008;100(9):672–9.

[77] Bleau AM, et al. PTEN/PI3K/Akt pathway regulates the side population phenotype and ABCG2 activity in glioma tumor stem-like cells. Cell Stem Cell 2009;4(3):226–35.

[78] Pang R, et al. A subpopulation of $CD26^+$ cancer stem cells with metastatic capacity in human colorectal cancer. Cell Stem Cell 2010;6(6):603–15.

[79] Lee TK, et al. $CD24^+$ liver tumor-initiating cells drive self-renewal and tumor initiation through STAT3-mediated NANOG regulation. Cell Stem Cell 2011;9(1):50–63.

[80] Hoey T, et al. DLL4 blockade inhibits tumor growth and reduces tumor-initiating cell frequency. Cell Stem Cell 2009;5(2):168–77.

[81] Willingham SB, et al. The CD47-signal regulatory protein alpha (SIRPa) interaction is a therapeutic target for human solid tumors. Proc Natl Acad Sci USA 2012.

CHAPTER 9

Cancer Stem Cells as New Therapeutic Targets for Ovarian Cancer

Y. Wang[1], F. Fang[1], S. Condello[2], D. Matei[2,3,4], K.P. Nephew[1,2,3]
[1]Indiana University, Bloomington, IN, United States; [2]Indiana University School of Medicine, Indianapolis, IN, United States; [3]Indiana University Melvin and Bren Simon Cancer Center, Indianapolis, IN, United States; [4]VA Roudebush Hospital, Indianapolis, IN, United States

CONTENTS

INTRODUCTION

Ovarian cancer (OC) causes more deaths than any other gynecologic malignancy. Five-year survival rates have only marginally improved over the past three decades, with progression to drug resistance remaining the major therapeutic barrier. Similar to a number of other carcinomas, recent reports suggest that ovarian tumors may exhibit a hierarchical organization of cell types, with tumor development and progression driven by "cancer stem cells (CSCs)" that are inefficiently targeted by conventional therapies. This chapter will focus on the CSC hypothesis, as it may relate to OC, examine reports of ovarian CSCs (OCSCs), and discuss potentially improved therapeutic strategies, based on the specific targeting of these tumor progenitors.

OVARIAN CANCER BIOLOGY AND PATHOLOGY

Globally, OC is the seventh leading cause of total cancer-related death, claiming 125,000 lives per year [1,2]. In the United States, an estimated 21,550 women will be diagnosed with, and more than 14,600 women die from, OC, in 2009 [3]. It is estimated that 1 woman in 70 will develop OC during her lifetime, and the lifetime risk of death for all women is 1 in 98 [1].

The current standard treatment, cytoreductive surgery followed by a platinum/taxane-based regimen, results in clinical complete remissions in approximately 70% of patients [4]. However, the majority of those initially responsive patients experience tumor recurrence associated with the development of chemoresistance, and platinum-resistant OC is uniformly fatal. A new paradigm explaining tumor relapse involves the persistence of "CSCs" which have been characterized in several solid tumors, including OC [5,6].

Cancer Stem Cells. http://dx.doi.org/10.1016/B978-0-12-803892-5.00009-7

Stem cells in adult epithelial tissue play a critical role in normal tissue turnover as well as cancer initiation, due to accumulation of genetic mutations [1,7]. The presence of resident stem cells in the epithelium that function in maintaining epithelium tissue homeostasis, repairing damaged tissue, and regeneration have been described and characterized in various organs, including the female reproductive tract [7]. Recently, Flesken-Nikitin et al. demonstrated that a stem cell niche resides in the hilum region of mouse ovary, the junction region of ovarian surface epithelium (OSE), mesothelium, and tubal epithelium, which is responsible for OSE regeneration during ovulation [8]. Cells in the junction region of OSE have been indicated to display stem cell properties with slowly cycling and expression of stemness markers [8]. However, presence of $Lgr5^+$ labeled stem cells in ovary reported by Ng et al. both in the OSE and tubal epithelia indicated these as a possible origin of OC [7,9]. Importantly, that study as well as Hirsch et al. in colon cancer [9] demonstrated that stem cells in stem niches are prone to malignant transformation followed by genetic mutation, including inactivation of *Trp53* and *Rb1* tumor-suppressor genes, signaling pathways that are frequently mutated in high-grade serous OC (HGSOC), suggesting their potential roles in the origin of OC [7,8]. In addition, microenvironment alterations also play a critical role in malignant transformation and growth of OC (reviewed in Ref. [10]).

ISOLATION AND CHARACTERIZATION OF OVARIAN CANCER STEM CELLS

The existence of CSCs was first demonstrated in hematologic malignancies [11] and, more recently, in solid tumors [12], including OC [5,13–16]. Putative OCSCs have now been isolated from a number of sources, including established OC cell lines, ascites, and primary tumors. Similar to the isolation of other tumor stem cells, enrichment of OCSCs relies on various phenotypes likely shared with normal stem cells, including the ability to form anchorage-independent spherical aggregates, express stem cell markers, undergo membrane efflux, display distinct surface proteins, form clones in stem cell cultured condition, and exhibit greatly enhanced tumor-forming ability [8].

Isolation of Putative Ovarian Cancer Stem Cells From Established Cell Cultures

A number of studies have now been performed demonstrating the presence of a subpopulation of tumorigenic stemlike cells in cultures of established cancer cell lines. While all of the aforementioned methods have been used to identify CSCs from cell cultures, the most widely used technique exploits the membrane efflux phenotype, by fluorescence-activated cell sorting (FACS) of cells capable of expelling a specific fluorophore ("side population" or SP cells), based on early studies of hematopoietic stem cells [17]. One of the most commonly used fluorescent reagents for SP isolation is the DNA-binding molecule, Hoechst 33342 [18], which has also been used to isolate and characterize OC SP cells from cell lines [19].

Isolation of Ovarian Cancer Stem Cells From Ascites

The first report (Bapat et al.) of the isolation and identification of OCSCs from OC patients described two ascites-derived clones able to form multiaggregate, anchorage-independent spheres in culture, and serially propagate xenograft tumors (ie, reisolation of stem cells capable of tumorigenesis in a newly engrafted animal) in nude mice that were histopathologically similar to their parental tumors [14]. A follow-up study by the same group demonstrated that in one of these ascites-derived clones, overexpression of mediators of the epithelial-to-mesenchymal transition (EMT, a facilitator of metastasis) associated with chemotherapy and radiotherapy resistance [20]. In another study, SP cells (see section "Isolation of Putative Ovarian Cancer Stem Cells From Established Cell Cultures") isolated from OC patient ascites were demonstrated (by immunohistochemistry) to express the stem cell markers OCT4, NANOG, STELLAR, and ABCG2/BCRP1, as compared to non-SP cells, while also exhibiting greater proliferation rates and tumor multiplicity in xenografted animals [21].

Isolation of Ovarian Cancer Stem Cells From Primary Ovarian Tumors

While sphere-forming and clonogenicity assays have been used to isolate CSCs from solid tumors, the use of cell surface markers or stem cell gene reporter assays has been employed most extensively. A number of stemness markers have been used to isolate CSCs from primary patient tumors and cancer cell lines, including OC (Table 9.1). Most of the "cluster of differentiation" (CD) cell surface markers were originally used for the identification of hematopoietic cells of distinct lineages and levels of differentiation [38]. Although these stemness markers are not defined as OCSCs-specific and there is a high degree of overlapping expression of those markers across a diverse types of cancer cells (reviewed in Ref. [22]), here we will focus on discussing the utilization of these stemness markers for OCSC identification, isolation, and characterization.

CD117. The *c-kit* protooncogene, encoding a tyrosine kinase receptor, is expressed in many normal and cancerous tissues and has also been used to isolate OCSCs from primary tumors. C-Kit kinase activity is induced by the binding of its ligand, stem cell factor (SCF), resulting in autophosphorylation of the receptor. Coexpression of *c-kit* and SCF has previously been examined in human OC tumors, in normal ovaries, and in cultured OSE cells [23]. Normal OSE cells expressed *SCF* but not *c-kit*; *c-kit* expression was, however, found in epithelial invaginations and inclusion cysts. While c-Kit kinase

Table 9.1 Candidate Ovarian Cancer Stem Cell (CSC) Markers

Type	Marker	Normal Function(s)	Proposed Ovarian Cancer (OC) Function(s)	Cancer Type	Reference(s)
Cell surface markers	CD117 (c-Kit)	Hematopoietic cell survival, proliferation, and differentiation	Cancer stemness, tumor proliferation, metastasis, and angiogenesis	OC Neuroblastoma Epidermoid osteosarcoma Rhabdomyosarcoma carcinoma Colon carcinoma Lung cancer Melanoma Liver cancer Leukemia Plasmacytoma Mammary adenocarcinoma Colorectal cancer	[5,22,23]
	CD44	Cell adhesion, hyaluronate degradation, and lymphocyte homing	Cancer stemness, mesothelium binding, Nanog activation, and cytoskeletal activation of MDR-1	Ovarian carcinoma acute myeloid leukemia	[5,22,24,25]
	CD133 (prominin-1)	Hematopoiesis, tissue development, and differentiation	Cancer stemness, metastasis, and angiogenesis	OC Glioblastoma Colon adenocarcinoma Hepatocellular carcinoma Glioblastoma	[13,15]
	CD24	Regulation of cell–cell and cell–matrix interactions. Regulation of metabolism, cell differentiation, migration, and apoptosis	Cancer stemness, metastasis, tumor formation, progression, and survival	OC Brain cancer Prostate cancer Breast cancer Lung cancer Leukemia Colorectal pancreatic cancer	[22,26]

		Lgr5	Maintenance of pluripotency, regulation of tissue repair, regeneration and homeostasis	Cancer stemness, tumor formation, proliferation, and migration	Stem cells of ovary and tubal epithelia OC (potential) Colorectal cancer Gastrointestinal tract cancer	[7,8,27]
Intracellular markers	Adaptor protein, expressed in cytoplasm	MyD88	Immune response, inflammation	Chemoresistance, proliferation	OC	[25]
	Detoxifying protein, expressed in cytoplasm, nucleus, or mitochondria	ALDH	Metabolism of aldehydes, alcohol oxidation, and differentiation (generation of retinoic acid)	Cancer stemness, tumor development	OC Breast cancers Colon cancers Liver cancers Myeloma cancers Acute myeloid leukemia Lung cancer	[22,28–30]
	Stress-inducible chaperone, expressed in endoplasmic reticulum, mitochondria, and plasma membrane	GRP78	Regulation of endoplasmic reticulum function, cell proliferation, apoptosis, and inflammation maintenance of stemness features of embryonic and mammary stem cells	Cancer stemness, angiogenesis, tumor development, metastasis, and invasion	OC Breast cancer Head and neck cancer	[31–36]
Transcription factors		LIN28	Stem cell self-renewal, maintenance of pluripotency	Cancer stemness, dedifferentiation	OC Breast cancer	[22,37]
		Oct4	Maintenance of pluripotency, stem cell self-renewal	Cancer stemness, dedifferentiation	OC Breast cancer Lung cancer Liver cancer	[21,22,37]

activity is well documented to be oncogenic, it was also reported that c-Kit expression is decreased in advanced stage disease, with c-Kit-negative patients having a significantly shorter disease-free survival time than c-Kit-positive patients [23]. These results suggest that c-Kit may play an early role in ovarian carcinogenesis, while loss of c-Kit expression associates with poor prognosis in later stage disease. Correspondingly, our group demonstrated that ovarian tumor cells coexpressing CD117 and the hyaluronate-binding protein CD44 (see the later discussion) isolated from serous ovarian adenocarcinomas exhibited numerous CSC properties, including formation of anchorage-independent, self-renewing spheres, self-renewal, and high tumorigenic potential (100 cells/mouse could form tumors, while 100,000 unsorted cells could not) [5]. $CD44^{+}/CD117^{+}$ cells could also serially propagate tumors identical to their original histology and expressed a number of genes associated with "stemness" (*BMI-1*, *SCF*, *OCT4*, *NES*, *NOTCH1*, and *NANOG*) [5]. Those highly tumorigenic cells were also found to express the membrane efflux pump ABCG2, while also demonstrating enhanced resistance to the conventional OC chemotherapies cisplatin or paclitaxel [5].

CD44. The hyaluronate receptor CD44, a single-chain transmembrane glycoprotein widely expressed in both epithelial and nonepithelial tissues, plays a role in numerous physiological processes, including cell–cell and cell–matrix interactions, cell adhesion, and cell migration [39]. CD44, however, has also been implicated in tumor progression, and its interaction with hyaluronan has been shown to play a role in the onset of drug resistance [40], and in OC progression CD44 expression has been correlated with the multidrug resistance proteins MDR1, MRP2, and the invasion mediator urokinase (uPA) [41]. Moreover, epidermal growth factor (EGF) family member heregulin activation of oncogenic ErbB2-ERK signaling was found to result in hyaluronan synthase phosphorylation/activation, with the subsequent upregulation of hyaluronan leading to CD44-mediated OC progression [42]. The hyaluronan–CD44 interaction has also been demonstrated to facilitate cytoskeletal protein binding to the multidrug resistance protein MDR1, augmenting drug efflux [42], and hyaluronan-based prodrugs against CD44 have demonstrated antitumor and antimetastasis activity in vivo [43]. Paradoxically, however, CD44 expression was found to correlate with well-differentiated, early-stage ovarian tumors, and greater survival [24]; the specific CD44 isoform that was analyzed may contribute to these differences.

CD133. In addition to CD44 and CD117, CD133 (prominin-1, formerly known as AC133), a plasma membrane glycoprotein, has now been described as a common CSC marker for numerous malignancies including OC [44]. It was reported that $CD133^{+}$ OC cell lines, primary tumor cells, and patient ascites-derived cells were more platinum-resistant than $CD133^{-}$ cells, in addition to forming more aggressive tumor grafts at lower inoculums [15]. In the latter study, CD133$^+$ cells sorted from primary ovarian carcinomas were more clonogenic in culture, and more proliferative, than CD133$^-$ cells. Moreover, CD133$^+$ cells were found significantly more frequently in ovarian carcinomas than in normal ovaries, benign ovarian tumors, or omental lesions [15], while also showing increased tumorigenic capacity and recapitulation of their original tumor phenotype [15]. CD133$^+$ cells were also shown to differentiate into CD133$^-$ cells, with CD133 silenced in progeny cells by DNA methylation [13]. In another study, however, while CD133$^+$ cells were found to interact with OCSCs, these cells were not tumorigenic, although they could facilitate tumor development by augmenting vasculogenesis (thus suggesting the CD133$^+$ cells to be OCSC-associated endothelial stem cells) [45]. Additionally, CD133 expression was found nonpredictive of patient response to treatment, time to progression (TTP), clinical prognosis, or overall survival (OS) [46]. Consequently, the precise role of this glycoprotein in ovarian tumors remains somewhat uncertain.

CD24. CD24 is a cell surface molecule upregulated in a large number of human malignancies, and in OC, its expression has been correlated with poor prognosis [47]. CD24 has also been used as a marker to identify solid tumor CSCs [48], and CD24$^+$ ovarian tumor cells were reported to possess various stemlike characteristics, including quiescence, chemoresistance, self-renewal, and differentiation [26]. In addition, low (5×10^3) numbers of CD24$^+$ cells were capable of xenograft formation in nude mice, while equal numbers of CD24$^-$ cells remained nontumorigenic [26]; CD24$^+$ cells were also found to overexpress the stem cell markers *NES*, *CTNNBIP1*, *BMI-1*, *OCT4*, *NOTCH1*, and *NOTCH4*, while underexpressing *CDH1*, as compared to CD24$^-$ cells [26].

LGR5. Leucine-rich-repeat-containing G-protein-coupled receptor 5 (LGR5), a seven transmembrane receptor protein, is a Wnt target gene that acts as receptor for the Wnt agonist R-spondin [9]. *Lgr5* has been reported as a resident stem cell marker in many organs, including intestine, skin hair follicle, kidney and ovary, all organs that require stem cells to repair tissue damage and regeneration throughout their life span [7,27]. Lgr5$^+$ epithelial stem cells also have been implicated as the origin of several types of malignancies, including colorectal cancer and gastrointestinal tract cancer. Recent findings on Lgr5$^+$ cell–driven tracing assay in mouse ovary models revealed that Lgr5$^+$ stem cells are constantly maintained in the adult OSE [7]. Early Lgr5$^+$ populations establish the OSE lineages in the ovary, as well as the epithelial cell lineage of the oviduct [7]. In the adult ovary, resident Lgr5$^+$ populations only maintain OSE homeostasis and ovulation-induced epithelial regeneration. The presence of Lgr5$^+$ cells with stem properties in ovary suggests that Lgr5$^+$ stem cells reside in two hypothesized OC origin sites, the OSE and the tubal epithelia, which further indicates that resident Lgr5$^+$ cells may be more vulnerable to transformation and contribute to the origin of the disease [7,8,27].

MyD88. Myeloid differentiation factor 88 (MyD88) is a critical component of the toll-like receptor pathway (associated with immune response) and an activator of the protooncogenic NF-κB signaling pathway [25]. In OC, Alvero et al. isolated $CD44^+/MyD88^+$ cells from solid ovarian tumors and ascites that demonstrated constitutive NF-κB activity, cytokine and chemokine production, a high capacity for DNA repair, and resistance to conventional chemotherapies [25]. Gene array analysis demonstrated that MyD88 was exclusively expressed in $CD44^+$ OC cells, while 10% of all genes examined were differentially expressed between $CD44^+$ and $CD44^-$ cells, including genes related to apoptosis, signal transduction, and cell differentiation [25].

ALDH. Aldehyde dehydrogenase (ALDH) enzymes comprised of 19 isoforms located in the cytoplasm, mitochondria, or nucleus [49], which are a family of detoxifying enzymes responsible for the oxidation of intracellular aldehydes, and were also reported to play a role in the early differentiation of stem cells through its oxidation of retinol to retinoic acid [50]. Recently, ALDH activity detectable through the Aldefluor assay has been well validated and is considered as a universal and robust CSC marker in a wide spectrum of cancers, including OC [6,49,51]. However, the expression of ALDH isoforms, which are responsible for the Aldefluor activity of ALDH, is cancer type dependent [49]. Several studies have indicated that the expression of ALDH1A1 isoforms is correlated to enzymatic activity of ALDH in OC [28,52]. $ALDH^+$ cells derived from OC cells have been observed to be resistant to cisplatin chemotherapy and display enhanced spheroid-forming ability (Fig. 9.1) and enhanced tumor-initiating capacity compared to $ALDH^-$ cells. In addition, upregulated expression of stem cell transcription factors (*SOX2*, *NANOG*, *OCT4*) was

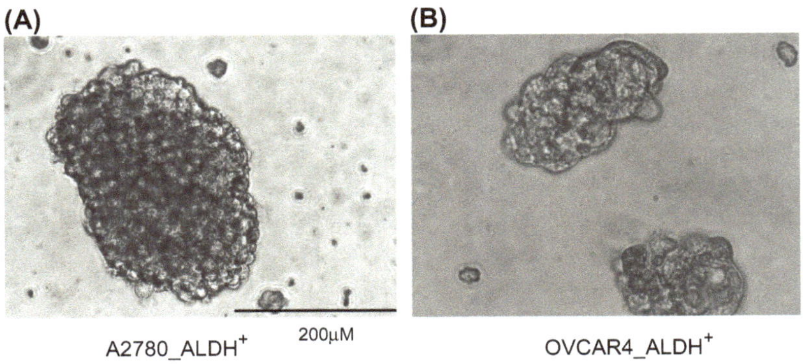

FIGURE 9.1

Self-renewal capacity of $ALDH^+$ ovarian cancer (OC) cells. Self-renewal spheroids formed by $ALDH^+$ cells derived from well-established OC cell lines, which have been cultured on the low-adhesion plates with stem cell medium for 7 days. (A) Spheroids formed by A2780_$ALDH^+$ cells; (B) OVCAR4_$ALDH^+$ cells.

observed in ALDH+ cells [6,29,51]. Taken together, ALDH+ cells appear to satisfy all criteria to be considered CSCs.

We and others have determined that the percentage of ALDH+ cells in OC cell lines and human tumors varies between 0.2% and 4% [6,29], and ALDH1 expression in OC correlates with worse survival and platinum resistance [16, 28, 29]. We further confirmed the stemlike properties of the ALDH+ OC cells by demonstrating increased expression of stem cell–associated genes (*NOTCH, ALDH1A1, BMI-1*), decreased expression levels of the differentiation-related genes *HOXA10* and *HOXA11* in ALDH+ compared to ALDH− cells [29]. Additionally, several other groups have now shown *ALDH1* expression to be a prognostic marker for a number of epithelial cancers [53,54]. Deng et al. [16] analyzed *ALDH1* expression in 24 types of normal tissues and a large collection of epithelial tumor specimens, in addition to a transgenic OC mouse model and murine OC cell lines. *ALDH1* expression, while minimal in OC cells and tumors, significantly associated with poor clinical outcomes in serous OC patients [16]. This finding, however, was contradicted by another study of 266 serous and 176 nonserous OC patients, in which *ALDH1* expression correlated with favorable prognosis [54]. These discrepancies may indicate that the prognostic value of *ALDH1* is tumor subtype specific [54].

GRP78. Glucose-regulated protein 78 (GRP78) is a stress-inducible chaperone, which belongs to the heat shock protein family (HSP) [31]. GRP78 resides

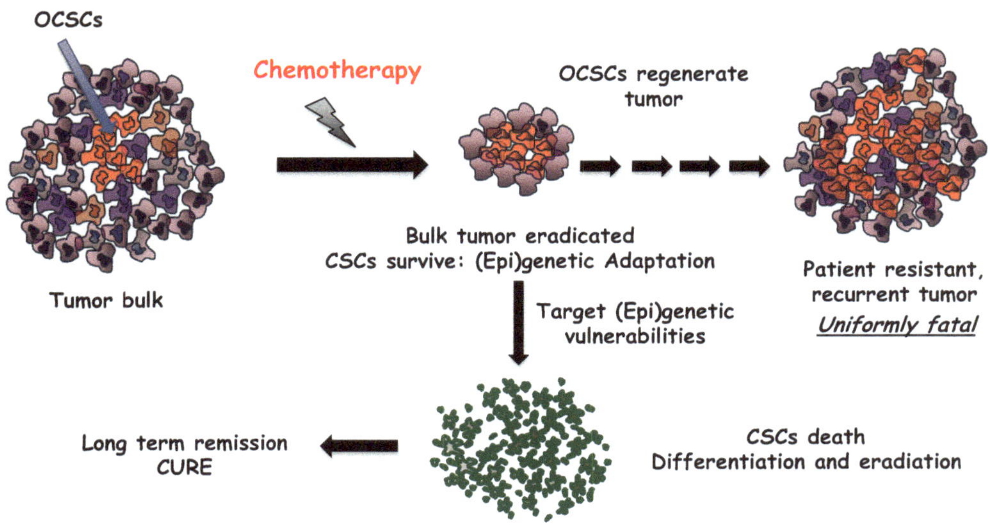

FIGURE 9.2
(Epi)genetic targeting of ovarian cancer stem cells (OCSCs). OCSCs elude the effects of traditional chemotherapy due to (epi)genetic adaptation and can reconstitute recurrent tumors which become recalcitrant to chemotherapy. Long-term remission or possibly a cure could be achieved by eliminating OCSCs by targeting their (epi)genetic vulnerabilities. *CSCs*, cancer stem cells.

mainly in the endoplasmic reticulum (ER) and mitochondria and also presents on plasma membrane of stem cells and malignant tumors [32]. Overexpression of GRP78 is widely indicated in a wide range of cancer and is associated with tumor initiation, growth, metastasis, and invasion [32], including epithelial OC [33]. Surface expression of GRP78 plays a crucial role in maintaining stemness features of embryonic stem cells and fetal and adult mammary stem cells [34]. GRP78 is involved in regulation of multiple stemness-associated signaling pathways including transforming growth factor β (TGF-β) and phosphatidylinositol 3-kinase (PI3K) [31]. Therefore, it is not surprising that GRP78 overexpression has been demonstrated to promote cancer stemness properties in cancer, including OC [35,36]. A recent study has shown that GRP78 expression was significantly increased in OC cells isolated from ascites associated with malignant OC. They found that $GRP78^+$ cells possess higher tumorigenic potential and stemness properties than $GRP78^-$ cells. Neutralizing GRP78 against the carboxy-terminal domain of GRP78 was able to reduce self-renewing ability of murine and human OC cells by suppressing GRP78/GSK3a-AKT/SNAI1 signaling [35]. Together, all these results indicate the potential role of GRP78 as novel CSC marker in OC and a promising therapeutic target for OCSCs.

OVARIAN CANCER STEM CELLS AND DRUG RESISTANCE

OCSCs are characterized by self-renewal and multipotency features, serial tumor transplantation or lineage-tracing assays should be applied followed by utilizing cell surface markers for CSC identification to further confirm stemness characteristic of those stem cell–like subpopulations [55]. In this regard, it has also been suggested that various CSC attributes can be conferred to normal/precancerous cells under extrinsic stimulation, such as chemotherapy. Our recent study demonstrated that while chemotherapy may succeed initially at decreasing the size and number of ovarian tumors, it leaves behind residual malignant cells, which we have shown is enriched in tumor progenitors or "OCSCs" [29]. Several other groups have published similar results. Abubaker et al. indicated that a short-term single treatment of cisplatin or paclitaxel leaves residual cells that are enriched in CSC-like traits, resulting in an increased metastatic potential of OC cells. Immunohistochemical studies on mouse tumors injected with chemotherapy showed treated residual cells displayed higher staining for the proliferative antigen Ki67, oncogenic CA125, epithelial E-cadherin as well as CSC markers such as Oct4 and CD117, compared to mice injected with control untreated cells [56]. Wintzell et al. also found that repeatedly treating SKOV-3 cells with low doses of cisplatin induced a stable population of "SKOV-3-R" showing increased motility, EMT, and expression of CSC markers CD117, CD44, and ALDH1 [57].

The cytokine interleukin (IL-6) and its signaling pathway are important regulators of self-renewal of CSCs in a broad spectrum of tumor types [58], and IL-6 may play a critical role in the conversion of non-CSCs to CSCs. In this regard, several studies have shown that differentiated cancer cells could be converted to stemlike cells. Kim et al. group demonstrated that the IL-6-JAK1-STAT3 signal transduction pathway plays an important role in the conversion of non-CSCs into CSCs through regulation of OCT4 gene expression in breast cancer [58]. Han et al. showed that IL-6 can convert non-ALDH$^+$ cells to ALDH$^+$ cells in prostate cancer cells and a STAT3 inhibitor inhibited the conversion. Moreover, blockade of STAT3 signaling significantly inhibited tumor initiating and tumor growth in both prostate cancer cell line xenograft model and patient-derived xenograft (PDX) tumor models [59].

In a related study, our group recently showed that in an OC model of platinum resistance, drug resistance positively correlates with a linear increase in the total genomic number of hypermethylated gene promoters, while drug sensitivity was subsequently restored by inhibitors of DNA methylation [60]. Another study showed that TGF-β induces non-CSC to CSC conversions in a basal breast cancer model by removal of the repressive H3K27me3 mark at the ZEB-1 promoter, allowing the transition of poised chromatin to active chromatin and transcription initiation [61]. These results provide additional support for the association of epigenome alterations with chemotherapy resistance and/or cancer stemness.

THERAPEUTIC APPROACHES FOR ERADICATING OVARIAN CANCER STEM CELLS

As a consequence of the CSC theory, it is hypothesized that tumors that initially undergo complete remission, but subsequently relapse to a refractory state such as OC, are more likely to possess CSCs than tumors that do not respond well to primary therapy Fig. 9.2. In this model, chemotherapeutics preferentially target the rapidly proliferating cells (presumably, CSC progeny cells) that comprise the bulk of the tumor, causing tumor regression, but fail to eradicate drug-resistant CSCs. Consequently, therapies are needed to target the small percentage of tumorigenic progenitors, in addition to the more rapidly proliferating, nontumorigenic progeny that comprise that bulk of the tumor [62]. Here we have categorized CSC-targeted therapies into four classifications: (1) elimination; (2) differentiation; (3) stem cell niche modification; and (4) epigenetic (Table 9.2).

Elimination Therapies Targeting Cancer Stemness-Related Pathways

PI3K/Akt signaling. One cascade upregulated in numerous solid cancers, possibly contributing to tumor initiation (and thus an attractive target for cancer therapeutics), is the phosphatidylinositol kinase-3 (PI3K)/Akt mitogenic

Table 9.2 Strategies Targeting CSCs in Ovarian Cancer

Category	Pathway	Drugs	Reference(s)
Elimination therapies	PI3K/Akt signaling	Daidzein-daunomycin, N-t-Boc-daidzein, CXCR1 antagonist	[63–66]
	Ephrin/Eph signaling	Anti-EFNA4-Ab-drug conjugate (PF-06647263)	[67]
	Sonic hedgehog pathway	Cyclopamine	[68,69]
	Notch pathway	Notch inhibitor(s)	[70]
	Wingless (Wnt) signaling	ZTM000990, PKF1158-310, anti-Wnt1 and anti-Wnt2-mAbs	[71,72]
	Aldehyde dehydrogenase isozymes	Disulfiram, aldi, A37	[52,73–75]
CSCs differentiation therapies	Embryo-like CSC phenotype, HOXA10, TWIST1	All-trans retinoic acid, histone deacetylase and DNA methyltransferase inhibitors	[5,29,76–79]
CSCs niche destruction therapies	Müllerian system	Müllerian inhibiting substance (MIS) adeno-associated virus–delivered human MIS [albumin leader Q425R MIS (LRMIS)]	[19,80,81]
Epigenetic therapies	Deoxycytosine methylation, histone deacetylation	Guadecitabine (SGI-110), decitabine, histone deacetylation inhibitors	[29,78,82–87]

signaling pathway [88]. Activation of the PI3K/AKT signaling pathway can upregulate the expression of inducers of EMT, including *EZH2*, *BMI-1*, and *SNAIL* and *SLUG* promote EMT and CSC features in human (endometrial) cancer [89]. Moreover, this pathway has also been implicated in regulating EMT-associated CSC enrichment in OC, and the PI3K/AKT pathway inhibitor LY294002 reduced expression of stemness genes in OC spheroids [90].

Two potential Akt signaling inhibitors, daidzein-daunomycin and N-t-Boc-daidzein, were derived from the promising cancer chemopreventative phytoestrogen daidzein [63,64]. Daidzein-daunomycin was reported to improve therapeutic response in an animal OC model [63], while N-t-Boc-daidzein could decrease the number of OCSCs isolated from patient ascites. N-t-Boc-daidzein was also found to elicit apoptosis of ascites-derived mature OC primary cell lines, in a dose- and time-dependent manner, due in part to the degradation of Akt [64]. In addition, Akt inhibitors were found effective in targeting $CD133^+$ cells in other cancers, including hepatocellular cancer [65]. In an impressive recent study of breast cancer, an antagonist of the IL-8 receptor CXCR1 was demonstrated to reduce the number of $ALDH^+$ (see section "Isolation of Ovarian Cancer Stem Cells From Primary Ovarian Tumors") breast CSCs, followed by massive apoptosis of the remaining bulk of the tumor; that antagonist was demonstrated to inhibit focal adhesion kinase (FAK) signaling through Akt [66]. Together, these reports strongly implicate the PI3K/Akt cascade in CSC maintenance and self-renewal.

Ephrin/Eph signaling. Ephrin receptors (Eph) are the largest family of receptor tyrosine kinases, which comprises 14 receptors [91]. The activation of Eph by interaction with cell surface ligands ephrins in the adjacent cells induces bidirectional signals that affect both receptor- and ephrin-expressing cells [92]. The Eph/ephrin system functions have been implicated to regulate signaling pathways that impact cell fate decisions during embryogenesis and adult tissue homeostasis. The aberrant expression of Eph receptors and ephrin ligands has been observed in many types of human cancers, including breast and ovaries cancers [93]. In addition, Eph/ephrin molecules also play a role in regulation of CSC self-renewal and tumorigenicity, thus targeting Eph and ephrins may hold promise for targeting on CSCs population in cancers [92]. Several types of ephrin/Eph signaling inhibitors have been examined in preclinical and clinical investigations [94], including synthetic peptides that inhibit Eph/ephrin interaction, Eph and ephrin exodomain fusion proteins that modulate Eph and ephrin activity and function, small molecule tyrosine kinase inhibitors, and mAbs-targeting Ephs. Recent evidence suggests the role of EFNA4 in regulation of cell fate decision of mammary epithelial cells and overexpression of EFNA4 have been observed in the CSCs population of triple-negative breast cancer (TNBC) and OC, suggesting EFNA4 as a therapeutic target of CSCs. A new drug, anti-EFNA4-antibody-drug conjugates (PF-06647263), has demonstrated antitumor activity in preclinical studies and is currently being evaluated in a Phase 1 clinical trial. This drug was also able to reduce CSC frequency and induce sustained tumor regressions in OC PDX tumor models [67]. Together, these results strongly support the role of ephrin/Eph signaling in OCSC maintenance, suggesting novel therapeutic target of CSCs.

Sonic hedgehog pathway. Another signal cascade implicated in playing a role in cancer stemness is the Sonic hedgehog (SHH) embryogenesis-associated pathway. SHH signaling, initiated by SHH binding to its receptor, Patched-1, is a crucial mediator of cell fate during early mammalian development [95]. However, SHH deregulation has been hypothesized to contribute to CSC self-renewal and therefore represents an attractive target for cancer therapy [96]. In support of such an approach, it was demonstrated that cyclopamine, a naturally occurring alkaloid also found to be specific SHH pathway inhibitor, strongly inhibited the proliferation and clonogenic growth of ovarian tumor cells in vitro, while also arresting ovarian tumor growth, in vivo [68]. However, another study demonstrated minimal SHH signaling in OC [69], leaving the specific role of this pathway in OC largely unresolved.

Notch. The Notch pathway is a cell–cell contact signaling cascade intimately involved in normal development and tissue renewal [97]. Signal transduction occurs when a surface Notch ligand, on one cell, activates its receptor on a contiguous cell, resulting in cleavage of the Notch intracellular domain (NICD) [97]. The NICD then relocates from the cytoplasm to the nucleus, resulting in gene transactivation via its interaction with the transcription factor C element-binding factor (CBF) [97]. However, Notch dysregulation has also been implicated in maintenance of the CSC phenotype, and a number of specific Notch inhibitors are currently in various phases of development [70]. In OC, various Notch pathway members are overexpressed in tumors, but not in adenomas [98]; correspondingly, while significant Notch signaling was also found in three OC cell lines and 76% of OC patient tumors [99]. Moreover, *Notch1* was found overexpressed in candidate OCSCs, as compared to the bulk population of tumor cells or OCSCs placed under differentiating conditions [5]. Together, these (and other) reports strongly implicate Notch in ovarian tumorigenesis and OCSC maintenance, making this signal an attractive therapeutic target.

Wingless (Wnt) signaling. The Wnt pathway is essential for embryonic morphogenesis and body axis specification, in addition to tissue homeostasis, due to its regulation of self-renewal of normal stem cells [100]. Signal propagation occurs upon the binding of Wnt ligand to its receptor, Frizzled, resulting in a cascade that leads to nuclear translocation of β-catenin, which upon binding to its transactivational cofactor, TCF, induces a number of protooncogenes (including *MYC*) [100]. Consequently, similar to other embryonic signaling pathways, Wnt dysregulation is also associated with carcinogenesis and tumor progression [100]. In one OC study, Rask et al. demonstrated increased expression of components of the Wnt pathway in malignant OC tumors, as compared to normal ovarian tissues [101]. Toward the targeting of Wnt signaling as an effective cancer therapy, two small molecules, ZTM000990 and PKF118-310, were identified in a high-throughput screen (based on the structure of the β-catenin/TCF complex) to target the canonical Wnt signaling cascade [71]. Additionally, anti-Wnt1 and anti-Wnt2 monoclonal antibodies were found to be potent inducers of apoptosis in mesothelioma and melanoma cells [72]. With further pharmacologic optimization, these small molecules or antibodies targeting the Wnt signaling pathway could represent effective OC therapeutics.

ALDH isozymes. The enzymatic activity of ALDH plays a role in maintaining stemness characteristic of CSCs [49], which could be regulated from both transcription and posttranslation level. In breast cancer cells, ALDH1A1 activity is inhibited by acetylation of lysine 353, which is regulated by acetyltransferase P300/CBP-associated factor and deacetylase sirtuin 2 (SIRT2). Acetylation of ALDH1A1 inhibited both the stem cell population and self-renewal properties in breast cancer. NOTCH signaling has been indicated to play a role in activation of ALDH1A1 activity through the induction of SIRT2 to promote breast CSCs [102], suggesting the potential to inhibit CSCs population by regulating ALDH enzymatic activity posttranslationally. However, the activity of ALDH also could be modulated at the transcription level. β-Catenin has been demonstrated to play a role in direct regulation of ALDH1A1 expression in the maintenance of OC spheroids, a characteristic associated with self-renewal capacity of stemlike cells. This finding indicated a new approach to target the OCSC population by inhibition of ALDH1A1 expression [52]. In addition, drugs directly inhibiting ALDH enzymatic activity are being developed including disulfiram and aldi, which showed synergistic effects on cytotoxicity in several types of cancer [73,74]. Disulfiram, an inhibitor of ALDH1 and ALDH2, was recently shown to eliminate crizotinib-resistant cancer cells by diminishing the protection exerted by ALDH1 against reactive oxygen species [75]. Other more specific ALDH1A1 inhibitors have been discovered and are being developed. For instance A37, a selective ALDH1A1 inhibitor was shown to block the growth of OC spheroids and inhibit β-catenin transcriptional activity associated with stem cells [52].

Therapies Inducing Differentiation of Ovarian Cancer Stem Cells

Another potential approach to CSC targeting is the use of differentiating agents, which presumably would alter the embryo-like CSC phenotype toward that of its normal, mature tissue; it is hypothesized that disruption of the aforementioned self-renewal pathways might serve this purpose. Various differentiating agents have now been examined with varying degrees of success, including dietary polyphenols and phytoestrogens, and vitamin D3. However, the agent best studied (and to date, the most successful) differentiating agent is all-trans retinoic acid (ATRA), which has demonstrated impressive effectiveness against acute promyelocytic leukemia, head/neck squamous carcinomas, and thyroid cancer [76]. In two serous OC cell lines, ATRA also altered cell morphology to induce differentiated epithelial cells, in addition to strongly inhibiting cell proliferation [77]. With regard to CSC differentiation, our group showed that OCSCs are more resistant to cisplatin and paclitaxel, but could be resensitized to both agents under differentiating conditions [5]. In a separate study, we also demonstrated that histone deacetylase (HDAC) inhibitors (a type of epigenetic therapy, see section "Epigenetic Therapies") induced morphological changes and epithelial differentiation markers in a platinum-resistant OC cell line [78]. In the recent study, we also showed that DNA methyltransferase (DNMT) inhibitor (see section "Epigenetic Therapies") reprogrammed OCSCs and resensitized them to platinum chemotherapy by derepressing expression of differentiation-associated gene, *HOXA10* [29]. Yin et al. reported the transcription factor TWIST1 also had the potential to drive differentiation of OCSC [79].

Other studies demonstrated differentiation of highly aggressive melanoma cells to normal melanocytes following plating of the melanoma cells atop an embryonic stem cell–derived extracellular matrix; the reciprocal placement of melanocytes onto a melanoma-derived microenvironment resulted in restoration of the aggressive malignant phenotype [103]. As normal differentiation is governed by epigenomic changes (see section "Epigenetic Therapies"), it is strongly believed that CSCs also possess a type of "epigenetic plasticity" capable of altering their degree of differentiation (and thus their malignant phenotype) [104]. Together, these results support that differentiation therapy, possibly facilitated by epigenetic therapies (see section "Epigenetic Therapies") of OCSCs, has strong potential as an effective therapeutic approach.

Destruction or Alteration of the Ovarian Cancer Stem Cell Niche

In vivo, stem cell self-renewal and differentiation are tightly controlled by a complex niche that physically harbors those cells in an anatomically well-defined location within a tissue (reviewed in Ref. [105]), and there is increasing evidence that the microenvironment regulates tissue specificity and contributes significantly to tumorigenesis (reviewed in Ref. [106]) and contributes the transformation and growth of OC (reviewed in Ref. [10]). The extracellular environment provides the structural platform necessary for cell growth and intercellular communication; analogously, various growth factors, and chemokines may enhance tumor cell proliferation and invasion [106]. Conversely, the tumor microenvironment may also stimulate production of antiangiogenic proteins and inhibitors of matrix metalloproteases that can obstruct tumorigenesis [107]. In brain cancer, including medulloblastoma and ependymoma, it was shown that CD133$^+$/Nestin$^+$ CSCs reside in a paravascular niche, and that inhibition of angiogenesis, via EGF signaling disruption or inhibition of vascular endothelial growth factor (VEGF), eradicated those self-renewing cells [108]. Moreover, as noted previously (section "Elimination Therapies Targeting Cancer Stemness-Related Pathways"), a number of studies have now demonstrated differentiation-associated "reprogramming" of aggressive melanoma and breast cancer cells, into normal epithelial cells, by culture in an embryonic stem cell–derived microenvironment [109]. Those studies demonstrate that even advanced stage cancers (presumably enriched in CSCs) exhibit a phenotypic "plasticity" for differentiation, governed by epigenomic changes. In particular, the embryonic microenvironmental signaling molecule responsible for melanoma cell reprogramming was later found to be an inhibitor of the embryonic morphogen Nodal, a member of the TGF-β family [109]. Nodal was later found to effect "vascular mimicry," formation of tubelike structures capable of perfusing the tumor [110] (thus possibly similar to brain CSC perivascular niche). In OC specifically, components of the secondary Müllerian system (paraovarian/paratubal cysts, rete ovarii, endosalpingiosis, endometriosis, and endomucinosis) may similarly provide a source of cells (or signaling molecules) allowing the different histologic types of ovarian malignancies [80]. Müllerian inhibiting substance (MIS) transduction signaling has been proposed to inhibit tumor of Müllerian origin by its physiological role in regression of the Müllerian structure in fetal development [111]. A recent study suggested MIS as a possible adjuvant to conventional OC chemotherapeutics that targets putative OCSCs, as MIS treatment strikingly inhibited proliferation of both SP and non-SP cells, while conventional chemotherapies primarily arrested only non-SP cells [19]. Moreover, using an adeno-associated virus (AAV)–delivered peptide engineered with modifications to human MIS [albumin leader Q425R MIS (LRMIS)] as a gene therapy approach in HGSOC, Pepin et al. demonstrated that AAV9-LRMIS monotherapy effectively inhibited OC spheroid formation and growth of chemoresistant serous adenocarcinoma ovarian PDX [81].

Epigenetic Therapies

Epigenetic alterations have been demonstrated to govern gene expression both in embryonic and tissue stem cells and thus likely play an important role in the tumorigenic potential (and differentiation prevention) in OCSCs. These alterations include methylation of deoxycytosine, as well as numerous modifications of histones, regulate distinct gene expression patterns that allow for specific cell or organotypic phenotypes [112]. It is also well established that epigenetics contributes significantly to ovarian tumorigenesis [105], likely in association with its intricate role in differentiation.

It is hypothesized that ovarian tumor chemotherapy resistance in OC results largely from the repression of tumor-suppressor genes (specifically, chemotherapy-response genes) by DNA methylation [113,114]. In contrast to DNA mutations and deletions, however, aberrant gene-repressive epigenetic modifications are potentially reversible, by epigenetic therapies, including inhibitors of DNMTs or various histone-modifying enzymes (reviewed in Ref. [114]). Although epigenetic monotherapies have shown little activity against solid tumors [115] (including OC [114]), preclinical studies of DNMT inhibitors by our group [60,116] and others [117] have demonstrated potent resensitization of drug-resistant OC cells and xenograft tumors to conventional chemotherapies, possibly by differentiation or apoptosis of CSCs [115]. As evidence suggests that DNMTs expression is increased in $ALDH^+$ cells, our group examined the effects of the second generation DNMT inhibitor Guadecitabine (SGI-110; 100 nM for 3 days) on OCSCs dissociated from cell lines and high-grade serous human tumors. Guadecitabine treatment was able to decrease the percentage of $ALDH^+$ cells in platinum-sensitive and platinum-resistant ovarian cells and primary ovarian tumors, indicating that low-dose Guadecitabine has the potential to target OCSCs [29]. We also found that $ALDH^+$ cells demonstrated greater sphere-forming ability and tumor-initiation ability than $ALDH^-$ cells, which consistent with their stemness features. Guadecitabine treatment alone markedly inhibited the spheroid-forming ability of both population in vitro and significantly inhibited tumor-forming ability of ALDH$^+$ cells in mice by targeting on ALDH$^+$ OC cell population. In ALDH$^+$ cells, Guadecitabine suppressed the expression of stemness-related genes and induced upregulation of differentiation-associated genes accompanied promoter region demethylation, consistent with Guadecitabine-hypomethylating effect.

We also found that DNA hypomethylation induced by Guadecitabine prevented tumor recurrence after maximal response to platinum therapy. Treatment with Guadecitabine decreased total tumor weight and reduced tumor-disassociated cells, spheroid-formation capability in vitro, consistent with inhibition of stem cell properties [29]. More than 62,000 methylation sites and 10,000 CpG islands were found to be significantly hypomethylated in Guadecitabine-treated tumors compared to controls and substantial global DNA hypomethylation induced by Guadecitabine in vivo was also demonstrated. Interestingly, it was also demonstrated that DNMT-inhibitor treatment was able to chemosensitize drug-resistant OC cells and xenografts by inducing expression of endogenous Wnt pathway inhibitor *SFRP5* (see section "Elimination Therapies Targeting Cancer Stemness-Related Pathways") [82]. Collectively these data support that DNMT inhibitors reprogram OCSCs preventing resistant tumors.

Based on the aforementioned studies demonstrating chemotherapy sensitization of resistant OC cells and tumors, several such combinatorial regimens are now being examined in cancer clinical trials [113,114]. Our group recently completed a Phase 1 trial (NCT00477386, Study ID 0704-07, www.clinicaltrials.gov) using a DNMT inhibitor, decitabine (Dacogen; Eisai, Inc., Tokyo, Japan) combined with the standard chemotherapeutic, carboplatin, hypothesizing that low-dose decitabine may derepress silenced tumor suppressors to chemosensitize platinum-resistant OCs [83]. In that study, a low dose (10 mg/m^2) of decitabine was administered for 5 consecutive days, followed by carboplatin (AUC5) administered 3 days later, with each treatment cycle lasting 28 days. That regimen demonstrated bioactivity in vivo, as assessed by decreased methylation of DNA repetitive elements (in patients' peripheral blood cells) and of specific genes (in patients' sera). Of ten patients, we observed one complete response, while four patients experienced stable disease for at least 6 months; interestingly the complete response did not occur prior to eight treatment cycles [83]. It is interesting that breast CSCs induce tumorigenesis only after a latency of 6 months, further supporting the idea that drug-associated hypomethylation of CSCs requires an extended period, based on the slow division those cells and their delayed response in other carcinomas [12]. This possibility is further reinforced by our follow-up Phase 2 trial demonstrating proof-of-principle for the concept that epigenetic modulation restores sensitivity to platinum in patients with platinum-resistant OC [84]. The decitabine–platinum combination resulted in significant clinical activity, as measured by high response rate and prolonged progression-free survival. We showed that these striking clinical effects were modulated through DNA hypomethylation and identified candidate genes and methylation profiles correlating with clinical outcome [84]. Gene expression profiling of patient biopsies identified specific pathways altered by decitabine and associated with platinum resensitization and clinical benefit in OC [85]. These findings justify further investigation of DNMT inhibitors/carboplatin combination in OC or other rationally designed epigenetic strategies in solid tumors. A very recent study of noncytotoxic DNMT1 depletion therapy in patients with myelodysplastic syndromes (MDS) suggested that noncytotoxic concentrations of decitabine produced p53-independent cell cycle exits by epigenetically driving proliferation-terminating differentiation genes in cancer cells [118]. While that clinical trial demonstrated that the current FDA-approved decitabine regimens could be redesigned to reduce cytotoxicity and increase S-phase exposure time for DNMT1 depletion and safely and effectively target P53 and apoptosis-deficient cancer in MDS, the results are clearly highly relevant to OC.

Since histone deacetylation is another transcriptional silencing mechanism in OC, HDAC inhibitors (HDACIs) can also relieve epigenetic gene repression; these agents also exert anticancer effects by inhibiting the deacetylation of nonhistone proteins [86]. Similar to DNA methylation inhibitors, HDACIs are most promising in combination with conventional agents, and studies by our group [78] and others [87] have demonstrated chemosensitizing effects on drug-resistant OC cells and tumors. HDACIs have also been effectively combined with a differentiating [119] or a "death receptor" ligand [120] in OC cell studies, similarly demonstrating additive or synergistic effects. It is also possible that HDACIs might also have direct effects on CSCs. In one study, a subpopulation of rapidly proliferating, drug-resistant lung cancer cells could be eliminated by HDACI treatment [121], while an HDACI/imatinib combination was demonstrated to target chronic myelogenous leukemia stem cells [122]. In other studies, an HDACI was found to block self-renewal and aggregation of breast cancer spheroids [123], while another HDACI suppressed expression of the stemness gene *NANOG* in embryonic carcinoma cells, resulting in loss of a stem cell "gene signature" [124]. Thus this class of epigenetic agents holds promise for the treatment of drug-resistant OC, possibly via their effects on OCSCs.

CONCLUSIONS

Based on several studies to date and similar to other epithelial tissues, both a dormant and a proliferative stem cell population exist in the ovary. These populations maintain normal tissue homeostasis, repair/replace damaged tissue and regenerate during ovulation [7,8]. It is highly likely that ovarian tumors, similar to numerous normal tissues and other solid cancers, possess a hierarchy of cell types, with tumor initiation, progression, and chemoresistance driven by slow-cycling dormant $ALDH^{+}$ labeled stem cell subpopulation(s) [28]. OC appears to follow a hierarchical model with rare stochastic events, as a branched differentiation hierarchy was recently demonstrated using a novel microfluidic single-cell culture method and differentiation capacity of four heterogeneous OC cell populations (defined by dual CSC markers ALDH and CD133) [125]. By promoting OCSC expansion and suppressing progenitor proliferation, bone morphogenetic protein 2 (BMP2) was indicated to play paradoxical role in these hierarchical differentiation cells [125].

Although CSC may arise from normal stem cells, an emerging model based upon recent findings in a number of laboratories is that hierarchically organized cancer cell populations are more plastic than previously appreciated, indicating that cancer cells may dedifferentiate and enter back into the CSC pool. In breast cancer EMT-inducing transcription factors, such as *SNAIL* or *TWIST*, were shown to facilitate reprogramming of the human mammary epithelial cells (HMLE) to induce $CD44^{+}/CD24^{-}$ neoplastic stem cells population [126]. Also in breast cancer, IL-6-JAK1-STAT3 signal transduction pathway played an important role in the conversion of non-CSCs into CSCs through upregulation of *OCT4* gene expression [58]. However, whether non-CSC to CSC conversion occurs in OC has not been fully explored.

Although numerous stemness markers have been used to identify, isolate, and characterize OCSCs from OC tissues, the specificity of those markers for OCSCs labeling remains to be established. Future OCSC purification studies may include combining cell surface markers with functional assays, such as stemness signaling pathway analysis or cell lineage-tracing assays.

Ovarian PDX has recently been developed that recapitulate the molecular and histological heterogeneity of human ovarian tumors, gene expression profiles, clinical response to chemotherapy, and the presence of OCSCs [127–129]. These studies strongly suggest that PDX will play a prominent role in the OCSC field, and in this regard PDX has been used for drug-targeting OCSC, for example, anti-ROR1 [30] therapies and AAV-delivered human MIS [81]. In addition, as a normal ovary stem cell population has been identified that likely contributes to the origin of OC, identification of genetic profiles involved in driving stem cell transformation could lead to development of early diagnostic biomarkers and OCSC-targeted therapies. Finally, as both genetic and epigenetic changes associated with stemness-related genes play critical roles in maintaining OCSCs, a better understanding of molecular pathways driving OCSCs behavior is critical for designing strategies that induce OCSC differentiation and/or eliminate OCSC perhaps by inhibiting specific pathways essential for OCSC maintenance.

Acknowledgment

The authors would like to thank the Ovarian Cancer Research Fund Alliance (PPDIU01.2011) for supporting this work.

List of Acronyms and Abbreviations

AAV Adeno-associated virus
ABC ATP-binding-cassette
ALDH Aldehyde dehydrogenase
ATRA All-trans retinoic acid
AUC Area under curve
BCRP Breakpoint cluster region pseudogene 1
BMI-1 B lymphoma Mo-MLV insertion region 1 homolog, polycomb complex protein
BMP2 Bone morphogenetic protein 2
CBF C element-binding factor
CD Cluster of differentiation
CD44 The hyaluronate receptor
CD113 Prominin-1
CD117 The *c-kit* protooncogene
CD147 Plasma membrane protein emmprin
CDH1 E-cadherin
CSC Cancer stem cells
CTNNBIP1 β-Catenin-interacting protein 1
DNMTi DNA methyltransferse inhibitors
DNMTs DNA methyltransferases
EGF Epidermal growth factor
EMT Epithelial-to-mesenchymal transition
ER Endoplasmic reticulum
FACS Fluorescence-activated cell sorting
FAK Focal adhesion kinase
FTE Fallopian tube epithelia
GRP78 Glucose-regulated protein 78
GSK3A Glycogen synthase kinase 3 α
HDACIs Histone deacetylase inhibitors
HGSEOC High-grade serous epithelium ovarian cancer
HMLE Human mammary epithelial cells
HSP Heat shock protein family
IFN Interferon
IL-6 Interleukin 6
LGR5 Leucine-rich-repeat-containing G-protein-coupled receptor 5
LRMIS Albumin leader Q425R MIS
MDR Multidrug resistance-associated transporter
MDS Myelodysplastic syndromes
MIS Müllerian inhibiting substance
MyD88 Myeloid differentiation factor 88
NES Nestin

NICD Notch intracellular domain
OC Ovarian cancer
OCSC Ovarian cancer stem cell
OCT4 Octamer-binding transcription factor 4
OS Overall survival
OSE Ovarian surface epithelium
PDX Patient-derived xenograft
PI3K Phosphatidylinositol 3-kinase
Rb1 Retinoblastoma protein 1
ROR1 Receptor-related 1
SCF Stem cell factor
SFRP5 Secreted frizzled-related protein 5
SHH Sonic hedgehog
SIRT2 Sirtuin 2
SP Side population
STIC Serous tubal intraepithelial carcinomas
TGF-β Transforming growth factor β
TNBCβ Triple-negative breast cancerβ
Trp53 Transformation-related protein 53
TTP Time to progression
uPA Urokinase
VEGF Vascular endothelial growth factor

References

[1] Dean M, Fojo T, Bates S. Tumour stem cells and drug resistance. Nat Rev Cancer 2005;5(4):275–84.

[2] Vaughan S, Coward JI, Bast RC, Berchuck A, Berek JS, Brenton JD, et al. Rethinking ovarian cancer: recommendations for improving outcomes. Nat Rev Cancer 2011;11(10):719–25.

[3] Jemal A, Siegel R, Ward E, Hao Y, Xu J, Thun MJ. Cancer statistics, 2009. CA Cancer J Clin 2009;59(4):225–49.

[4] Armstrong DK, Bundy B, Wenzel L, Huang HQ, Baergen R, Lele S, et al. Intraperitoneal cisplatin and paclitaxel in ovarian cancer. N Engl J Med 2006;354(1):34–43.

[5] Zhang S, Balch C, Chan MW, Lai HC, Matei D, Schilder JM, et al. Identification and characterization of ovarian cancer-initiating cells from primary human tumors. Cancer Res 2008;68(11):4311–20.

[6] Silva IA, Bai S, McLean K, Yang K, Griffith K, Thomas D, et al. Aldehyde dehydrogenase in combination with CD133 defines angiogenic ovarian cancer stem cells that portend poor patient survival. Cancer Res 2011;71(11):3991–4001.

[7] Ng A, Tan S, Singh G, Rizk P, Swathi Y, Tan TZ, et al. Lgr5 marks stem/progenitor cells in ovary and tubal epithelia. Nat Cell Biol 2014;16(8):745–57.

[8] Flesken-Nikitin A, Hwang CI, Cheng CY, Michurina TV, Enikolopov G, Nikitin AY. Ovarian surface epithelium at the junction area contains a cancer-prone stem cell niche. Nature 2013;495(7440):241–5.

[9] Hirsch D, Barker N, McNeil N, Hu Y, Camps J, McKinnon K, et al. LGR5 positivity defines stem-like cells in colorectal cancer. Carcinogenesis 2014;35(4):849–58.

[10] Bowtell DD, Bohm S, Ahmed AA, Aspuria PJ, Bast Jr RC, Beral V, et al. Rethinking ovarian cancer II: reducing mortality from high-grade serous ovarian cancer. Nat Rev Cancer 2015;15(11):668–79.

[11] Bonnet D, Dick JE. Human acute myeloid leukemia is organized as a hierarchy that originates from a primitive hematopoietic cell. Nat Med 1997;3(7):730–7.

[12] Dalerba P, Cho RW, Clarke MF. Cancer stem cells: models and concepts. Annu Rev Med 2007;58:267–84.

[13] Baba T, Convery PA, Matsumura N, Whitaker RS, Kondoh E, Perry T, et al. Epigenetic regulation of CD133 and tumorigenicity of CD133+ ovarian cancer cells. Oncogene 2009;28(2):209–18.

[14] Bapat SA, Mali AM, Koppikar CB, Kurrey NK. Stem and progenitor-like cells contribute to the aggressive behavior of human epithelial ovarian cancer. Cancer Res 2005;65(8):3025–9.

[15] Curley MD, Therrien VA, Cummings CL, Sergent PA, Koulouris CR, Friel AM, et al. CD133 expression defines a tumor initiating cell population in primary human ovarian cancer. Stem Cells 2009;27(12):2875–83.

[16] Deng S, Yang X, Lassus H, Liang S, Kaur S, Ye Q, et al. Distinct expression levels and patterns of stem cell marker, aldehyde dehydrogenase isoform 1 (ALDH1), in human epithelial cancers. PLoS One 2010;5(4):e10277.

[17] Kim M, Turnquist H, Jackson J, Sgagias M, Yan Y, Gong M, et al. The multidrug resistance transporter ABCG2 (breast cancer resistance protein 1) effluxes Hoechst 33342 and is overexpressed in hematopoietic stem cells. Clin Cancer Res 2002;8(1):22–8.

[18] Goodell MA, Brose K, Paradis G, Conner AS, Mulligan RC. Isolation and functional properties of murine hematopoietic stem cells that are replicating in vivo. J Exp Med 1996;183(4):1797–806.

[19] Szotek PP, Pieretti-Vanmarcke R, Masiakos PT, Dinulescu DM, Connolly D, Foster R, et al. Ovarian cancer side population defines cells with stem cell-like characteristics and mullerian inhibiting substance responsiveness. Proc Natl Acad Sci USA 2006;103(30):11154–9.

[20] Kurrey NK, Jalgaonkar SP, Joglekar AV, Ghanate AD, Chaskar PD, Doiphode RY, et al. Snail and slug mediate radioresistance and chemoresistance by antagonizing p53-mediated apoptosis and acquiring a stem-like phenotype in ovarian cancer cells. Stem Cells 2009;27(9):2059–68.

[21] Hu L, McArthur C, Jaffe RB. Ovarian cancer stem-like side-population cells are tumourigenic and chemoresistant. Br J Cancer 2010;102(8):1276–83.

[22] Hong IS, Lee HY, Nam JS. Cancer stem cells: the 'Achilles heel' of chemo-resistant tumors. Recent Pat Anticancer Drug Discov 2015;10(1):2–22.

[23] Tonary AM, Macdonald EA, Faught W, Senterman MK, Vanderhyden BC. Lack of expression of c-KIT in ovarian cancers is associated with poor prognosis. Int J Cancer 2000;89(3):242–50.

[24] Sillanpaa S, Anttila MA, Voutilainen K, Tammi RH, Tammi MI, Saarikoski SV, et al. CD44 expression indicates favorable prognosis in epithelial ovarian cancer. Clin Cancer Res 2003;9(14):5318–24.

[25] Alvero AB, Chen R, Fu HH, Montagna M, Schwartz PE, Rutherford T, et al. Molecular phenotyping of human ovarian cancer stem cells unravels the mechanisms for repair and chemoresistance. Cell Cycle 2009;8(1):158–66.

[26] Gao MQ, Choi YP, Kang S, Youn JH, Cho NH. CD24(+) cells from hierarchically organized ovarian cancer are enriched in cancer stem cells. Oncogene 2010;29(18):2672–80.

[27] Capel B. Ovarian epithelium regeneration by Lgr5(+) cells. Nat Cell Biol 2014;16(8):743–4.

[28] Meng E, Mitra A, Tripathi K, Finan MA, Scalici J, McClellan S, et al. ALDH1A1 maintains ovarian cancer stem cell-like properties by altered regulation of cell cycle checkpoint and DNA repair network signaling. PLoS One 2014;9(9):e107142.

[29] Wang Y, Cardenas H, Fang F, Condello S, Taverna P, Segar M, et al. Epigenetic targeting of ovarian cancer stem cells. Cancer Res 2014;74(17):4922–36.

[30] Zhang S, Cui B, Lai H, Liu G, Ghia EM, Widhopf 2nd GF, et al. Ovarian cancer stem cells express ROR1, which can be targeted for anti-cancer-stem-cell therapy. Proc Natl Acad Sci USA 2014;111(48):17266–71.

[31] Lee AS. Glucose-regulated proteins in cancer: molecular mechanisms and therapeutic potential. Nat Rev Cancer 2014;14(4):263–76.

[32] Miao YR, Eckhardt BL, Cao Y, Pasqualini R, Argani P, Arap W, et al. Inhibition of established micrometastases by targeted drug delivery via cell surface-associated GRP78. Clin Cancer Res 2013;19(8):2107–16.

[33] Huang LW, Lin CY, Lee CC, Liu TZ, Jeng CJ. Overexpression of GRP78 is associated with malignant transformation in epithelial ovarian tumors. Appl Immunohistochem Mol Morphol 2012;20(4):381–5.

[34] Spike BT, Kelber JA, Booker E, Kalathur M, Rodewald R, Lipianskaya J, et al. CRIPTO/GRP78 signaling maintains fetal and adult mammary stem cells ex vivo. Stem Cell Rep 2014;2(4):427–39.

[35] Mo L, Bachelder RE, Kennedy M, Chen PH, Chi JT, Berchuck A, et al. Syngeneic murine ovarian cancer model reveals that ascites enriches for ovarian cancer stem-like cells expressing membrane GRP78. Mol Cancer Ther 2015;14(3):747–56.

[36] Li B, Cheng XL, Yang YP, Li ZQ. GRP78 mediates radiation resistance of a stem cell-like subpopulation within the MCF-7 breast cancer cell line. Oncol Rep 2013;30(5):2119–26.

[37] Peng S, Maihle NJ, Huang Y. Pluripotency factors Lin28 and Oct4 identify a sub-population of stem cell-like cells in ovarian cancer. Oncogene 2010;29(14):2153–9.

[38] Zola H, Swart B, Nicholson I, Aasted B, Bensussan A, Boumsell L, et al. CD molecules 2005: human cell differentiation molecules. Blood 2005;106(9):3123–6.

[39] Lesley J, Hyman R, Kincade PW. CD44 and its interaction with extracellular matrix. Adv Immunol 1993;54:271–335.

[40] Anttila MA, Tammi RH, Tammi MI, Syrjanen KJ, Saarikoski SV, Kosma VM. High levels of stromal hyaluronan predict poor disease outcome in epithelial ovarian cancer. Cancer Res 2000;60(1):150–5.

[41] Chen H, Hao J, Wang L, Li Y. Coexpression of invasive markers (uPA, CD44) and multiple drug-resistance proteins (MDR1, MRP2) is correlated with epithelial ovarian cancer progression. Br J Cancer 2009;101(3):432–40.

[42] Bourguignon LY, Peyrollier K, Gilad E, Brightman A. Hyaluronan-CD44 interaction with neural Wiskott-Aldrich syndrome protein (N-WASP) promotes actin polymerization and ErbB2 activation leading to beta-catenin nuclear translocation, transcriptional up-regulation, and cell migration in ovarian tumor cells. J Biol Chem 2007;282(2):1265–80.

[43] Auzenne E, Ghosh SC, Khodadadian M, Rivera B, Farquhar D, Price RE, et al. Hyaluronic acid-paclitaxel: antitumor efficacy against CD44(+) human ovarian carcinoma xenografts. Neoplasia 2007;9(6):479–86.

[44] Yin S, Li J, Hu C, Chen X, Yao M, Yan M, et al. CD133 positive hepatocellular carcinoma cells possess high capacity for tumorigenicity. Int J Cancer 2007;120(7):1444–50.

[45] Kusumbe AP, Mali AM, Bapat SA. CD133-expressing stem cells associated with ovarian metastases establish an endothelial hierarchy and contribute to tumor vasculature. Stem Cells 2009;27(3):498–508.

[46] Ferrandina G, Martinelli E, Petrillo M, Prisco MG, Zannoni G, Sioletic S, et al. CD133 antigen expression in ovarian cancer. BMC Cancer 2009;9:221.

[47] Kristiansen G, Denkert C, Schluns K, Dahl E, Pilarsky C, Hauptmann S. CD24 is expressed in ovarian cancer and is a new independent prognostic marker of patient survival. Am J Pathol 2002;161(4):1215–21.

[48] Li C, Heidt DG, Dalerba P, Burant CF, Zhang L, Adsay V, et al. Identification of pancreatic cancer stem cells. Cancer Res 2007;67(3):1030–7.

[49] Marcato P, Dean CA, Pan D, Araslanova R, Gillis M, Joshi M, et al. Aldehyde dehydrogenase activity of breast cancer stem cells is primarily due to isoform ALDH1A3 and its expression is predictive of metastasis. Stem Cells 2011;29(1):32–45.

[50] Kastan MB, Schlaffer E, Russo JE, Colvin OM, Civin CI, Hilton J. Direct demonstration of elevated aldehyde dehydrogenase in human hematopoietic progenitor cells. Blood 1990;75(10):1947–50.

[51] Yasuda K, Torigoe T, Morita R, Kuroda T, Takahashi A, Matsuzaki J, et al. Ovarian cancer stem cells are enriched in side population and aldehyde dehydrogenase bright overlapping population. PLoS One 2013;8(8):e68187.

[52] Condello S, Morgan CA, Nagdas S, Cao L, Turek J, Hurley TD, et al. Beta-catenin-regulated ALDH1A1 is a target in ovarian cancer spheroids. Oncogene 2015;34(18): 2297–308.

[53] Ginestier C, Hur MH, Charafe-Jauffret E, Monville F, Dutcher J, Brown M, et al. ALDH1 is a marker of normal and malignant human mammary stem cells and a predictor of poor clinical outcome. Cell Stem Cell 2007;1(5):555–67.

[54] Chang B, Liu G, Xue F, Rosen DG, Xiao L, Wang X, et al. ALDH1 expression correlates with favorable prognosis in ovarian cancers. Mod Pathol 2009;22(6):817–23.

[55] Rycaj K, Tang DG. Cell-of-origin of cancer versus cancer stem cells: assays and interpretations. Cancer Res 2015;75(19):4003–11.

[56] Abubaker K, Latifi A, Luwor R, Nazaretian S, Zhu H, Quinn MA, et al. Short-term single treatment of chemotherapy results in the enrichment of ovarian cancer stem cell-like cells leading to an increased tumor burden. Mol Cancer 2013;12:24.

[57] Wintzell M, Lofstedt L, Johansson J, Pedersen AB, Fuxe J, Shoshan M. Repeated cisplatin treatment can lead to a multiresistant tumor cell population with stem cell features and sensitivity to 3-bromopyruvate. Cancer Biol Ther 2012;13(14):1454–62.

[58] Kim SY, Kang JW, Song X, Kim BK, Yoo YD, Kwon YT, et al. Role of the IL-6-JAK1-STAT3-Oct-4 pathway in the conversion of non-stem cancer cells into cancer stem-like cells. Cell Signal 2013;25(4):961–9.

[59] Han Z, Wang X, Ma L, Chen L, Xiao M, Huang L, et al. Inhibition of STAT3 signaling targets both tumor-initiating and differentiated cell populations in prostate cancer. Oncotarget 2014;5(18):8416–28.

[60] Li M, Balch C, Montgomery JS, Jeong M, Chung JH, Yan P, et al. Integrated analysis of DNA methylation and gene expression reveals specific signaling pathways associated with platinum resistance in ovarian cancer. BMC Med Genomics 2009;2:34.

[61] Chaffer CL, Marjanovic ND, Lee T, Bell G, Kleer CG, Reinhardt F, et al. Poised chromatin at the ZEB1 promoter enables breast cancer cell plasticity and enhances tumorigenicity. Cell 2013;154(1):61–74.

[62] Jordan CT. Cancer stem cells: controversial or just misunderstood? Cell Stem Cell 2009;4(3):203–5.

[63] Somjen D, Katzburg S, Nevo N, Gayer B, Hodge RP, Renevey MD, et al. A daidzein-daunomycin conjugate improves the therapeutic response in an animal model of ovarian carcinoma. J Steroid Biochem Mol Biol 2008;110(1–2):144–9.

[64] Green JM, Alvero AB, Kohen F, Mor G. 7-(O)-Carboxymethyl daidzein conjugated to N-t-Boc-hexylenediamine: a novel compound capable of inducing cell death in epithelial ovarian cancer stem cells. Cancer Biol Ther 2009;8(18):1747–53.

[65] Ma S, Lee TK, Zheng BJ, Chan KW, Guan XY. CD133+ HCC cancer stem cells confer chemoresistance by preferential expression of the Akt/PKB survival pathway. Oncogene 2008;27(12):1749–58.

[66] Ginestier C, Liu S, Diebel ME, Korkaya H, Luo M, Brown M, et al. CXCR1 blockade selectively targets human breast cancer stem cells in vitro and in xenografts. J Clin Invest 2010;120(2):485–97.

[67] Damelin M, Bankovich A, Park A, Aguilar J, Anderson W, Santaguida M, et al. Anti-EFNA4 calicheamicin conjugates effectively target triple-negative breast and ovarian tumor-initiating cells to result in sustained tumor regressions. Clin Cancer Res 2015;21(18):4165–73.

[68] Bhattacharya R, Kwon J, Ali B, Wang E, Patra S, Shridhar V, et al. Role of hedgehog signaling in ovarian cancer. Clin Cancer Res 2008;14(23):7659–66.

[69] Yang L, He J, Huang S, Zhang X, Bian Y, He N, et al. Activation of hedgehog signaling is not a frequent event in ovarian cancers. Mol Cancer 2009;8:112.

[70] Pannuti A, Foreman K, Rizzo P, Osipo C, Golde T, Osborne B, et al. Targeting notch to target cancer stem cells. Clin Cancer Res 2010;16(12):3141–52.

[71] Lepourcelet M, Chen YN, France DS, Wang H, Crews P, Petersen F, et al. Small-molecule antagonists of the oncogenic Tcf/beta-catenin protein complex. Cancer Cell 2004;5(1):91–102.

[72] Barker N, Clevers H. Mining the Wnt pathway for cancer therapeutics. Nat Rev Drug Discov 2006;5(12):997–1014.

[73] Khanna M, Chen CH, Kimble-Hill A, Parajuli B, Perez-Miller S, Baskaran S, et al. Discovery of a novel class of covalent inhibitor for aldehyde dehydrogenases. J Biol Chem 2011;286(50):43486–94.

[74] Lin J, Haffner MC, Zhang Y, Lee BH, Brennen WN, Britton J, et al. Disulfiram is a DNA demethylating agent and inhibits prostate cancer cell growth. Prostate 2011;71(4):333–43.

[75] Raha D, Wilson TR, Peng J, Peterson D, Yue P, Evangelista M, et al. The cancer stem cell marker aldehyde dehydrogenase is required to maintain a drug-tolerant tumor cell subpopulation. Cancer Res 2014;74(13):3579–90.

[76] Sell S. Stem cell origin of cancer and differentiation therapy. Crit Rev Oncol Hematol 2004;51(1):1–28.

[77] Caliaro MJ, Marmouget C, Guichard S, Mazars P, Valette A, Moisand A, et al. Response of four human ovarian carcinoma cell lines to all-trans retinoic acid: relationship with induction of differentiation and retinoic acid receptor expression. Int J Cancer 1994;56(5):743–8.

[78] Yang YT, Balch C, Kulp SK, Mand MR, Nephew KP, Chen CS. A rationally designed histone deacetylase inhibitor with distinct antitumor activity against ovarian cancer. Neoplasia 2009;11(6):552–63. 3 p following 63.

[79] Yin G, Alvero AB, Craveiro V, Holmberg JC, Fu HH, Montagna MK, et al. Constitutive proteasomal degradation of TWIST-1 in epithelial-ovarian cancer stem cells impacts differentiation and metastatic potential. Oncogene 2013;32(1):39–49.

[80] Dubeau L. The cell of origin of ovarian epithelial tumours. Lancet Oncol 2008;9(12):1191–7.

[81] Pepin D, Sosulski A, Zhang L, Wang D, Vathipadiekal V, Hendren K, et al. AAV9 delivering a modified human Mullerian inhibiting substance as a gene therapy in patient-derived xenografts of ovarian cancer. Proc Natl Acad Sci USA 2015;112(32):E4418–27.

[82] Su HY, Lai HC, Lin YW, Liu CY, Chen CK, Chou YC, et al. Epigenetic silencing of SFRP5 is related to malignant phenotype and chemoresistance of ovarian cancer through Wnt signaling pathway. Int J Cancer 2010;127(3):555–67.

[83] Fang F, Balch C, Schilder J, Breen T, Zhang S, Shen C, et al. A phase I and pharmacodynamic study of decitabine in combination with carboplatin in patients with recurrent, platinum-resistant, epithelial ovarian cancer. Cancer 2010:4043–53.

[84] Matei D, Fang F, Shen C, Schilder J, Arnold A, Zeng Y, et al. Epigenetic resensitization to platinum in ovarian cancer. Cancer Res 2012;72(9):2197–205.

[85] Fang F, Zuo Q, Pilrose J, Wang Y, Shen C, Li M, et al. Decitabine reactivated pathways in platinum resistant ovarian cancer. Oncotarget 2014;5(11):3579–89.

[86] Minucci S, Pelicci PG. Histone deacetylase inhibitors and the promise of epigenetic (and more) treatments for cancer. Nat Rev Cancer 2006;6(1):38–51.

[87] Strait KA, Warnick CT, Ford CD, Dabbas B, Hammond EH, Ilstrup SJ. Histone deacetylase inhibitors induce G2-checkpoint arrest and apoptosis in cisplatinum-resistant ovarian cancer cells associated with overexpression of the Bcl-2-related protein Bad. Mol Cancer Ther 2005;4(4):603–11.

[88] Vivanco I, Sawyers CL. The phosphatidylinositol 3-kinase AKT pathway in human cancer. Nat Rev Cancer 2002;2(7):489–501.

[89] Dong P, Konno Y, Watari H, Hosaka M, Noguchi M, Sakuragi N. The impact of micro RNA-mediated PI3K/AKT signaling on epithelial-mesenchymal transition and cancer stemness in endometrial cancer. J Transl Med 2014;12:231.

[90] Luo X, Dong Z, Chen Y, Yang L, Lai D. Enrichment of ovarian cancer stem-like cells is associated with epithelial to mesenchymal transition through an miRNA-activated AKT pathway. Cell Prolif 2013;46(4):436–46.

[91] Lisabeth EM, Falivelli G, Pasquale EB. Eph receptor signaling and ephrins. Cold Spring Harb Perspect Biol 2013;5(9).

[92] Chen J, Song W, Amato K. Eph receptor tyrosine kinases in cancer stem cells. Cytokine Growth Factor Rev 2015;26(1):1–6.

[93] Xi HQ, Wu XS, Wei B, Chen L. Eph receptors and ephrins as targets for cancer therapy. J Cell Mol Med 2012;16(12):2894–909.

[94] Boyd AW, Bartlett PF, Lackmann M. Therapeutic targeting of EPH receptors and their ligands. Nat Rev Drug Discov 2014;13(1):39–62.

[95] Chari NS, McDonnell TJ. The sonic hedgehog signaling network in development and neoplasia. Adv Anat Pathol 2007;14(5):344–52.

[96] Ischenko I, Seeliger H, Schaffer M, Jauch KW, Bruns CJ. Cancer stem cells: how can we target them? Curr Med Chem 2008;15(30):3171–84.

[97] Kopan R, Ilagan MX. The canonical Notch signaling pathway: unfolding the activation mechanism. Cell 2009;137(2):216–33.

[98] Hopfer O, Zwahlen D, Fey MF, Aebi S. The Notch pathway in ovarian carcinomas and adenomas. Br J Cancer 2005;93(6):709–18.

[99] Rose SL, Kunnimalaiyaan M, Drenzek J, Seiler N. Notch 1 signaling is active in ovarian cancer. Gynecol Oncol 2010;117(1):130–3.

[100] Klaus A, Birchmeier W. Wnt signalling and its impact on development and cancer. Nat Rev Cancer 2008;8(5):387–98.

[101] Rask K, Nilsson A, Brannstrom M, Carlsson P, Hellberg P, Janson PO, et al. Wnt-signalling pathway in ovarian epithelial tumours: increased expression of beta-catenin and GSK3beta. Br J Cancer 2003;89(7):1298–304.

[102] Zhao D, Mo Y, Li MT, Zou SW, Cheng ZL, Sun YP, et al. NOTCH-induced aldehyde dehydrogenase 1A1 deacetylation promotes breast cancer stem cells. J Clin Invest 2014;124(12):5453–65.

[103] Seftor EA, Brown KM, Chin L, Kirschmann DA, Wheaton WW, Protopopov A, et al. Epigenetic transdifferentiation of normal melanocytes by a metastatic melanoma microenvironment. Cancer Res 2005;65(22):10164–9.

[104] Feinberg AP. Phenotypic plasticity and the epigenetics of human disease. Nature 2007;447(7143):433–40.

[105] Kobel Sal RP. High-throughput methods to define complex stem cell niches. Biotechniques 2010;48(4):14.

[106] Bissell MJ, Labarge MA. Context, tissue plasticity, and cancer: are tumor stem cells also regulated by the microenvironment? Cancer Cell 2005;7(1):17–23.

[107] Hamano Y, Zeisberg M, Sugimoto H, Lively JC, Maeshima Y, Yang C, et al. Physiological levels of tumstatin, a fragment of collagen IV alpha3 chain, are generated by MMP-9 proteolysis and suppress angiogenesis via alphaV beta3 integrin. Cancer Cell 2003;3(6):589–601.

[108] Calabrese C, Poppleton H, Kocak M, Hogg TL, Fuller C, Hamner B, et al. A perivascular niche for brain tumor stem cells. Cancer Cell 2007;11(1):69–82.

[109] Hendrix MJ, Seftor EA, Seftor RE, Kasemeier-Kulesa J, Kulesa PM, Postovit LM. Reprogramming metastatic tumour cells with embryonic microenvironments. Nat Rev Cancer 2007;7(4):246–55.

[110] McAllister JC, Zhan Q, Weishaupt C, Hsu MY, Murphy GF. The embryonic morphogen, Nodal, is associated with channel-like structures in human malignant melanoma xenografts. J Cutan Pathol 2010;37(Suppl. 1):19–25.

[111] Stephen AE, Pearsall LA, Christian BP, Donahoe PK, Vacanti JP, MacLaughlin DT. Highly purified mullerian inhibiting substance inhibits human ovarian cancer in vivo. Clin Cancer Res 2002;8(8):2640–6.

[112] Jablonka E, Lamb MJ. The changing concept of epigenetics. Ann NY Acad Sci 2002;981: 82–96.

[113] Balch C, Matei D, Huang TH-M, Nephew KP. Role of epigenomics in ovarian and endometrial cancers. Epigenomics 2011;2(3):419–47.

[114] Matei DE, Nephew KP. Epigenetic therapies for chemoresensitization of epithelial ovarian cancer. Gynecol Oncol 2010;116(2):195–201.

[115] Jones PA, Baylin SB. The epigenomics of cancer. Cell 2007;128(4):683–92.

[116] Fang F, Munck J, Tang J, Taverna P, Wang Y, Miller DF, et al. The novel, small-molecule DNA methylation inhibitor SGI-110 as an ovarian cancer chemosensitizer. Clin Cancer Res 2014;20(24):6504–16.

[117] Steele N, Finn P, Brown R, Plumb JA. Combined inhibition of DNA methylation and histone acetylation enhances gene re-expression and drug sensitivity in vivo. Br J Cancer 2009;100(5):758–63.

[118] Saunthararajah Y, Sekeres M, Advani A, Mahfouz R, Durkin L, Radivoyevitch T, et al. Evaluation of noncytotoxic DNMT1-depleting therapy in patients with myelodysplastic syndromes. J Clin Invest 2015;125(3):1043–55.

[119] Zuco V, Benedetti V, De Cesare M, Zunino F. Sensitization of ovarian carcinoma cells to the atypical retinoid ST1926 by the histone deacetylase inhibitor, RC307: enhanced DNA damage response. Int J Cancer 2010;126(5):1246–55.

[120] Park SJ, Kim MJ, Kim HB, Sohn HY, Bae JH, Kang CD, et al. Trichostatin A sensitizes human ovarian cancer cells to TRAIL-induced apoptosis by down-regulation of c-FLIPL via inhibition of EGFR pathway. Biochem Pharmacol 2009;77(8):1328–36.

[121] Sharma SV, Lee DY, Li B, Quinlan MP, Takahashi F, Maheswaran S, et al. A chromatin-mediated reversible drug-tolerant state in cancer cell subpopulations. Cell 2010;141(1) :69–80.

[122] Zhang B, Strauss AC, Chu S, Li M, Ho Y, Shiang KD, et al. Effective targeting of quiescent chronic myelogenous leukemia stem cells by histone deacetylase inhibitors in combination with imatinib mesylate. Cancer Cell 2010;17(5):427–42.

[123] Robertson FM, Woodward WA, Pickei R, Ye Z, Bornmann W, Pal A, et al. Suberoylanilide hydroxamic acid blocks self-renewal and homotypic aggregation of inflammatory breast cancer spheroids. Cancer 2010;116(Suppl. 11):2760–7.

[124] You JS, Kang JK, Seo DW, Park JH, Park JW, Lee JC, et al. Depletion of embryonic stem cell signature by histone deacetylase inhibitor in NCCIT cells: involvement of Nanog suppression. Cancer Res 2009;69(14):5716–25.

[125] Choi YJ, Ingram PN, Yang K, Coffman L, Iyengar M, Bai S, et al. Identifying an ovarian cancer cell hierarchy regulated by bone morphogenetic protein 2. Proc Natl Acad Sci USA 2015;112(50):E6882–8.

[126] Mani SA, Guo W, Liao MJ, Eaton EN, Ayyanan A, Zhou AY, et al. The epithelial-mesenchymal transition generates cells with properties of stem cells. Cell 2008;133(4):704–15.

[127] Dobbin ZC, Katre AA, Steg AD, Erickson BK, Shah MM, Alvarez RD, et al. Using heterogeneity of the patient-derived xenograft model to identify the chemoresistant population in ovarian cancer. Oncotarget 2014;5(18):8750–64.

[128] Ricci F, Bizzaro F, Cesca M, Guffanti F, Ganzinelli M, Decio A, et al. Patient-derived ovarian tumor xenografts recapitulate human clinicopathology and genetic alterations. Cancer Res 2014;74(23):6980–90.

[129] Weroha SJ, Becker MA, Enderica-Gonzalez S, Harrington SC, Oberg AL, Maurer MJ, et al. Tumorgrafts as in vivo surrogates for women with ovarian cancer. Clin Cancer Res 2014;20(5):1288–97.

SECTION 3

CSC Features and Mechanisms

CHAPTER 10

Understanding Cancer Cells of Origin in Cutaneous Tumors

H. Moon, L.R. Donahue, A.C. White
Cornell University, Ithaca, NY, United States

INTRODUCTION

Benign and malignant epithelial tumors arise from normal epithelial cells harboring oncogenic somatic mutations and epigenetic changes [1]. Although the genetic basis of tumor development and progression continues to gain increased depth and understanding, the cells which have the unique propensity to efficiently develop into cancerous epithelial cells, once a sufficient number of cancer-causing mutations have accumulated, remain poorly understood. Nonmelanoma skin cancer, which includes cutaneous basal cell and squamous cell carcinoma (SCC), is particularly well studied with regard to cancer cells of origin (CCOs) and tumor initiation. In this chapter, using nonmelanoma skin cancer models as our focus, we will review our current understanding of the adult stem cell hierarchy of the epidermis, and look at previous studies aimed at elucidating the cell intrinsic and cell extrinsic requirements for cancer initiation from their CCOs, as well as the role of adult stem cells in the process of tumor initiation.

Cancer is a complex and heterogeneous disease, even within specific tumor subtypes in a tissue or organ of interest, which obscures the ability to identify the cellular origin of each cancer [2–4]. Moreover, since cancer development is a transformative process, the final histopathological characteristics of a cancer may no longer resemble the initiating cells' morphology. Thus, to determine the early cellular and molecular events of tumor initiation, it is critical to define the essential factors influencing tumor initiation. First, it is necessary to identify the specific cellular origin of each cancer. In many cases, adult stem cells appear to serve as CCOs. Secondly, the concurrent physiological stresses (extrinsic factors) combined with genetic mutations (intrinsic factors) that initiate malignant transformation from the cellular origin of a tumor need to be fully elucidated. In some cases, pathological conditions causing inflammation or injury may act as a stimulus for cancer development. Finally, changes within the CCO niche and microenvironment should be determined. Modulation in

CONTENTS

Cancer Stem Cells. http://dx.doi.org/10.1016/B978-0-12-803892-5.00010-3

the CCO niche may significantly contribute to either facilitating or suppressing tumor formation, similar to the effects of tumor microenvironment within a tumor (ie, cancer-associated fibroblasts regulating tumor growth and metastasis). Taken together, this understanding can lead to improved prevention, the development of biomarkers for early detection, and effective therapeutic strategies for patients who are at high risk of developing cancers.

CANCER CELLS OF ORIGIN VERSUS CANCER STEM CELLS

While CCOs and cancer stem cells (CSCs) may be potentially linked, supporting data is currently lacking. CCO(s) are the target cell or cells which accumulate the necessary mutations and/or epigenetic alterations needed for cancer development and are subjected to the necessary conditions that allow for initiation of a new tumor [5,6] (Fig. 10.1). CSCs, in contrast, denote the cells existing within established tumors that have self-renewal capabilities and the ability to generate new tumors by serial transplantation [7] (Fig. 10.1). CSCs are also thought to be the cells that resist chemotherapy and are at the foundation of relapse [8,9]. A direct relationship between CCOs and CSCs, however, has not been clearly defined. Furthermore, it is possible that CSCs are dynamically generated and can emerge from distinct cell populations within a tumor, descended from CCOs, after malignant epithelial cell transformation. For example, one well-known cancerous event, epithelial-to-mesenchymal

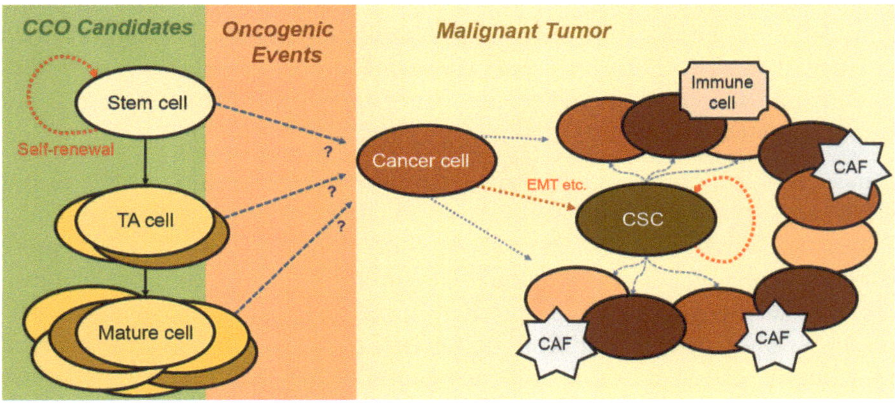

FIGURE 10.1 Cancer cells of origin (CCOs) and cancer stem cell (CSC).
Theoretically, all cell lineages from stem cells to transit-amplifying cells (TA cells) and differentiated, mature cells can be the cellular origin of each cancer. Following oncogenic events, the CCO will form a malignant tumor. While CCOs refers to the cellular origin of cancer cells, CSCs can arise from existing cancer cells within a tumor through various molecular mechanisms such as epithelial-to-mesenchymal transition (EMT). Tumor propagation from CSCs and disease progression are accelerated by surrounding tissues, especially tumor-associated immune cells and cancer-associated fibroblasts (CAFs).

transition (EMT), has been suggested to be a major mechanism for malignant epithelial cells to acquire stemlike characteristics, thus potentially facilitating the generation of new CSCs within existing tumors [10,11]. As lineage tracing improves, future studies may better determine the relationship between these two distinct populations [12,13].

Apart from the unknown lineage relationship between CCOs and CSCs, these distinct identities are sometimes confused, and terminologies are used incorrectly. However, CCOs and CSCs should be considered separately. Adult stem cells are now thought to be compelling candidates for CCOs in many types of cancers, as they are long lived, multi- or unipotent, have significant self-renewal potential and can differentiate into a number of cell types [14]. During normal homeostasis, adult stem cells frequently self-renew and serve to repair and replenish tissues throughout the life of an organism [15]. Since CSCs have stemlike characteristics and behaviors within tumors, CSCs can be easily confused with CCOs, as CCOs are often derived from adult stem cells. Established tumors also appear to contain a hierarchical organization, in which CSCs are able to generate distinct subpopulations [16,17]. The conceptual and terminological similarities generate confusion in distinguishing between CCOs and CSCs, with further discrepancies emerging from the use of terms such as tumor-initiating cells and tumor-propagating cells. In this chapter, we provide insight into the current state of CCO research, particularly with respect to the most common forms of cancer, cutaneous tumors.

IDENTIFYING CANCER CELLS OF ORIGIN

To identify CCOs, it is crucial to understand the normal cell fate of adult stem cells and their terminally differentiated, mature, progeny. Adult stem cells are found in many types of organs and contribute to tissue homeostasis in the organ in which they reside. For example, skin tissues contain several distinct adult stem and/or progenitor cells which may differentially contribute to skin homeostasis [18–20]. In general, many adult stem cell populations are found in a quiescent, or resting, state. However, upon normal physiological activation (ie, natural hair cycling in the skin) or in the case of physiological stress such as injuries, quiescent stem cells give rise to their committed progenitors, often called transit-amplifying (TA) cells [21]. These TA cells are known to be multilineage restricted progenitors (Fig. 10.1), which quickly divide and differentiate into terminally differentiated cells. Terminally differentiated cells are usually mature cells that perform the necessary roles required for proper organ function. Notwithstanding some exceptions, many of our major organs prove the model of cellular hierarchies from adult stem cells to terminally differentiated, mature cells. Therefore, the cellular hierarchy can provide a framework to identify which cells can initiate tumor formation.

Hypothetically, all cell lineages, including adult stem, TA and mature cells, can acquire oncogenic mutations and have the potential to develop tumors (Fig. 10.1). To test tumorigenic potential, earlier in vitro and ex vivo studies had been used extensively. Using cell surface markers, specific cell populations can be isolated using fluorescence-activated cell sorting (FACS), allowing nonnative, fusion genes to then be inserted into the cells of interest for malignant transformation. Through the introduction of transgenes, isolated cells were tested for the potential to develop into malignant cells through transgene activation of known tumor-associated genes. For example, a leukemogenic mixed lineage leukemia (MLL) fusion gene has been artificially introduced in normal hematopoietic stem cells to determine its potential to transform wild-type cells into leukemia cells [22]. Similarly, these potential tumor-causing genes have been tested in downstream descendants. By comparing differing but developmentally similar cell types, the capabilities to transform into malignant cells provided the initial ideas regarding the cellular origins of cancers. However, despite their experimental convenience, in vitro and ex vivo experimental systems do not completely recapitulate actual physiological in vivo situations including a natural tissue environment. Furthermore, isolated cells often lose their innate in vivo characteristics when moved to an in vitro setting, which represents a significant limitation to this system.

In contrast, there is a significant advantage to test malignancy potential in in vivo genetically engineered mouse models (GEMMs). GEMMs allow for the differential targeting of individual cell populations, such as stem cells or their descendants, without perturbation of their normal physiological microenvironments. Cell type–specific promoters can provide efficient spatial control in targeting distinct cell populations. Additionally, temporal control of gene targeting can be achieved with cell type–specific promoters coupled with inducible Cre recombinase alleles, such as those fused with modified estrogen (ER) or progesterone (PR) receptors. In these cases, addition of tamoxifen to activate ER, or mifepristone to activate PR, will induce activation of Cre recombinase (Fig. 10.2A). This technique harbors a significant advantage, since it facilitates preferential targeting in adult mice. Since most cancers arise during adulthood, this provides more clinical relevance for studying cancer development in GEMMs.

Cre recombinase alleles, in combination with reporter genes, allow for genetic lineage-tracing experiments in mice, in which the initial cell with Cre activity and all descendant cells are labeled [5,12]. Inducible fluorescent proteins, such as yellow fluorescent protein (YFP) or tdTomato, or β-galactosidase via lacZ expression, are commonly used example of lineage-tracing alleles. This system typically employs a fluorescent or LacZ cassette knocked-in to the ubiquitously expressed *Rosa* locus. A stop sequence in front of the fluorescent cassette prevents its expression without Cre activation. Cre activity removes this stop sequence, which is flanked by *lox P* sites (flox), thus permanently labeling the target cell and all of its descendants (Fig. 10.2B). Hence, this in vivo system is

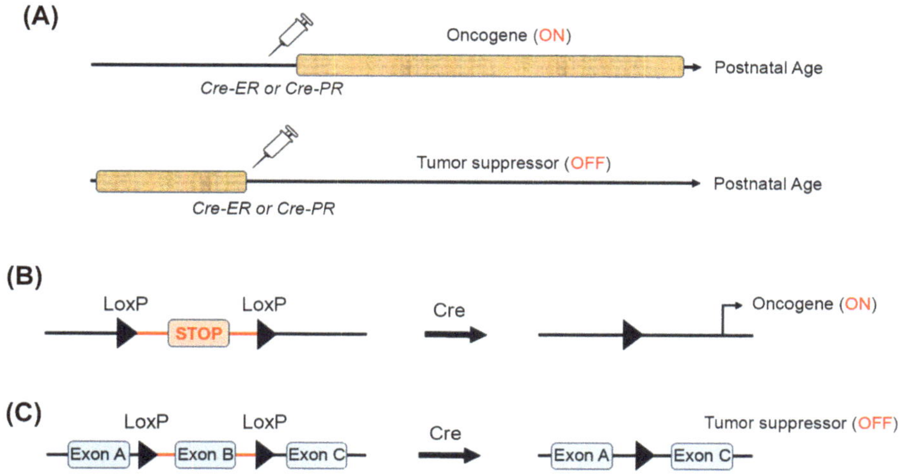

FIGURE 10.2 Genetic approaches using Cre-Lox recombination mechanisms.
(A) Expression timing of oncogenic alleles or floxed tumor suppressor alleles: The activity of Cre recombinase can be controlled through modifications by the estrogen receptor (ER) or progesterone receptor (PR). The administration of the receptor antagonists, tamoxifen or mifepristone, can activate the translocation of Cre to the nucleus for excision activity. This facilitates temporal specificity of Cre-mediated genetic modifications. (B) Overexpression of oncogenes: Through the nuclear translocation of Cre, Cre recombinase can remove the STOP cassette ahead of oncogenes. This process then induces the conditional expression of oncogenic alleles. (C) Deletion of tumor suppressors: Similar to overexpression, Cre recombinase can excise essential exons of tumor suppressors, which leads to silencing tumor suppressors.

one of the best methods to identify stem cell niches and their roles in tissue homeostasis in many tissues and organs, principally with an intact native surrounding tissue microenvironment. Similarly, in addition to lineage tracing, cell type–specific promoter-driven Cre recombinase expression also allows for the conditional expression, overexpression, or deletion of prospective oncogenes or tumor suppressors in precise cells or tissues of interest, provided a suitable promoter can be identified. As aforementioned, expression of a genetically engineered oncogene of interest can employ a knockin to the endogenous locus with the addition of a loxp-Stop-loxp (LSL) sequence upstream of the translation start site (Fig. 10.2B). Likewise, to determine the necessity of potential tumor suppressors, engineering loxp sequences to flank critical exons will effectively delete these genes in a conditional manner (Fig. 10.2C).

These techniques have been used in recent years to significantly advance the field of CCOs. Notably, by targeting distinct cell populations within a tissue or organ of interest with identical tumorigenic combinations, it appears that distinct cancer subtypes are defined by (1) the mutational profile, (2) the unique cell that undergoes transformation, and (3) the cell extrinsic conditions that enable tumor initiation.

STEM CELL NICHES OF THE EPIDERMIS

Structures of the Skin

As the primary protective barrier between an organism and its environment, the skin is a complex organ composed of a number of layers. These include a stratified epithelial layer that extends and surrounds the hair follicles, a mesenchymal dermis, and a connective subcutis. The dermal layer is found beneath the epidermis and is composed of nerve and connective tissues including fibroblasts and immune cells [23]. The epidermis is composed of five layers: the stratum basale, stratum spinosum, stratum granulosum, stratum lucidum, and stratum corneum [18,24]. The innermost basal layer is located on a basement membrane and contains highly proliferative columnar-shaped cells. These cells express high levels of cytokeratins 5 and 14 (*K5* and *K14*) and are essential in maintaining epidermal keratinocytes during homeostasis, due to their ability to undergo terminal differentiation during upward stratification (Fig. 10.3A). This differentiation process creates the different cell types that compose the remainder of the epidermis. Once the cells of the basal layer begin the process of terminal differentiation, they cease to express *K5* and *K14*, and begin to express cytokeratins 1 and 10 (*K1* and *K10*). The stratum spinosum is found directly above the basal layer. This layer has a large number of desmosomes and keratin filaments, giving these cells their spiny shape. As the cells continue upward into the stratum granulosum, they begin to lose all membrane bound organelles and become squamous cells. The outermost layer, the stratum corneum, is made up entirely of highly keratinized squames that eventually flake away from the surface of the skin and are replaced as the process of differentiation continues.

In addition to the keratinocytes of the interfollicular epidermis and hair follicles, several other types of cells are also found within the epidermis. Melanocytes, the cells that provide pigmentation to the skin and hair, are found within the basal layer and within the hair follicles. Merkel cells and Langerhans' cells assist in touch sensory and immune function, respectively [25,26].

Structures such as hair follicles, sebaceous and sweat glands, and nails are referred to as epidermal appendages. All of these structures are present in the epidermis, but originate well within the dermis. Of these appendages, the hair follicle is of particular interest in studies of focused on epidermal stem cell biology.

Stem Cell Niches of the Hair Follicle

The hair follicle contains several subpopulations of stem cells which contribute to full regeneration of hair follicles and to repair of the interfollicular epidermis in wounding situations [20,27]. The hair follicle is a self-renewing appendage that cycles through a three-step process of regeneration and rest [28]. When in a state of rest (telogen), the hair follicle is small and contains bulge stem cells, the secondary hair germ, and a sebaceous gland (Fig. 10.3B). During the growth

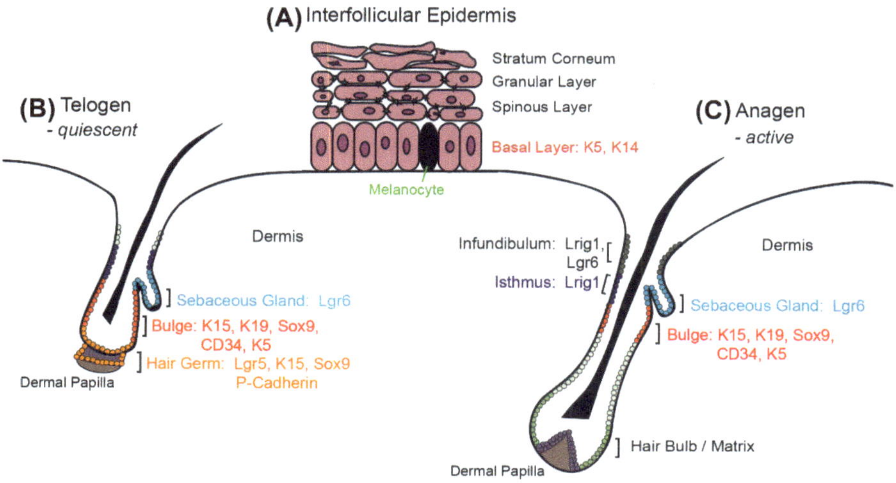

FIGURE 10.3 Stem cell populations of the skin.
(A) The interfollicular epidermis is composed of four to five layers, and the basal stem/progenitor cells maintain epidermal homeostasis. (B and C) The hair follicle undergoes morphogenic changes as it transitions through the hair cycle. While in telogen (B), most cells located below the oil-secreting sebaceous gland are stem cells, though there is some heterogeneity in marker expression. Located at the bottom of the telogen hair follicle, the lower bulge and hair germ are thought to be the first to reconstitute the anagen follicle upon onset of the hair cycle. During the transition to anagen (C), the cells within the bulge expand to form the hair bulb, which will in turn engulf the dermal papilla and generate the rapidly proliferating transit-amplifying cells of the hair follicle matrix.

period (anagen), the stem cells regenerate all the lineages of the full hair follicle. The anagen hair bulb extends downward through the dermis and engulfs the dermal papilla (Fig. 10.3C); a group of mesenchymal cells that provide the necessary signals for the construction of the follicle shaft and inner root sheath from the rapidly proliferating TA cells of its surrounding matrix [28,29]. When anagen is complete, extension of the hair shaft ceases and a destructive process (catagen) begins, in which most hair follicle cells undergo apoptosis, leaving behind the sebaceous gland and the bulge stem cell population.

Elegant studies have definitively shown that hair follicle stem cells reside in the bulge [30,31]. Through label retaining assays based on pulse/chase experiments using bromo-deoxyuridine (BrdU) incorporation and green fluorescent protein (GFP)-labeled histone H2B (H2B-GFP), two landmark papers found that the majority of label-retaining cells reside within the hair follicle bulge [30,31]. These studies showed that the cells within the bulge are slow cycling and label retaining and fully form all of the downstream lineages of the hair follicle during its growth phase.

To gain further insight into the stem cell niches within the hair follicles, many studies have identified markers of hair follicle stem cells. Furthermore, genetic alleles have been developed to exploit these markers to direct expression of Cre recombinase in genetic loss of function and gain of function studies. K5, previously mentioned as being a marker of the proliferative epidermal basal layer, is also expressed throughout the hair follicle, including within the bulge cells. In recent years, however, distinct stem cell niches have been identified within the K5 positive cell layers [19,20]. The murine bulge shows unique expression of the cell surface marker CD34, the transcription factor Sox9, and cytokeratins 15 (K15) and 19 (K19) [32–36]. Leucine-rich repeat containing G protein-coupled receptor 5 (Lgr5) is a well-documented marker of intestinal stem cells that is also highly expressed in the bulge and outer root sheath of hair follicles [37]. The Lgr5 positive cell population is found in the lower bulge and hair germ during telogen, and throughout the lower outer root sheath during anagen. Lgr5-expressing cells are thought to be the first cells that migrate to the outer root sheath during the onset of anagen, while the upper bulge cells remain in the stem cell niche. Leucine-rich repeat containing G protein-coupled receptor 6 (Lgr6), closely related to Lgr5, marks another stem cell population located directly above the hair bulge, which does not overlap with Lgr5 [38]. In cell transplant studies using immunodeficient nude mice, both Lgr5 and Lgr6 positive cells were shown to successfully form full hair follicles. The Lgr6 positive population was additionally shown to be capable of migrating outside of its niche to regenerate all keratinocyte varieties within the epidermis to maintain tissue homeostasis in response to physiological stresses such as a wounding event [37,38]. However, under normal homeostasis, Lgr6-expressing cells appear to mostly act as a stem cell reservoir for the sebaceous gland. Finally, leucine-rich repeats and immunoglobulin-like domain protein 1 (Lrig1) marks the stem cell niche of the hair follicle above the sebaceous glands and has the capability to generate all cell types found within the epidermis [39].

In sum, an unprecedented amount of work focusing on the roles and markers of individual subpopulations of stem cells within the epidermis and hair follicle has led to the development of a myriad of genetically modified mice that are able to directly target these individual populations. This resource of genetic tools has led to extensive testing and continued exploration into identifying the CCOs for epidermal cancers such as basal cell carcinoma (BCC) and SCC.

CANCER CELLS OF ORIGIN IN CUTANEOUS TUMORS

Cancer Cells of Origin for Cutaneous Basal Cell Carcinoma

Nonmelanoma skin cancer is the most common form of human cancer. Nonmelanoma skin cancers include both cutaneous basal cell and SCC at an incidence of over 3 million new cases worldwide per year [24]. Major risk factors for nonmelanoma skin cancers include exposure to ultraviolet (UV) radiation and fair skin [40]. It is estimated that 80% of all new nonmelanoma skin cancers diagnosed are BCC [41,42]. Though mortality for BCC is low, surgical excision of advanced BCCs can lead to severe disfigurement [24].

CCOs for cutaneous BCC have been long postulated to be found at adult stem cells of the hair follicle due to the many similar characteristics between BCC and hair follicles [43]. In recent years, the ability to test this directly has emerged; numerous cell-specific targeting *Cre* alleles which direct expression to the stem, TA, and differentiated cells intrinsic to the hair follicle and the skin have been developed, based on the unique markers found in these cellular subpopulations. Furthermore, BCC appears to be primarily caused by mutations in the Hedgehog (Hh) pathway and several genetic techniques have been developed to cause aberrant activation of this pathway [44]. Together, this has led to an intense effort from a number of groups to determine the CCOs for BCC (Fig. 10.4A and B) [45,48,49]. These efforts serve to exemplify how GEMMs, combined with lineage tracing, can significantly inform on the CCOs, as well as the potential experimental caveats that need to be considered in interpreting these results as a whole.

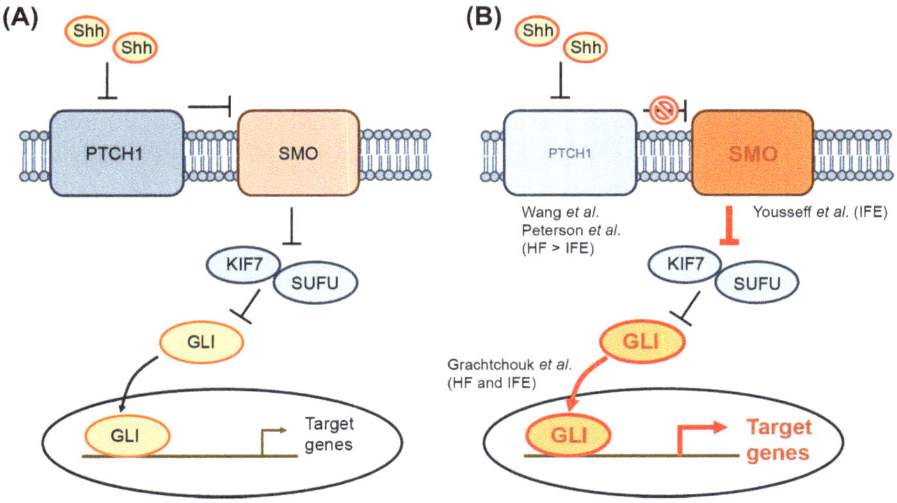

FIGURE 10.4 Hh signaling and basal cell carcinoma (BCC) tumor formation.
(A) Hh signaling is well controlled during homeostasis. Smoothened (SMO) and its downstream signaling can be activated through the inhibition of Patched 1 (PTCH1) by the Shh ligand. The inhibition of negative regulators, kinesin-like protein (KIF7) and suppressor of fused homolog (SUFU), can activate the nuclear translocation of Gli transcription factors. (B) Aberrant activation of Hh signaling is often found in BCC. Several groups used genetic modifications to define CCOs in BCC. Wang et al. [45] and Peterson et al. [46] found that hair follicle (HF) stem cells rather than stem cells within interfollicular epidermis (IFE) are capable of BCC formation via loss of *Ptch1*. On the other hand, Youssef et al. found IFE stem cells are able to form BCC through oncogenic mutant Smo allele [47]. Grachtchouk et al. demonstrated that multiple stem cells within HF and IFE are able to initiate BCC through constant activation of the GLI2 transcription factor [48].

In the first description of potential CCOs for BCC using these techniques, Youssef et al. tested the ability of differing compartments of the epidermis to generate BCC using the ubiquitously expressing *Rosa26*-SmoM2 allele [49]. This genetic tool can cause constitutive Hh signaling via an activated form of the G protein-coupled receptor Smoothened (SmoM2) (Fig. 10.4B). First, by directing SmoM2 expression throughout the epidermis and hair follicle, *K14-CreER*; SmoM2 mice demonstrated macroscopic tumors resembling human nodular BCCs. Through careful clonal analysis and lineage tracing, BCCs in *K14-CreER*; SmoM2 mice appeared to arise from the interfollicular epidermis and infundibulum, rather than the previously postulated origin at the hair follicle [43]. Additionally, many of the BCCs in this model arose from the skin of the ear or tail. To then directly test whether the hair follicles are capable to act as BCC CCOs, Youssef et al. further utilized an extensive number of the *Cre* alleles that targeted individual subpopulations within the hair follicle, including matrix progenitor cells (*Shh-CreER*; SmoM2) and hair follicle bulge stem cells (*K15-CrePR*; SmoM2, *K15-CreER*; SmoM2 and *K19-CreER*; SmoM2). Intriguingly, they found that perturbed Hh signaling did not induce BCC formation from bulge stem cells or from matrix TA cells, and therefore suggested that the BCC CCO is located at the interfollicular epidermis and, to a much lesser extent, the infundibulum. Furthermore, activating oncogenic SMO in the skin of the paws, which are devoid of hair follicles, also led to BCC transformation, which provided additional support to the contention that the interfollicular epidermal basal cells are the CCOs for these SMO-induced nodular BCC-like tumors (Fig. 10.4B). In a follow-up study, this group further showed that during the early stages of activated Smo-mediated BCC formation, cells of the interfollicular epidermis undergo reprogramming to cells that resemble embryonic hair follicle cells on a transcriptomic level, thereby providing a mechanistic rationale as to why marker expression within BCCs resemble hair follicles [47].

In contrast to findings shown by this group, other studies reported conflicting results for the CCOs in mouse models of BCC-like tumors [45,48]. Wang and colleagues induced active Hh-driven BCC-like tumors by partially removing PTCH1 function (Fig. 10.4B) using $Ptch1^{+/-}$ (heterozygous loss of *Ptch1*) mice [45]. Previous reports had shown that $Ptch1^{+/-}$ mice are highly susceptible to BCC development when mice are exposed to well-known risk factors such as UV light and ionizing radiation exposure [50,51]. Using an established ionizing radiation method as a mutagenic insult in $Ptch1^{+/-}$ mice, the fate of *K15-CreER* and *K14-CreER* coupled with reporter gene allele, *Rosa26-LSL-YFP*, was monitored [45]. This YFP lineage trace from *K15-CrePR*+ bulge stem cells showed that the predominant CCOs in BCCs were found in hair follicles, rather than the K14 expressing basal progenitor cells. In an additional contrast to oncogenic SMO driven BCCs [49], PTCH1 suppression mediated BCCs arising from the dorsal skin, but not ears or tails. The differing results were potentially due to nondetectable endogenous Smo expression in the ear, tails, and interfollicular epidermis in *Ptch1*$^{+/-}$ mice. However, additional loss of *TP53* in *Ptch1*$^{+/-}$ mice accelerated BCC development in both the interfollicular epidermis and hair follicles, potentially through induction of Smo in these regions, which normally do not appear to express Smo. Furthermore, BCCs arising from *Ptch1*$^{+/-}$ hair follicles were of differing BCC subtype to those found arising from activated Smo. This study indicated that multiple stem/progenitor cells could be CCOs in BCCs, but the efficiency of tumor initiation may be changed under the different oncogenic loads, even within the Hh signaling pathway (Fig. 10.4B).

In a third report on CCOs for BCC, Grachtchouk et al. reported that multiple origins including both hair follicles and the interfollicular epidermis are involved in BCC tumor formation [48]. Interestingly, the conclusions drawn from this paper indicated that the hair follicle is the CCO for the nodular BCC subtype, whereas the interfollicular epidermis is the CCO of the superficial BCC subtype. In this study, inducible expression of the Hh pathway was achieved through the use of the activated mutant GLI2 (GLI2ΔN) via *Rosa26-LSL-rtTA; tetO-GLI2ΔN*. This allowed for strong activation of downstream target genes without modulation of the upstream Hh signaling components PTCH1 and SMO (Fig. 10.4B). Using *K15-CrePR; Rosa26-LSL-rtTA; tetO-GLI2ΔN* (*iK15; rtTA; GLI2ΔN*) and *Lgr5-CreER; Rosa26-LSL-rtTA; tetO-GLI2ΔN* (*iLgr5; rtTA; GLI2ΔN*), it was shown that nodular BCC-like tumors more efficiently arose from K15 and Lgr5 expressing hair follicle stem cells, most notably from the hair germ (Fig. 10.4B). Furthermore, while *K14-rtTA; tetO-GLI2ΔN* could induce superficial BCC-like tumors, short-term induction was less efficient. In summary, these multiple studies illustrate the power of mouse genetics to discern CCOs, but also show the need for critical assessment of the target gene utilized to model the cancer subtype of interest, as well as determining the strength of the promoter strategy used to express Cre recombinase.

Downstream Factors and Niche Requirements Necessary for Hh Pathway-Mediated BCC Formation

The previous studies demonstrated the importance of the Hh signaling pathway in multiple potential CCOs in BCC. Numerous Hh pathway inhibitors are under clinical study for treatment of BCC [52]. Vismodegib, an inhibitor of Smo, has been approved for treatment of locally advanced or metastatic BCC [53]. Though this inhibitor works well for most patients, early reports indicate resistance to Vismodegib occurs in as much as 20% of treated patients [53]. Current work is underway to determine mechanisms of resistance [54].

Studies that model the initiation of cancer from the CCOs have unique translational potential via the development of preventative and early treatment strategies through the identification of factors downstream of Hh signaling necessary for BCC development. These targets have significant potential as additive therapies in combination to Hh inhibitors. Using these mouse models, Wnt signaling through β-catenin has been reported as necessary for aberrantly activated SMO-mediated BCC formation [47]. Here, genetic and pharmacological inhibition of Wnt signaling effectively prevented BCC-like tumors from initiating. Even more recently, Larsimont et al. demonstrated the requirement of *Sox9* expression to initiate BCC formation [55]. This group reported that Sox9 controls BCC tumorigenesis through regulation of self-renewal in SmoM2-mediated CCOs, and that Sox9 is required for the long-term maintenance of BCC transformed cells. Most importantly, they further found that conditional deletion of *Sox9* at the CCO significantly blocks Hh-mediated BCC formation in both *K14-CreER*; *Rosa26*-SmoM2-*YFP*; *Sox9-flox/flox* and *K14-CreER*; *Ptch1-flox/flox*; *Sox9-flox/flox* mouse models. These studies demonstrate the ability to identify necessary downstream factors for tumor development, which holds translational potential, by modeling tumor initiation from CCOs in GEMMs.

Cell extrinsic factors derived from the stem cell niche may have potential in contributing to the CCO identity and to tumor initiation. The microenvironment within or surrounding the stem cell niche consists of myriad of differing cell types such as fibroblasts, immune cells, and other stem cell populations. Within a tumor, a supportive, and sometimes necessary, role of cancer-associated fibroblasts has been well described [56]. Similarly, many types of cells and their secreted proteins (such as proinflammatory cytokines) resident to the stem cell niche microenvironment may also regulate tumor initiation, and CCOs may strongly or partially require their physiological involvement. Recently, Peterson and colleagues reported that although cells within the interfollicular epidermis are generally not responsive to the oncogenic load via loss of *Ptch1*, BCC-like tumors do arise from the interfollicular epidermis within the touch dome epithelia niche. Interestingly, they also found that surgical denervation significantly suppressed the tumorigenic potential of these touch domes [46]. This study serves to demonstrate that CCO identity can be heavily influenced by external cellular influences such as cutaneous sensory nerves, and by unique niches such as the touch dome, which provides a particularly susceptible environment to BCC initiation.

Taken together, it appears that *Ptch1* loss of function mediates nodular BCCs arising from the hair follicle, whereas activated SMO mediates superficial BCCs arising from the interfollicular epidermis. Furthermore, though *Ptch1* mutant-mediated BCC does not typically arise from the interfollicular epidermis, the nerve associated touch domes of the interfollicular epidermis represent a unique BCC susceptible population.

Cell Extrinsic Factors Regulating Cancer Cells of Origin in Hh Pathway-Driven Basal Cell Carcinoma

Identification of the environmental causes and their roles in tumorigenesis via CCOs will be invaluable in developing preventative strategies in many cancers.

The most well-known risk factor for BCC is UV light exposure and chronic or intermittent sunburn, as BCCs most frequently occur on the skin of the face, head, and neck [40,50]. Furthermore, risk for many types of cancers including BCCs has been long associated with injury and inflammation [57] which may regulate adult stem cells and CCOs in BCC. Uncovering whether acute or chronic inflammation could be a strong physiological activator in tumor initiation, and the role of injury and inflammation in dictating the CCO, represents a significant experimental challenge.

Adult stem cells of the hair follicle can respond to a wounding injury by assisting in repair and reepithelization of the interfollicular epidermis [27,58]. To test whether epidermal injury and inflammation affect the identity of BCC CCOs as a cell extrinsic tumor initiator, Wong and Reiter used *K15-CrePR*; SmoM2 mice under wounding and nonwounding conditions [59]. Similar to the findings by Youssef et al. [49], tumor initiation was not efficient in this mouse model. However, when epidermal wounds were induced in *K15-CrePR*; SmoM2 mice, significant superficial BCC-like tumors were observed at the induced wound sites. This finding suggests that though hair follicle stem cells may not actively initiate tumors under normal homeostasis, an additional stimulus such as injury and inflammation can promote tumorigenesis from alternative CCOs. In other words, cell extrinsic factors can change the status of CCOs from resting, but poised, to active. An additional study by Kasper and colleagues also supported the importance of injury as an extrinsic factor in BCC tumorigenesis [60]. To induce Hh pathway-driven tumor formation, the group used genetic alleles that overexpressed *human Gli1* and cell-specific deletion of *Ptch1*. Specifically, *K5tTA; TRE-Gli1*, *K5-CreER; Ptch1-flox/flox*, and *Lgr5-CreER; Ptch1-flox/flox* mice were used to induce BCC, and similar to Wong et al., wounding accelerated BCC-like tumor formation at the wound sites. These independent studies demonstrated the potential association between cancer initiation and injury, in terms of the status of CCOs and locale of tumor development.

Cancer Cells of Origin in Cutaneous Squamous Cell Carcinoma

Cutaneous SCC is the second most commonly diagnosed nonmelanoma skin cancer diagnosed, with 700,000 new cases expected annually in the United States [61]. Like BCC, major risk factors are high lifetime exposure to UV radiation and fair skin. In contrast to BCC, SCC can metastasize and metastatic recurrence harbors a poor survival rate [24,62]. Therefore, identifying the cellular origin of SCC, together with defining the early events of tumor initiation, can establish better preventative strategies.

As we mentioned previously, BCC was thought to arise from hair follicles due to the similarity in morphological and molecular features, but CCO modeling in mice revealed that multiple origins, including the interfollicular epidermis and the hair follicles, are sufficient to initiate BCC-like tumors [45,46,48,49,59,60]. SCC has been thought to arise from differentiated cells of the interfollicular epidermal layers, due to the squamous appearance in SCCs. Like BCC, however, recent studies have indicated that hair follicle stem cells can, in fact, act significantly as SCC CCOs [63,64]. Numerous reports have determined that *Ras* and *p53* mutations are found in both human and murine SCCs [65,66]. Based on previous findings, two groups tested which epidermal cells can efficiently develop squamous tumors when a combination of mutant oncogenic *Kras* (*LSL Kras-G12D*) and/or deletion of *p53* (*p53 flox/flox*) is used [63,64]. Surprisingly, cutaneous papillomas and epidermal cysts developed from hair follicle stem cells in *K15-CrePR; LSL Kras-G12D* and *K19-CreER; LSL Kras-G12D* mice. The contribution of hair follicle stem cells to the initiation of cutaneous papillomas is also supported by a recent study by Li et al. which utilizes a chemical carcinogen, dimethylbenzanthracene, tetradecanoylphobol-acetate (DMBA/TPA), to initiate tumorigenesis in a *K15-CrePR* mouse line [67]. On the other hand, the similar pathologic changes were not observed when hair follicle TA cells were targeted in *Shh-CreER; LSL Kras-G12D* animals. In further studies, cutaneous SCCs were formed from hair follicle stem cells when mutant *Kras-G12D* was combined with *p53* deletion in *K15-CreER; LSL Kras-G12D; p53 flox/flox* and *K19-CreER; LSL Kras-G12D; p53 flox/flox* mice [63,64]. Similarly, no SCC tumor formation was observed arising from the TA cells in *Shh-CreER; LSL Kras-G12D; p53 flox/flox* mice, suggesting hair follicle stem cells can act as CCOs in cutaneous SCC (Fig. 10.5A). These independent reports demonstrated that cutaneous SCCs can originate from hair follicle stem cells whereas SCCs cannot originate from the hair follicle stem cell descendants, the TA cells. The reason why the TA cells have limited potential to form tumors in this model remains unclear. However, understanding how these two very closely related cell subpopulations have completely different susceptibilities to tumor formation can provide clues on mechanisms of potential tumor prevention.

Suppressed Tumor Initiation via Adult Stem Cell Quiescence

Although it is recognized that genetic and epigenetic alterations cause cancer, and that the CCO can dictate tumor subtype, the stimulus needed to initiate cellular proliferation is not yet clear. Indeed, though a CCO may harbor the tumorigenic load sufficient for tumorigenesis, tumor initiation sometimes has a long latency period. With regard to BCC development, wounding has been noted as a stimulus that can facilitate the development of BCC, and influence the CCO identity [59,60]. During wounding, hair follicle stem cells are activated to assist in repair of the interfollicular epidermis. Thus, this process seems to cause hair follicle stem cells to gain receptiveness to the oncogenic stimulus.

Hair follicles cycle through a state of stem cell quiescence during telogen, and state of stem cell activity and proliferation at the onset of anagen. It has also

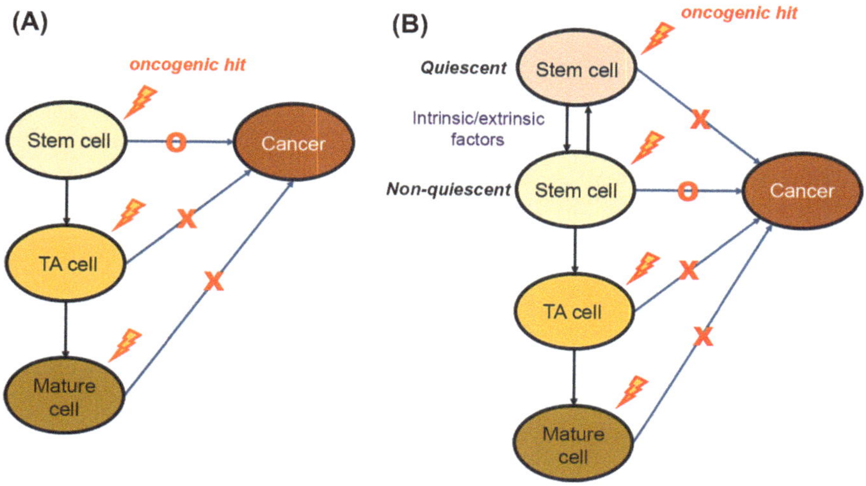

FIGURE 10.5 Traditional model of cancer cells of origin (CCOs) in cutaneous squamous cell carcinomas (SCCs), and the requirement of stem cell activation for SCC tumor initiation.
(A) It has been found that hair follicle stem cells can serve as the cellular origin of cutaneous squamous tumors under oncogenic events of mutant *Kras* with loss of *p53*, but transit-amplifying (TA) and differentiated cells do not show potential for tumor formation [63,64]. (B) In addition to the identification of CCOs, it has also been found that quiescent hair follicle stem cells with the same oncogenic events are unable to form the tumor. Quiescent adult stem cells require cell intrinsic and extrinsic factors to be activated to initiate the tumor formation [68].

long been known that depilation (removal of the hair shafts) can initiate hair follicle stem cell activation during telogen, which causes resting hair follicles to undergo stem cell activation and anagen induction. The mechanism through which depilation is able to activate stem cells is not well understood, though the recruitment of immune cells and activated transforming growth factor (TGF) signaling in the epidermal stem cells may have a role [69]. In one model of BCC, depilation-mediated anagen induction was shown to significantly accelerate BCC-like tumor formation in mice, suggesting a relationship between adult stem cell activity and tumor initiation [48]. Additionally, the loss of *p53* was found to accelerate BCC-like tumor formation [45]. Stem cell maintenance in hematopoietic stem cells is known to be regulated, at least in part, by *p53* [70]. Therefore, it is also possible that the status of *p53* expression in the cellular origin of BCC can affect the maintenance of the stem cell quiescence. Taken together, these results may indicate that adult stem cell populations within a quiescent state may be less able to develop BCC-like tumors, even when harboring mutations sufficient to cause aberrant activation of Hh signaling.

Intriguingly, a recent study revealed that expression of oncogenic *Kras* coupled with loss of tumor-suppressor *p53* in hair follicle stem cells was not able to initiate cutaneous SCC formation during stem cell quiescence, whereas SCCs efficiently form following the onset of stem cell activation when these mutations are present (Fig. 10.5B) [68]. The murine hair cycle is synchronized and well defined during early life. Mice at around 50 days postnatal show telogen hair follicles, with the next anagen generally beginning at approximately 12 weeks of age [28,71]. Using the physiologically synchronized hair cycle, this study investigated the relative ability for SCC CCOs to initiate tumors within the context of stem cell quiescence and activation. Notably, SCC formation occurred only during induction of anagen, whereas quiescent CCOs were completely resistant to SCC formation. This suggests that CCOs require additional mechanisms to initiate cancers, even when they express well-known mutant genes and epigenetic changes. Furthermore, the natural ability for quiescent CCOs may inform on potential strategies of prevention and early treatment. In addition to the finding that stem cell quiescence can act as a natural tumor suppressor, it was also revealed that loss of PTEN is sufficient to facilitate cutaneous SCC formation from quiescent CCOs. Furthermore, since this finding has been published, a similar mechanism has been shown in which quiescent muscle satellite cells are also able to suppress soft tissue sarcoma formation [72]. Taken together, these findings indicate the combined requirement of both cell intrinsic and extrinsic factors in the tumorigenesis from their CCOs.

MELANOCYTIC TUMOR INITIATION

While many studies found the essential role of adult stem cells in nonmelanoma cutaneous tumor development, cutaneous melanoma has not been well studied for their CCOs and their related factors. Although the incidence of melanoma is relatively low compared to BCC and SCC (approximately 132,000 cases worldwide per year), its annual incidence and mortality rates are gradually and continuously increasing [73]. Melanoma is usually considered the deadliest skin cancer, since it is associated with a higher aggressiveness, rapid progression, and frequent metastasis. Melanoma causes death in about 65,000 patients per year [24]. Similar to nonmelanoma cutaneous tumors, it has been suggested that long-term regular use of sunscreen may efficiently prevent melanoma development [74].

Melanocyte stem cells are known to be located within the hair follicle within the epidermal hair follicle stem cell niche, and their physiological cycles and functional activities are closely linked to each other [75]. Furthermore, melanocyte stem cells respond to physiological or environmental factors such as wounds and UVB radiation, similar to epidermal stem cells [76]. Therefore, it will be important to identify the role of melanocyte stem cells or melanocyte progenitors in the cellular origin of cutaneous melanoma. However, understanding the roles of melanocyte stem cells in melanoma initiation may still require more intensive future studies, due to lack of cell-specific mouse models, particularly for melanocyte stem cells or TA cells. Various -omics techniques, such as single-cell transcriptomics [77] and single-cell epigenomics [78,79], will enable the elucidation of better candidates to design new mouse models. Furthermore, new models will be established more quickly and efficiently through the advent of the revolutionary CRISPR/Cas9 gene editing techniques [80–82].

CONCLUDING REMARKS AND FUTURE DIRECTIONS

In the nonmelanoma cutaneous tumors BCC and SCC, it is evident that in most cases, epidermal stem cell populations are more susceptible to tumor initiation relative to their progeny, the TA cells, and mature differentiated cells. However, it is also clear from BCC CCO studies that the mutational load selected to modulate the appropriate signaling pathways implicated in a cancer subtype can differentially affect the CCO identity. Taken together, these experimental approaches provide an improved understanding of which cells can initiate tumors and also generates a framework that be applied for defining the mechanisms of how CCOs induce the first proliferative step to forming a nonmelanoma skin tumor. However, since current nonmelanoma skin cancer studies are mostly based on Cre-mediated tumorigenic systems, oncogenic combinations concurrently occur. Therefore, the sequential progression of cancer through its CCO is still not well defined. To understand the mechanisms of oncogenic hits and bona fide disease progression, additional genetic technologies will need to be employed to understand how CCOs may be shaped by earlier mutations that allow them to become more susceptible to tumorigenesis. Additionally, the development of a valid in vitro or grafting system to test these results with human skin will be very important in future studies. Unfortunately, technical limitations currently prevent the generation of mutations in particular cells of human hair follicles grafted to mouse skin.

Other organ systems have employed similar studies as those described earlier to probe CCOs. Prostate cancer exemplifies another cancer type with a well-developed body of literature exploring potential CCOs [83]. The prostate contains both basal and luminal cells, which during homeostasis appear to maintain their own populations in a unipotent manner [84]. However, in some cases basal cells can produce luminal cells. Studies exploring prostate cells of origin using both human-grafted cells and genetically modified mice with lineage tracing have produced conflicting results, dependent on the model system used and the tumorigenic event employed. Recently, it has been proposed that both luminal and basal cells can act as a CCO, but the predominance of data suggesting a luminal origin may be due to a mechanism in which basal cells that are capable of initiating prostate cancer undergo a basal to luminal cell differentiation prior to tumor initiation [85,86]. This distinction from a basal "cell of mutation" to a luminal "CCO" is an intriguing proposal that should be taken into account in nonmelanoma skin cancer and other CCO studies in the future [83].

While CCOs and CSCs need to be considered as independent entities, many of the molecular mechanisms that regulate adult stem cell biology and their transformation into CCOs may also be important in regulation of CSCs and their role in tumor maintenance, or vice versa. Additional work will help establish the similarities and dissimilarities among the mechanisms of hierarchal organization descendant from adult stem cells, CCOs and CSCs. Furthermore, though the relationship between adult stem cells and CCOs is becoming better understood, the role between CCOs and CSCs is not clear. In fact, it remains to be determined if there is a relationship between these two cell populations, or if CSCs can arise from another source altogether.

List of Acronyms and Abbreviations

BCC Basal cell carcinoma
BrdU Bromo-deoxyuridine, which is incorporated into DNA and can be used to detect label-retaining cells
CCO Cancer cell of origin
CSC Cancer stem cell
EMT Epithelial-to-mesenchymal transition
ER Estrogen receptor, contextually found attached to Cre recombinase
FACS Fluorescence-activated cell sorting
H2B-GFP GFP (green fluorescent protein) labeling of histone H2B; a method of lineage tracing and detecting label-retaining cells
Hh The development-associated Hedgehog pathway
GEMM Genetically engineered mouse model
LSL Loxp-Stop-loxp site found in genetically engineered knockin systems which allows for the conditional expression of a gene of interest
MLL Mixed lineage leukemia; a gene located on chromosome 11, often found with mutations in leukemia patients
PR Progesterone receptor, contextually found attached to Cre recombinase
TGF Transforming growth factor signaling
UV Ultraviolet (light, radiation)
YFP Yellow fluorescent protein

References

[1] Hanahan D, Weinberg RA. Hallmarks of cancer: the next generation. Cell 2011;144(5):646–74.

[2] Marusyk A, Tabassum DP, Altrock PM, Almendro V, Michor F, Polyak K. Non-cell-autonomous driving of tumour growth supports sub-clonal heterogeneity. Nature 2014;6(2):54–8.

[3] Gerlinger M, Rowan AJ, Horswell S, Larkin J, Endesfelder D, Gronroos E, et al. Intratumor heterogeneity and branched evolution revealed by multiregion sequencing. N Engl J Med 2012:883–92.

[4] Burrell RA, McGranahan N, Bartek J, Swanton C. The causes and consequences of genetic heterogeneity in cancer evolution. Nature 2013;501(7467):338–45.

[5] Blanpain C. Tracing the cellular origin of cancer. Nat Cell Biol 2013;15(2):126–34.

[6] Visvader JE. Cells of origin in cancer. Nature 2011;469(7330):314–22.

[7] Clarke MF, Dick JE, Dirks PB, Eaves CJ, Jamieson CHM, Jones DL, et al. Cancer stem cells–perspectives on current status and future directions: AACR Workshop on cancer stem cells. Cancer Res 2006;66(19):9339–44.

[8] Koch U, Krause M, Baumann M. Cancer stem cells at the crossroads of current cancer therapy failures. Semin Cancer Biol 2010:1–9.

[9] Alison MR, Lim SM, Nicholson LJ. Cancer stem cells: problems for therapy? J Pathol 2011;223(2):147–61.

[10] Ye X, Weinberg RA. Epithelial–Mesenchymal plasticity: a central regulator of cancer progression. Trends Cell Biol 2015;25(11):675–86.

[11] Mitra A, Mishra L, Li S. EMT, CTCs and CSCs in tumor relapse and drug-resistance. Oncotarget 2015;6(13):10697–711.

[12] Blanpain C, Simons BD. Unravelling stem cell dynamics by lineage tracing. Nat Rev Mol Cell Biol 2013;14(8):489–502.

[13] Schepers AG, Snippert HJ, Stange DE, van den Born M, van Es JH, van de Wetering M, et al. Lineage tracing reveals Lgr5+ stem cell activity in mouse intestinal adenomas. Science 2012;337(6095):730–5.

[14] White AC, Lowry WE. Refining the role for adult stem cells as cancer cells of origin. Trends Cell Biol 2014;25(1):11–20.

[15] Fuchs E, Chen T. A matter of life and death: self-renewal in stem cells. EMBO Rep 2012;14(1):39–48.

[16] McGranahan N, Swanton C. Biological and therapeutic impact of intratumor heterogeneity in cancer evolution. Cancer Cell 2015;27(1):15–26.

[17] Almendro V, Marusyk A, Polyak K. Cellular heterogeneity and molecular evolution in cancer. Annu Rev Pathol 2013;8(1):277–302.

[18] Hsu Y-CC, Li L, Fuchs E. Emerging interactions between skin stem cells and their niches. Nat Med 2014;20(8):847–56.

[19] Rompolas P, Greco V. Stem cell dynamics in the hair follicle niche. Semin Cell Dev Biol 2014;25–26:34–42.

[20] Arwert EN, Hoste E, Watt FM. Epithelial stem cells, wound healing and cancer. Nat Rev Cancer March 2012;12:170–80.

[21] Hsu Y-C, Li L, Fuchs E. Transit-amplifying cells orchestrate stem cell activity and tissue regeneration. Cell 2014;157(4):935–49.

[22] Cozzio A, Passegué E, Ayton PM, Karsunky H, Cleary ML, Weissman IL. Similar MLL-associated leukemias arising from self-renewing stem cells and short-lived myeloid progenitors. Genes Dev 2003;17(24):3029–35.

[23] Ross M, Pawlina W. Histology: a text and atlas. In: Histology: a text and atlas. 2004. p. 184–91.

[24] Simoes MCF, Sousa JJS, Pais AACC. Skin cancer and new treatment perspectives: a review. Cancer Lett 2015;357(1):8–42.

[25] Maksimovic S, Nakatani M, Baba Y, Nelson AM, Marshall KL, Wellnitz SA, et al. Epidermal Merkel cells are mechanosensory cells that tune mammalian touch receptors. Nature 2014;509(7502):617–21.

[26] Kissenpfennig A, Henri S, Dubois B, Laplace-Builhé C, Perrin P, Romani N, et al. Dynamics and function of Langerhans cells in vivo: dermal dendritic cells colonize lymph node areas distinct from slower migrating Langerhans cells. Immunity 2005;22(5):643–54.

[27] Ito M, Liu Y, Yang Z, Nguyen J, Liang F, Morris RJ, et al. Stem cells in the hair follicle bulge contribute to wound repair but not to homeostasis of the epidermis. Nat Med 2005;11(12):1351–4.

[28] Müller-Röver S, Handjiski B, Van Der Veen C, Eichmu S, Foitzik K, Mckay IA, et al. A comprehensive guide for the accurate classification of murine hair follicles in distinct hair cycle stages. J Invest Dermatol 2001;117:3–15.

[29] Driskell RR, Clavel C, Rendl M, Watt FM. Hair follicle dermal papilla cells at a glance. J Cell Sci 2011;124(8):1179–82.

[30] Cotsarelis G, Sun TT, Lavker RM. Label-retaining cells reside in the bulge area of pilosebaceous unit: implications for follicular stem cells, hair cycle, and skin carcinogenesis. Cell 1990;61(7):1329–37.

[31] Tumbar T, Guasch G, Greco V, Blanpain C, Lowry WE, Rendl M, et al. Defining the epithelial stem cell niche in skin. Science 2004;303(5656):359–63.

[32] Trempus CS, Morris RJ, Bortner CD, Cotsarelis G, Faircloth RS, Reece JM, et al. Enrichment for living murine keratinocytes from the hair follicle bulge with the cell surface marker CD34. J Invest Dermatol 2003;120(4):501–11.

[33] Vidal VPI, Chaboissier M-C, Lützkendorf S, Cotsarelis G, Mill P, Hui C-C, et al. Sox9 is essential for outer root sheath differentiation and the formation of the hair stem cell compartment. Curr Biol 2005;15(15):1340–51.

[34] Morris RJ, Liu Y, Marles L, Yang Z, Trempus C, Li S, et al. Capturing and profiling adult hair follicle stem cells. Nat Biotechnol 2004;22(4):411–7.

[35] Commo S, Gaillard O, Bernard BA. The human hair follicle contains two distinct K19 positive compartments in the outer root sheath: a unifying hypothesis for stem cell reservoir? Differentiation 2000;66(4–5):157–64.

[36] Means AL, Xu Y, Zhao A, Ray KC, Gu G. A CK19(CreERT) knockin mouse line allows for conditional DNA recombination in epithelial cells in multiple endodermal organs. Genesis 2008;46(6):318–23.

[37] Jaks V, Barker N, Kasper M, van Es JH, Snippert HJ, Clevers H, et al. Lgr5 marks cycling, yet long-lived, hair follicle stem cells. Nat Genet 2008;40(11):1291–9.

[38] Snippert HJ, Haegebarth A, Kasper M, Jaks V, van Es JH, Barker N, et al. Lgr6 marks stem cells in the hair follicle that generate all cell lineages of the skin. Science 2010;327(5971):1385–9.

[39] Jensen KB, Collins CA, Nascimento E, Tan DW, Frye M, Itami S, et al. Lrig1 expression defines a distinct multipotent stem cell population in mammalian epidermis. Cell Stem Cell 2009;4(5):427–39.

[40] Rubin AI, Chen EH, Ratner D. Basal-cell carcinoma. N Engl J Med 2005;353(21):2262–9.

[41] Rogers HW, Weinstock MA, Feldman SR, Coldiron BM. Incidence estimate of nonmelanoma skin cancer (keratinocyte carcinomas) in the US population, 2012. JAMA Dermatol 2015;151(10):1081–6.

[42] Gandhi SA, Kampp J. Skin cancer epidemiology, detection, and management. Med Clin North Am 2015;99(6):1323–35.

[43] Donovan J. Review of the hair follicle origin hypothesis for basal cell carcinoma. Dermatol Surg 2009;35(9):1311–23.

[44] Teglund S, Toftgård R. Hedgehog beyond medulloblastoma and basal cell carcinoma. Biochim Biophys Acta 2010;1805(2):181–208.

[45] Wang GY, Wang J, Mancianti M-L, Epstein EH. Basal cell carcinomas arise from hair follicle stem cells in Ptch1(+/−) mice. Cancer Cell 2011;19(1):114–24.

[46] Peterson SC, Eberl M, Vagnozzi AN, Belkadi A, Veniaminova NA, Verhaegen ME, et al. Basal cell carcinoma preferentially arises from stem cells within hair follicle and mechanosensory niches. Cell Stem Cell 2015;16(4):400–12.

[47] Youssef KK, Lapouge G, Bouvrée K, Rorive S, Brohée S, Appelstein O, et al. Adult interfollicular tumour-initiating cells are reprogrammed into an embryonic hair follicle progenitor-like fate during basal cell carcinoma initiation. Nat Cell Biol 2012;14(11):1–14.

[48] Grachtchouk M, Pero J, Yang SH, Ermilov AN, Michael LE, Wang A, et al. Basal cell carcinomas in mice arise from hair follicle stem cells and multiple epithelial progenitor populations. J Clin Invest 2011;121(5):1768–81.

[49] Youssef KK, Van Keymeulen A, Lapouge G, Beck B, Michaux C, Achouri Y, et al. Identification of the cell lineage at the origin of basal cell carcinoma. Nat Cell Biol March 2010;12(3):299–305.

[50] Armstrong BK, Kricker A. The epidemiology of UV induced skin cancer. J Photochem Photobiol B October 2001;63(1–3):8–18.

[51] Moan J, Grigalavicius M, Baturaite Z, Dahlback A, Juzeniene A. The relationship between UV exposure and incidence of skin cancer. Photodermatol Photoimmunol Photomed 2014:1–10 (Cm).

[52] Amakye D, Jagani Z, Dorsch M. Unraveling the therapeutic potential of the Hedgehog pathway in cancer. Nat Med 2013;19(11):1410–22.

[53] Sekulic A, Migden MR, Oro AE, Dirix L, Lewis KD, Hainsworth JD, et al. Efficacy and safety of Vismodegib in advanced basal-cell carcinoma. N Engl J Med 2012;366(23):2171–9.

[54] Sharpe HJ, Pau G, Dijkgraaf GJ, Basset-Seguin N, Modrusan Z, Januario T, et al. Genomic analysis of smoothened inhibitor resistance in basal cell carcinoma. Cancer Cell 2015;27(3):327–41.

[55] Larsimont J-C, Youssef KK, Sánchez-Danés A, Sukumaran V, Defrance M, Delatte B, et al. Sox9 controls self-renewal of oncogene targeted cells and links tumor initiation and invasion. Cell Stem Cell 2015:60–73.

[56] Erez N, Truitt M, Olson P, Hanahan D. Cancer-associated fibroblasts are activated in incipient Neoplasia to orchestrate tumor-promoting inflammation in an NF-κB-dependent manner. Cancer Cell 2010;17(2):135–47.

[57] Loeb LA, Harris CC. Advances in chemical carcinogenesis: a historical review and prospective. Cancer Res September 01, 2008;68(17):6863–72.

[58] Blanpain C, Fuchs E. Plasticity of epithelial stem cells in tissue regeneration. Science 2014;344(6189):1242281.

[59] Wong SY, Reiter JF. Wounding mobilizes hair follicle stem cells to form tumors. Proc Natl Acad Sci USA March 2011;108(10):4093–8.

[60] Kasper M, Jaks V, Are A, Bergström Å, Schwäger A, Barker N, et al. Wounding enhances epidermal tumorigenesis by recruiting hair follicle keratinocytes. Proc Natl Acad Sci USA 2011;108(10):4099–104.

[61] Karia PS, Han J, Schmults CD. Cutaneous squamous cell carcinoma: estimated incidence of disease, nodal metastasis, and deaths from disease in the United States, 2012. J Am Acad Dermatol June 2013;68(6):957–66.

[62] Mullen JT, Feng L, Xing Y, Mansfield PF, Gershenwald JE, Lee JE, et al. Invasive squamous cell carcinoma of the skin: defining a high-risk group. Ann Surg Oncol July 2006;13(7):902–9.

[63] White AC, Tran K, Khuu J, Dang C, Cui Y, Binder SW, et al. Defining the origins of Ras/p53-mediated squamous cell carcinoma. Proc Natl Acad Sci USA 2011;108(18):7425–30.

[64] Lapouge G, Youssef KK, Vokaer B, Achouri Y, Michaux C, Sotiropoulou PA, et al. Identifying the cellular origin of squamous skin tumors. Proc Natl Acad Sci USA 2011;108(18):7431–6.

[65] Nassar D, Latil M, Boeckx B, Lambrechts D, Blanpain C. Genomic landscape of carcinogen-induced and genetically induced mouse skin squamous cell carcinoma. Nat Med 2015;21(8):946–54.

[66] Balmain A, Yuspa SH. Milestones in skin carcinogenesis: the biology of multistage carcinogenesis. J Invest Dermatol 2014;134(e1):E2–7.

[67] Li S, Park H, Trempus CS, Gordon D, Liu Y, Cotsarelis G, et al. A keratin 15 containing stem cell population from the hair follicle contributes to squamous papilloma development in the mouse. Mol Carcinog 2013;52(10):751–9.

[68] White AC, Khuu JK, Dang CY, Hu J, Tran KV, Liu A, et al. Stem cell quiescence acts as a tumour suppressor in squamous tumours. Nat Cell Biol 2014;16(1):99–107.

[69] Chen C-CC, Wang L, Plikus MV, Jiang TX, Murray PJ, Ramos R, et al. Organ-level quorum sensing directs regeneration in hair stem cell populations. Cell 2015;161(2):277–90.

[70] Liu Y, Elf SE, Miyata Y, Sashida G, Liu Y, Huang G, et al. p53 regulates hematopoietic stem cell quiescence. Cell Stem Cell 2009;4(1):37–48.

[71] Geyfman M, Plikus MV, Treffeisen E, Andersen B, Paus R. Resting no more: re-defining telogen, the maintenance stage of the hair growth cycle. Biol Rev Camb Philos Soc 2015;90(4):1179–96.

[72] Van Mater D, Añó L, Blum JM, Webster MT, Huang W, Williams N, et al. Acute tissue injury activates satellite cells and promotes sarcoma formation via the HGF/c-MET signaling pathway. Cancer Res 2015;75(3):605–14.

[73] Siegel RL, Miller KD, Jemal A. Cancer statistics, 2015. CA Cancer J Clin January 2015;65(1):5–29.

[74] Green AC, Williams GM, Logan V, Strutton GM. Reduced melanoma after regular sunscreen use: randomized trial follow-up. J Clin Oncol January 20, 2011;29(3):257–63.

[75] Rabbani P, Takeo M, Chou W, Myung P, Bosenberg M, Chin L, et al. Coordinated activation of Wnt in epithelial and melanocyte stem cells initiates pigmented hair regeneration. Cell June 2011;145(6):941–55.

[76] Chou WC, Takeo M, Rabbani P, Hu H, Lee W, Chung YR, et al. Direct migration of follicular melanocyte stem cells to the epidermis after wounding or UVB irradiation is dependent on Mc1r signaling. Nat Med June 2013;19(7):924–9.

[77] Shalek AK, Satija R, Adiconis X, Gertner RS, Gaublomme JT, Raychowdhury R, et al. Single-cell transcriptomics reveals bimodality in expression and splicing in immune cells. Nature 2013;498(7453):236–40.

[78] Cusanovich DA, Daza R, Adey A, Pliner HA, Christiansen L, Gunderson KL, et al. Epigenetics. Multiplex single-cell profiling of chromatin accessibility by combinatorial cellular indexing. Science 2015;348(6237):910–4.

[79] Buenrostro JD, Wu B, Litzenburger UM, Ruff D, Gonzales ML, Snyder MP, et al. Single-cell chromatin accessibility reveals principles of regulatory variation. Nature 2015;523(7561):486–90.

[80] Cho SW, Kim S, Kim JM, Kim J-S. Targeted genome engineering in human cells with the Cas9 RNA-guided endonuclease. Nat Biotechnol 2013;31(3):230–2.

[81] Horvath P, Barrangou R. CRISPR/Cas, the immune system of bacteria and archaea. Science 2010;327(5962):167–70.

[82] Cong L, Ran FA, Cox D, Lin S, Barretto R, Habib N, et al. Multiplex genome engineering using CRISPR/Cas systems. Science 2013;339(6121):819–23.

[83] Lee SH, Shen MM. Cell types of origin for prostate cancer. Curr Opin Cell Biol 2015; 37:35–41.

[84] Wang ZA, Mitrofanova A, Bergren SK, Abate-Shen C, Cardiff RD, Califano A, et al. Lineage analysis of basal epithelial cells reveals their unexpected plasticity and supports a cell of origin model for prostate cancer heterogeneity. Nat Cell Biol 2013;15(3):274–83.

[85] Stoyanova T, Cooper AR, Drake JM, Liu X, Armstrong AJ, Pienta KJ, et al. Prostate cancer originating in basal cells progresses to adenocarcinoma propagated by luminal-like cells. Proc Natl Acad Sci USA 2013;110(50):20111–6.

[86] Lu TL, Huang YF, You LR, Chao NC, Su FY, Chang JL, et al. Conditionally ablated Pten in prostate basal cells promotes basal-to-luminal differentiation and causes invasive prostate cancer in mice. Am J Pathol 2013;182(3):975–91.

CHAPTER 11

Asymmetric Division of Cancer Stem Cells

C. Petritsch[1], X. Shen[2]
[1]*University of California San Francisco, San Francisco, CA, United States;* [2]*Duke University, Durham, NC, United States*

INTRODUCTION

Cancer cells divide excessively to multiply their numbers. Cancer cell division was often assumed to be largely symmetric, producing a homogeneous population of progeny within the tumor, although distinct mutations may be inherited due to genomic instability at the time of mitosis. However, the discovery of cancer stem cells (CSCs), and the realization that solid tumors often maintain heterogeneous and hierarchical structures reminiscent of the tissues they originated from have prompted researchers to reexamine this assumption.

In contrast to symmetric division, which result in identical progeny, asymmetric cell division (ACD) generates progeny, which carry distinct features immediately after cytokinesis. A cell undergoes an ACD to generate daughter cells, each of which can divide symmetrically to amplify its number or asymmetrically to further increase diversity and establish a hierarchy of cell types (lineages). There are two types of ACD: extrinsic and intrinsic ACD. In *Drosophila*, ovarian stem cell ACD represents the extrinsic type, where two daughter cells are initially identical but receive external signaling cues from the niche, which differ depending on the position of the progeny and which stimulate distinct fates. In contrast, *Drosophila* neuroblast (NB) ACD represent the classic, intrinsic type, during which the dividing cell establishes polarity prior to division and allocates regulatory factors, called cell fate determinants (CFDs), differentially between the two daughter cell compartments. The two daughter cells of an NB will always acquire distinct fates [1,2] (Fig. 11.1). The imbalance of CFDs causes bifurcation of the regulatory (and often epigenetic) states in the two daughter cells, leading to distinct cell fates. ACD is an evolutionary conserved mechanism existent in unicellular and multicellular, prokaryotic and eukaryotic organisms [3]. We will first review the functional implications of ACD and the different ACD strategies employed throughout evolution and among

CONTENTS

Cancer Stem Cells. http://dx.doi.org/10.1016/B978-0-12-803892-5.00011-5

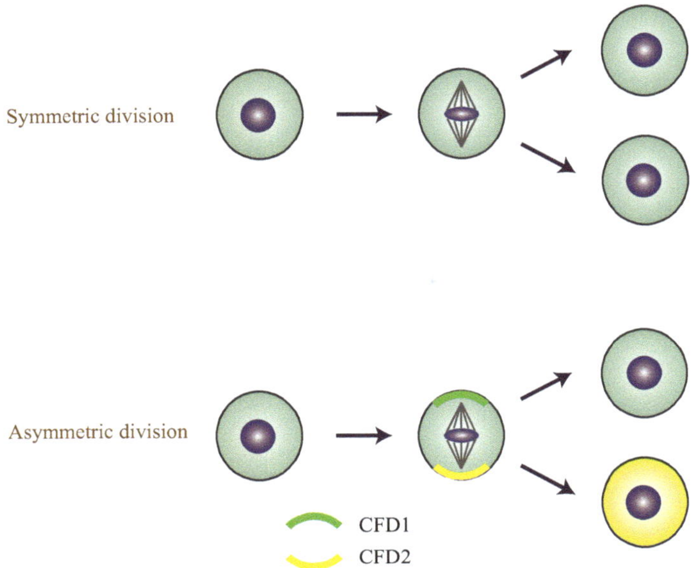

FIGURE 11.1 Schematic illustration of symmetric versus asymmetric division. Asymmetric polarization of cell fate determinants (CFDs) leads to asymmetric cell fates.

tissues in multicellular organisms. We will then outline the molecular mechanism for cell polarization, centrosome localization/spindle orientation and cell fate determination in *Drosophila* NBs, which exemplify the best-studied model system of classic, intrinsic ACD. In the remainder of the chapter, we will outline the cell division mode of CSCs in several cancer types in relation to that of tissue stem and progenitor cells in the corresponding normal tissue.

ASYMMETRIC CELL DIVISION IN UNICELLULAR ORGANISMS

Examination of unicellular organisms such as bacteria and yeast suggests that ACD can contribute to survival of those species by inducing cell fate switches, thereby generating functional diversity and by regulating aging (Fig. 11.2). One example of ACD generating functional diversity is the bacterium *Caulobacter crescentus*. The *Caulobacter* mother cell grows a stalk to attach to a surface and divides asymmetrically to produce a swarmer daughter cell that swims away to start new colony [4]. Another example of ACD-associated cell fate switching occurs in budding yeast, *Saccharomyces cerevisiae* [3]. The mother yeast cell buds off a young daughter cell that maintains her original mating type, while the mother cell switches her mating type. ACD is also used as a survival strategy to increase diversity in response to stress. Under starvation, the bacterium *Bacillus subtilis* turns on ACD for sporulation, producing a small dormant spore and thereby increasing the chance of survival [3]. ACD regulates aging

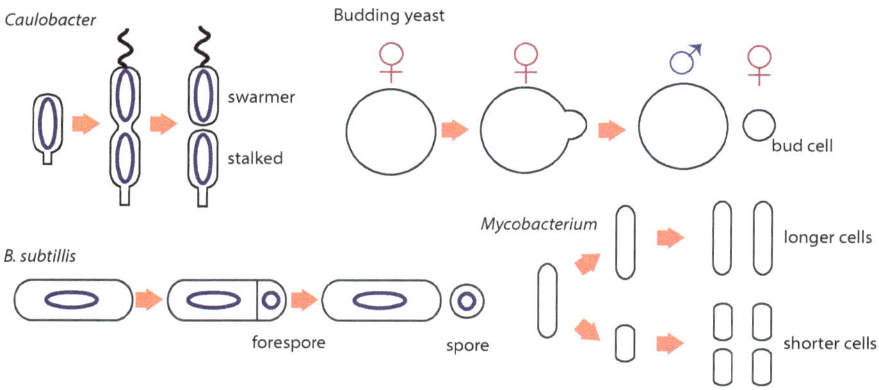

FIGURE 11.2 Schematic illustration of representative asymmetric cell divisions of unicellular organisms.
Caulobacter, *Saccharomyces cerevisiae* (budding yeast), *Bacillus subtilis*, and *Mycobacterium* are included.

by preferentially segregating damaged proteins or DNA strands more prone to mutation into one daughter cell, to allow the other daughter cell to remain "young." Both *Caulobacter* and budding yeast ACDs retain "aging" factors in the mother cell, which has a limited division cycles and upon each division loses its ability to reproduce [5]. Therefore, the need to generate diversity while protecting cells from damage associated with aging can be balanced through ACD. In fact, many species that are considered symmetrically dividing in terms of morphology may still possess intrinsic asymmetry. Studies have shown that *Escherichia coli* preferentially segregates damaged molecules, and the mother cell gradually loses its growth potential and becomes more prone to death [6]. In contrast, in *Mycobacterium* such as *Mycobacterium tuberculosis*, the daughter cell inherits the old pole and divides faster. This mechanism generates cells with different elongation rates and increases heterogeneity in a deterministic way, which may have been an evolutionary strategy for *M. tuberculosis* to adapt to its infection cycle and host environment [7]. From the previous studies, it is conceivable that most species possess a certain degree of asymmetry. Even though symmetric divisions may amplify cell number transiently, ACD allows for diversification and preferential aging, which protects cells from extinction and allows tissues to become optimally adapted to functional demands.

ASYMMETRIC CELL DIVISION IN MULTICELLULAR ORGANISMS

In theory, stem and progenitor cells achieve their dual function in three different ways. They can divide symmetrically to generate two stem cells (symmetric self-renewing divisions) or two differentiating cells (symmetric differentiating divisions). Lastly, stem and progenitor cells can undergo ACD whereby cell fate is determined during mitoses and one daughter cell remains a stem cell and self-renews, whereas the other daughter cell takes on a path of differentiation, often first becoming expanding progenitor cells and eventually becoming terminally differentiated. Through ACD, the number of stem cell remains constant, preventing either depletion of stem cells, which deprives the tissue of regeneration capability, or overproliferation of stem cells that leads to dysplasia (Fig. 11.3). ACD plays an even more prominent role in multicellular organisms where it generates diversity during development and achieves tissue homeostasis in the adult organism [8]. To date, the cell division mode of a few types of tissue stem and progenitor cells has been characterized and of those examples little is known about how cell division mode changes during development and during aging.

Aside from *Drosophila* NBs, evidence of classic intrinsic ACD has been observed in neural stem cells (NSCs) of the developing mouse cortex [9] and in oligodendrocyte progenitor cells (OPCs) in the adult brain [10,11]. For example, in developing telencephalon, asymmetric division prevails among progenitor cells during later stages of neurogenesis in the proliferative ventricular zone [12]. In neuroepithelial cells, asymmetric inheritance of a cellular structure such as the basal projection in embryonic NSCs or the apical plasma membrane in neuroepithelial cells [13] has been associated with daughter cells acquiring distinct cell fate during or shortly after cell division. In the OPC, proteins such as NG2, EGFR, GPR17, and PDGFRα have been shown to distribute unequally between daughter cells [10,11]. In a broader sense, the unequal distribution of CFD may not always be required or may not be the determining factor for the ultimate fate decision. Two daughter cells can acquire distinct fates because they were placed in distinct microenvironments right after birth. This occurs, for example, in basal epithelial cells of the developing epidermis [14,15], which divide either within a plane or reorient their division plane and divide perpendicular to the plane. Such "extrinsic asymmetric" divisions rely on similar polarity and spindle orientation mechanisms as those described for *Drosophila* NBs. The developing epidermis may be an example for a tissue where the role of CFD asymmetry is secondary to

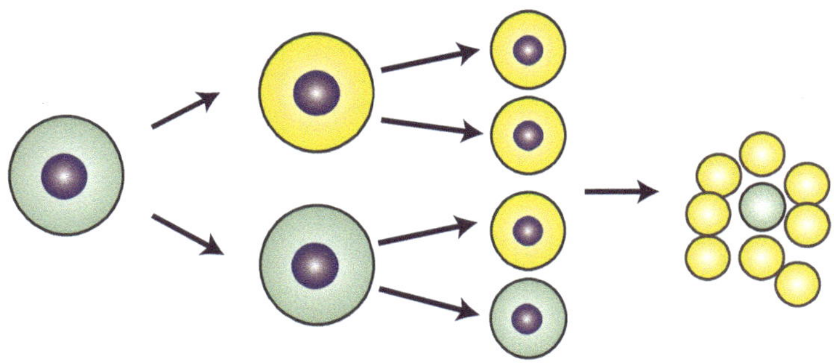

FIGURE 11.3 Asymmetric cell division increases the cell population while keeping the number of self-renewing cells stable.

the positional information for cell fate decisions. Due to the increasing complexity of mammalian tissue, many adult tissue stem cells may depend on a niche formed by supporting cells for their self-renewal [16], in addition to asymmetric inheritance of CFDs. The niche provides yet an alternative, spatial scheme to ACD for keeping stem cell numbers relatively constant. Although it is a common belief that ACD helps avoid accumulation of DNA damage, recent computational models suggest an alternative theory that, in rapidly cycling tissues, symmetric division may also reduce the risk of accumulating heritable DNA damage over the lifetime via extinction of damaged stem cell clones [17]. Like *Drosophila* germline stem cells, the niche may enable some mammalian stem cells to undergo extrinsic ACD, during which the stem cell first undergo symmetric division, and the daughter cell detached from the niche will lose external signaling cues for self-renewal and start to differentiate (Fig. 11.4). ACD has also been described in the mammary epithelium [18] and the hematopoietic compartment. ACD is also the mechanism for erythroblast enucleation, a process to extrude the nucleus [19].

Detailed mechanistic studies of ACD in invertebrate model organisms *Caenorhabditis elegans* and *Drosophila* [1,2] have identified molecular regulators of ACD, the majority of which are conserved in the mammalian genome. The zygote of *C. elegans* divides asymmetrically into a larger anterior AB and a smaller posterior P1 daughter cell (Fig. 11.5). At subsequent stages, ACD is triggered recurrently during embryo development to generate diverse cell lineages. *Drosophila* NB is another well-studied example for ACD. Type 1 NB divides asymmetrically into an NB and a ganglion mother cell (GMC), which divides into neurons (Fig. 11.5). One prominent example of an evolutionary conserved ACD module is the PAR complex. The PAR protein complex controls spindle orientation and division plane formation in worms, flies, and vertebrates.

DROSOPHILA MELANOGASTER NEUROBLASTS—A MODEL SYSTEM FOR CLASSIC, INTRINSIC ASYMMETRIC CELL DIVISIONS

Drosophila NBs undergo intrinsic ACD by establishing apicobasal polarity and aligning the mitotic spindle along the axis of polarity. During metaphase, the

FIGURE 11.4 Schematic illustration of extrinsic asymmetric cell division.
The daughter cell attached to the niche remains as stem cell, while the daughter cell detached from the niche differentiates.

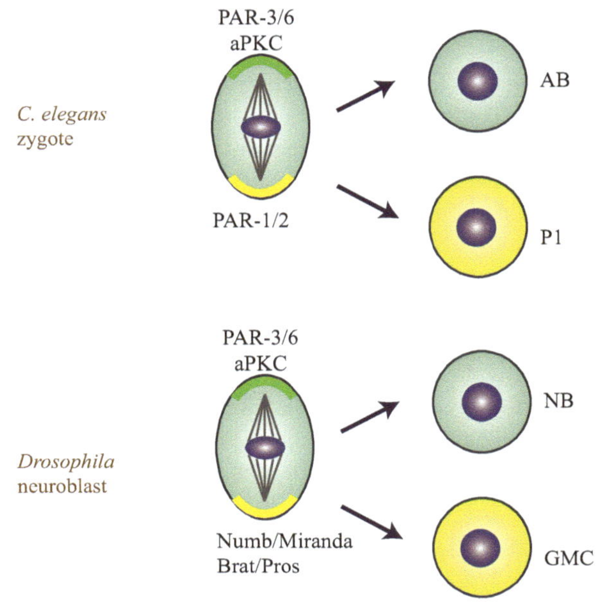

FIGURE 11.5 Classic asymmetric cell division (ACD) models.
Cell fate determinants in *Caenorhabditis elegans zygote* and *Drosophila* neuroblast ACD.

spindle acquires size asymmetry and differentiation-promoting CFD localize to the basal pole, and subsequently segregate to the smaller daughter cell during cytokinesis [20]. This process is repeated for several rounds of ACD, to generate each time another NB and a more restricted progenitor called GMC. At early stages of larval development, and after a period of quiescence, NBs reenter the cell cycle and continue to divide asymmetrically to produce GMCs, either directly (type I NBs) or via intermediate progenitors (type II NBs) [21–23].

Establishing Polarity

The first generation of embryonic NBs delaminates from a polarized neuroectoderm from which it inherits apically positioned Bazooka (Baz or Par3) protein. In prophase, Baz/Par3 serves as a crystallization point for the apical polarity complex, which consists of active Rho GTPase Cdc42 [24], atypical protein kinase C (aPKC), and the aPKC inhibitory subunit Par6 [24,25]. The adaptor protein Inscuteable (Insc) binds to the Par complex [26] and links it to a second apical complex consisting of partner of Insc (Pins) [27], the heterotrimeric G protein-coupled subunits Gαι and Gβγ and the coiled–coiled domain protein mushroom body-defective (Mud). Pins-dependent heterotrimer formation of Gαι/βγ activates G protein signaling in a transmembrane receptor-independent manner [28] and in the absence of nucleotide exchange [29].

In metaphase, a phosphorylation/dephosphorylation cascade catalyzed by Aurora A kinase (AurAK) and protein phosphatase 2A (PP2A) activates aPKC and restricts it to the apical cortex in larval NB (for review see Gómez-López, Lerner, & Petritsch, 2013) [30]. The activation of G protein signaling through Pins occurs cell intrinsically and not only stabilizes apical polarity, but also positions the nascent mitotic spindle along the apicobasal axis and determines its size asymmetry.

The Ras-like small GTPase Rap1 through the Ral guanine nucleotide exchange factor Rgl, Ral, and the PDZ domain protein Canoe regulate apical localization of Mud and to a lesser extent of Baz/Par3 and aPKC. Coimmunoprecipitation experiments find that Ral is in a complex with aPKC and Par6. Consequently, basal CFDs are mislocalized in Ral mutants and the mitotic spindle is misoriented in metaphase. Basal CFDs acquire a normal localization in telophase of Rgl1 mutant NBs due to a process called telophase rescue [31]. Thus, Rap1-Rgl-Ral signaling is an additional cell intrinsic regulator pathway that synergistically cooperates with the apical complexes to coordinate cell polarity and spindle orientation to achieve ACD of NBs.

Mitotic Spindle Orientation and Spindle Asymmetry

Proper apicobasal spindle positioning is crucial for the switch from symmetric to asymmetric division in NBs and it is achieved by a dynamic cross talk between the two apical complexes with centrosomes and astral microtubules. Recent data reveal an important function of the centrosome in providing polarity cues, which precede those provided by the Par complex in second generation NBs [32]. While the apical polarity components are downregulated in interphase, the centrosome is rather stable at the apical cortex and therefore crucial in providing spatial memory for proper spindle axis formation in subsequent, postdelaminated NB divisions [33–35]. After each NB division, the centrosome consists of a newer "daughter" and an older "mother" centriole. The centrioles contain pericentriolar material (PCM), which is a group of proteins with microtubule organizing activity. In interphase, the centrosome migrates toward and anchors at the apical cell cortex and splits shortly after. While the mother centriole sheds PCM, moves away from the apical through the cytoplasm and localizes near the basal pole, the daughter centriole retains PCM and remains at the apical pole [35] (for review see Conduit, 2013) [36]. The mother centriole only expresses PCM and matures after reaching the basal pole. Loss-of-function, time-lapse, and pharmacologic inhibitor studies revealed that Polo kinase plays a crucial role in centrosome asymmetry. Polo localizes to the interphase centrosome but is subsequently shed from the mother centriole in a Bld10/Cep135-dependent manner. Bld10/Cep135 is important for asymmetric centriole inheritance and centrosome positioning [37]. Polo stably localizes to the daughter centriole and phosphorylates the

Drosophila homolog of human Centrobin. Centrobin (Cnb) phosphorylation is essential for the formation of aster microtubules and is therefore a key driver of centrosome anchoring in interphase [38].

Taken together, centrosome asymmetry is important for correct centrosome positioning and the daughter centrosome is "dominant" over the mother centrosome because it anchors to the apical cortex and orients the mitotic spindle in prophase. However, despite failed centrosome asymmetry, Bld10/Cep135 mutant NBs undergo seemingly normal ACD with intact cortical polarity and proper apical–basal spindle orientation [37]. These experiments point to the presence of several backup mechanisms which ensure that ACD occur properly even in the absence of centrosome asymmetry. This begs the question whether centrosome asymmetry is important for ACD and associated cell fate decisions. Noteworthy, mammalian radial glial cells also segregate their centrosomes nonrandomly during ACD and centrosomal components control cell fate [39]. Future functional studies of centrosome regulators will reveal the short- and long-term consequences of defective centrosome asymmetry for ACD and cell fate decisions in neural progenitors but also other mammalian tissue stem and progenitor cells.

In mitosis, the centriole duplicates and the daughter centrosome generates large aster microtubules, which bind and link Mud to Pins/Gαi to form the second apical complex and to align the mitotic spindle along the apicobasal axis [40–42]. The MAGUK disc large (Dlg) binds to kinesin Khc-73 at the microtubule plus ends [43,44] and thereby maintains apical cortical polarity as well as proper spindle orientation. Spindle positioning is linked to global control of mitosis by Polo and AurA kinases [45–47].

Noteworthy, in *Drosophila* NB ACD, the difference in cell fate is accompanied by a difference in cell size, which in turn is accompanied by unequal spindle size. Time-lapse imaging of NBs expressing Dlg-GFP showed that the apical but not the basal cortex extends significantly in anaphase, which leads to a significant size asymmetry [48]. Simultaneously, the apical mitotic spindle elongates the central spindle, aster microtubules and increases the size of the spindle pole on the apical side, which leads to a significant size asymmetry between the apical and basal spindle [49]. As a consequence, the site where the microtubules of the apical and basal central spindle overlap is shifted toward the basal side. This site determines the position of the cleavage furrow and therefore a large (NB) and a small (GMC) daughter are produced during cytokinesis. Mitotic spindle asymmetry is controlled by the WD40 protein lethal giant larvae (Lgl), Dlg, and Scribble [50] as well as Pins and heterotrimeric G proteins [48,51].

Loss-of-function and time-lapse studies recently revealed that the site of the cleavage furrow formation is selected not only by the activity of the "centralspindlin" complex, which includes kinesin Pavarotti, the RACGAP50 Tumbleweed and the RhoGEF Pebble [52], but also by a second, spindle-independent mechanism [53]. The latter involves the localization to the basal cortex of early and late furrow proteins Pavarotti, Anillin, and Myosin II in late anaphase in a Pins-dependent manner. This process can induce a furrow near the basal cortex even in the absence of a mitotic spindle [54].

Most of the molecules that control mitotic spindle and size asymmetry are conserved in the mammalian genome and a few show functional overlap with their invertebrate homologs. For example, the abnormal spindle microcephaly-associated asp protein localizes to the mitotic spindle pole in *Drosophila* NB and maintains spindle orientation. In mammalian embryonic NSCs or radial glial (RG) cells, the asp homolog Aspm is required for mitotic spindle assembly and positioning as well as mitotic progression. Aspm promotes symmetric divisions in RG cells and its loss is associated with increased asymmetric divisions (for review see Ref. [55]). Human ASPM has been identified as a molecular target of oncogenic EGFR signaling in glioma [56]. Its expression is upregulated in human medulloblastoma [57] and Aspm knockout slows growth of a murine medulloblastoma model [58]. Future studies will reveal the detailed mechanism by which Aspm inhibits ACD. Given the tight link of centrosome/spindle regulation and cell cycle control [59], and the high frequency of spindle and centrosomal abnormalities in tumor cells, it will not come as a surprise if regulators of mitotic spindle positioning, spindle and centrosome asymmetry will be identified as key players in cancer.

Cell Fate Decision Control

Cell fate in *Drosophila* NBs is controlled by the unequal segregation of CFDs Numb, Staufen, Prospero, and Brat to the GMC. Numb is an evolutionary conserved protein primarily known for its ability to inhibit the Notch signaling pathway, and the adaptor protein Miranda (Mira). This process depends on an intact actin cytoskeleton [60] and two myosin motors, MyoII and MyoVI, which operate downstream of the apical complex [61].

In interphase aPKC and Par6 form a complex with the WD40 protein lethal giant larvae (Lgl). Loss-of-function studies have demonstrated that Lgl and aPKC have a mutually antagonistic relationship. Lgl restricts aPKC localization at the apical cortex while aPKC phosphorylates Lgl in an AurA-Par6-dependent mechanism, which triggers Lgl exchange for Baz/Par3 in the Par complex and releases Lgl from the apical cortex [62,63]. Assembly of Par3/Par6/aPKC changes aPKC substrate specificity, allowing it to phosphorylate Numb and releasing it from the apical cortex [62,64]. Phosphorylated Numb then localizes to the basal side of the cell [64] assisted by the adaptor protein Partner of Numb (PON) [65], whose activity is largely dependent on its phosphorylation by Polo kinase [45]. Once in the GMC, Numb prevents self-renewal, predominantly by antagonizing Notch signaling [46].

Similar to Numb, Mira polarization has been proposed to occur as a consequence of direct aPKC phosphorylation [62,66]. However, in *aurA* mutant NBs, which display delocalized aPKC, Mira distribution is unaltered [46,47], implying the existence of alternative mechanisms acting upstream or in parallel to aPKC. Additional mechanisms of Mira localization have been identified and include the anaphase-promoting complex/cyclosome (APC) which ubiquitylates Mira at the C-terminus and directs it toward the basal cortex. Ubiquitylation targets proteins for degradation by the proteasome. However, engineered loss of proteasome-mediated degradation is not responsible for Mira mislocalization in NBs. This points to a proteasome-independent mechanism of ubiquitylation, which reportedly could include kinase activation, vesicle trafficking, DNA repair, and gene silencing [67]. Notably, Mira interacts with MyoVI in vesicle-like structures and MyoVI is required for basal Mira localization [61]. It remains to be determined if ubiquitylation is required for MyoVI/Mira interactions. In addition, time-lapse analyses coupled with FRAP have demonstrated that basal localization of Mira is in part achieved by passive diffusion [68]. Thus, basal localization of Mira is regulated by multiple mechanisms that include the apical polarity complex with Lgl, Myosins, and the APC.

Mira binds to and directs the asymmetric localization of distinct CFDs, including the transcriptional regulator Prospero (Pros) [69,70], the double-stranded RNA-binding protein Staufen (Stau) [71,72] and the NHL-domain protein brain tumor (Brat) [73]. Stau binds to *pros* mRNA, through its 3′ untranslated regions (3′ UTRs), and localizes it to the GMC [74,75], which itself does not transcribe the *pros* gene [75]. In the GMC, Prospero activates a neurogenic program [76,77] and represses expression of genes associated with stem cell fate and cell cycle progression [78]. Brat, on the other hand, promotes cell cycle exit, represses NB fate and inhibits cell growth [73,79,80], at least in part, by posttranscriptionally inhibiting the expression of the transcription factor dMyc [73]. In certain APC mutants, Mira is mislocalized to a pericentrosomal region and this prevents basal localization of Staufen, Prospero, and Brat [67].

As a result of unequal segregation of basal CFDs, the apical NB daughter continues to proliferate, while the basal GMC daughter is committed to differentiate (Fig. 11.1). The specification of NB and GMC fates is irreversible and spontaneous dedifferentiation of a GMC into an NB has never been observed in wild-type *Drosophila*.

ASYMMETRIC CELL DIVISION AND CANCER

As recent reports show that cancer risk among tissues are largely correlated with the number of stem cell division, the potential role of stem and progenitor cells in initiating tumor growth is increasingly appreciated. Moreover, CSCs have been demonstrated to maintain tumors and regrow them after treatment

[20]. ACD has been observed in CSCs of many types of cancer, including in glioma, myeloid leukemia, breast cancer, colon cancer, lung, and gastric cancer. In some normal tissue counterparts, stem cells may undergo ACD, suggesting that the ACD mechanisms in CSCs may have been inherited from normal stem cells or may have been reactivated in established tumors. The first evidence that ACD may have a tumor-suppressive effect on stem cells was provided by the discovery that *Drosophila* larval NBs divide more symmetrically and overproliferate in situ when carrying loss-of-function mutants in key regulators of ACD, including *lgl, prospero, numb*, and *brat*. In further support of a growth-inhibitory effect of ACD, brain tissue transplants from larvae mutant for *lgl* and *dlg*, or for *pins* as well as CFDs *numb, mira, pros*, or *brat*, grew expansively in the fly host abdomen and invaded host tissue. Ectopic expression of a membrane-tethered mutant aPKC, which localizes uniformly around the cortex in larval NBs leads to overproliferation (for a detailed review see Ref. [30]). A recent study uncovered that Lgl is required for the maintenance of GMC/progenitor fate in the type I and II NB lineages; *lgl* mutant cells revert back to NBs fate presumably due to increased cortical aPKC kinase activity, which leads to insufficient Numb activity and hence increased Notch activity in GMC/progenitors [81]. Interestingly, a complete deletion of ACD as seen in *dlg-Gb13F* double-mutant NBs cells leads to low levels of nuclear Pros expression, and mutant cells either undergo symmetric differentiation or symmetric self-renewal depending on the developmental stage [82]. Thus, when CFDs fail to distribute asymmetrically, daughter cells do not acquire proper differentiated fate, presumably because differentiation-promoting factors are not present at sufficiently high levels. ACD-defective cells in some instances rather revert back to a stemlike state with increased proliferation properties and aberrant self-renewal, the extent of which depends on the developmental context and probably the microenvironment.

Suppression of ACD is likely to maintain CSC in their stemlike fate and to promote symmetric self-renewing divisions. In progressive colorectal cancer (CRC), CSCs undergo ACD at early stages of tumorigenesis, while ACD tends to be silenced in later stage, more undifferentiated tumors, wherein CSCs almost undergo symmetric division exclusively. Therefore, in certain context, ACD's capability of generating diversity may still be important to tumor cell survival under growth-limiting or stress conditions, eg, by hypoxia, inflammation, or chemotherapy, in a way analogous to *B. subtilis* [82a,82b,158] (Fig. 11.6). One possible interpretation of the dynamic switch from ACD to symmetric divisions is that CSCs evolve to be more autonomous as the tumor becomes more independent of the microenvironment and consequently, the tissue hierarchy of the original tissue is erased.

It is worth noting that ACD in cancer cells can be driven not only by stem cell–related mechanisms but also by mechanisms related to abnormalities in DNA structures or mitotic machinery. Oncogenes in cancer cells can be amplified by

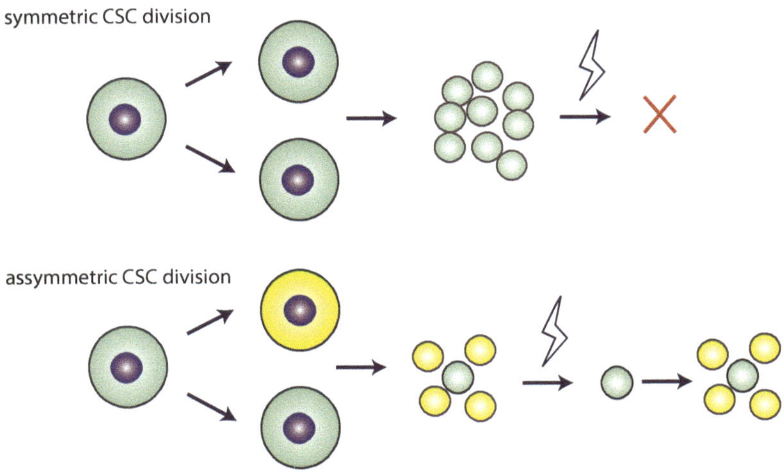

FIGURE 11.6 Asymmetric cell division may help tumor resistance to chemo- or radiation therapy.
CSC, Cancer stem cell.

paired acentric chromatin bodies called double minute chromosomes (DMs), which are circular extrachromosomal DNA that undergo replication but lack centromere. Because mitotic spindle cannot attach to the centromere-less DNA fragments, DMs can be asymmetrically segregated between two daughter cells [83]. Bridge-fusion-breakage cycle is another mechanism of gene amplification. Fusion of telomeres erodes chromosome ends to create a dicentromeric chromosome, which might be broken apart by the mechanical force of mitotic spindle pull, causing the two daughter cells to inherent asymmetrical sets of genome [84]. Missegregation of chromosomes due to abnormality in centromere number/function or other mechanisms such as mitotic checkpoints can also cause asymmetric segregation of genome between the daughter cells and generate aneuploidy cells [85,86].

Later, we will discuss ACD in CSCs of multiple tumor types in relation to their normal tissue stem and progenitor cell counterparts.

Colorectal Cancer

CRC is a leading cause of cancer death in the United States and worldwide [87]. Colon cancer stem cells (CCSCs), also called colon cancer–initiating cells (CCICs), isolated from primary CRCs are important for CRC tumorigenesis and initiating metastasis [88–90]. In contrast to commonly used CRC cell lines, CCSC serially maintain and propagate large tumors with the histological, pathological, and molecular markers of the primary CRC from which they were derived. CCSC purified from tumors can self-renew and serially form new xenografts while the non-CCSC containing cell fractions cannot. CCSC are highly resistant to current CRC chemotherapy and cytotoxic chemotherapy increases the percentage of CCSC in CRCs [91,92].

CCSCs form xenograft tumors that largely maintain the heterogeneous cellular morphology and differentiation hierarchy of their original human CRC tumors upon xenografting in immunodeficient mice, unlike more homogeneous xenograft tumors formed by common CRC lines that almost exclusively undergo symmetric division [93]. This raised the possibility that ACD may contribute to the ability of CCSC to create cellular diversity. Furthermore, CCSC regulate cell fate decision of self-renewal versus differentiation by Notch and Wnt signaling pathways both of which are downstream targets of polarizing CFDs such as Numb. Indeed, asymmetric Numb distribution was observed in 24% of CCSC divisions, while the majority still undergoes symmetric division [94]. Knockdown of ID1 and ID3, which belong to a family of homologous helix-loop-helix transcriptional regulatory factors, suppressed asymmetric Numb segregation. A more detailed study using various CCSC markers and time-lapse microscopy with a fluorescent Notch reporter showed that imbalance of Notch signaling indeed gives rise to daughter cell fate asymmetry—the Notch^{high} daughter cell continues expressing CCSC markers for self-renewal, while the Notch^{low} daughter cell switched to differentiation markers [95]. Numb is one of the CFDs that regulates CCSC asymmetry as Numb segregation is associated with Notch signaling and cell fate asymmetry.

Interestingly, microRNAs and long non-coding RNAs (lncRNAs) can also act as CFD to regulate CCIC asymmetry. MicroRNAs are small noncoding RNAs that silence gene expression by binding to the 3′ UTRs of target mRNAs, inhibiting their translation or marking them for degradation [96]. Imbalance of the microRNA miR-34a, which targets Notch for suppression, leads to daughter cell fate asymmetry [95]. Silencing of miR-34a inhibits ACD and promotes symmetric self-renewal, increasing CCIC proliferation in both sphere cultures in vitro and xenograft tumors in vivo. By regulating asymmetry, miR-34a creates Notch bimodality in the cell population to generate distinct cell lineages. miR-34a tends to be silenced in tumors from late-stage (III and IV) CRC patients, which promotes symmetric division of these more autonomous cells and gives rise to more homogeneous and undifferentiated tumors (Fig. 11.7). Apparently miR-34a is not the only microRNA that can regulate ACD. In Snail-expressing CCICs, miR-146 targets Numb to stabilize the Wnt signaling effector β-catenin and promote symmetric division. Wnt in turn upregulates Snail, and miR-146, forming a self-sustaining feedback loop. Inhibition of the Snail-miR-146-β-catenin axis reduces symmetric division, tumorigenicity and cetuximab resistance [97]. Furthermore, a lncRNA, Lnc34a, has been shown to initiate miR-34a asymmetry via epigenetic regulation, revealing another layer of regulatory complexity by non-coding RNA [97a].

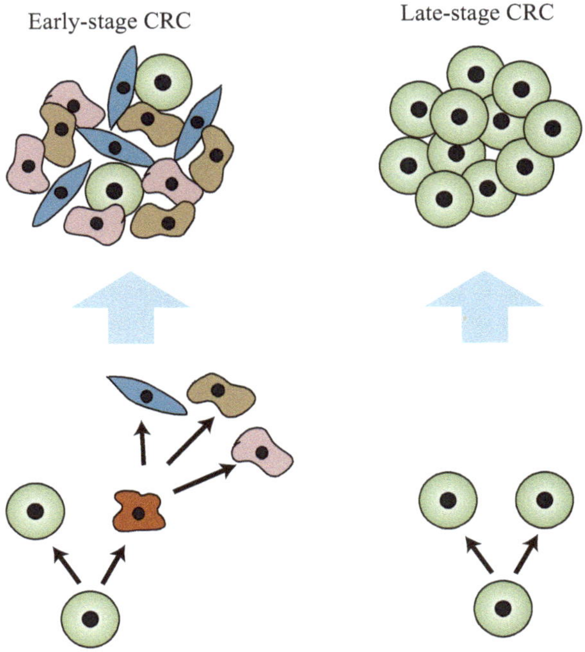

FIGURE 11.7 Asymmetric cell division (ACD) silencing in late-stage colorectal cancer (CRC). Tumor progression decreases ACD and increases symmetric self-renewal, giving rise to more undifferentiated tumors.

Studies so far suggest that CCSC cell fate symmetry is largely regulated through Notch and Wnt signaling, both of which are essential for CCSC self-renewal. Multiple upstream CFDs target Notch and Wnt signaling to regulate ACD, and their relationship and contexts still await elucidation (Fig. 11.8).

The origin of the ACD mechanism in CCSC is still somewhat a mystery. One would have naturally assumed that such mechanisms originate from normal stem cells. The CCSC markers CD44, CD133, ALDH1, Lgr5, EphB2, and Bmi1 have all been shown to mark the stem cell compartment at the base of the intestinal crypt, with various specificity. Both Notch and Wnt signaling are essential for the self-renewal of intestinal and colon stem cells, consistent with their roles in CCSCs. However, whether intestinal and colon stem cells undergo ACD is still up to debate. Earlier reports showed that slow-cycling stem cells around the +4 position undergo ACD exclusively to protect their number and genomic integrity [98–100]. However, more recent work identified Lgr5$^+$ crypt base columnar (CBC) cells as actively cycling stem cells, and they perform symmetric division while competing with each other in a neutral drift process [101,102] (Fig. 11.9). It is worth noting that, contrary to a common misconception, neutral drift proves the existence of symmetric division

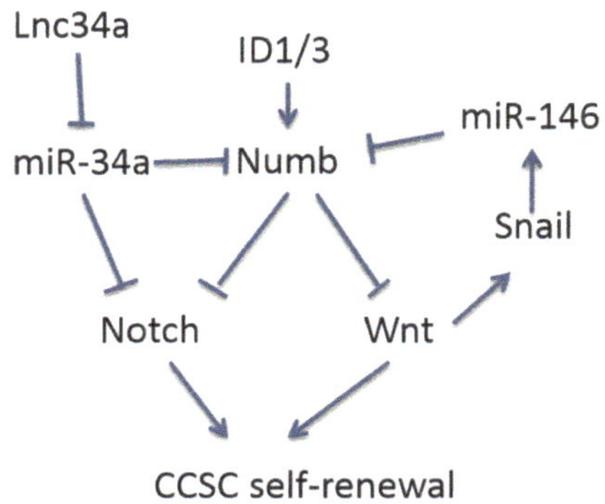

FIGURE 11.8 Regulators involved in colon cancer stem cell (CCSC) asymmetric cell division.

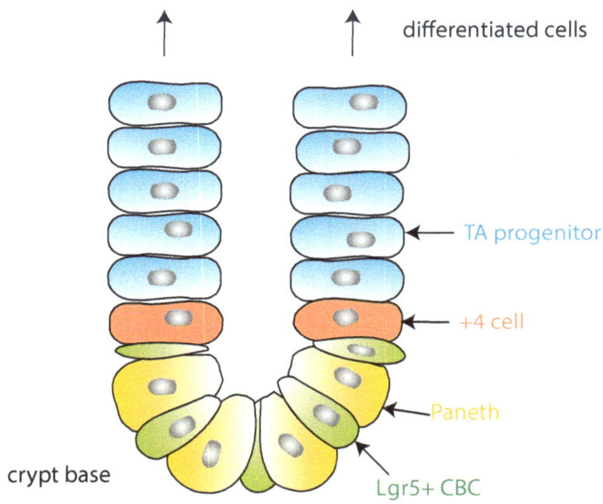

FIGURE 11.9 Intestinal crypt.

Lgr5+ crypt base columnars (CBCs) (stem cells) and Paneth cells form a stem cell niche at the base of the crypt. Putative quiescent stem or progenitor cells locate around the +4 location. CBCs divide and migrate upward while differentiating into trans-amplifying (TA) progenitor cells and undergoing ~8 rounds of division in the TA compartment.

but does not exclude ACD, which may happen at a lower rate or activated at a particular developmental stage [103]. Emerging evidence suggests that stem cells may turn on ACD in stress conditions or during tissue repair after damage [158]. Furthermore, fast-cycling Lgr5+ CBC cells and slow-cycling +4 Bmi1+ cells are capable of direct inter-convertion via ACD, especially under inflammatory conditions, which has also been observed in CCSCs [82a,82b]. Therefore, the capability of CCSCs to undergo ACD may have originated from a stress response mechanism in normal colon stem cells.

Brain Tumors

The most frequent type of brain tumor is glioma, which can be further divided in three classes, ependymoma, oligodendroglioma, and astrocytoma. Oligodendroglioma are frequently diagnosed as low-grade (WHO grade II) tumors but despite their slow growth and chemosensitivity they inevitably progress to a higher grade (WHO grade III). Astrocytoma on the other hand are intrinsically therapy resistant and arise either from a grade II tumor which rapidly progress to grade III and IV tumors, or are first diagnosed as grade IV tumors, which are referred to as glioblastoma (GBM). Despite aggressive standard treatment, consisting of surgery, radiation and temozolomide therapy, GBM patients have median survival of less than 14 months [104]. The genetic and phenotypic heterogeneity poses a major challenge for the effective treatment of GBM. Intratumoral heterogeneity manifests as the presence of phenotypically distinct tumor-propagating cells, which have been identified in studies examining the tumorigenicity of xenotransplanted cells sorted from GBM surgical specimen. Tumor-propagating cells in GBM similar to those in other tumors display stem cell features, such as expression of stem cell markers, aberrant self-renewal and impaired differentiation and are therefore commonly referred to as CSC. CSCs are common in grade III astrocytoma and GBM [105–107] and present albeit at lower frequency in anaplastic, high-grade oligodendrogliomas [108]. Human NSCs, activated adult murine NSCs [109] and GBM CSC express similar markers, including Nestin and CD133/Prominin-1 (Prom1). $CD133^+$ CSC are of particular interest because they were shown to be resistant to radiation [110,111] which has led to the speculation that they contribute to therapy resistance and recurrence of GBM. In support of the role of CSCs in therapy resistance and recurrence, a $Nestin^+$ neural progenitor population in a murine GBM model survives treatment with the alkylating agent temozolomide, while the tumor bulk shrinks. Moreover, the $Nestin^+$ population significantly contributes to tumor regrowth [112]. In contrast, NG2 positivity that is associated with OPCs has been shown to identify tumor-propagating cells that respond well to chemotherapy [10,113].

Molecular analyses have recently found that the cysteine-rich polypeptide progranulin upregulates CD133 expression and other stemness genes in addition to DNA repair genes [114]. In addition to self-renewal and multipotentiality, $CD133^+$ CSCs have the ability to undergo ACD. Single-cell analyses of dividing $CD133^+$ GBM cells have revealed that CD133 protein itself forms polarized cortical crescents and distributes unequally in the progeny [115]. The rates of ACD of $CD133^+$ cells are rather low but still exceed that of $NG2^+$ OPC-like cells in corresponding GBM cultures. Flow cytometric analyses show that activity of the mammalian Polo kinase homolog Plk1 is increased in $CD133^+$ GBM CSC compared with their $CD133^-$ counterparts. Plk1 localizes to the centrosome as expected but also colocalizes with CD133 to polarized, cortical crescents.

Pharmacologic inhibitor studies showed that there is a mutual dependence of polarized $CD133^{+}$ localization, ACD, Plk1 activity and mitotic progression. These data support the existence of a mitotic polarity/ACD checkpoint in this therapy-resistant CSC population [116]. *Drosophila* Polo kinase regulates asymmetric localization of CFDs and polo mutant NBs develop hyperplasia, which points to a conserved polarity-establishing mechanism in CSCs. What might be the effect of $CD133^{+}$ asymmetric localization on cell fate has yet to be elucidated, and further studies are needed to determine if spindle orientation plays a role in CSCs ability to maintain their tumorigenic potential.

CD133/Prom1 is a pentaspan transmembrane glycoprotein that is specifically sorted to microvilli and plasma membrane protrusions and localizes to membrane microdomains that are released into the extracellular space [117]. CD133/Prom1 also segregates asymmetrically in hematopoietic stem cells (HSCs) and progenitor cells, suggesting a broader and common role in somatic stem and progenitor cells [118]. So far, described phenotypes for genetic depletion of *Prom1* in mice are restricted to retinal degeneration [119]. Therefore, the significance of the asymmetric distribution of CD133/Prom1 in stem cells and CSCs remains unclear. CD133 activates the PI3K/Akt signaling pathway, which is frequently upregulated in glioma, by physically interacting with the PI3K regulatory subunit p85 and thereby promotes CSC self-renewal and tumor-initiating potential [120]. In addition to Plk1, the CFD Numb has been reported to asymmetrically segregate to $CD133^{high}$-expressing GBM cells [121]. Future studies will determine if CD133 itself is a direct substrate for Plk1 kinase activity.

The ability to enter relative quiescence is thought to contribute to therapy resistance of CSC but is poorly understood on a molecular level. *Drosophila* NBs, for example, divide asymmetrically during embryonic development and eventually become quiescent at late embryonic stages and reenter the cell cycle at different larval stages in a neuroanatomical and cell type–specific manner. Temporal restricted low-level expression of nuclear Prospero in late embryonic NBs is required and sufficient to induce quiescence while maintaining NB fate. Noteworthy, high levels of nuclear Prospero on the other hand induce cell cycle arrest, differentiation, and GMC daughter cell fate [122] which suggests that ACD-related CFDs functional output depends on expression levels and the cellular and developmental context. It is interesting that Prospero plays this role because it is known to be expressed mainly during ACD, which provides a mechanism by which asymmetrically dividing NBs can enter quiescence rapidly. Mammalian adult NSCs in the subventricular zone are largely quiescent [123], and CSCs are partially slow-dividing [124] but whether Prospero homolog Prox1 induces quiescence in these cells remains to be determined. A recent study shows that $CD133^{+}$ GBM cells proliferate less and divide asymmetric at higher rates, while also showing decreased sensitivity to the antiproliferative effects of pharmacological

MAPK pathway inhibition, compared to their $CD133^-$ counterparts. These data suggest that asymmetric dividing CSC evade BRAF/MAPK-directed therapy due to their lower proliferation rate [116].

It is attractive to speculate that pharmacological inhibition of CSC-specific regulators of ACD will increase symmetric, self-renewing divisions of $CD133^+$ CSC and might thereby sensitize these cells to chemotherapy. We conclude that investigations into the mechanism of ACD may aid in the identification of approaches that support a more durable antitumor effect.

In contrast to GBM, the majority of oligodendroglioma have tumor propagating cells (TPCs) with progenitor-like rather than stem cell features [10,113]. Examination of $NG2^+$ TPCs isolated from human surgical glioma specimens from patients prior to treatment revealed that these cells divide predominantly symmetrically [10]. It is also feasible that an $NG2^+$ TPC has limited self-renewal potential and may eventually differentiate, which may explain the slower tumor growth of oligodendroglioma. Oligodendroglioma TPCs presumably expand the tumor by nonhierarchical, symmetric divisions; and symmetric divisions correlate with chemosensitivity. It is attractive to speculate that $NG2^+$ TPC show increased chemosensitivity because they undergo symmetrical, self-renewing divisions and thereby continuously proliferate and do not enter quiescence. Understanding cell division mode and why some populations of cells acquire and/or retain ACD, while others expand symmetrically, may provide explanations as to why distinct cancer cell types are unequally affected by treatment.

The cellular origin of glioma is best studied in genetically engineered mouse models, which are also suitable to address the major open question whether CSCs have a common predecessor, such as an NSC or OPC, or whether they equally arise from progenitor or more differentiating cells. Such a direct lineage relationship of the CSCs and TPCs and their nonneoplastic counterpart is suggested by their biological similarities. Studies with genetically engineered mouse models of glioma have suggested that both NSCs and OPCs can act as cellular origin of astrocytoma [125–128]. These data suggest that CSCs are equally likely to directly descend from transformed NSC and arise de novo from differentiating cells. The latter possibility is further supported by the additional observation that high-grade astrocytoma can arise from mature functional cells [129]. Mammalian embryonic and adult-activated NSC as well as OPC undergo ACD. ACD in mammalian embryonic NSC has been well characterized and it is regulated in part by conserved mechanisms involving mammalian Par3 and Numb, the mammalian Brat homolog Trim32, Inscuteable, the mammalian Pins homolog LGN, the mammalian Mud homolog NuMA as well as Plk1. Moreover, OPC undergo ACD but the mechanism for cell polarity and CFD distribution is much less understood than that of NSC (for review see Ref. [30]). It is interesting to note that astrocytes, for example, divide mostly symmetrically [130], which indirectly suggests that there might be instances when CSCs acquire cell polarity and asymmetric division modes de novo.

Hematologic Malignancies

The potential role of ACD in hematologic malignancies, especially leukemia, has been the subject of investigation [131]. Expression of the translocated fusion oncogene NUP98-HOXA9, commonly associated with acute myelogenous leukemia (AML) and the blast crisis phase of chronic myelogenous leukemia (CML), promotes symmetric self-renewal division and reduces ACD [132].

Proper regulation of ACD relies on the proper alignment of spindle orientation. Loss of Lis1 causes random spindle orientation, which significantly increases the frequency of ACD of leukemia stem cells [133]. The increase in differentiation is associated with better survival for CML and AML. The cytoskeletal protein Llgl1 regulates HSC fitness and it is also downregulated in AML, and its loss is associated with poor survival [134].

Notch signaling also seems to play a role in regulating division symmetry of leukemia-initiating cells. As CML advances to blast crisis, the CFD Numb, which promotes Notch degradation, is downregulated, while Musashi, the RNA-binding protein that inhibits Numb, is upregulated [135]. Musashi is overexpressed in human AML cell lines and correlates with poor survival in AML patients [136]. Musashi also synergize with the oncoprotein BCR-ABL1 to initiate aggressive leukemia. Silencing of Musashi or ectopic expression of Numb deters leukemia progression in mouse models [135].

The mechanisms regulating ACD of leukemia stem cells likely originates from HSCs or progenitor cells. Hematopoietic precursors undergo both symmetric and asymmetric division, which has been shown in mice carrying a Notch reporter that indicates differentiation [132]. Numb can localize in one of the daughter cells during certain HSC divisions and causes asymmetric cell fate. An in vivo RNAi knockdown screen identified polarity proteins Pard6a and Prkcz as well as CFD Musashi as factors that are important for HSC self-renewal. Their knockdown led to more differentiation and marked decrease in HSC repopulation [137]. Genetic deletion of Musashi and Lis1 reduced HSC self-renewal, while genetic deletion of the spindle orientation regulator Llgl1 enhanced HSC repopulation and engraftment capacity [133–135].

Breast Cancer

The highly proliferative breast cancer cell line, MCF7, can divide asymmetrically to provide slow-dividing, "G0-like" progeny [138]. These "G0-like" cells have low levels of reactive oxygen species, MKI67, MCM2, H3K9me2, and AKT and high levels of Notch signaling. It turned out that these breast cancer cells suppress AKT signaling asymmetrically during mitosis to produce the slowly cycling daughter cell, and the ability to undergo ACD seems to be a latent but general property which all cancer cells rather than a discrete subpopulation can acquire. Further studies revealed that a β1-integrin/FAK/mTORC2/AKT1 signaling axis triggers fast-cycling cancer cells to divide asymmetrically to produce a fast-cycling and a slow-cycling AKT^{low}, G0-like daughter cells [139]. AKT^{low} cancer cells in human breast tumors are highly resistant to prolonged combination chemotherapy of adriamycin, cyclophosphamide, and paclitaxel, which further enrich this slow-proliferating population. Therefore, ACD can contribute to cancer cell survival despite slowing down their growth.

Another study of 12 human breast cancer cells found basal-like and claudin-low subtypes are more likely to asymmetrically segregate template DNA during cell division [140]. Breast cancer cell lines with the capability of asymmetric DNA template segregation have higher invasion and migration abilities. $CD44^+$/$CD24^-$ stemlike cells contain a subpopulation of cells that asymmetrically segregates template DNA, although symmetric self-renewal is enriched in the stemlike population.

A study using transgenic ErbB2 breast cancer mouse model demonstrates that CSCs have higher rates of symmetric division than normal mammary stem cells [141]. Furthermore, loss of p53 in normal mammary stem cells promotes symmetric division and inhibits ACD. As a consequence, p53 knockout mammary stem cells self-renew and proliferate at rates similar to that of CSCs. Pharmacological reactivation of p53 function is associated with restoration of ACD and reduction of tumor grown in the p53-null mammary glands. Another study also confirmed that p53 loss leads to expansion of both luminal and basal stem/progenitor cells [142]. Therefore, loss of p53 promotes symmetric division and contributes to tumor growth.

Mechanisms of ACD in breast CSCs likely originated from normal mammary stem cells. Murine mammary glands contain self-renewing epithelial stem cells [18,143]. The slow-cycling stem cells could be labeled using [^{3}H]-thymidine (^{3}HTdR) and retain their label [144]. These cells asymmetrically segregate template DNA during division, retaining original ^{3}HTdR-labeled template DNA strands in the daughter stem cells and passing newly synthesized 5BrdU-labeled DNA to their progeny. Another study showed that mammary stem cells which were identified as slow-cycling population based on their ability to retain lipophilic fluorescent dye PKH26, predominantly undergo ACD [141]. On the other hand, a population of fast-cycling, K5-expressing basal stem or progenitor cells colocalize with label-retaining stem cells and undergo ACD that generate luminal cells [145].

Expression patterns of Numb, Musashi1, and Notch1 suggest that the Notch signaling pathway may regulate ACD of mammary stem cells [146]. Aurora A kinase regulates mitotic spindle orientation in a Notch-dependent manner to determine basal versus luminal cell fate in the mammary epithelium [147]. Furthermore, Slit2/Robo1 signaling acts through Inscuteable (mInsc) and Snail to regulate the choice between asymmetric and symmetric division of basal and luminal stem cells [18]. The Par3-like polarity protein (Par3L) is also essential for mammary stem cell maintenance, though whether it regulates ACD is still unclear [148]. Aging seems to affect bipotent basal stem cells and cause tissue atrophy [149]. Deletion of metalloproteinase inhibitors Timp1 and Timp3 activates Notch to disrupt mitotic spindle orientation and ACD, which enriches the stem/progenitor pool to prevent tissue regression. Interestingly, stemlike cells in cultures of immortalized human mammary epithelial cells asymmetrically segregate aged mitochondria. As a result the stemlike daughter cells receive fewer old mitochondria [150]. Disruption of this asymmetric mitochondria apportioning mechanism leads to loss of stem cell characteristics in the progeny.

Taken together, the regulation of asymmetric division of breast cancer stemlike cells and normal mammary stem cells is regulated by similar mechanisms. Loss of p53 by mutation or constitutive activation of Notch increases symmetric division at the expense of ACD, which contributes to tumor stem cell expansion. On the other hand, ACD of mammary CSCs generates a subpopulation of more quiescent and chemoresistant cells that survive chemotherapy [138,139]. Therefore, ACD seems to be a double-edged sword that balances tumor growth with adaptation to changes in the tumor microenvironment.

Other Cancers

ACD has also been reported in other cancer types. $CD133^+$ CSCs in lung cancer cell lines and primary cultures undergo ACD and asymmetrically segregate their template DNA into the $CD133^+$, CSC daughter cell [151]. The frequency of ACD is modulated by extrinsic factors, including cell–cell contact, cell density, hypoxia, and serum deprivation. Highly metastatic $CXCR4^+$ gastric CSCs are also capable of ACD [152]. Treatment with TGF-β induces ACD and promotes differentiation, which enhances anticancer effects of docetaxel treatment. This synergistic effect even worked on dormant gastric CSCs. Human thyroid cancers contain a population of label-retaining CSCs that divide template DNA both symmetrically and asymmetrically [153]. ACD of these CSCs seems to increase diversity and enhances resistance to chemotherapy. ACD has also been reported in the prostate. In developing prostatic epithelia, while luminal cells exclusively undergo symmetrical divisions, basal cells divide both symmetrically and asymmetrically [154]. The different cell fates contribute to a hierarchy of prostate epithelial cell lineages. In the prostate epithelial hierarchy from both normal and tumor tissues, ACD retains monoallelic regulation of TMPRSS2 transcription, which is relaxed in differentiated epithelial cells [155]. ACD of a side population of CSC-like cells in ovarian cancer cell lines has also been reported [156]. The change from asymmetric to symmetric division also appears to be a mechanism associated with oral carcinogenesis [157].

CONCLUDING REMARKS AND FUTURE DIRECTIONS

ACD has been reported in a variety of cancers, often associate with CSCs. ACD has been mainly identified through CFDs, polarity proteins, CSC markers, activation of fluorescence signaling or lineage-tracing reporters, mitotic activity, young versus aged mitochondria, or template versus de novo synthesized DNA strands. CSC ACD seems to originate from parallel mechanisms in normal stem cells, although the possibility that CSCs can turn on ACD mechanisms de novo, especially under environmental cues or stress (eg, chemotherapy or hypoxia), cannot be excluded. In general, ACD promotes more differentiation, hierarchy, and diversity in the population, and abolishment of ACD in favor of symmetric self-renewal contributes to proliferation and undifferentiated tumors overall. However, ACD may also enhance tumor survival and resistance to therapy in certain context by generating more quiescent cells.

Many open questions about ACD and cancer remain to be addressed. First and foremost, it is unclear how ACD originates. Do cancer cells activate ACD de novo, or do they inherit ACD as direct descendants from mutated tissue stem or progenitor cells? For example, CCSCs have been consistently reported to undergo ACD, but normal proliferating $LGR5^{+}$ intestinal and colon stem cells are thought to largely undergo symmetric division in the niche, which suggests that CCSCs activate ACD de novo. However, recent work suggests that inflammation and oncogenic mutation can trigger normal stem cells to switch to ACD, hence supporting the latter hypothesis [158]. Second, the frequency with which ACD occur in vivo has yet to be determined for the majority of tissue types. Moreover, ACD relevance to tumor physiology is still much up for debate. Frequency of ACD can vary greatly between cancer types and stages, but ACD is often less frequent than symmetric cell division, raising the possibility that ACD may not always play a tumor-suppressor role and can potentially enhance drug resistance. It is worth noting that ACD does not have to be limited to stem- or progenitor-like cancer cells. As cancer cells do not terminally differentiate or stop dividing as normal tissue cells do, it is very likely that differentiated cancer cells can also undergo ACD, and we simply lack the markers to distinguish the daughter cells. Third, mechanisms of ACD still need to be better elucidated. Are ACD mechanisms shared among different cancer types or at least among individual tumors of the same cancer type, or are they idiosyncratic? The former is likely to be true if cancer ACD originates from normal stem cells, while the latter is more likely to be true if cancer cells activate ACD mechanisms de novo. This question of general applicability has important clinical implications with regard to the feasibility of developing therapeutics to target ACD mechanisms in cancer.

List of Acronyms and Abbreviations

ACD Asymmetric cell division
CCSC Colon cancer stem cell
CFD Cell fate determinant
CRC Colorectal cancer
CSC Cancer stem cell

References

[1] Neumuller RA, Knoblich JA. Dividing cellular asymmetry: asymmetric cell division and its implications for stem cells and cancer. Genes Dev 2009;23(23):2675–99.

[2] Knoblich JA. Mechanisms of asymmetric stem cell division. Cell 2008;132(4):583–97.

[3] Li R. The art of choreographing asymmetric cell division. Dev Cell 2013;25(5):439–50.

[4] Childers WS, Xu Q, Mann TH, Mathews II, Blair JA, Deacon AM, et al. Cell fate regulation governed by a repurposed bacterial histidine kinase. PLoS Biol 2014;12(10):e1001979.

[5] Mogk A, Bukau B. Mitochondria tether protein trash to rejuvenate cellular environments. Cell 2014;159(3):471–2.

[6] Lindner AB, Madden R, Demarez A, Stewart EJ, Taddei F. Asymmetric segregation of protein aggregates is associated with cellular aging and rejuvenation. Proc Natl Acad Sci USA 2008;105(8):3076–81.

[7] Aldridge BB, Fernandez-Suarez M, Heller D, Ambravaneswaran V, Irimia D, Toner M, et al. Asymmetry and aging of mycobacterial cells lead to variable growth and antibiotic susceptibility. Science 2012;335(6064):100–4.

[8] Inaba M, Yamashita YM. Asymmetric stem cell division: precision for robustness. Cell Stem Cell 2012;11(4):461–9.

[9] Noctor SC, Martinez-Cerdeno V, Ivic L, Kriegstein AR. Cortical neurons arise in symmetric and asymmetric division zones and migrate through specific phases. Nat Neurosci 2004;7(2):136–44.

[10] Sugiarto S, Persson AI, Munoz EG, Waldhuber M, Lamagna C, Andor N, et al. Asymmetry-defective oligodendrocyte progenitors are glioma precursors. Cancer Cell 2011;20(3):328–40.

[11] Boda E, Di Maria S, Rosa P, Taylor V, Abbracchio MP, Buffo A. Early phenotypic asymmetry of sister oligodendrocyte progenitor cells after mitosis and its modulation by aging and extrinsic factors. Glia 2015;63(2):271–86.

[12] Haydar TF, Ang Jr E, Rakic P. Mitotic spindle rotation and mode of cell division in the developing telencephalon. Proc Natl Acad Sci USA 2003;100(5):2890–5.

[13] Kosodo Y, Roper K, Haubensak W, Marzesco AM, Corbeil D, Huttner WB. Asymmetric distribution of the apical plasma membrane during neurogenic divisions of mammalian neuroepithelial cells. EMBO J 2004;23(11):2314–24.

[14] Lechler T, Fuchs E. Asymmetric cell divisions promote stratification and differentiation of mammalian skin. Nature 2005;437(7056):275–80.

[15] Kulukian A, Fuchs E. Spindle orientation and epidermal morphogenesis. Philos Trans R Soc Lond B Biol Sci 2013;368(1629):20130016.

[16] Morrison SJ, Spradling AC. Stem cells and niches: mechanisms that promote stem cell maintenance throughout life. Cell 2008;132(4):598–611.

[17] McHale PT, Lander AD. The protective role of symmetric stem cell division on the accumulation of heritable damage. PLoS Comput Biol 2014;10(8):e1003802.

[18] Ballard MS, Zhu A, Iwai N, Stensrud M, Mapps A, Postiglione MP, et al. Mammary stem cell self-renewal is regulated by Slit2/Robo1 signaling through SNAI1 and mINSC. Cell Rep 2015;13(2):290–301.

[19] Ji P, Jayapal SR, Lodish HF. Enucleation of cultured mouse fetal erythroblasts requires Rac GTPases and mDia2. Nat Cell Biol 2008;10(3):314–21.

[20] Knoblich JA. Asymmetric cell division: recent developments and their implications for tumour biology. Nat Rev Mol Cell Biol 2010;11(12):849–60.

[21] Bello BC, Izergina N, Caussinus E, Reichert H. Amplification of neural stem cell proliferation by intermediate progenitor cells in *Drosophila* brain development. Neural Dev 2008;3:5.

[22] Boone JQ, Doe CQ. Identification of *Drosophila* type II neuroblast lineages containing transit amplifying ganglion mother cells. Dev Neurobiol 2008;68(9):1185–95.

[23] Bowman SK, Rolland V, Betschinger J, Kinsey KA, Emery G, Knoblich JA. The tumor suppressors Brat and Numb regulate transit-amplifying neuroblast lineages in *Drosophila*. Dev Cell 2008;14(4):535–46.

[24] Atwood SX, Chabu C, Penkert RR, Doe CQ, Prehoda KE. Cdc42 acts downstream of Bazooka to regulate neuroblast polarity through Par-6 aPKC. J Cell Sci 2007;120(Pt 18):3200–6.

[25] Peterson FC, Penkert RR, Volkman BF, Prehoda KE. Cdc42 regulates the Par-6 PDZ domain through an allosteric CRIB-PDZ transition. Mol Cell 2004;13(5):665–76.

[26] Wodarz A, Ramrath A, Grimm A, Knust E. Drosophila atypical protein kinase C associates with Bazooka and controls polarity of epithelia and neuroblasts. J Cell Biol 2000;150(6):1361–74.

[27] Schaefer M, Shevchenko A, Shevchenko A, Knoblich JA. A protein complex containing Inscuteable and the Gα-binding protein Pins orients asymmetric cell divisions in *Drosophila*. Curr Biol 2000;10(7):353–62.

[28] Willard FS, Kimple RJ, Siderovski DP. Return of the GDI: the GoLoco motif in cell division. Annu Rev Biochem 2004;73:925–51.

[29] Yu F, Wang H, Qian H, Kaushik R, Bownes M, Yang X, et al. Locomotion defects, together with Pins, regulates heterotrimeric G-protein signaling during *Drosophila* neuroblast asymmetric divisions. Genes Dev 2005;19(11):1341–53.

[30] Gomez-Lopez S, Lerner RG, Petritsch C. Asymmetric cell division of stem and progenitor cells during homeostasis and cancer. Cell Mol Life Sci 2014;71(4):575–97.

[31] Carmena A, Makarova A, Speicher S. The Rap1-Rgl-Ral signaling network regulates neuroblast cortical polarity and spindle orientation. J Cell Biol 2011;195(4):553–62.

[32] Januschke J, Gonzalez C. The interphase microtubule aster is a determinant of asymmetric division orientation in *Drosophila* neuroblasts. J Cell Biol 2010;188(5):693–706.

[33] Rebollo E, Sampaio P, Januschke J, Llamazares S, Varmark H, Gonzalez C. Functionally unequal centrosomes drive spindle orientation in asymmetrically dividing *Drosophila* neural stem cells. Dev Cell 2007;12(3):467–74.

[34] Siller KH, Doe CQ. Spindle orientation during asymmetric cell division. Nat Cell Biol 2009;11(4):365–74.

[35] Januschke J, Llamazares S, Reina J, Gonzalez C. *Drosophila* neuroblasts retain the daughter centrosome. Nat Commun 2011;2:243.

[36] Conduit PT. The dominant force of Centrobin in centrosome asymmetry. Nat Cell Biol 2013;15(3):235–7.

[37] Singh P, Ramdas Nair A, Cabernard C. The centriolar protein Bld10/Cep135 is required to establish centrosome asymmetry in *Drosophila* neuroblasts. Curr Biol 2014;24(13):1548–55.

[38] Januschke J, Reina J, Llamazares S, Bertran T, Rossi F, Roig J, et al. Centrobin controls mother-daughter centriole asymmetry in *Drosophila* neuroblasts. Nat Cell Biol 2013;15(3):241–8.

[39] Wang X, Tsai JW, Imai JH, Lian WN, Vallee RB, Shi SH. Asymmetric centrosome inheritance maintains neural progenitors in the neocortex. Nature 2009;461(7266):947–55.

[40] Izumi Y, Ohta N, Hisata K, Raabe T, Matsuzaki F. *Drosophila* Pins-binding protein Mud regulates spindle-polarity coupling and centrosome organization. Nat Cell Biol 2006;8(6):586–93.

[41] Siller KH, Cabernard C, Doe CQ. The NuMA-related Mud protein binds Pins and regulates spindle orientation in *Drosophila* neuroblasts. Nat Cell Biol 2006;8(6):594–600.

[42] Bowman SK, Neumuller RA, Novatchkova M, Du Q, Knoblich JA. The *Drosophila* NuMA Homolog Mud regulates spindle orientation in asymmetric cell division. Dev Cell 2006;10(6):731–42.

[43] Siegrist SE, Doe CQ. Microtubule-induced Pins/Galphai cortical polarity in *Drosophila* neuroblasts. Cell 2005;123(7):1323–35.

[44] Siegrist SE, Doe CQ. Microtubule-induced cortical cell polarity. Genes Dev 2007;21(5):483–96.

[45] Wang H, Ouyang Y, Somers WG, Chia W, Lu B. Polo inhibits progenitor self-renewal and regulates Numb asymmetry by phosphorylating Pon. Nature 2007;449(7158):96–100.

[46] Wang H, Somers GW, Bashirullah A, Heberlein U, Yu F, Chia W. Aurora-A acts as a tumor suppressor and regulates self-renewal of *Drosophila* neuroblasts. Genes Dev 2006;20(24): 3453–63.

[47] Lee CY, Andersen RO, Cabernard C, Manning L, Tran KD, Lanskey MJ, et al. *Drosophila* Aurora-A kinase inhibits neuroblast self-renewal by regulating aPKC/Numb cortical polarity and spindle orientation. Genes Dev 2006;20(24):3464–74.

[48] Connell M, Cabernard C, Ricketson D, Doe CQ, Prehoda KE. Asymmetric cortical extension shifts cleavage furrow position in *Drosophila* neuroblasts. Mol Biol Cell 2011;22(22): 4220–6.

[49] Kaltschmidt JA, Davidson CM, Brown NH, Brand AH. Rotation and asymmetry of the mitotic spindle direct asymmetric cell division in the developing central nervous system. Nat Cell Biol 2000;2(1):7–12.

[50] Albertson R, Doe CQ. Dlg, Scrib and Lgl regulate neuroblast cell size and mitotic spindle asymmetry. Nat Cell Biol 2003;5(2):166–70.

[51] Fuse N, Hisata K, Katzen AL, Matsuzaki F. Heterotrimeric G proteins regulate daughter cell size asymmetry in *Drosophila* neuroblast divisions. Curr Biol 2003;13(11):947–54.

[52] D'Avino PP. How to scaffold the contractile ring for a safe cytokinesis–lessons from Anillin-related proteins. J Cell Sci 2009;122(Pt 8):1071–9.

[53] Roth M, Roubinet C, Ifflander N, Ferrand A, Cabernard C. Asymmetrically dividing *Drosophila* neuroblasts utilize two spatially and temporally independent cytokinesis pathways. Nat Commun 2015;6:6551.

[54] Cabernard C, Prehoda KE, Doe CQ. A spindle-independent cleavage furrow positioning pathway. Nature 2010;467(7311):91–4.

[55] Chavali PL, Putz M, Gergely F. Small organelle, big responsibility: the role of centrosomes in development and disease. Philos Trans R Soc Lond B Biol Sci 2014;369(1650).

[56] Horvath S, Zhang B, Carlson M, Lu KV, Zhu S, Felciano RM, et al. Analysis of oncogenic signaling networks in glioblastoma identifies ASPM as a molecular target. Proc Natl Acad Sci USA 2006;103(46):17402–7.

[57] Vulcani-Freitas TM, Saba-Silva N, Cappellano A, Cavalheiro S, Marie SK, Oba-Shinjo SM, et al. ASPM gene expression in medulloblastoma. Childs Nerv Syst 2011;27(1):71–4.

[58] Williams SE, Garcia I, Crowther AJ, Li S, Stewart A, Liu H, et al. Aspm sustains postnatal cerebellar neurogenesis and medulloblastoma growth. Development 2015;142(22):3921–32.

[59] Fu J, Hagan IM, Glover DM. The centrosome and its duplication cycle. Cold Spring Harb Perspect Biol 2015;7(2):a015800.

[60] Kraut R, Chia W, Jan LY, Jan YN, Knoblich JA. Role of inscuteable in orienting asymmetric cell divisions in *Drosophila*. Nature 1996;383(6595):50–5.

[61] Petritsch C, Tavosanis G, Turck CW, Jan LY, Jan YN. The *Drosophila* myosin VI Jaguar is required for basal protein targeting and correct spindle orientation in mitotic neuroblasts. Dev Cell 2003;4(2):273–81.

[62] Wirtz-Peitz F, Nishimura T, Knoblich JA. Linking cell cycle to asymmetric division: Aurora-A phosphorylates the Par complex to regulate Numb localization. Cell 2008;135(1):161–73.

[63] Betschinger J, Mechtler K, Knoblich JA. The Par complex directs asymmetric cell division by phosphorylating the cytoskeletal protein Lgl. Nature 2003;422(6929):326–30.

[64] Smith CA, Lau KM, Rahmani Z, Dho SE, Brothers G, She YM, et al. aPKC-mediated phosphorylation regulates asymmetric membrane localization of the cell fate determinant Numb. EMBO J 2007;26(2):468–80.

[65] Lu B, Rothenberg M, Jan LY, Jan YN. Partner of Numb colocalizes with Numb during mitosis and directs Numb asymmetric localization in *Drosophila* neural and muscle progenitors. Cell 1998;95(2):225–35.

[66] Atwood SX, Prehoda KE. aPKC phosphorylates Miranda to polarize fate determinants during neuroblast asymmetric cell division. Curr Biol 2009;19(9):723–9.

[67] Slack C, Overton PM, Tuxworth RI, Chia W. Asymmetric localisation of Miranda and its cargo proteins during neuroblast division requires the anaphase-promoting complex/cyclosome. Devepment 2007;134(21):3781–7.

[68] Erben V, Waldhuber M, Langer D, Fetka I, Jansen RP, Petritsch C. Asymmetric localization of the adaptor protein Miranda in neuroblasts is achieved by diffusion and sequential interaction of Myosin II and VI. J Cell Sci 2008;121(Pt 9):1403–14.

[69] Shen CP, Jan LY, Jan YN. Miranda is required for the asymmetric localization of Prospero during mitosis in *Drosophila*. Cell 1997;90(3):449–58.

[70] Ikeshima-Kataoka H, Skeath JB, Nabeshima Y, Doe CQ, Matsuzaki F. Miranda directs Prospero to a daughter cell during *Drosophila* asymmetric divisions. Nature 1997;390(6660):625–9.

[71] Shen CP, Knoblich JA, Chan YM, Jiang MM, Jan LY, Jan YN. Miranda as a multidomain adapter linking apically localized Inscuteable and basally localized Staufen and Prospero during asymmetric cell division in *Drosophila*. Genes Dev 1998;12(12):1837–46.

[72] Matsuzaki F, Ohshiro T, Ikeshima-Kataoka H, Izumi H. Miranda localizes Staufen and Prospero asymmetrically in mitotic neuroblasts and epithelial cells in early *Drosophila* embryogenesis. Development 1998;125(20):4089–98.

[73] Betschinger J, Mechtler K, Knoblich JA. Asymmetric segregation of the tumor suppressor brat regulates self-renewal in *Drosophila* neural stem cells. Cell 2006;124(6):1241–53.

[74] Li P, Yang X, Wasser M, Cai Y, Chia W. Inscuteable and Staufen mediate asymmetric localization and segregation of prospero RNA during *Drosophila* neuroblast cell divisions. Cell 1997;90(3):437–47.

[75] Broadus J, Fuerstenberg S, Doe CQ. Staufen-dependent localization of prospero mRNA contributes to neuroblast daughter-cell fate. Nature 1998;391(6669):792–5.

[76] Chu-Lagraff Q, Wright DM, McNeil LK, Doe CQ. The prospero gene encodes a divergent homeodomain protein that controls neuronal identity in *Drosophila*. Dev Suppl 1991;(Suppl. 2):79–85.

[77] Doe CQ, Chu-LaGraff Q, Wright DM, Scott MP. The prospero gene specifies cell fates in the *Drosophila* central nervous system. Cell 1991;65(3):451–64.

[78] Choksi SP, Southall TD, Bossing T, Edoff K, de Wit E, Fischer BE, et al. Prospero acts as a binary switch between self-renewal and differentiation in *Drosophila* neural stem cells. Dev Cell 2006;11(6):775–89.

[79] Lee CY, Wilkinson BD, Siegrist SE, Wharton RP, Doe CQ. Brat is a Miranda cargo protein that promotes neuronal differentiation and inhibits neuroblast self-renewal. Dev Cell 2006;10(4):441–9.

[80] Bello B, Reichert H, Hirth F. The brain tumor gene negatively regulates neural progenitor cell proliferation in the larval central brain of *Drosophila*. Development 2006; 133(14):2639–48.

[81] Haenfler JM, Kuang C, Lee CY. Cortical aPKC kinase activity distinguishes neural stem cells from progenitor cells by ensuring asymmetric segregation of Numb. Dev Biol 2012;365(1):219–28.

[82] Kitajima A, Fuse N, Isshiki T, Matsuzaki F. Progenitor properties of symmetrically dividing *Drosophila* neuroblasts during embryonic and larval development. Dev Biol 2010;347(1):9–23.

[82a] Srinivasan T, Walters J, Bu P, Than EB, Tung KL, Chen KY, et al. NOTCH signaling regulates asymmetric Cell Fate of Fast- and Slow-Cycling Colon Cancer Initiating Cells. Cancer Research 2016. http://dx.doi.org/10.1158/0008-5472.CAN-15-3198.

[82b] Srinivasan T, Bich ET, Bu P, Tung K-L, Chen K-Y, Augenlicht L, et al. Notch signalling regulates asymmetric division and inter-conversion between lgr5 and bmi1 expressing intestinal stem cells. Scientific Reports 2016;6(26069). http://dx.doi.org/10.1038/srep26069.

[83] Kanda T, Sullivan KF, Wahl GM. Histone-GFP fusion protein enables sensitive analysis of chromosome dynamics in living mammalian cells. Curr Biol 1998;8(7):377–85.

[84] Smith KA, Stark MB, Gorman PA, Stark GR. Fusions near telomeres occur very early in the amplification of CAD genes in Syrian hamster cells. Proc Natl Acad Sci USA 1992; 89(12):5427–31.

[85] Santaguida S, Amon A. Short- and long-term effects of chromosome mis-segregation and aneuploidy. Nat Rev Mol Cell Biol 2015;16(8):473–85.

[86] Weaver BA, Silk AD, Montagna C, Verdier-Pinard P, Cleveland DW. Aneuploidy acts both oncogenically and as a tumor suppressor. Cancer Cell 2007;11(1):25–36.

[87] Jemal A. Cancer statistics. CA A Cancer J Clin 2007;57:43–66.

[88] Ricci-Vitiani L, Lombardi DG, Pilozzi E, Biffoni M, Todaro M, Peschle C, et al. Identification and expansion of human colon-cancer-initiating cells. Nature 2007;445(7123):111–5.

[89] Dalerba P, Dylla SJ, Park IK, Liu R, Wang X, Cho RW, et al. Phenotypic characterization of human colorectal cancer stem cells. Proc Natl Acad Sci USA 2007;104(24):10158–63.

[90] O'Brien CA, Pollett A, Gallinger S, Dick JE. A human colon cancer cell capable of initiating tumour growth in immunodeficient mice. Nature 2007;445(7123):106–10.

[91] Hoey T, Yen W, Axelrod F, Basi J, Dylla S, Clarke M, et al. DLL4 blockade inhibits tumor growth and reduces tumor-initiating cell frequency. Cell Stem Cell 2009;5:168–77.

[92] Dylla SJ, Beviglia L, Park IK, Chartier C, Raval J, Ngan L, et al. Colorectal cancer stem cells are enriched in xenogeneic tumors following chemotherapy. PLoS One 2008;3(6):e2428.

[93] Bu P, Chen KY, Chen JH, Wang L, Walters J, Shin YJ, et al. A microRNA miR-34a-regulated bimodal switch targets Notch in colon cancer stem cells. Cell Stem Cell 2013; 12(5):602–15.

[94] O'Brien CA, Kreso A, Ryan P, Hermans KG, Gibson L, Wang Y, et al. ID1 and ID3 regulate the self-renewal capacity of human colon cancer-initiating cells through p21. Cancer Cell 2012;21(6):777–92.

[95] Bu P, Chen KY, Lipkin SM, Shen X. Asymmetric division: a marker for cancer stem cells in early stage tumors? Oncotarget 2013;4(7):948–9.

[96] Pauli A, Rinn JL, Schier AF. Non-coding RNAs as regulators of embryogenesis. Nat Rev Genet 2011;12(2):136–49.

[97] Hwang WL, Jiang JK, Yang SH, Huang TS, Lan HY, Teng HW, et al. MicroRNA-146a directs the symmetric division of Snail-dominant colorectal cancer stem cells. Nat Cell Biol 2014;16(3):268–80.

[97a] Wang L, Bu P, Ai Y, Srinivasan T, Chen HJ, Xiang K, et al. A long non-coding RNA targets microRNA miR-34a to regulate colon cancer stem cell asymmetric division. eLife 2016. http://dx.doi.org/10.7554/eLife.14620.

[98] Potten CS, Owen G, Booth D. Intestinal stem cells protect their genome by selective segregation of template DNA strands. J Cell Sci 2002;115(Pt 11):2381–8.

[99] Quyn AJ, Appleton PL, Carey FA, Steele RJ, Barker N, Clevers H, et al. Spindle orientation bias in gut epithelial stem cell compartments is lost in precancerous tissue. Cell Stem Cell 2010;6(2):175–81.

[100] Goulas S, Conder R, Knoblich JA. The par complex and integrins direct asymmetric cell division in adult intestinal stem cells. Cell Stem Cell 2012;11(4):529–40.

[101] Snippert HJ, van der Flier LG, Sato T, van Es JH, van den Born M, Kroon-Veenboer C, et al. Intestinal crypt homeostasis results from neutral competition between symmetrically dividing Lgr5 stem cells. Cell 2010;143(1):134–44.

[102] Lopez-Garcia C, Klein AM, Simons BD, Winton DJ. Intestinal stem cell replacement follows a pattern of neutral drift. Science 2010;330(6005):822–5.

[103] Itzkovitz S, Blat IC, Jacks T, Clevers H, van Oudenaarden A. Optimality in the development of intestinal crypts. Cell 2012;148(3):608–19.

[104] Stupp R, Hegi ME, Mason WP, van den Bent MJ, Taphoorn MJ, Janzer RC, et al. Effects of radiotherapy with concomitant and adjuvant temozolomide versus radiotherapy alone on survival in glioblastoma in a randomised phase III study: 5-year analysis of the EORTC-NCIC trial. Lancet Oncol 2009;10(5):459–66.

[105] Singh SK, Clarke ID, Terasaki M, Bonn VE, Hawkins C, Squire J, et al. Identification of a cancer stem cell in human brain tumors. Cancer Res 2003;63(18):5821–8.

[106] Singh SK, Hawkins C, Clarke ID, Squire JA, Bayani J, Hide T, et al. Identification of human brain tumour initiating cells. Nature 2004;432(7015):396–401.

[107] Hemmati HD, Nakano I, Lazareff JA, Masterman-Smith M, Geschwind DH, Bronner-Fraser M, et al. Cancerous stem cells can arise from pediatric brain tumors. Proc Natl Acad Sci USA 2003;100(25):15178–83.

[108] Kelly JJ, Blough MD, Stechishin OD, Chan JA, Beauchamp D, Perizzolo M, et al. Oligodendroglioma cell lines containing t(1;19)(q10;p10). Neuro Oncol 2010;12(7):745–55.

[109] Luo Y, Coskun V, Liang A, Yu J, Cheng L, Ge W, et al. Single-cell transcriptome analyses reveal signals to activate dormant neural stem cells. Cell 2015;161(5):1175–86.

[110] Bao S, Wu Q, McLendon RE, Hao Y, Shi Q, Hjelmeland AB, et al. Glioma stem cells promote radioresistance by preferential activation of the DNA damage response. Nature 2006;444(7120):756–60.

[111] Tamura K, Aoyagi M, Wakimoto H, Ando N, Nariai T, Yamamoto M, et al. Accumulation of CD133-positive glioma cells after high-dose irradiation by Gamma Knife surgery plus external beam radiation. J Neurosurg 2010;113(2):310–8.

[112] Chen J, Li Y, Yu TS, McKay RM, Burns DK, Kernie SG, et al. A restricted cell population propagates glioblastoma growth after chemotherapy. Nature 2012;488(7412):522–6.

[113] Persson AI, Petritsch C, Swartling FJ, Itsara M, Sim FJ, Auvergne R, et al. Non-stem cell origin for oligodendroglioma. Cancer Cell 2010;18(6):669–82.

[114] Bandey I, Chiou SH, Huang AP, Tsai JC, Tu PH. Progranulin promotes Temozolomide resistance of glioblastoma by orchestrating DNA repair and tumor stemness. Oncogene 2015;34(14):1853–64.

[115] Lathia JD, Hitomi M, Gallagher J, Gadani SP, Adkins J, Vasanji A, et al. Distribution of CD133 reveals glioma stem cells self-renew through symmetric and asymmetric cell divisions. Cell Death Dis 2011;2:e200.

[116] Lerner RG, Grossauer S, Kadkhodaei B, Meyers I, Sidorov M, Koeck K, et al. Targeting a Plk1-controlled polarity checkpoint in therapy-resistant glioblastoma-propagating cells. Cancer Res 2015;75(24):5355–66.

[117] Sykes AM, Huttner WB. Prominin-1 (CD133) and the cell biology of neural progenitors and their progeny. Adv Exp Med Biol 2013;777:89–98.

[118] Bauer N, Fonseca AV, Florek M, Freund D, Jaszai J, Bornhauser M, et al. New insights into the cell biology of hematopoietic progenitors by studying prominin-1 (CD133). Cells Tissues Organs 2008;188(1–2):127–38.

[119] Zacchigna S, Oh H, Wilsch-Brauninger M, Missol-Kolka E, Jaszai J, Jansen S, et al. Loss of the cholesterol-binding protein prominin-1/CD133 causes disk dysmorphogenesis and photoreceptor degeneration. J Neurosci 2009;29(7):2297–308.

[120] Wei Y, Jiang Y, Zou F, Liu Y, Wang S, Xu N, et al. Activation of PI3K/Akt pathway by CD133-p85 interaction promotes tumorigenic capacity of glioma stem cells. Proc Natl Acad Sci USA 2013;110(17):6829–34.

[121] Jiang X, Xing H, Kim TM, Jung Y, Huang W, Yang HW, et al. Numb regulates glioma stem cell fate and growth by altering epidermal growth factor receptor and Skp1-Cullin-F-box ubiquitin ligase activity. Stem Cells 2012;30(7):1313–26.

[122] Lai SL, Doe CQ. Transient nuclear Prospero induces neural progenitor quiescence. Elife 2014;3.

[123] Ponti G, Obernier K, Guinto C, Jose L, Bonfanti L, Alvarez-Buylla A. Cell cycle and lineage progression of neural progenitors in the ventricular-subventricular zones of adult mice. Proc Natl Acad Sci USA 2013;110(11):E1045–54.

[124] Deleyrolle LP, Harding A, Cato K, Siebzehnrubl FA, Rahman M, Azari H, et al. Evidence for label-retaining tumour-initiating cells in human glioblastoma. Brain 2011;134(Pt 5):1331–43.

[125] Marumoto T, Tashiro A, Friedmann-Morvinski D, Scadeng M, Soda Y, Gage FH, et al. Development of a novel mouse glioma model using lentiviral vectors. Nat Med 2009;15(1):110–6.

[126] Alcantara Llaguno S, Chen J, Kwon CH, Jackson EL, Li Y, Burns DK, et al. Malignant astrocytomas originate from neural stem/progenitor cells in a somatic tumor suppressor mouse model. Cancer Cell 2009;15(1):45–56.

[127] Alcantara Llaguno SR, Wang Z, Sun D, Chen J, Xu J, Kim E, et al. Adult lineage-restricted CNS progenitors specify distinct glioblastoma subtypes. Cancer Cell 2015;28(4):429–40.

[128] Liu C, Sage JC, Miller MR, Verhaak RG, Hippenmeyer S, Vogel H, et al. Mosaic analysis with double markers reveals tumor cell of origin in glioma. Cell 2011;146(2):209–21.

[129] Friedmann-Morvinski D, Bushong EA, Ke E, Soda Y, Marumoto T, Singer O, et al. Dedifferentiation of neurons and astrocytes by oncogenes can induce gliomas in mice. Science 2012;338(6110):1080–4.

[130] Ge WP, Miyawaki A, Gage FH, Jan YN, Jan LY. Local generation of glia is a major astrocyte source in postnatal cortex. Nature 2012;484(7394):376–80.

[131] Bajaj J, Zimdahl B, Reya T. Fearful symmetry: subversion of asymmetric division in Cancer development and progression. Cancer Res 2015;75(5):792–7.

[132] Wu M, Kwon HY, Rattis F, Blum J, Zhao C, Ashkenazi R, et al. Imaging hematopoietic precursor division in real time. Cell Stem Cell 2007;1(5):541–54.

[133] Zimdahl B, Ito T, Blevins A, Bajaj J, Konuma T, Weeks J, et al. Lis1 regulates asymmetric division in hematopoietic stem cells and in leukemia. Nat Genet 2014;46(3):245–52.

[134] Heidel FH, Bullinger L, Arreba-Tutusaus P, Wang Z, Gaebel J, Hirt C, et al. The cell fate determinant Llgl1 influences HSC fitness and prognosis in AML. J Exp Med 2013;210(1):15–22.

[135] Ito T, Kwon HY, Zimdahl B, Congdon KL, Blum J, Lento WE, et al. Regulation of myeloid leukaemia by the cell-fate determinant Musashi. Nature 2010;466(7307):765–8.

[136] Kharas MG, Lengner CJ, Al-Shahrour F, Bullinger L, Ball B, Zaidi S, et al. Musashi-2 regulates normal hematopoiesis and promotes aggressive myeloid leukemia. Nat Med 2010;16(8):903–8.

[137] Hope KJ, Cellot S, Ting SB, MacRae T, Mayotte N, Iscove NN, et al. An RNAi screen identifies Msi2 and Prox1 as having opposite roles in the regulation of hematopoietic stem cell activity. Cell Stem Cell 2010;7(1):101–13.

[138] Dey-Guha I, Wolfer A, Yeh AC, Albeck JG, Darp R, Leon E, et al. Asymmetric cancer cell division regulated by AKT. Proc Natl Acad Sci USA 2011;108(31).

[139] Dey-Guha I, Alves CP, Yeh AC, Salony, Sole X, Darp R, et al. A mechanism for asymmetric cell division resulting in proliferative asynchronicity. Mol Cancer Res 2015;13(2):223–30.

[140] Liu W, Jeganathan G, Amiri S, Morgan KM, Ryan BM, Pine SR. Asymmetric segregation of template DNA strands in basal-like human breast cancer cell lines. Mol Cancer 2013;12(1):139.

[141] Cicalese A, Bonizzi G, Pasi CE, Faretta M, Ronzoni S, Giulini B, et al. The tumor suppressor p53 regulates polarity of self-renewing divisions in mammary stem cells. Cell 2009;138(6):1083–95.

[142] Chiche A, Moumen M, Petit V, Jonkers J, Medina D, Deugnier MA, et al. Somatic loss of p53 leads to stem/progenitor cell amplification in both mammary epithelial compartments, basal and luminal. Stem Cells 2013;31(9):1857–67.

[143] Visvader JE, Stingl J. Mammary stem cells and the differentiation hierarchy: current status and perspectives. Genes Dev 2014;28(11):1143–58.

[144] Smith GH. Label-retaining epithelial cells in mouse mammary gland divide asymmetrically and retain their template DNA strands. Development 2005;132(4):681–7.

[145] Rios AC, Fu NY, Lindeman GJ, Visvader JE. In situ identification of bipotent stem cells in the mammary gland. Nature 2014;506(7488):322–7.

[146] Clarke RB, Anderson E, Howell A, Potten CS. Regulation of human breast epithelial stem cells. Cell Prolif 2003;36(Suppl. 1):45–58.

[147] Regan JL, Sourisseau T, Soady K, Kendrick H, McCarthy A, Tang C, et al. Aurora A kinase regulates mammary epithelial cell fate by determining mitotic spindle orientation in a Notch-dependent manner. Cell Rep 2013;4(1):110–23.

[148] Huo Y, Macara IG. The Par3-like polarity protein Par3L is essential for mammary stem cell maintenance. Nat Cell Biol 2014;16(6):529–37.

[149] Jackson HW, Waterhouse P, Sinha A, Kislinger T, Berman HK, Khokha R. Expansion of stem cells counteracts age-related mammary regression in compound Timp1/Timp3 null mice. Nat Cell Biol 2015;17(3):217–27.

[150] Katajisto P, Dohla J, Chaffer CL, Pentinmikko N, Marjanovic N, Iqbal S, et al. Stem cells. Asymmetric apportioning of aged mitochondria between daughter cells is required for stemness. Science 2015;348(6232):340–3.

[151] Pine SR, Ryan BM, Varticovski L, Robles AI, Harris CC. Microenvironmental modulation of asymmetric cell division in human lung cancer cells. Proc Natl Acad Sci USA 2010;107(5):2195–200.

[152] Fujita T, Chiwaki F, Takahashi RU, Aoyagi K, Yanagihara K, Nishimura T, et al. Identification and characterization of CXCR4-positive gastric Cancer stem cells. PLoS One 2015;10(6):e0130808.

[153] Ma R, Minsky N, Morshed SA, Davies TF. Stemness in human thyroid cancers and derived cell lines: the role of asymmetrically dividing cancer stem cells resistant to chemotherapy. J Clin Endocrinol Metab 2014;99(3):E400–9.

[154] Wang J, Zhu HH, Chu M, Liu Y, Zhang C, Liu G, et al. Symmetrical and asymmetrical division analysis provides evidence for a hierarchy of prostate epithelial cell lineages. Nat Commun 2014;5:4758.

[155] Polson ES, Lewis JL, Celik H, Mann VM, Stower MJ, Simms MS, et al. Monoallelic expression of TMPRSS2/ERG in prostate cancer stem cells. Nat Commun 2013;4:1623.

[156] Boesch M, Zeimet AG, Reimer D, Schmidt S, Gastl G, Parson W, et al. The side population of ovarian cancer cells defines a heterogeneous compartment exhibiting stem cell characteristics. Oncotarget 2014;5(16):7027–39.

[157] Gonzalez-Moles MA, Plaza-Campillo J, Ruiz-Avila I, Herrera P, Bravo M, Gil-Montoya JA. Asymmetrical proliferative pattern loss during malignant transformation of the oral mucosa. J Oral Pathol Med 2014;43(7):507–13.

[158] Bu P, Wang L, Chen KY, Srinivasan T, Murthy PKL, Varanko AK, et al. A miR-34a-Numb feed-forward loop triggered by inflammation regulates asymmetric stem cell division in intestine and colon cancer. Cell Stem Cell 2016;18(2):189–202. http://dx.doi.org/10.1016/j.stem.2016.01.006.

CHAPTER 12

Metastasis and Metastatic Cells: A Historical Perspective and Current Analysis

K. Rycaj[1], D.G. Tang[1,2,3]
[1]University of Texas MD Anderson Cancer Center, Smithville, TX, United States; [2]Tongji University School of Medicine, Shanghai, China; [3]University of Texas MD Anderson Cancer Center, Houston, TX, United States

CONTENTS

INTRODUCTION

Cancer is a complicated disease that coevolves with the host. Primary tumors are responsible for only about 10% of deaths from cancer, whereas the remaining 90% of patients are overcome with metastasis. The growth of metastasis research in the past decade has yielded information about the tumor-intrinsic and tumor-extrinsic mechanisms that dictate metastatic behaviors, the molecular and cellular basis underlying the distinct courses of metastatic progression in different cancers and what renders metastatic cancer refractory to available therapies. What remains is the integration of such new knowledge into an improved, metastasis-oriented oncological drug development strategy to stop the development of metastatic disease at every stage of progression [1]. In this chapter, we provide a historical perspective of the metastatic cascade and key cellular determinants (Fig. 12.1), mainly cancer stem cells (CSCs), circulating tumor cells (CTCs), and disseminated tumor cells (DTCs). We also discuss the epithelial–mesenchymal transition (EMT) program and its malignant phenotypes associated with metastasis in several different cancers.

THE METASTATIC CASCADE

Metastasis causes more than 90% of cancer-related deaths but strikingly still remains poorly understood [2]. Metastases are formed by cancer cells that have left the primary tumor mass and have traveled, via blood and lymphatic vessels, to other locations in the body (Fig. 12.1). This process of cancer metastasis consists of many stochastic events that are dependent on both intrinsic properties of the tumor cells and the responses of the microenvironment (Fig. 12.1). Indeed, many of the complex cell-biological events are orchestrated

Cancer Stem Cells. http://dx.doi.org/10.1016/B978-0-12-803892-5.00012-7

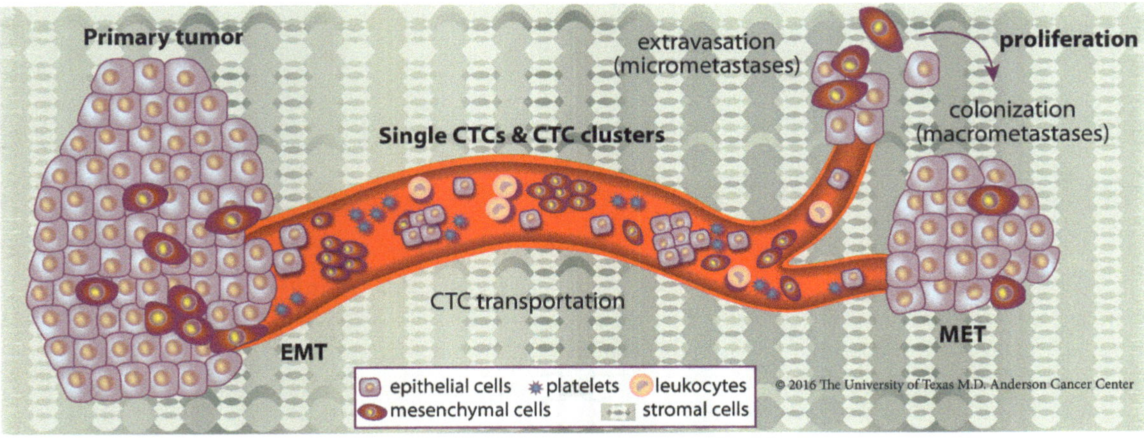

FIGURE 12.1 **The metastatic cascade via hematogenous dissemination.**
Depicted on the left is a primary epithelial tumor in which most tumor cells present as epithelial although there are also some mesenchymal cells, derived presumably through the process of EMT. Both epithelial and mesenchymal tumor cells in the primary tumor can be shed into the bloodstream, either as individual cells or tumor cell clusters. Tumor cells in the blood may also form clusters (aggregates) with blood cells such as platelets and lymphocytes. Both tumor cell–tumor cell and tumor cell–platelet clusters have been shown to be more metastatic than single tumor cells. In the end organ microvasculature, tumor cells extravasate into the parenchyma, colonize, and grow into a secondary macrometastasis under the right conditions. A process of mesenchymal–epithelial transition (MET) is thought to occur before the formation of metastatic colonies. The tumor microenvironment, including stromal cells, modulates this process. *CTCs*, circulating tumor cells; *EMT*, epithelial-to-mesenchymal transition.

by molecular pathways operating within the cancer cells as well as cell-nonautonomous interactions between cancer cells and stroma and stromal cells [3]. Our ability to effectively treat cancer is largely dependent on our capacity to inhibit the process of metastasis. Intuitively, the first goal should be determining how and when precursors of overt metastasis arise.

There are two general models that explain the genesis of metastatic cells in primary tumors (Fig. 12.2). The first predicts that highly metastatic tumor cell variants preexist in the parental population (Fig. 12.2, left) [4], and this view was supported by evidence from cancer cell transplantation experiments in mice [5]. Parental tumors were defined as heterogeneous masses that contained highly metastatic tumor cell variants (clones), and the metastatic ability was conferred by genetic mutations that confer these rare cells in a clone, the metastatic capability (Fig. 12.2, left), although the specific mutations that drove metastasis remained unknown at the time. The second model employed to explain metastasis is clonal evolution or clonal selection [6–10]. In the mid-1970s, the classical clonal evolutionary theory of cancer was advocated that most neoplasms arise from the transformation of a single normal cell, and tumor progression results from acquired genetic variability within the original clones allowing sequential selection of more aggressive sublines [6]. The

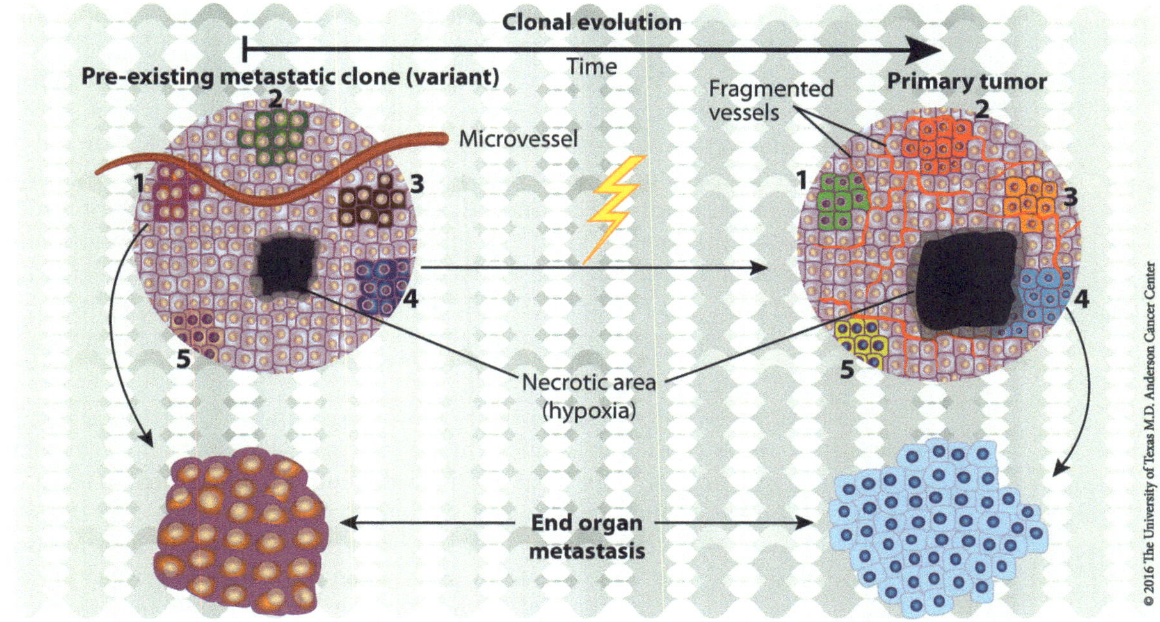

FIGURE 12.2 Two general models explaining the genesis of metastatic cells in primary tumors. The first (left) proposes that highly metastatic tumor cell variants (clones) preexist in the parental population. In the example shown, there are five tumor clones, one of which (ie, clone 1) is close to the vasculature and possesses much higher metastatic potential than the rest of the clones. Thus, clone 1 represents the preexistent metastatic clone. The metastasis in the end organ derived from clone 1 cells may be genetically similar to the parental clone 1 cells but phenotypically different (indicated by different colors) due to the microenvironmental influences. The second model (right) predicts that primary tumors progress towards metastasis via acquisition of multiple genetic alterations, ie, clonal selection (or evolution). The model suggests that none of the five clones in the early-stage primary tumor will be metastatic. However, time-related genetic mutations and/or epigenetic alterations would render a nonmetastatic clone 4 now highly metastatic (right). Depicted in the right is the close association of clone 4 cells with hypoxic necrotic area, and hypoxia has been reported to induce mesenchymal-like cancer stem cells that are highly invasive and metastatic. Depicted in this figure are also time-dependent changes in tumor vasculature, which, in early-stage tumors, is comprised of relatively normal vessels but becomes highly disorganized and leaky in advanced tumors (ie, tumor neoangiogenesis).

progression from early neoplastic lesion to metastasis is an extremely long process that involves the emergence of new mutations and clonal selection (Fig. 12.2, right). Comparative exome sequencing data has shown that it is unlikely that random mutations induce metastasis, but instead the driving forces are probably multiple genomic and epigenomic alterations that provide selective advantages to specific mutations [9,10].

Metastasis consists of several stochastic events including primary tumor formation, local invasion, intravasation (gaining access to the blood or lymphatic system), transport in the circulation (interaction with platelets, lymphocytes, and other blood components), arrest in microvessels of various organs, extravasation, formation of micrometastasis, and finally colonization and subsequent formation of macrometastasis [11] (Fig. 12.1). This process is called the "invasion-metastasis cascade" [11], in which there exists a dynamic cross talk between tumors and their microenvironment: tumor cells secrete many cytokines that activate stromal fibroblasts and induce the recruitment of immune cells to the tumor, whereas signals provided by the local environment (stromal, inflammatory, and immune cells) can promote the ability of tumor cells to invade and metastasize [12]. Indeed, during early tumor development stage, chronic inflammation and the local microenvironment shift to a growth-promoting state. Before tumor cells breach the basement membrane, they also stimulate angiogenesis on the stromal side of the membrane via angiogenic factors [11].

Once breaking away from the primary tumor mass and in the stroma, tumor cells gain access to the blood and lymphatic vessels, supplying the cells with nutrients and oxygen. This local invasion depends largely on the release of several proteases, which remodel the extracellular matrix (ECM). By degrading the basement membrane, invasive tumor cells harvest growth and survival factors that have been sequestered by attachment to the tissue architecture. In addition, the mechanics of intravasation are likely to be strongly influenced by the structural features of tumor-associated blood vessels [3]. Through a variety of mechanisms, tumor cells stimulate the formation of new blood vessels via neoangiogenesis. In contrast to blood vessels present in normal tissues, the neovasculature generated by tumor cells is prone to leakiness [13]. The weak interactions between adjacent endothelial cells that form the tumor-associated microvasculature and the absence of extensive pericyte coverage likely facilitate intravasation [14].

After intravasation, the CTCs, individually or in clusters with other tumor cells or with various host cells such as platelets and leukocytes, travel to other areas of the body via the venous and arterial circulation (Fig. 12.1). CTCs must survive a variety of stresses to reach distant sites (described later). Experiments suggest that cancer cells use specific cell surface receptors such as integrins to initially adhere to arterioles and capillaries in certain tissues; however, far more evidence suggests that simple physical trapping within small vessels determines the final destination of these cancer cells [11]. Ultimately, the capacity of CTCs to pass through endothelial walls influences the organ tropism of tumors. Once at these destinations, cancer cells may cross from vessel lumen into the tissue parenchyma by penetrating the endothelial cell and pericyte layers that separate vessel lumina from the stromal microenvironment, a process known as extravasation. Once in the parenchyma of the tissues, metastasizing cancer cells may form micrometastases or small clumps of disseminated cancer cells (Fig. 12.1). The extravasated cancer cells must first survive in the foreign microenvironment, which is influenced by many factors including the types of stromal cells, ECM constituents, available growth factors and cytokines, and even the microarchitecture of the tissue itself [3].

Finally, colonization occurs once these micrometastases expand into large macroscopic metastases (Fig. 12.1) [11]. The micrometastases may persist via two mechanisms. First, the DTCs may be largely quiescent, with their proliferation at metastatic sites greatly impaired due to incompatibilities with the foreign microenvironments [15]. Alternatively, the cancer cells in these micrometastases may proliferate continuously. However, their net increase in overall number may not occur due to the counterbalancing effects of a high apoptotic rate [3]. Success in metastatic colonization is also likely influenced by the ability of founder cells to possess high self-renewal capacity as seen in CSCs (see later discussion). Xenograft serial transplantation studies involving several tumor types lend support to this model, although the applicability of these findings to all types of human malignancies continues to be debated [16,17].

One question that remains unanswered is whether the metastatic potential of tumor cells is entirely shaped at the primary tumor site or also by signaling events occurring during the intravascular transit of tumor cells. Recent evidence suggests that the metastatic potential of tumor cells is not only shaped at the primary tumor but can also be modulated during their transit through the bloodstream. For example, transforming growth factor-beta (TGF-β1) is often elevated in the plasma of patients with malignant tumors. In normal cells, TGF-β, acting through its signaling pathway, stops the cell cycle at the G1 stage to stop proliferation, induce differentiation, or promote apoptosis. However, in cancer, this pathway is mutated causing cancer cells and surrounding stromal cells (fibroblasts) to proliferate. Both cells increase their production of TGF-β. TGF-β released in the serum acts on the surrounding stromal cells, immune cells, endothelial and smooth-muscle cells and causes immunosuppression and angiogenesis [18]. Indeed, TGFβ is known to promote metastasis by enhancing EMT and invasiveness in primary carcinomas [19]. Platelets in the bloodstream may also contribute to the physical interaction between CTCs and vascular endothelial cells by supporting the stable adhesion to endothelium and/or transmigration of tumor cells out of the vasculature. Local platelet activation could promote the migration of inflammatory cells, enhancing tumor stroma formation. Alternatively, tumor cell–associated platelets could facilitate tumor cell metastasis by preventing interactions between tumor cells and innate immune cells [20]. In terms of metastatic potential, a study has shown that platelet–tumor cell interactions are sufficient to prime tumor cells for subsequent metastasis [21]. Specifically, platelet-derived TGFβ and direct platelet–tumor cell contacts synergistically activate the TGFβ/Smad and NF-κB pathways in cancer cells, resulting in their transition to an invasive mesenchymal-like phenotype and enhanced metastasis in vivo [21].

The variable tendencies of different tumors to metastasize are unknown. Tumors in certain tissues have high probabilities of metastasizing, whereas others do not. Also, physical dissemination of cancer cells does not necessarily correlate with colonization, as many patients with disseminated micrometastases never progress [22,23]. Another major unresolved issue involves metastatic tropism, the preferential seeding of colonies in particular tissues located in distant locations. This could be explained by the "seed and soil" theory that states that metastasis depends on cross talk between selected cancer cells and specific organ microenvironments. The seed and soil hypothesis applies to the metastatic patterns of certain metastasis such as preferential metastatic spread of prostate and breast cancer cells to the bone. On the other hand, we also know that seeding and the subsequent establishment of metastatic niches are dependent on circulation patterns, extravasation barriers, and survival on arrival [11]. Indeed, solid tumors display dramatic variation in the pattern of metastasis. Many cancers spread to multiple organs such as triple-negative breast cancers, skin melanomas, lung cancers, and renal carcinoma, and, as mentioned previously, the specific hydrodynamics in different capillaries in different organs plays a large role in the ultimate distribution and colonization of these tumor cells [11]. Aggressive tumors release thousands of cancer cells in the circulation each day [24–26]; however, most patients develop only few metastases. This low rate of success in forming metastases is sometimes termed metastatic inefficiency.

Additionally, the metastatic tumor progression timeline can vary dramatically between tumor types and patients. For example, locally invasive cancers, such as glioblastomas, form distant metastases only rarely [27], whereas medulloblastomas frequently metastasize [28]. Lung and pancreatic cancers are frequently associated with metastasis at the time of diagnosis [29,30] whereas breast and prostate cancers (PCa) are not [31,32].

CANCER STEM CELLS AND METASTASIS: PROSTATE CANCER AS AN EXAMPLE

Tumor relapse and metastasis are the primary causes of poor survival rates in patients with advanced cancer, and one of the causes is likely due to the persistence of CSCs. As early as 1937, Furth and Kahn successfully transplanted leukemia with a single mouse leukemic cell [33], showing the first evidence of stem cell-like cancer cells, now termed CSCs. The human leukemic stem cells (LSCs) were first reported in the late 1990s [34]. Subsequently, CSCs have been identified in human solid tumors such as the breast [35] and brain [36]. These CSCs are defined by two characteristics: self-renewal and multipotency. Therefore, an established clinical tumor is sustained by CSCs that can generate, intraclonally, both tumorigenic and nontumorigenic cells [37].

Elucidating the mechanisms underlying PCa metastasis, especially with respect to the involvement of prostate CSCs (PCSCs), is critical for effective therapeutic intervention. It has been shown that PTEN loss or PI3K/AKT signaling pathway activation correlates with PCa progression and metastasis [38]. However, in preclinical murine models, deletion of *Pten* alone fails to mimic the significant metastatic burden that frequently accompanies the end stage of human disease. To identify additional pathway alterations that cooperate with PTEN loss in PCa progression, one group surveyed tissue microarrays and found that the RAS/MAPK pathway is significantly elevated in both primary and metastatic lesions [39]. Although RAS activation alone cannot initiate PCa development, it significantly accelerated progression caused by PTEN loss, accompanied by EMT and macrometastasis with 100% penetrance [39]. Interestingly, a novel stem/progenitor subpopulation with mesenchymal characteristics was isolated from the compound mutant prostates, and, of importance, inhibition of RAS/MAPK signaling in this population significantly reduced the metastatic progression initiated from transplanted stem/progenitor cells [39]. Thus, *Pten* loss and RAS/MAKP activation cooperate to promote EMT and metastasis initiated from PCSCs [39].

During the EMT program, a single epithelial cell undergoes significant alterations [ie, in cell surface marker expression, morphology, transcription factor (TF) expression, transcriptome, etc] when transitioning into a mesenchymal state and again when reverting back into an epithelial cell (see detailed discussion later). During this transition, the ability to adopt phenotypic changes, such as acquiring stem-like characteristics, would presumably provide cells with survival advantages necessary for distant metastasis [40,41]. Thus, the EMT program may not only enable cancer cells to physically disseminate from primary tumors but also can confer on such cells the self-renewal capability that is crucial to their subsequent clonal expansion at sites of dissemination [42–45]. This leads to the suggestion that both epithelial and mesenchymal cells within this program can be considered as a CSC, such that the CSC attribute is an "activity" rather than a defined state. A study [46] highlights this concept. The authors developed an in vivo system that allows tracking of the dynamic EMT program and isolation of cells from a PCa mouse model that have either completed (mesenchymal-like) or are transitioning through an EMT [46]. The EMT and mesenchymal-like tumor cells displayed enhanced stemness and invasive characteristics, compared with epithelial tumor cells, and displayed an enriched tumor-initiating capacity and were able to regenerate epithelial glandular structures in vivo. Most important, only epithelial and EMT tumor cells, not mesenchymal-like tumor cells, could form macrometastases [46]. This suggests that both mesenchymal and epithelial states contribute differentially to different stages of PCa development and progression, and EMT tumor cells exist with plasticity that contributes to multiple stages of the metastatic cascade.

It was shown that the formation of autonomic nerve fibers in the prostate gland regulates PCa development and dissemination in mouse models [47]. Tumors in this study were infiltrated by parasympathetic cholinergic fibers that promoted cancer dissemination. Whether and how the sympathetic and parasympathetic nervous systems regulate PCSCs remains an interesting and open question. In a study, another group performed gene expression profiling of PCSCs derived from DU145 human PCa cell cultures to identify factors involved in metastatic progression [48]. Their studies revealed that contactin 1 (CNTN1), a neural cell adhesion protein, promotes PCa progression and metastasis. CNTN1 knockdown reduced PCSC-mediated tumor initiation, whereas CNTN1 overexpression enhanced PC cell invasion in vitro and promoted xenograft tumor formation and lung metastasis in vivo [48]. Thus this study implicates proneural molecules in PCSCs and PCa metastasis, and this represents a fascinating new area to explore.

Two studies also highlight how dynamic interactions among epithelial, self-renewal, and mesenchymal gene programs determine the plasticity of epithelial tumor-initiating cells (TICs)/CSCs [49,50]. One group characterized two human cellular models derived from prostate and bladder cancer cell lines to better understand the relationship between TICs and EMT programs [49]. Tumor subpopulations that expressed a strong epithelial gene program were enriched in highly metastatic TICs. The second population with mesenchymal traits was deprived of TICs. Constitutive overexpression of the TF Snai1 in the epithelial/TIC-enriched populations engaged a mesenchymal gene program and suppressed their self-renewal and metastatic phenotypes. Conversely, knockdown of EMT factors in the mesenchymal-like PCa cell subpopulation caused a gain in epithelial features and properties of TICs [49]. The second study [50] involved breast cancer stem cells (BCSCs) that have been suggested to mediate metastasis, are resistant to radiation and chemotherapy, and contribute to relapse. The authors show that BCSCs exist in distinct mesenchymal-like (EMT) and epithelial-like [mesenchymal–epithelial transition (MET)] states [50]. The mesenchymal-like BCSCs characterized as CD24(−)CD44(+) were found to be primarily quiescent and localized at the tumor invasive front, whereas epithelial-like BCSCs expressed aldehyde dehydrogenase (ALDH), were proliferative and, were located more centrally. Based on these results, authors proposed that the plasticity of BCSCs that allows them to transition between EMT- and MET-like states endows these cells with the capacity for tissue invasion, dissemination, and growth at metastatic sites [50]. Overall, the relationship between CSCs and EMT is dynamic and cell type- and context-dependent.

CIRCULATING TUMOR CELLS AND METASTASIS

CTCs are cells that are shed from primary tumors and have intravasated into the bloodstream and are thus crucial in improving our understanding of the metastatic cascade (Fig. 12.1) [51]. CTCs were first reported in 1869 in a patient with advanced metastatic cancer [52]. CTCs are heterogeneous and can be broadly classified into three categories according to morphology and phenotypic traits: epithelial, mesenchymal, and intermediate (ie, transitioning from epithelial to mesenchymal). CTCs are extraordinarily rare, roughly one CTC per billion normal blood cells in the circulation of patients with advanced cancer, because the vast majority of CTCs die in the bloodstream, likely due to several causes such as anoikis, as these cells have lost adhesion-dependent survival signals. CTCs captured in xenograft PCa models have highlighted the importance of decreased adhesiveness, resistance to anoikis, and resistance to apoptosis in the survival of these cells [53,54]. Those that do survive have an inherent aspect of anoikis resistance, and IGFR1 has been evinced to support the survival of these CTCs by inhibiting the anoikis cascade [55]. Metastatic inefficiency, due to significant loss (death) of CTCs, is also caused by the shear force of the bloodstream. Hydrodynamic shear forces in the circulation are extreme in small vessel and tear cells apart. Furthermore, the majority of CTCs do not survive the harsh conditions when infiltrating distant organs due to the inability of adapting to new microenvironments. Cancer cells adapt to these new and inhospitable microenvironments via several strategies. For example, the late occurrences of brain metastasis in the clinic indicate that CTCs have a high rate of attrition particularly in the brain, due to two critical microenvironmental determinants, ie, the ability of CTCs to escape reactive stromal signals and to associate with capillary blood vessels known as vascular cooption [56].

Of note, CTCs may persist at distant sites and survive in host tissues for long periods of time. This period between diagnosis of a primary tumor and detection of metastatic lesions is referred to as "metastatic latency." Alternatively, the period between the arrival of cancer cells at distant locations till their colonization of the organ is named "dormancy." In breast cancer and PCa, CTCs disseminate long before they acquire the capabilities of metastatic colonization. The biology of metastatic latency and reactivation is largely unknown, at least partially due to the lack of suitable experimental systems that would model this aspect of metastasis. However, work in mouse models of breast cancer provides insights into the kind of differentiation signals and stromal interactions that may be involved [57,58].

CTCs hold the key to understanding the biology of metastasis; however, their rarity makes isolation and characterization a major technological challenge. Attempts have been made to capture CTCs based on cell surface markers, size, density, electromechanical characteristics, tumor-specific epitopes, nucleic acid-based detection, and high-throughput microscopic scanning approaches (reviewed in Ref. [59]). However, characterization of CTCs that induce metastasis remains obscure because platforms that capture CTCs are not all-inclusive. The Veridex CellSearch platform, the only test approved by the US FDA, relies

Table 12.1 CTC Biomarkers

Cancer	Phenotypic Markers	Capture Antibody	References
CRPC	$CK8^+$, $CK18^+$, $CK19^+$, $CD45^-$	$EpCAM^+$	[77]
Breast	$CD44^+$, $CD47^+$, MET^+	$EpCAM^+$	[25]
Breast	$HER2^+$,$EGFR^+$, $HPSE^+$, $Notch1^+$	$EpCAM^-$	[84]
Breast	$HER2^+$,$EGFR^+$	$EpCAM^+$	[68]
SCLC	$CK8^+$, $CK18^+$, $CK19^+$	$EpCAM^+$	[76]
Colon	$CK8^+$, $CK18^+$, $CK19^+$, $CD45^-$	$EpCAM^+$	[126]
GBM	$Nestin^+$, $Telomerase^+$	$EpCAM^-$	[127]
Sarcoma	CSV^+	$EpCAM^-$	[128]
Melanoma	$MCAM^+$/$MUC18^+$/$CD146^+$, $ABCB5^+$, $nestin^+$, $CD133^+$, $CD271^+$		[129]

ABCB5, *ATP-binding cassette subfamily B member 5;* CRPC, *castration-resistant prostate cancer;* CSV, *cell-surface vimentin;* EGFR, *human epithelial growth factor receptor;* HER2, *human epidermal growth factor receptor 2;* HPSE, *heparanase;* MCAM, *melanoma cell adhesion molecule.*

solely on the use of an antibody against the epithelial cell adhesion molecule (EpCAM; Table 12.1). Therefore, CTCs that are undetectable by this antibody or are $EpCAM^-$ are ignored. CellSearch is also unable to capture CTCs in 30–35% of metastatic breast cancer patients [60–62]. New approaches to identify and characterize CTCs are discussed in specific studies below. Beyond this, the goal of the field is to demonstrate that human CTCs can cause new tumors and be characterized as organ-specific homing subsets.

In one new approach, a series of microfluidic devices have been developed that have the advantage of low shear, yet high throughput, allowing interrogation of unprocessed whole blood, providing highly enriched and unfixed CTCs that are suitable for detailed molecular analysis [26,63,64]. In a study, both the herringbone (^{HB}CTC-Chip) and ^{neg}CTC-iChip were used along with in vivo fluorescence-activated cell sorting and next generation RNA sequencing to characterize CTCs [65]. Both the metastatic propensity between CTC clusters versus single cells and the origin of CTC clusters were experimentally tested using patients with metastatic breast cancer and mouse tumor models. Data from mouse models revealed that CTC clusters are derived from oligoclonal clumps of primary tumor cells and not from the proliferation of a single tumor cell in the vasculature [65]. These endogenous CTC clusters, defined as 2-50 cancer cells, have increased metastatic potential compared to single CTCs and have been detected within the circulation of patients with metastatic epithelial cancers [66–68]. Specifically, CTC clusters constitute only 2–5% of all CTC events detected in the circulation, but their dramatically elevated metastatic potential (23–50 times that of single CTCs) contributes to approximately half of all metastatic lesions in orthotopic breast cancer models [65]. Additionally, RNA sequencing of human breast CTC clusters revealed plakoglobin as a key mediator of tumor cell clustering and thus greatly contributes to the metastatic spread of cancer [65]. Finally, continuous presence of CTC clusters was found to be associated with an adverse clinical outcome. Thus, the identification of specific transcripts that enhance the metastatic potential of tumor cells may enable therapeutic strategies to suppress the blood-borne spread of cancer.

Tight vascular wall barriers, unfavorable conditions for survival in distant organs, and a rate-limiting acquisition of organ colonization functions are just some of the impediments to the formation of distant metastasis by the CTCs [69]. Typically, cancer cells leave the primary tumor and seed metastasis in distant organs. However, since the neovasculature of tumors is leaky, in theory it would be easier for CTCs to reenter their tumor of origin after circulating. One group thus proposed a new concept of "tumor self-seeding," in which injected tagged human cancer cell lines may colonize an existing tumor deposit, with the newly recruited tumor cells conferring increased aggressiveness to the existing tumor [70]. Indeed, this self-seeding can accelerate tumor growth, angiogenesis, and stromal recruitment through seed-derived factors including the chemokine CXCL1. This tumor self-seeding was found to be a general phenomenon in experimental models of breast carcinoma, colon carcinoma, and malignant melanoma. Unlike colonization of distant organs, self-seeding requires little, if any, additional adaptation of CTCs to the recipient microenvironment. However, self-seeding does select for cancer cell populations that are more aggressive than the bulk population of the primary tumor. Self-seeding requires tumor-derived attraction signals which may include IL-6 and IL-8 as tumor-derived attractants of CTCs in breast carcinoma and melanoma models [70].

Breast cancer primarily metastasizes to the bone, lung, liver, and brain and accounts for the vast majority of cancer-related deaths in women [69]. Until recently, the existence and phenotype of metastasis-initiating cells among CTCs had not been experimentally demonstrated. To clarify this population, one group developed a xenograft assay to show that primary human luminal breast CTCs contain cells that initiate metastasis to bone, lung, and liver in mice. These metastasis-initiating cells were found to be likely present with a low frequency among highly heterogeneous CTC populations, and metastasis-initiating cells from luminal breast cancer patients demonstrated an $EPCAM^{+}$ $CD44^{+}CD47^{+}MET^{+/-}$ phenotype [25]. Indeed, in patient samples, the frequency of $CD44^{+}CD47^{+}MET^{+}$ cells correlated with clinical progression and was also a better indicator of patient overall survival and metastatic burden than bulk CTCs [25]. As CD44, CD46, and c-MET are well-established phenotypic markers for CSCs, these studies suggest that at least some CTCs may be CSCs.

Other studies have also validated the existence of stem cell-like CTCs, which have the ability to self-renew, clonally expand, and initiate tumors [25,71–73]. Wnt signaling contributes to the reprogramming and maintenance of CSC states that are activated by EMT. In a study, CTCs captured via a microfluidic device from an endogenous mouse pancreatic cancer model were subjected to single-molecule RNA sequencing, identifying Wnt2 as a candidate gene enriched in CTCs [74]. Expression of WNT2 in pancreatic cancer cells suppressed anoikis, enhanced anchorage-independent sphere formation, and increased metastatic propensity in vivo. CTCs from human samples also revealed enrichment for WNT signaling in 5 out of 11 cases [74]. In addition to the Wnt pathway, future molecular analysis efforts of CTCs may be useful in identifying other therapeutic targets to prevent metastasis.

CTCs are highly prevalent in patients with small cell lung cancer (SCLC) compared to other cancers and CTC number is of prognostic significance [75]. CTCs from patients with either chemosensitive or chemorefractory SCLC were found to be tumorigenic in immune-compromised mice [76]. Additionally, the resultant CTC-derived explants mirrored the donor patient's response to platinum and etoposide chemotherapy and thus represent clinical SCLC [76]. The authors also reported the first genomic comparison of single CTCs directly isolated from patient blood, and the resultant matched CTC-derived explants obtained following transplant into mice. This comparison revealed considerable similarity between the two, suggesting that CTCs from a patient with extensive-stage SCLC are largely homogeneous [76]. Thus, these CTC-derived explant models generated from minimally invasive clinical samples provide a model to study SCLC biology and search for new therapeutics.

Previous studies have identified PCa-specific molecular alterations in CTCs such as androgen receptor (AR) gene amplification, PTEN loss, and TMPRSS2:ERG fusion confirming the malignant origin of CTCs [67,77,78]. In a study, isolated CTCs from castration resistant prostate cancer (CRPC) patients followed longitudinally over the course of taxane chemotherapy were utilized, and the authors observed a decrease in AR intensity and a significant correlation between AR cytoplasmic localization and clinical response, suggesting that the changes in AR subcellular localization can be used as a predictive biomarker of response [79]. Indeed, CTCs captured from peripheral blood have been shown to predict disease outcome and therapy response in cancer patients [60,80–83]. A study interrogating candidate genes in breast CTCs derived from a patient with breast cancer suggested that coexpression of EpCAM, CD44, CD47, and MET identifies a subset with increased metastatic capacity [25]. Only a small number of candidate genes (346) were identified with significantly divergent expressions, including transcriptional regulators (XBP1), signaling molecules (AGR2 and HER3), and plakoglobin.

CTCs are a heterogeneous population. To investigate biomarkers expressed by CTCs and their ability to metastasize to the brain, a study characterized CTCs isolated from peripheral blood mononuclear cells of patients with breast cancer [84]. In EpCAM-negative CTCs, the authors identified a potential signature of brain metastasis comprising "brain metastasis selected markers (BMSMs)" $HER2^+/EGFR^+/HPSE^+/Notch1^+$. CTC lines expressing the BMSM signature were highly invasive and capable of generating brain and lung metastases when xenografted in nude mice [84]. Additionally, increased brain metastatic capabilities, frequency, and quantitation were detected in $EpCAM^-$ CTCs overexpressing the BMSM signature [84].

After a short period of circulation in the blood, CTCs lodge in distant organs and are termed DTCs. DTCs have mainly been detected and characterized in the bone marrow, because this organ is more easily accessible through needle aspiration than other organs such as liver, lung, or brain. Although studies suggest that early DTCs rarely transform into metastasis-initiator cells [85,86], DTCs have the potential to provide novel insights into the process of dissemination. Indeed, there are numerous reports on the prognostic impact of DTCs, which are commonly detected by the histogenetic markers, EpCAM or cytokeratins (CKs), in lymph nodes and bone marrow, respectively. In contrast to CTCs, which are mostly detected and studied in patients with metastases [24], DTCs are clearly relevant in early disease stages [87–89]. In a study, a combined genome and transcriptome analysis of single cells revealed that DTCs might express hematopoietic transcripts, and the results point to an unexpected transcriptome plasticity of epithelial cancer cells in bone marrow [90]. In breast cancer, there is a significant correlation between the detection of CTCs and DTCs [91]. Specifically, CTCs and extravasated DTCs in the bone marrow of patients with breast cancer have been shown to correlate with increased metastatic burden, aggressive disease, and a decreased time to relapse [87].

THE ROLE OF EPITHELIAL-TO-MESENCHYMAL TRANSITION IN METASTASIS

The majority of cancers occur in epithelial tissues, producing carcinomas. The majority of epithelial tissues are comprised of relatively thin sheets of epithelial cells that sit atop layers of stroma. Separating the two is an ECM known as the basement membrane [11]. The physical dissemination of cancer cells from the primary tumor to distant tissues and the subsequent adaptation of these cells to foreign tissue microenvironments that results in successful colonization constitute the two major phases of metastasis [92]. Both the EMT and the subsequent MET have been implicated in the development of clinical metastases [11]. The EMT program allows stationary and polarized epithelial cells to undergo multiple biological changes that enable them to disrupt cell–cell adherence, lose apical–basal polarity, dramatically remodel the cytoskeleton, and acquire mesenchymal characteristics such as enhanced migratory capacity, invasiveness, elevated resistance to apoptosis, and greatly increased production of ECM components. Finally at distant sites, their extravasation and subsequent MET allows growth of micrometastatic lesions into macroscopic tumors [92]. Thus, the motility and invasiveness of cancer cells that undergo EMT favor dispersion to distant tissues whereas the MET is needed to start metastatic colonization and expansion. Recent evidence from experimental systems shows that cancer cells that underwent an EMT for metastatic dissemination must undergo the reverse process. In a model of squamous cell carcinoma, expression of the EMT master regulator Twist enhanced cancer cell dissemination, but the formation of overt metastases required a loss of Twist expression and reacquisition of epithelial traits [93]. It is this balance between proliferation and motility that enables metastasis.

A fundamental question regarding the initiation of metastasis is whether cancer cells break free from their primary tumor via active invasion or a passive shedding of cells due to compromised tumor vasculature. Several models have been proposed to elucidate the trigger of metastasis initiation; however, the debate continues [94–96]. The leading hypothesis supports the former and involves EMT. The EMT process has been implicated in many epithelial cancers including breast [97], prostate [98], and colorectal cancer [99–101]. In a study of human breast cancer, mesenchymal markers indicative of EMT were expressed within the cancer cells comprising CTC clusters, implicating a role for EMT in the bloodborne dissemination of human breast cancer [68]. Serial CTC monitoring in 11 patients suggested an association of mesenchymal CTCs with disease progression. Notably, there was a striking association between expression of mesenchymal markers and clusters of CTCs, rather than single CTCs [68].

EMT facilitates metastasis and is thought to generate CSCs and contribute to therapy resistance. Therefore, the EMT pathway is of great therapeutic interest in the treatment of cancer and could be targeted either to prevent tumor dissemination or to eradicate existing metastatic cancer cells in patients with more advanced diseases [102]. There are several potential stages to pharmacologically target EMT events in cancer: inhibiting EMT induction, targeting the mesenchymal cancer cells, or inhibiting MET. With regard to EMT induction, many signals from the tumor microenvironment, such as reduced oxygen tension, cytokines, and growth factors, can be targeted [103–106]. One limitation could be that resistance to these therapies may rapidly emerge if other EMT-inducing pathways remain intact, as has been observed in lung carcinoma studies [107]. In another study, molecular chaperone Hsp27 has been shown to drive EMT in PCa, and conversely its attenuation reverses EMT and decreases cell migration, invasion, and matrix metalloproteinase activity [108]. Most importantly, Hsp27 inhibition via OGX-427, an antisense therapy currently in phase

2 trials, led to a decreased number of CTCs in patients with metastatic CRPC in a phase 1 clinical trial. Additionally, decreases in CTC counts in patients with PCa undergoing treatment are associated with improved survival [109].

Inhibition of mesenchymal regulators such as N-cadherin and fibronectin may be a mechanism to target the mesenchymal phenotype of metastasizing tumor cells and to halt metastasis. For example, targeting of N-cadherin using a monoclonal antibody approach has been shown to inhibit PCa cell invasion and reduce metastasis formation [110]. Targeting EMT-associated TFs may also be an option. Metastatic cells are regulated by three core groups of transcriptional regulators: the Snail zinc-finger family of TFs, SNAI1, and SNAI2 (SLUG) [111–113], the zinc-finger E-box-binding homeobox family proteins ZEB1 and ZEB2 (SIP1) [114], and the basic helix-loop-helix (bHLH) family of TFs, TWIST1, TWIST2, and E12/E47 [115–117]. All these master TFs transcriptionally repress E-cadherin and induce EMT. For example, inhibition of Twist in a highly metastatic mammary carcinoma cell line reduces the frequency of lung metastases in a mouse model of breast cancer and reduced the number of CTCs in this model [115]. Another option that has gained recent attention is the development of miRNA-based therapies [118]. For example, AC1MMYR2, a specific small molecule inhibitor of miR-21 production, reduces EMT markers and inhibits invasion of glioblastoma, breast cancer, and gastric cancer cells [119]. Finally, the stage of reversion of mesenchymal-like cells to a more differentiated, epithelial-like state, ie, MET, can be targeted as studies have shown the need for MET for metastatic capacity [93,120]. However, the main dilemma with this approach is that agents that inhibit the mesenchymal state could accelerate epithelialization of disseminated deposits and, thus, drive the formation of metastases; whereas agents that target MET could promote mesenchymalization and therapy resistance in premetastatic cancers.

Despite preponderant data linking EMT to metastasis, challenges have been raised against the role of EMT as a crucial effector of cancer metastasis. Current understanding of EMT-induced metastasis is largely derived from in vitro studies in human cancer cell lines. But the inherent plasticity of the process and the limited clinical evidence supporting the occurrence of EMT in tumor specimens [68,121] have led to skepticism about EMT being the predominant mechanism governing the early steps of metastasis. Two studies provide evidence that epithelial tumor cells do not need to transition to a mesenchymal-cell state to form a metastasis but that the process does contribute to drug resistance [122]. In one study, an EMT lineage-tracing system was utilized to survey the fate of epithelial tumor cells transitioning to a mesenchymal state, from the cells' inception and dissemination through the bloodstream to their exit from blood vessels and metastatic growth in mice using a spontaneous breast-to-lung metastasis model [123]. To do this, the authors monitored the expression of green fluorescent protein (GFP) as a proxy for the expression of the genes that encode fibroblast-specific protein 1 (FSP1) or vimentin, which are triggered when epithelial tumor cells switch to a mesenchymal state. The green fluorescence persists in the progeny of these cells well after they revert to an epithelial fate. Authors show that lung metastasis mainly consists of non-EMT tumor cells that maintain their epithelial phenotype. Indeed, the vast majority of metastatic tumors were not derived from the mesenchymal-switched cells expressing GFP but from disseminating epithelial cells. Interestingly, EMT cells significantly contribute to recurrent lung metastasis formation after chemotherapy [123]. A second group examined the role of EMT in pancreatic ductal adenocarcinoma (PDAC) by generating mouse models of PDAC with deletion of Snail or Twist, two key TFs responsible for EMT [124]. They found that EMT suppression in the primary tumor does not alter the emergence of invasive PDAC, systemic dissemination, or metastasis. However, EMT induces chemoresistance in PDAC [124]. These two findings challenge the prevailing hypothesis that EMT is a key element in the metastatic dissemination of epithelial cancers, and they point to a distinct role of this cell-fate transition in enhancing cancer-cell survival during drug treatment.

CONCLUDING REMARKS

Metastasis is the ultimate challenge in our effort to fight cancer as a life-threatening disease. The mechanisms responsible for the seeding into the bloodstream of tumors cells, followed by their survival in the circulations, and ultimately invasion and proliferation in distant tissues are all the subject of intent investigations. In human tumors, such studies have been limited by the difficulty in accessing different stages of the metastatic cascade. As a result, conventional therapeutic strategies targeting primary tumor cells have limited success in preventing and treating metastasis. There is also a need for better biomarkers for monitoring efficacy of antimetastasis treatments. Another issue is the discordance in mutational status of target molecules in metastasis as compared to primary tumors [125]. This may explain why metastasis is more resistant to current treatment as compared to primary tumors. Thus, additional mechanisms that render metastasis more resistant to current treatment must be explored [1]. Advanced technologies in genomic interrogation and cellular imaging and better experimental models are needed to allow more in-depth tracing and characterization of metastatic tumor cells and their environment to uncover the vulnerabilities in metastatic cells. Further understanding of the mechanisms regulating epithelial–mesenchymal plasticity and the relationship between EMT, MET, CTCs, and CSCs will help to unveil novel therapies.

List of Acronyms and Abbreviations

ALDH Aldehyde dehydrogenase
AR Androgen receptor
CK Cytokeratin

CNTN1 Contactin 1
CRPC Castration-resistant prostate cancer
CSCs Cancer stem cells
CTCs Circulating tumor cells
DTCs Disseminated tumor cells
ECM Extracellular matrix
EMT Epithelial–mesenchymal transition
LSCs Leukemic stem cells
PCa Prostate cancer
PCSCs Prostate cancer stem cells
PDAC Pancreatic ductal adenocarcinoma
PTEN Phosphatase and tensin homolog
SCLC Small cell lung cancer
TGF-β Transforming growth factor-β
TICs Tumor-initiating cells

Acknowledgments

The work in the authors' lab was supported, in part, by grants from the NIH R01-CA155693, Department of Defense W81XWH-13-1-0352 and W81XWH-14-1-0575, CPRIT RP120380, and MDACC Center for Cancer Epigenetics (DGT). K. Rycaj was supported, in part, by NIH postdoctoral fellowship 5T32CA009480. We thank Ms. J. Holcomb for assistance in generating the figures.

References

[1] Wan L, Pantel K, Kang Y. Tumor metastasis: moving new biological insights into the clinic. Nat Med 2013;19(11):1450–64.

[2] Brabletz T, Lyden D, Steeg PS, Werb Z. Roadblocks to translational advances on metastasis research. Nat Med 2013;19(9):1104–9.

[3] Valastyan S, Weinberg RA. Tumor metastasis: molecular insights and evolving paradigms. Cell 2011;147(2):275–92.

[4] Fidler IJ, Kripke ML. Metastasis results from preexisting variant cells within a malignant tumor. Science 1977;197(4306):893–5.

[5] Kripke ML, Gruys E, Fidler IJ. Metastatic heterogeneity of cells from an ultraviolet light-induced murine fibrosarcoma of recent origin. Cancer Res 1978;38(9):2962–7.

[6] Nowell PC. The clonal evolution of tumor cell populations. Science 1976;194(4260):23–8.

[7] Bernards R, Weinberg RA. A progression puzzle. Nature 2002;418(6900):823.

[8] Greaves M, Maley CC. Clonal evolution in cancer. Nature 2012;481(7381):306–13.

[9] Jones S, Chen WD, Parmigiani G, Diehl F, Beerenwinkel N, Antal T, et al. Comparative lesion sequencing provides insights into tumor evolution. Proc Natl Acad Sci USA 2008;105(11):4283–8.

[10] Vanharanta S, Massague J. Origins of metastatic traits. Cancer Cell 2013;24(4):410–21.

[11] Weinberg RA. The biology of cancer. 2nd ed. 2014. p. 641–722.

[12] Joyce JA, Pollard JW. Microenvironmental regulation of metastasis. Nat Rev Cancer 2009;9(4):239–52.

[13] Carmeliet P, Jain RK. Principles and mechanisms of vessel normalization for cancer and other angiogenic diseases. Nat Rev Drug Discov 2011;10(6):417–27.

[14] Gupta GP, Nguyen DX, Chiang AC, Bos PD, Kim JY, Nadal C, et al. Mediators of vascular remodelling co-opted for sequential steps in lung metastasis. Nature 2007;446(7137):765–70.

[15] Chambers AF, Groom AC, MacDonald IC. Dissemination and growth of cancer cells in metastatic sites. Nat Rev Cancer 2002;2(8):563–72.

[16] Shackleton M, Quintana E, Fearon ER, Morrison SJ. Heterogeneity in cancer: cancer stem cells versus clonal evolution. Cell 2009;138(5):822–9.

[17] Clevers H. The cancer stem cell: premises, promises and challenges. Nat Med 2011;17(3): 313–9.

[18] Blobe GC, Schiemann WP, Lodish HF. Role of transforming growth factor beta in human disease. N Engl J Med 2000;342(18):1350–8.

[19] Oft M, Heider KH, Beug H. TGFbeta signaling is necessary for carcinoma cell invasiveness and metastasis. Curr Biol 1998;8(23):1243–52.

[20] Palumbo JS, Talmage KE, Massari JV, La Jeunesse CM, Flick MJ, Kombrinck KW, et al. Platelets and fibrin(ogen) increase metastatic potential by impeding natural killer cell-mediated elimination of tumor cells. Blood 2005;105(1):178–85.

[21] Labelle M, Begum S, Hynes RO. Direct signaling between platelets and cancer cells induces an epithelial-mesenchymal-like transition and promotes metastasis. Cancer Cell 2011;20(5):576–90.

[22] Talmadge JE, Fidler IJ. AACR centennial series: the biology of cancer metastasis: historical perspective. Cancer Res 2010;70(14):5649–69.

[23] McGowan PM, Kirstein JM, Chambers AF. Micrometastatic disease and metastatic outgrowth: clinical issues and experimental approaches. Future Oncol 2009;5(7):1083–98.

[24] Allard WJ, Matera J, Miller MC, Repollet M, Connelly MC, Rao C, et al. Tumor cells circulate in the peripheral blood of all major carcinomas but not in healthy subjects or patients with nonmalignant diseases. Clin Cancer Res 2004;10(20):6897–904.

[25] Baccelli I, Schneeweiss A, Riethdorf S, Stenzinger A, Schillert A, Vogel V, et al. Identification of a population of blood circulating tumor cells from breast cancer patients that initiates metastasis in a xenograft assay. Nat Biotechnol 2013;31(6):539–44.

[26] Nagrath S, Sequist LV, Maheswaran S, Bell DW, Irimia D, Ulkus L, et al. Isolation of rare circulating tumour cells in cancer patients by microchip technology. Nature 2007;450(7173):1235–9.

[27] Lun M, Lok E, Gautam S, Wu E, Wong ET. The natural history of extracranial metastasis from glioblastoma multiforme. J Neurooncology 2011;105(2):261–73.

[28] Wu X, Northcott PA, Dubuc A, Dupuy AJ, Shih DJ, Witt H, et al. Clonal selection drives genetic divergence of metastatic medulloblastoma. Nature 2012;482(7386):529–33.

[29] Werner J, Combs SE, Springfeld C, Hartwig W, Hackert T, Buchler MW. Advanced-stage pancreatic cancer: therapy options. Nat Rev Clin Oncol 2013;10(6):323–33.

[30] Feld R, Rubinstein LV, Weisenberger TH. Sites of recurrence in resected stage I non-small-cell lung cancer: a guide for future studies. J Clin Oncol 1984;2(12):1352–8.

[31] Lim E, Metzger-Filho O, Winer EP. The natural history of hormone receptor-positive breast cancer. Oncology 2012;26(8). 688–94, 696.

[32] Popiolek M, Rider JR, Andren O, Andersson SO, Holmberg L, Adami HO, et al. Natural history of early, localized prostate cancer: a final report from three decades of follow-up. Eur Urol 2013;63(3):428–35.

[33] Furth J, Kahn M. The transmission of leukemia of mice with a single cell. Cancer Res 1937:276–82.

[34] Bonnet D, Dick JE. Human acute myeloid leukemia is organized as a hierarchy that originates from a primitive hematopoietic cell. Nat Med 1997;3(7):730–7.

[35] Al-Hajj M, Wicha MS, Benito-Hernandez A, Morrison SJ, Clarke MF. Prospective identification of tumorigenic breast cancer cells. Proc Natl Acad Sci USA 2003;100(7):3983–8.

[36] Bao S, Wu Q, McLendon RE, Hao Y, Shi Q, Hjelmeland AB, et al. Glioma stem cells promote radioresistance by preferential activation of the DNA damage response. Nature 2006;444(7120):756–60.

[37] Rycaj K, Tang DG. Cell-of-origin of cancer versus cancer stem cells: assays and interpretations. Cancer Res 2015;75.

[38] Taylor BS, Schultz N, Hieronymus H, Gopalan A, Xiao Y, Carver BS, et al. Integrative genomic profiling of human prostate cancer. Cancer Cell 2010;18(1):11–22.

[39] Mulholland DJ, Kobayashi N, Ruscetti M, Zhi A, Tran LM, Huang J, et al. Pten loss and RAS/MAPK activation cooperate to promote EMT and metastasis initiated from prostate cancer stem/progenitor cells. Cancer Res 2012;72(7):1878–89.

[40] Mani SA, Guo W, Liao MJ, Eaton EN, Ayyanan A, Zhou AY, et al. The epithelial-mesenchymal transition generates cells with properties of stem cells. Cell 2008;133(4):704–15.

[41] Thiery JP, Acloque H, Huang RY, Nieto MA. Epithelial-mesenchymal transitions in development and disease. Cell 2009;139(5):871–90.

[42] Brabletz T, Jung A, Spaderna S, Hlubek F, Kirchner T. Opinion: migrating cancer stem cells – an integrated concept of malignant tumour progression. Nat Rev Cancer 2005;5(9):744–9.

[43] Tirino V, Camerlingo R, Bifulco K, Irollo E, Montella R, Paino F, et al. TGF-beta1 exposure induces epithelial to mesenchymal transition both in CSCs and non-CSCs of the A549 cell line, leading to an increase of migration ability in the $CD133^+$ A549 cell fraction. Cell Death Dis 2013;4:e620.

[44] Giordano A, Gao H, Anfossi S, Cohen E, Mego M, Lee BN, et al. Epithelial-mesenchymal transition and stem cell markers in patients with HER2-positive metastatic breast cancer. Mol Cancer Ther 2012;11(11):2526–34.

[45] Chang L, Graham PH, Hao J, Ni J, Bucci J, Cozzi PJ, et al. Acquisition of epithelial-mesenchymal transition and cancer stem cell phenotypes is associated with activation of the PI3K/Akt/mTOR pathway in prostate cancer radioresistance. Cell Death Dis 2013;4:e875.

[46] Ruscetti M, Quach B, Dadashian EL, Mulholland DJ, Wu H. Tracking and functional characterization of epithelial-mesenchymal transition and mesenchymal tumor cells during prostate cancer metastasis. Cancer Res 2015;75(13):2749–59.

[47] Magnon C, Hall SJ, Lin J, Xue X, Gerber L, Freedland SJ, et al. Autonomic nerve development contributes to prostate cancer progression. Science 2013;341(6142):1236361.

[48] Yan J, Ojo D, Kapoor A, Lin X, Pinthus JH, Aziz T, et al. Neural cell adhesion protein CNTN1 promotes the metastatic progression of prostate cancer. Cancer Res 2016;76.

[49] Celia-Terrassa T, Meca-Cortes O, Mateo F, de Paz AM, Rubio N, Arnal-Estape A, et al. Epithelial-mesenchymal transition can suppress major attributes of human epithelial tumor-initiating cells. J Clin Invest 2012;122(5):1849–68.

[50] Liu S, Cong Y, Wang D, Sun Y, Deng L, Liu Y, et al. Breast cancer stem cells transition between epithelial and mesenchymal states reflective of their normal counterparts. Stem Cell Rep 2014;2(1):78–91.

[51] Pantel K, Speicher MR. The biology of circulating tumor cells. Oncogene 2015;35.

[52] Ashworth TR. A case of cancer in which cells similar to those in the tumors were seen in the blood after death. Aust Med J 1869;14:146–9.

[53] Berezovskaya O, Schimmer AD, Glinskii AB, Pinilla C, Hoffman RM, Reed JC, et al. Increased expression of apoptosis inhibitor protein XIAP contributes to anoikis resistance of circulating human prostate cancer metastasis precursor cells. Cancer Res 2005;65(6):2378–86.

[54] Howard EW, Leung SC, Yuen HF, Chua CW, Lee DT, Chan KW, et al. Decreased adhesiveness, resistance to anoikis and suppression of GRP94 are integral to the survival of circulating tumor cells in prostate cancer. Clin Exp Metastasis 2008;25(5):497–508.

[55] Sachdev D, Zhang X, Matise I, Gaillard-Kelly M, Yee D. The type I insulin-like growth factor receptor regulates cancer metastasis independently of primary tumor growth by promoting invasion and survival. Oncogene 2010;29(2):251–62.

[56] Kienast Y, von Baumgarten L, Fuhrmann M, Klinkert WE, Goldbrunner R, Herms J, et al. Real-time imaging reveals the single steps of brain metastasis formation. Nat Med 2010;16(1):116–22.

[57] Gao H, Chakraborty G, Lee-Lim AP, Mo Q, Decker M, Vonica A, et al. The BMP inhibitor Coco reactivates breast cancer cells at lung metastatic sites. Cell 2012;150(4):764–79.

[58] Lu X, Mu E, Wei Y, Riethdorf S, Yang Q, Yuan M, et al. VCAM-1 promotes osteolytic expansion of indolent bone micrometastasis of breast cancer by engaging alpha4beta1-positive osteoclast progenitors. Cancer Cell 2011;20(6):701–14.

[59] Yu M, Stott S, Toner M, Maheswaran S, Haber DA. Circulating tumor cells: approaches to isolation and characterization. J Cell Biol 2011;192(3):373–82.

[60] Cristofanilli M, Budd GT, Ellis MJ, Stopeck A, Matera J, Miller MC, et al. Circulating tumor cells, disease progression, and survival in metastatic breast cancer. N Engl J Med 2004;351(8):781–91.

[61] Alix-Panabieres C, Schwarzenbach H, Pantel K. Circulating tumor cells and circulating tumor DNA. Annu Rev Med 2012;63:199–215.

[62] Mego M, De Giorgi U, Dawood S, Wang X, Valero V, Andreopoulou E, et al. Characterization of metastatic breast cancer patients with nondetectable circulating tumor cells. Int J Cancer 2011;129(2):417–23.

[63] Ozkumur E, Shah AM, Ciciliano JC, Emmink BL, Miyamoto DT, Brachtel E, et al. Inertial focusing for tumor antigen-dependent and -independent sorting of rare circulating tumor cells. Sci Transl Med 2013;5(179):179ra47.

[64] Stott SL, Hsu CH, Tsukrov DI, Yu M, Miyamoto DT, Waltman BA, et al. Isolation of circulating tumor cells using a microvortex-generating herringbone-chip. Proc Natl Acad Sci USA 2010;107(43):18392–7.

[65] Aceto N, Bardia A, Miyamoto DT, Donaldson MC, Wittner BS, Spencer JA, et al. Circulating tumor cell clusters are oligoclonal precursors of breast cancer metastasis. Cell 2014;158(5):1110–22.

[66] Cho EH, Wendel M, Luttgen M, Yoshioka C, Marrinucci D, Lazar D, et al. Characterization of circulating tumor cell aggregates identified in patients with epithelial tumors. Phys Biol 2012;9(1):016001.

[67] Stott SL, Lee RJ, Nagrath S, Yu M, Miyamoto DT, Ulkus L, et al. Isolation and characterization of circulating tumor cells from patients with localized and metastatic prostate cancer. Sci Transl Med 2010;2(25):25ra3.

[68] Yu M, Bardia A, Wittner BS, Stott SL, Smas ME, Ting DT, et al. Circulating breast tumor cells exhibit dynamic changes in epithelial and mesenchymal composition. Science 2013;339(6119):580–4.

[69] Nguyen DX, Bos PD, Massague J. Metastasis: from dissemination to organ-specific colonization. Nat Rev Cancer 2009;9(4):274–84.

[70] Kim MY, Oskarsson T, Acharyya S, Nguyen DX, Zhang XH, Norton L, et al. Tumor self-seeding by circulating cancer cells. Cell 2009;139(7):1315–26.

[71] Aktas B, Tewes M, Fehm T, Hauch S, Kimmig R, Kasimir-Bauer S. Stem cell and epithelial-mesenchymal transition markers are frequently overexpressed in circulating tumor cells of metastatic breast cancer patients. Breast Cancer Res 2009;11(4):R46.

[72] Theodoropoulos PA, Polioudaki H, Agelaki S, Kallergi G, Saridaki Z, Mavroudis D, et al. Circulating tumor cells with a putative stem cell phenotype in peripheral blood of patients with breast cancer. Cancer Lett 2010;288(1):99–106.

[73] Raimondi C, Gradilone A, Naso G, Vincenzi B, Petracca A, Nicolazzo C, et al. Epithelial-mesenchymal transition and stemness features in circulating tumor cells from breast cancer patients. Breast Cancer Res Treat 2011;130(2):449–55.

[74] Yu M, Ting DT, Stott SL, Wittner BS, Ozsolak F, Paul S, et al. RNA sequencing of pancreatic circulating tumour cells implicates WNT signalling in metastasis. Nature 2012;487(7408): 510–3.

[75] Krebs MG, Sloane R, Priest L, Lancashire L, Hou JM, Greystoke A, et al. Evaluation and prognostic significance of circulating tumor cells in patients with non-small-cell lung cancer. J Clin Oncol 2011;29(12):1556–63.

[76] Hodgkinson CL, Morrow CJ, Li Y, Metcalf RL, Rothwell DG, Trapani F, et al. Tumorigenicity and genetic profiling of circulating tumor cells in small-cell lung cancer. Nat Med 2014;20(8):897–903.

[77] Shaffer DR, Leversha MA, Danila DC, Lin O, Gonzalez-Espinoza R, Gu B, et al. Circulating tumor cell analysis in patients with progressive castration-resistant prostate cancer. Clin Cancer Res 2007;13(7):2023–9.

[78] Attard G, Swennenhuis JF, Olmos D, Reid AH, Vickers E, A'Hern R, et al. Characterization of ERG, AR and PTEN gene status in circulating tumor cells from patients with castration-resistant prostate cancer. Cancer Res 2009;69(7):2912–8.

[79] Darshan MS, Loftus MS, Thadani-Mulero M, Levy BP, Escuin D, Zhou XK, et al. Taxane-induced blockade to nuclear accumulation of the androgen receptor predicts clinical responses in metastatic prostate cancer. Cancer Res 2011;71(18):6019–29.

[80] Cohen SJ, Punt CJ, Iannotti N, Saidman BH, Sabbath KD, Gabrail NY, et al. Relationship of circulating tumor cells to tumor response, progression-free survival, and overall survival in patients with metastatic colorectal cancer. J Clin Oncol 2008;26(19):3213–21.

[81] de Bono JS, Scher HI, Montgomery RB, Parker C, Miller MC, Tissing H, et al. Circulating tumor cells predict survival benefit from treatment in metastatic castration-resistant prostate cancer. Clin Cancer Res 2008;14(19):6302–9.

[82] Gasent Blesa JM, Alberola Candel V, Esteban Gonzalez E, Vidal Martinez J, Gisbert Criado R, Provencio Pulla M, et al. Circulating tumor cells in breast cancer: methodology and clinical repercussions. Clin Transl Oncol 2008;10(7):399–406.

[83] Goodman Jr OB, Fink LM, Symanowski JT, Wong B, Grobaski B, Pomerantz D, et al. Circulating tumor cells in patients with castration-resistant prostate cancer baseline values and correlation with prognostic factors. Cancer Epidemiol Biomarkers Prev 2009;18(6):1904–13.

[84] Zhang L, Ridgway LD, Wetzel MD, Ngo J, Yin W, Kumar D, et al. The identification and characterization of breast cancer CTCs competent for brain metastasis. Sci Transl Med 2013;5(180):180ra48.

[85] Husemann Y, Geigl JB, Schubert F, Musiani P, Meyer M, Burghart E, et al. Systemic spread is an early step in breast cancer. Cancer Cell 2008;13(1):58–68.

[86] Sanger N, Effenberger KE, Riethdorf S, Van Haasteren V, Gauwerky J, Wiegratz I, et al. Disseminated tumor cells in the bone marrow of patients with ductal carcinoma in situ. Int J Cancer 2011;129(10):2522–6.

[87] Braun S, Vogl FD, Naume B, Janni W, Osborne MP, Coombes RC, et al. A pooled analysis of bone marrow micrometastasis in breast cancer. N Engl J Med 2005;353(8):793–802.

[88] Stoecklein NH, Hosch SB, Bezler M, Stern F, Hartmann CH, Vay C, et al. Direct genetic analysis of single disseminated cancer cells for prediction of outcome and therapy selection in esophageal cancer. Cancer Cell 2008;13(5):441–53.

[89] Weckermann D, Polzer B, Ragg T, Blana A, Schlimok G, Arnholdt H, et al. Perioperative activation of disseminated tumor cells in bone marrow of patients with prostate cancer. J Clin Oncol 2009;27(10):1549–56.

[90] Guzvic M, Braun B, Ganzer R, Burger M, Nerlich M, Winkler S, et al. Combined genome and transcriptome analysis of single disseminated cancer cells from bone marrow of prostate cancer patients reveals unexpected transcriptomes. Cancer Res 2014;74(24):7383–94.

[91] Bednarz-Knoll N, Alix-Panabieres C, Pantel K. Clinical relevance and biology of circulating tumor cells. Breast Cancer Res 2011;13(6):228.

[92] Hanahan D, Weinberg RA. Hallmarks of cancer: the next generation. Cell 2011;144(5):646–74.

[93] Tsai JH, Donaher JL, Murphy DA, Chau S, Yang J. Spatiotemporal regulation of epithelial-mesenchymal transition is essential for squamous cell carcinoma metastasis. Cancer Cell 2012;22(6):725–36.

[94] Ledford H. Cancer theory faces doubts. Nature 2011;472(7343):273.

[95] Tarin D, Thompson EW, Newgreen DF. The fallacy of epithelial mesenchymal transition in neoplasia. Cancer Res 2005;65(14):5996–6000. Discussion-1.

[96] Molnar B, Ladanyi A, Tanko L, Sreter L, Tulassay Z. Circulating tumor cell clusters in the peripheral blood of colorectal cancer patients. Clin Cancer Res 2001;7(12):4080–5.

[97] Burgess DJ. Breast cancer: circulating and dynamic EMT. Nat Rev Cancer 2013;13(3):148.

[98] Mak P, Leav I, Pursell B, Bae D, Yang X, Taglienti CA, et al. ERbeta impedes prostate cancer EMT by destabilizing HIF-1alpha and inhibiting VEGF-mediated snail nuclear localization: implications for Gleason grading. Cancer Cell 2010;17(4):319–32.

[99] Zhu W, Cai MY, Tong ZT, Dong SS, Mai SJ, Liao YJ, et al. Overexpression of EIF5A2 promotes colorectal carcinoma cell aggressiveness by upregulating MTA1 through C-myc to induce epithelial-mesenchymal transition. Gut 2012;61(4):562–75.

[100] Cao H, Xu E, Liu H, Wan L, Lai M. Epithelial-mesenchymal transition in colorectal cancer metastasis: a system review. Pathol Res Pract 2015;211(8):557–69.

[101] Pena C, Garcia JM, Silva J, Garcia V, Rodriguez R, Alonso I, et al. E-cadherin and vitamin D receptor regulation by SNAIL and ZEB1 in colon cancer: clinicopathological correlations. Hum Mol Genet 2005;14(22):3361–70.

[102] Davis FM, Stewart TA, Thompson EW, Monteith GR. Targeting EMT in cancer: opportunities for pharmacological intervention. Trends Pharmacol Sci 2014;35(9):479–88.

[103] Polyak K, Weinberg RA. Transitions between epithelial and mesenchymal states: acquisition of malignant and stem cell traits. Nat Rev Cancer 2009;9(4):265–73.

[104] Lester RD, Jo M, Montel V, Takimoto S, Gonias SL. uPAR induces epithelial-mesenchymal transition in hypoxic breast cancer cells. J Cell Biol 2007;178(3):425–36.

[105] Yadav A, Kumar B, Datta J, Teknos TN, Kumar P. IL-6 promotes head and neck tumor metastasis by inducing epithelial-mesenchymal transition via the JAK-STAT3-SNAIL signaling pathway. Mol Cancer Res 2011;9(12):1658–67.

[106] Lo HW, Hsu SC, Xia W, Cao X, Shih JY, Wei Y, et al. Epidermal growth factor receptor cooperates with signal transducer and activator of transcription 3 to induce epithelial-mesenchymal transition in cancer cells via up-regulation of TWIST gene expression. Cancer Res 2007;67(19):9066–76.

[107] Thomson S, Petti F, Sujka-Kwok I, Epstein D, Haley JD. Kinase switching in mesenchymal-like non-small cell lung cancer lines contributes to EGFR inhibitor resistance through pathway redundancy. Clin Exp Metastasis 2008;25(8):843–54.

[108] Shiota M, Bishop JL, Nip KM, Zardan A, Takeuchi A, Cordonnier T, et al. Hsp27 regulates epithelial mesenchymal transition, metastasis, and circulating tumor cells in prostate cancer. Cancer Res 2013;73(10):3109–19.

[109] Scher HI, Jia X, de Bono JS, Fleisher M, Pienta KJ, Raghavan D, et al. Circulating tumour cells as prognostic markers in progressive, castration-resistant prostate cancer: a reanalysis of IMMC38 trial data. Lancet Oncol 2009;10(3):233–9.

[110] Tanaka H, Kono E, Tran CP, Miyazaki H, Yamashiro J, Shimomura T, et al. Monoclonal antibody targeting of N-cadherin inhibits prostate cancer growth, metastasis and castration resistance. Nat Med 2010;16(12):1414–20.

[111] Batlle E, Sancho E, Franci C, Dominguez D, Monfar M, Baulida J, et al. The transcription factor snail is a repressor of E-cadherin gene expression in epithelial tumour cells. Nat Cell Biol 2000;2(2):84–9.

[112] Cano A, Perez-Moreno MA, Rodrigo I, Locascio A, Blanco MJ, del Barrio MG, et al. The transcription factor snail controls epithelial-mesenchymal transitions by repressing E-cadherin expression. Nat Cell Biol 2000;2(2):76–83.

[113] Hajra KM, Chen DY, Fearon ER. The SLUG zinc-finger protein represses E-cadherin in breast cancer. Cancer Res 2002;62(6):1613–8.

[114] Comijn J, Berx G, Vermassen P, Verschueren K, van Grunsven L, Bruyneel E, et al. The two-handed E box binding zinc finger protein SIP1 downregulates E-cadherin and induces invasion. Mol Cell 2001;7(6):1267–78.

[115] Yang J, Mani SA, Donaher JL, Ramaswamy S, Itzykson RA, Come C, et al. Twist, a master regulator of morphogenesis, plays an essential role in tumor metastasis. Cell 2004;117(7):927–39.

[116] Fang X, Cai Y, Liu J, Wang Z, Wu Q, Zhang Z, et al. Twist2 contributes to breast cancer progression by promoting an epithelial-mesenchymal transition and cancer stem-like cell self-renewal. Oncogene 2011;30(47):4707–20.

[117] Perez-Moreno MA, Locascio A, Rodrigo I, Dhondt G, Portillo F, Nieto MA, et al. A new role for E12/E47 in the repression of E-cadherin expression and epithelial-mesenchymal transitions. J Biol Chem 2001;276(29):27424–31.

[118] Ling H, Fabbri M, Calin GA. MicroRNAs and other non-coding RNAs as targets for anticancer drug development. Nat Rev Drug Discov 2013;12(11):847–65.

[119] Shi Z, Zhang J, Qian X, Han L, Zhang K, Chen L, et al. AC1MMYR2, an inhibitor of dicer-mediated biogenesis of Oncomir miR-21, reverses epithelial-mesenchymal transition and suppresses tumor growth and progression. Cancer Res 2013;73(17):5519–31.

[120] Ocana OH, Corcoles R, Fabra A, Moreno-Bueno G, Acloque H, Vega S, et al. Metastatic colonization requires the repression of the epithelial-mesenchymal transition inducer Prrx1. Cancer Cell 2012;22(6):709–24.

[121] Ye X, Weinberg RA. Epithelial-mesenchymal plasticity: a central regulator of cancer progression. Trends Cell Biol 2015;25(11):675–86.

[122] Maheswaran S, Haber DA. Cell fate: transition loses its invasive edge. Nature 2015;527(7579):452–3.

[123] Fischer KR, Durrans A, Lee S, Sheng J, Li F, Wong ST, et al. Epithelial-to-mesenchymal transition is not required for lung metastasis but contributes to chemoresistance. Nature 2015;527(7579):472–6.

[124] Zheng X, Carstens JL, Kim J, Scheible M, Kaye J, Sugimoto H, et al. Epithelial-to-mesenchymal transition is dispensable for metastasis but induces chemoresistance in pancreatic cancer. Nature 2015;527(7579):525–30.

[125] Stoecklein NH, Klein CA. Genetic disparity between primary tumours, disseminated tumour cells, and manifest metastasis. Int J Cancer 2010;126(3):589–98.

[126] Cayrefourcq L, Mazard T, Joosse S, Solassol J, Ramos J, Assenat E, et al. Establishment and characterization of a cell line from human circulating colon cancer cells. Cancer Res 2015;75(5):892–901.

[127] Macarthur KM, Kao GD, Chandrasekaran S, Alonso-Basanta M, Chapman C, Lustig RA, et al. Detection of brain tumor cells in the peripheral blood by a telomerase promoter-based assay. Cancer Res 2014;74(8):2152–9.

[128] Satelli A, Mitra A, Cutrera JJ, Devarie M, Xia X, Ingram DR, et al. Universal marker and detection tool for human sarcoma circulating tumor cells. Cancer Res 2014;74(6):1645–50.

[129] Rodic S, Mihalcioiu C, Saleh RR. Detection methods of circulating tumor cells in cutaneous melanoma: a systematic review. Crit Rev Oncol Hematol 2014;91(1):74–92.

CHAPTER 13

Cancer Stem Cells: Metastasis and Evasion From the Host Immune Responses

A.S. Chopra[1,a], X. Liu[2,a], H. Liu[2,3]

[1]University of Connecticut, Storrs, CT, United States; [2]Case Western Reserve University, Cleveland, OH, United States; [3]Northwestern University, Chicago, IL, United States

CONTENTS

CANCER STEM CELLS, METASTASIS, AND HOST IMMUNITY: AN OVERVIEW

Metastasis, the spreading of cancer cells from the primary tumor site to distant organs, is responsible for over 90% of cancer-related deaths [1]. Table 13.1 shows the 5-year survival rate for the five most common cancers (by incidence rate) in the United States alone. The 5-year survival rate for these cancers is very low, contributing to the overall increased cancer-related deaths.

Cancer stem cells (CSCs) represent a subset of tumor cells that have the ability to self-renew and differentiate and are necessary for tumorigenesis and tumor progression [2,3]. Several lines of evidence suggest a critical role of CSCs in enhancing the metastatic potential of tumors by promoting migration and extravasation through an epithelial-to-mesenchymal (EMT) process [4–6]. CSCs, first identified in hematopoietic malignancies [7], have distinct characteristics that likely facilitate migration of tumor cells away from the primary tumor. For instance, coexpression of cell surface markers, CD133 and CD44, is considered a good predictive marker for metastatic CSCs in pancreatic ductal adenocarcinoma, whereas coexpression of CD133/CD44/CD166 on CSCs has been associated with poor clinical outcomes for colorectal cancers [8,9]. In addition to their self-renewal properties, subsets of CSCs may be mobile, invasive, and display resistance to host immune responses. Moreover, studies suggest a negative association between incidence of cancer and host immune responses [10–12].

The objective of this chapter is to evaluate the role of CSCs in establishing cancer metastasis and to propose the mechanisms that CSCs and cancer cells

[a] These authors equally contributed to the work.

Cancer Stem Cells. http://dx.doi.org/10.1016/B978-0-12-803892-5.00013-9

Table 13.1 5-Year Survival Rates of Individuals Diagnosed With Top Five Metastatic Cancers (by Incidence) in the United States

Metastatic Cancer	5-Year Survival (%)
Bladder	15
Breast	22
Colorectal	6
Endometrial	18
Renal	8

may utilize to escape targeted host immune responses. This chapter reviews the current literature on CSCs and the role of CSCs in promoting metastasis. The chapter describes, in brief, the characteristics of circulating tumor cells (CTCs) and metastatic CSCs (MCSCs), in addition to their value for the development of novel therapeutic agents. This chapter also discusses potential mechanisms that CSCs employ to evade and suppress the host immune responses and the current knowledge of the innate and adaptive immune and inflammatory responses that may be enhanced to target CSCs for elimination. Understanding the role of CSCs in cancer progression and metastasis is valuable from a therapeutic perspective and may lead to the development of effective strategies to specifically and completely eliminate cancer cells and CSCs, thereby improving the overall quality of life in cancer patients and reducing the costs and burden associated with cancer.

CANCER STEM CELLS AND METASTASIS

The importance of CSCs in tumor initiation has been well established in leukemia and a variety of solid tumors enumerated in different chapters throughout this book. However, the role of CSCs in metastasis has not been well defined with only partial experimental evidence to support the hypotheses. In this section, we review the current understanding of the role of CSCs in metastasis and their potential as therapeutic targets against cancer metastasis.

Metastatic Cancer Stem Cells: Concept and Origin

MCSCs, theoretically, are a subgroup of CSCs that display both stem cell-like properties and are capable of contributing to overall cancer metastasis. The concept of "migrating cancer stem cell" was first introduced by Brabletz et al. [13]. In their model, they separated the CSCs into stationary and mobile CSCs (migrating CSCs)—stationary CSCs cannot disseminate, but a subpopulation of migrating CSCs can disseminate via blood or lymphatic vessels and generate a metastatic mass at a new location. This heterogeneity within the CSC compartment was first identified in pancreatic cancers where the CSC population is defined by CD133 expression. However, among these $CD133^+$ populations, only the subpopulation expressing CXCR4 ($CD133^+$ $CXCR4^+$) is directly responsible for metastasis [14]. In vivo, only migrating CSC population ($CD133^+$ $CXCR4^+$) can give rise to CTCs, detectable in the portal vein followed by liver metastasis [14]. In 2006, Balic et al. first linked metastasis to CSCs by showing that a majority of early disseminated cancer cells in the bone marrow from breast cancer patients possess a putative stem cell phenotype [15], suggesting that CSCs might be the "lethal seeds" capable of reinitiating growth to form metastases. Although the presence of these cells in the bone marrow is associated with a poor prognosis, almost 50% of patients with micrometastases in their bone marrow did not develop apparent macrometastases during a 10-year follow-up, suggesting some disseminated cancer cells (DTCs) may not have stem cell properties or remain quiescent to reinitiate a tumor growth. Baccelli et al. proposed the "metastasis-initiating cell" (MIC) concept in which the MICs are functionally distinguished from CSCs by their metastatic capacity, meaning that MICs must be able to disseminate, survive in the systemic circulation, extravasate to the metastatic site, and grow in the new environment [16]. However, not every DTC has the metastasis-initiating potential; recently, Oskarsson et al. improved the model for MCSCs [17]. They used the term "metastatic stem cell" referring to any DTC that is capable of reinitiating macroscopic tumor growth in a distant tissue [17].

The origin of MCSCs remains elusive and may depend on the type of cancer. One possibility is that subpopulations of disseminated non-MCSCs may regain tumor-initiating capacity due to phenotypic plasticity, such as EMT, once exposed to the appropriate microenvironment. Several studies have demonstrated that EMT can induce a stem cell-like property of self-renewal and an increase in the CSC population [18,19] (Fig. 13.1A). However, the mechanism by which EMT links to CSCs is unclear; EMT may not necessarily be required for metastasis under certain circumstances as shown in two lineage tracing mouse studies of breast cancer [20] and pancreatic cancer [21]. Alternatively, MCSCs may already exist within subpopulations of CSCs in the primary tumor [17,22], which is supported by the evidence that the cell population capable of generating metastases in vivo can be isolated from primary tumor cells using stem cell markers [5,23]. Another line of evidence, supporting the preexistence of MCSCs in the tumor, is provided by Dieter et al. who found three distinct types of CSCs in colon cancers—extensively self-renewing long-term tumor-initiating cells (LT-TICs), tumor transient amplifying cells (T-TACs), and delayed contributing tumor-initiating cells (DC-TICs) [24]. Among them, LT-TICs are the only subpopulation that is able to contribute to metastasis [24] (Fig. 13.1B). Yet, some of the other studies hypothesize cell fusion as a mechanism for the MCSCs establishment [25,26]. In this notion, cancer cells may either fuse with other cancer cells or with tumor tissue–associated cells, such as

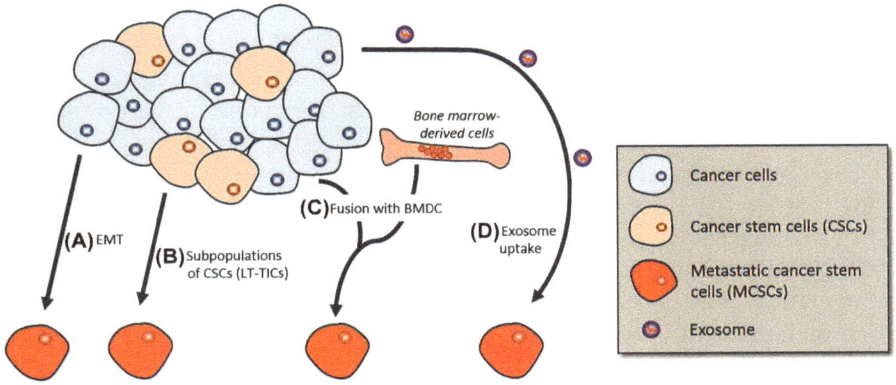

FIGURE 13.1 The origin of MCSCs.

(A) Some primary non-CSCs may regain the tumor-initiating capacity through phenotypic plasticity (EMT) once exposed to the appropriate microenvironment or arrested in the premetastatic niche. (B) MCSCs may already exist within subpopulations of CSCs (LT-TICs) in the primary tumor. (C) Cancer cells fuse with migratory BMDCs to generate hybrids, which can metastasize to another site and establish secondary cancer. (D) Cancer cells may internalize extracellular vesicles or exosomes that are transporting RNAs, lipids, and proteins from CSCs, stem cells, and/or myeloid cells (monocytes and neutrophils). *BMDC*, bone marrow-derived cell; *EMT*, epithelial-to-mesenchymal.

macrophages, bone marrow-derived cells (BMDCs), or tissue stem cells. These hybrids are thought to undergo "nuclear reprogramming" or metastasize to another site to maintain their cancer-initiating properties [27,28] (Fig. 13.1C). Most of the published cell fusion studies are based on analyses of limited markers and a small number of events, lacking DNA content analysis at a single cell level. Ding et al. found that breast cancer cells fused with macrophages gained $CD44^{+}/CD24^{-/low}$ phenotype and EMT-associated genes, along with enhanced tumorigenicity and metastatic ability in nonobese diabetic/severe combined immunodeficient (NOD/SCID) mice [29]. Another hypothesis suggests that cancer cells or stromal cell-derived extracellular vesicles, especially exosomes, may serve as mediators of cell-to-cell communication and potentially transmit stem cell phenotypes to recipient cancer cells, facilitating tumorigenesis and metastasis [30] (Fig. 13.1D). However, there is no direct evidence demonstrating that exosomes can modulate the fate of cancer cells to facilitate their conversion or transformation into MCSCs.

Markers for Metastatic Cancer Stem Cells

Human MCSCs have not been prospectively identified to date. However, accumulating evidence indicates that MCSCs might be found within subpopulations of CSCs, which can initiate metastasis. In 2006, Balic et al. provided the first evidence suggesting the existence of putative stem-like phenotype ($CD44^{+}$ $CD24^{-/low}$) within the DTC in bone marrow of early breast cancer patients [15].

Table 13.2 Markers for Metastatic Cancer Stem Cells (MCSCs) by Tumor Type

Tumor Type	MCSC Markers
Breast cancer	ALDH$^+$ CD44$^+$ CD24$^-$ [32] EPCAM$^+$ CD44$^+$ CD47$^+$ MET$^+$ [16]
Colorectal cancer	CD44v6$^+$ [33], CD26$^+$ [23]
Pancreatic cancer	CD133$^+$ CXCR4$^+$ [14]
Liver cancer	CD45$^-$ CD90$^+$ [37]

Later, Liu et al. found that CD44$^+$ breast CSCs are metastatic in patient-derived xenografts (PDXs) and develop spontaneous lung metastases [5]. An alternative marker proposed for the breast CSCs is aldehyde dehydrogenase (ALDH). Intracardiac injection of ALDH$^+$ cells but not ALDH$^-$ cells into NOD/SCID mice generated multiple metastases at distinct organs (bone, muscle, lung, and soft tissue), indicating that the metastatic ability is predominantly mediated by CSCs contained in the ALDH$^+$ population [31]. When combining the ALDH$^+$ and CD44$^+$ CD24$^{-/low}$ markers to sort CSCs from different breast cancer cell lines, Croker et al. found that only ALDH$^+$ CD44$^+$ CD24$^-$ cells can grow into large pulmonary metastases, indicating that the metastatic ability was enhanced in this CSC-enriched population [32]. However, ALDH expression in CSCs and human tumor metastasis may also be context and cancer subtype dependent.

The evidence suggesting the existence of MCSCs has also been observed in other cancers (Table 13.2). In pancreatic cancers, CD133$^+$ CSCs have the ability to initiate tumor development; however, only the subpopulation of CD133$^+$ CXCR4$^+$ CSCs are essential for liver metastasis [14]. Depletion of CD133$^+$ CXCR4$^+$ CSC pool virtually abrogated the metastatic phenotype of pancreatic tumors without affecting their tumorigenic potential. In colorectal cancer, subpopulation of CD26$^+$ cells were uniformly present in both the primary tumor and the liver metastasis [23]. Isolated CD26$^+$ cells, but not CD26$^-$ cells, led to development of distant metastasis when injected into mouse cecal wall [23]. Furthermore, the presence of CD26$^+$ CSCs in the primary tumor was predictive of the development of metastasis during clinical follow-up [23]. In contrast, patients without CD26$^+$ cells did not develop distant metastasis during follow-up [23]. Recently, Todaro et al. found that CD44v6 was expressed in all colorectal CSCs, which is required for the metastatic potential of CSCs [33]. CD44v6$^-$ colorectal progenitor cells, on the other hand, had no potential to enhance the metastasis of CSCs, but when treated with cytokines, these cells acquired CD44v6 expression and metastatic capacity [33]. This evidence indicates that the analysis of CSC subsets in the primary tumor can be utilized to predict metastasis. Development of therapeutic strategies targeting CSCs and MCSCs will be essential in diagnosing and preventing cancer metastasis.

Circulating Tumor Cells and Metastatic Cancer Stem Cells

Cancer cells that detach from the original tumor and circulate through the bloodstream or lymphatics are called CTCs, which are considered responsible for the development of metastases. The presence of CTCs in a patient's blood is associated with poor prognosis. Millions of cells are shed from a tumor every day (approximately 4×10^6 cells/g primary tumor) [34]; however, only a minority of the CTCs may form a secondary tumor. This has led to the hypothesis that only a small proportion of CTCs with stem cell-like properties allows them to form a tumor at a distant site. Recent studies indeed have shed light on the relationship of CTCs, MCSCs, and metastasis.

CTCs isolated from peripheral blood of patients frequently exhibit EMT features and putative stem cell phenotypes. Yu et al. characterized EMT in CTCs from breast cancer patients. They found that mesenchymal cells were highly enriched in CTCs (M^+ CTCs) and associated with disease progression [35]. Interestingly, the increase in M^+ CTCs was accompanied by the appearance of CTC clusters, which are strongly positive for mesenchymal markers but not epithelial markers [35]. Another study demonstrated that 66.7% of breast cancer patients show a putative stem cell/progenitor phenotype (35.2% $CD44^+$ $CD24^{-/low}$) or (17.7% $ALDH1^{high}$ $CD24^{-/low}$) in CTCs [36]. Similarly, a subgroup of CSCs ($CD45^-$ $CD90^+$) is detectable in 90% of blood samples from liver cancer patients. $CD45^-$ $CD90^+$ CSCs isolated from the circulation of liver cancer patients generate tumors in a second and even subsequent third batch of immunodeficient mice [37].

However, direct experimental evidence is lacking regarding whether these "stem cells" in CTCs can lead to metastasis. A study characterized a subpopulation of MCSCs ($EPCAM^+$ $CD44^+$ $CD47^+$ MET^+) in the blood of patients with metastatic luminal breast cancer [16]. These cells were correlated with a lower overall survival and increased metastatic potential in xenograft assays compared with bulk epithelial cell adhesion molecule-positive ($EPCAM^+$) cells. In addition to single CTCs, the CTC clusters/aggregates are frequently observed in cancer patients and correlate with poor prognosis [38]. Although rare in the circulation compared to single CTCs, CTC clusters show increased metastasis potential [39]. Although Yu's finding [35] suggested that CTC clusters may possess EMT properties that may be linked to stem cell properties, further research will be needed to determine the stem cell markers and signaling in clustered CTCs. A better understanding of the CTC clusters may identify a novel and potential target for metastatic cancers.

Dormant Metastatic Cancer Stem Cells

Surgical resection of primary tumors may not completely prevent recurrence. Many cancer patients suffer from metastatic relapse years after the removal of the primary tumor, increasing the risk of cancer-related mortalities. One common risk for cancer relapse is the reactivation of dormant CSCs and MCSCs. Increasing evidence suggests that once the disseminated tumor cells settle in the target organ, they may sustain CSC-like properties in a dormant state, resisting conventional therapies and becoming the source of cancer recurrence. During the dormant period, a subset of these dormant cells or MCSCs may be reactivated and give rise to metastatic tumors. Therefore, understanding the mechanisms that allow cells to enter dormancy and/or be reactivated will provide new insights into the development of strategies to target dormant MCSCs for elimination.

Liao et al. proposed three different strategies to target MCSCs—direct targeting of CSCs via surface markers, targeting self-renewal and differentiation pathways, or interrupting both CSC niche and premetastatic niche [40]. However, these strategies do not specifically target dormant MCSCs. Given that dormant MCSCs are resistant to conventional antiproliferative therapies, "awakening" dormant cells has been proposed to facilitate killing. However, evidence suggests that awakened dormant tumor cells seem to promote tumor recurrence and may worsen patient outcome in the absence of effective cytotoxic therapies [41]. Thus, maintaining dormant state and/or killing dormant CSCs directly using innovative approaches are therapeutically more attractive than attempting to wake them from dormancy [42,43]. This notion is supported by a study showing that the quiescent leukemia stem cell (LSC) pool in chronic myeloid leukemia (CML) can be gradually eradicated by the glitazones, agonists of peroxisome proliferator–activated receptor-γ (PPARγ) [44]. A further mechanistic study indicated that glitazones decreased expression of signal transducer and activator of transcription 5 (STAT5), hypoxia-inducible factor-2α (HIF2α), and CREB-binding protein/p300-interacting transactivator with Asp/Glu-rich C-terminal domain, 2 (CITED2), key guardians of the quiescence and stemness of CML LSCs [44].

To achieve the goal of keeping dormant CSCs quiescent, a better understanding of the biology of cancer cell dormancy, including the molecular mechanisms and the signaling pathways that lead to dormancy, is necessary. In a review, Sosa et al. summarized the markers that might dictate the state of DTCs [45]. Proliferative markers include increased expression of COCO, fibronectin, periostin (POSTN), transforming growth factor-β1 (TGF-β1), and phosphorylated ERK (p-ERK), as well as low expression of phosphorylated p38 (p-p38). Meanwhile, dormancy markers include low expression of p-ERK as well as high expression of p-p38, TGF-βR3, bone morphogenetic protein receptor type 2 (BMPR2), TGF-β2, and BMPs [45]. Therefore, strategies aimed at maintaining dormancy and/or suppressing proliferative signaling could theoretically result in preventing tumor recurrence. For example, a gain-of-function cDNA screen revealed that Coco, a secreted antagonist of TGF-β ligands, was able to induce the exit from dormancy of breast cancer cells and proliferation by blocking paracrine BMP signaling [46]. Alternatively, targeting dormancy-specific survival mechanisms could potentially eradicate MCSCs. C-Src activation is associated with late onset bone metastasis in breast cancer. A mechanistic study revealed that Src supports the survival of dormant breast cancer cells in the bone marrow by activating Akt in responses to chemokine ligand 12 (CXCL12) and tumor necrosis factor (TNF)-related apoptosis inducing ligand (TRAIL) [47], suggesting that c-Src could be a target to kill dormant CSCs.

Ghajar proposed that a direct targeting of the dormant niche may be used to prevent metastasis [43]. Once the DTCs reach a target tissue, these cells likely undergo proliferative arrest (dormancy) in the niche due to deprivation of appropriate adhesive and signaling interactions. Therefore, this study raises the possibility that we can target these cells by altering their dormant niche. One strategy to target dormant niche may be to sensitize dormant DTCs to cytotoxic treatments, which might be achieved by targeting molecules that tether DTCs to the niche or confer therapeutic resistance within the niche [43]. CXCL12 and osteopontin, which are derived from the osteoblastic and the perivascular niches (PVNs), have the major role in mediating chemoresistance of LSCs [48]. Antagonists to receptors for the niche constituents CXCL12 (CXCR4 antagonists) and osteopontin (CD44 and integrin $\alpha_4\beta_1$ antagonists) have been employed to treat refractory acute myeloid leukemia (AML) patients. The efficacy of CXCR4 inhibitors to sensitize refractory AML cells to chemotherapy is now being tested in clinical trials [48]. PVNs support MCSCs by supplying attachment, oxygen, nutrients, and paracrine factors from the activated endothelium. Endothelial-derived thrombospondin-1 (TSP-1) in PVNs can induce sustained breast cancer cell quiescence. However, this suppressive cue was lost in the sprouting neovasculature in which tumor-promoting factors TGF-β1 and POSTN are rich [49], suggesting that stable microvasculature constitutes a dormant niche, whereas the sprouting neovasculature sparks micrometastatic outgrowth. Infiltrating tumor cells also need to induce stromal POSTN expression in the secondary target organs, such as the lung, to initiate colonization [50]. POSTN is required for cancer stem cell maintenance and blocking its function prevents metastasis. Taken together, a direct disruption of supportive niches of MCSCs may be a novel strategy for the treatment of metastasis. A better characterization of the niche constituents and signaling cross talk between MCSCs and their niches is needed in the future studies.

THE INTERPLAY BETWEEN IMMUNE RESPONSES AND CANCER

Recent success of anticytotoxic T lymphocyte antigen 4 (anti-CTLA-4) and anti-programmed death ligand 1 (anti-PD-L1) antibodies against subsets of metastatic melanoma and lung cancers have led insight into immunomodulation as a potential method against more aggressive cancers [51]. Elimination of cancer and CSCs while minimizing adverse effects on surrounding normal tissue requires effective and precise targeting of CSCs, their self-renewal and differentiation pathways in CSCs, and/or premetastatic niches. Immune cells are able to identify cancer cells and CSCs through expression of cell surface markers. However, cancer cells and CSCs may employ various methods to not only dampen immune response but also modulate such responses to the support tumor growth and development. Although a plethora of evidence demonstrate a reduction in immune activity by cancer cells, little evidence exists to support the role of CSCs in immune modulation. In this section, we will focus on the published literature on cancer cells and discuss the immune-modulating potential of CSCs.

Acute inflammatory and immune responses can suppress cancer growth and progression through direct cytotoxic effects on cancer cells. Attempts to employ immune and inflammatory stimulants for cancer treatment date back to the 19th century [52,53]. William Coley was one of the pioneers in using bacterial toxins to successfully treat patients with sarcomas. Alternatively, chronically inflamed tissues have long been noted to spawn neoplastic lesions [54,55]. Rudolf Virchow observed the presence of leukocytes within tumors, provisioning the first evidence of an association between immune response and tumor development [56].

Transition of the immune and inflammatory response from an acute, cytotoxic response to a chronic, nonresolving inflammation is likely to contribute to these divergent effects. Cancer cells in the premalignant tissues may develop resistance to the apoptotic ligands generated by immune and inflammatory cells, which may support metastatic progression. Cancer cells use inflammatory cytokines to promote cell proliferation and survival in the tumor microenvironment. Moreover, altered expression of surface markers in cancer cells help evade phagocytosis and promote cell proliferation and metastasis [57,58]. In addition, cancer cells and CSCs, in close association with macrophages, activate a positive feedback signaling to induce intravasation and promote metastasis [59,60]. Finally, once in the blood vessels, metastatic CTCs activate a coagulation cascade and platelets to survive the shearing forces of the bloodstream and avoid detection by the immune cells in the bloodstream [61,62].

Overall, the immune responses against cancer cells exhibit conflicting roles at various stages of tumor progression and cancer cell metastasis. The role of immune cells and inflammatory responses in tumor initiation and metastasis is not well understood, including how cancer cells and CSCs modulate immune attacks to dampen the apoptotic signaling and how immune cells regulate CSCs and MCSCs in tumor progression. Understanding how CSCs and MCSCs modulate immune signals to promote tumorigenesis is important for developing effective tumor-preventive and immunotherapeutic agents. Moreover, understanding the mechanisms by which metastatic cancer cells or MCSCs survive immune surveillance may help reduce the incidence of secondary tumors. Specifically targeting the CSCs and the MCSCs via alteration of innate immune responses holds promise in the field of cancer therapy. In this section of the chapter, we review the literature on the role of the immune system in tumorigenesis and metastasis as well as the current knowledge of how cancer cells evade immune responses.

Innate and Adaptive Immune Responses Against Tumor Cells

The immune and acute inflammatory responses have the potential to suppress cancer growth and progression through the death-inducing capacity of the infiltrating immune cells. Natural killer (NK) cells, antigen-presenting cells (APCs), and subclasses of T and B lymphocytes can mount an immune response and, thus, are generally considered tumor suppressive [63]. The potentiation of the adaptive immune response through recognition of tumor-associated antigens (TAAs) by $CD8^+$ cytotoxic T cells (CTLs) can induce cell death of antigenic cancer cells using perforin/granzyme and CD95L (FasL) [64–66]. Cells of the innate immune response also have anticancer activities. NK cells can induce cell death through both the perforin/granzyme and the Fas pathways, and their presence in tumors has been associated with improved patient survival [58,67]. Moreover, activation of NK cells by Nlrp3 inflammasome has been shown to suppress colorectal cancer metastasis to the liver [68]. The importance of the immune responses in cancer suppression is evident from a greater risk of certain cancers in immunocompromised individuals. Studies suggest that mice deficient in the development or function of $CD8^+$ CTLs, $CD4^+$ T_h1 helper T cells, and/or NK cells exhibit higher incidence of tumor development [69,70]. Despite the expression of TAAs, EMT and autophagy induction make tumor cells elusive to the host immune system, resulting in poor clinical outcomes of cancer treatment and most cancer vaccines [71].

In addition to the cytotoxic effects of the T cells, a number of cytokines, including TNF, and TRAIL, have been implicated in tumor suppression; TNF was first identified as a mediator of tumor necrosis in animals treated with endotoxin [72–74] and was originally envisioned as a potential anticancer therapeutic. However, the transition of the inflammatory response from an acute to a chronic, nonresolving inflammation and increased cancer cell resistance to apoptotic ligands generated by immune cells pose an impediment to the elimination of cancer cells by the immune system. Acute inflammation and presence of TNF during early stages of tumorigenesis may suppress tumor growth, but higher levels of TNF in the tumor microenvironment are associated with tumor promotion via the NF-κB transcription pathway. Such a transition, mediated by chronic inflammation, can facilitate tumor initiation, promotion, and metastatic progression [57]. In addition, transplantation experiments revealed that weakly immunogenic cancer cells, generated under high immune response selection by "immunoediting," are more capable, than their strongly immunogenic counterparts, in proliferating in an immunocompetent environment [75]. Together, these data support the hypothesis that CSCs may be a part of the weakly immunogenic subpopulation of cancer cells, enabling them to survive and proliferate under high immune and inflammatory cell infiltration of the tumor. Developing approaches that promote the anticancer actions of the immune and inflammatory response may have important applications in cancer treatment and prevention.

Mechanisms of Evading the Host Immune Response

Evasion of immune surveillance is an emerging hallmark of cancers. The mechanisms that cancer cells employ to modulate the immune response are not fully understood. Nevertheless, these mechanisms of immune evasion can largely be classified into three subcategories: (1) alteration of the cancer cells to avoid killing by the host immune response, (2) modulation of the immune response in the tumor microenvironment to aid in tumor progression, and (3) implementation of immunosuppressive methods to suppress apoptotic immune response and to promote tumor growth as well as cancer metastases. CSCs have been proposed to evade the host immune responses. However, the hypothesis needs to be appropriately tested in immune-competent models.

Alteration of Cancer Cells and Cancer Stem Cells

Both the innate and adaptive immune responses aim to eliminate tumor cells. Inflammation, one of the first innate immune responses, plays an important role in activating apoptotic signaling of tumor cells. Various inflammatory cytokines, such as TNF, produced by macrophages, have been shown to induce tumor necrosis and are associated with better clinical outcomes [76]. However, as noted earlier, chronic and nonresolving inflammation along with resistance of cancer cells to apoptotic activity of the inflammatory response can lead to tumor promotion. TNF and other cytokines (such as interleukin-6 (IL-6)) have been implicated in both tumor regression and tumor promotion. Interaction of these cytokines with their corresponding receptors on the tumor cells can mediate both apoptotic and cell survival response. For instance, expression of the proinflammatory IL-6 has been implicated in cancer progression (Fig. 13.2). Furthermore, activated prostate epithelial stem cells have been hypothesized to acquire growth and survival advantage by expressing one or more of the proinflammatory cytokines, such as IL-6 [77]. Activation of the prostate CSCs is thought to initiate self-propagating autocrine signaling with positive feedback from IL-6, even in the absence of inflammation. Second, suppression of the apoptotic responses by

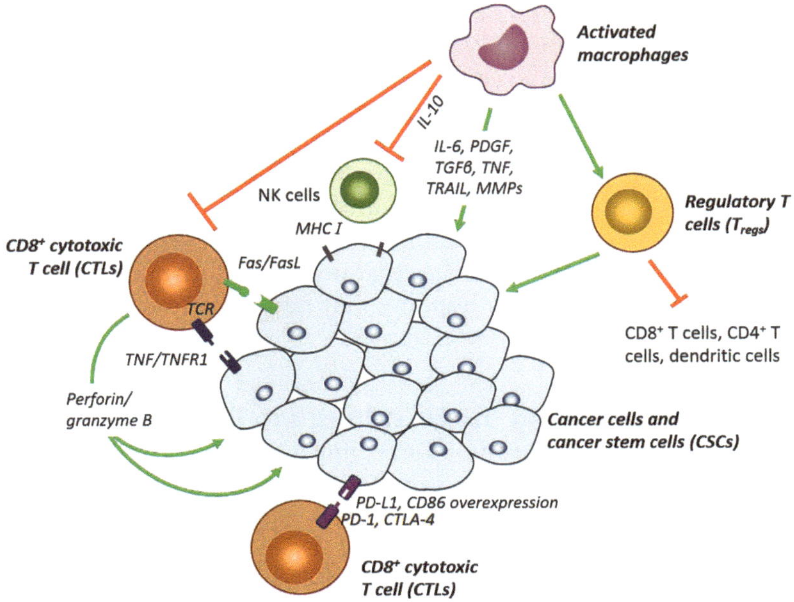

FIGURE 13.2 Innate and adaptive host immune responses against cancer cells and CSCs and some of the mechanisms of evasion.

CSCs and cancer cells evade detection by apoptotic receptor/ligand interactions through elimination of cell surface expression of tumor antigens (death signaling pathways such as FasL). Cancer cells or CSCs maintain threshold levels of antigen presentation (MHC Class I) to avoid lysis by natural killer (NK) cells and may also recruit macrophages that suppress a slew of immune responses. Activated macrophages suppress the activity of CTLs, antigen presentation by dendritic cells, secrete IL-6, TGF-β, TNF, and other prosurvival and growth signaling molecules. These activated macrophages also secrete IL-10 (antiinflammatory) to suppress the activity of NK cells on cancer cells. Moreover, activated macrophages recruit regulatory T cells into the tumor microenvironment for immunosuppression, and in turn, promote growth and survival of cancer cells as well as CSCs. Finally, cancer cells and CSCs have been demonstrated to overexpress CD8+ CTL costimulatory signals (CTLA-4 and PD-L1) that have an immunosuppressive effect. *MHC*, major histocompatibility complex.

the cancer cells favors the cell survival arm of the cytokine–receptor interaction. In addition, growing evidence points toward interaction of TNF with more than one receptor (TNFR1 and TNFR2) for its activity. Although TNF primarily binds to TNFR1 to mediate both apoptosis and cell survival [78,79], evidence indicates that cancer cells may preferentially increase the expression of TNFR2 in both mice and humans; TNFR2 exclusively signals for cell survival and proliferation via NF-κB pathway [80]. Hamilton et al. showed an increase in the expression of TNFR2 in AOM/DSS mouse model of colitis-associated cancers [81]. Moreover, TNFR1 and TNFR2 knockdown studies have demonstrated the protective effect of TNFR1 and tumor-promoting role of TNFR2 in the presence of TNF [81].

Additionally, the synthesis of prostaglandins by inflammatory cells expressing inducible cyclooxygenase 2 (COX-2) may further aid in the metastatic progression [82] and chemoresistance [83]. Kurtova et al. [83] demonstrated a novel mechanism of CSC repopulation in bladder urothelial carcinoma xenografts in response to gemcitabine and cisplatin. The study showed that PGE_2 release promoted the repopulation of CSCs even in the presence of the chemotherapeutic and contributed to chemoresistance in the xenografts. The authors further showed that PGE_2-neutralizing antibodies and the addition of celecoxib, a selective COX-2 inhibitor effectively abrogated chemoresistance and repopulation of CSCs in the xenografts [83]. Collectively, these studies shed new insights on the progressive development of chemoresistance and may aid in the development of novel adjunct therapies to reduce proliferation and recurrence of cancer cells by targeting CSCs.

Several other components of the immune response may also elicit apoptotic induction in cancer cells, including APCs, T cells, and NK cells. Presentation of the TAAs on major histocompatibility complexes (MHCs) and their recognition by APCs allows for the recruitment of $CD8^+$ T cells that induce apoptotic responses in cancer cells and enhance the inflammatory signaling in the tumor microenvironment [84] (Fig. 13.2). $CD8^+$ T cells play a crucial role in the host defense against malignancies in both mice and humans [71]. Identification of TAAs by the $CD8^+$ T lymphocytes has allowed for the development of specific immune responses in immunotherapeutic strategies. Cancer cells have been shown to reduce the cell surface expression of these antigens and/or that of MHC-I, to avoid detection by CTLs. However, cancer cells maintain threshold levels of self-antigen presentation on their cell surfaces to avoid detection and extrinsic apoptosis activation by NK cells [85]. Notably, this potent immune response is often dampened by tumor-derived overexpression of immunosuppressive cytokines, such as TGF-β, IL-10, and IL-23 [86] (Fig. 13.2). Furthermore, cancer cells and CSCs may actively suppress immune responses using costimulatory signals [87]. Tumor cells have been shown to overexpress CD86 and PD-L1 to inhibit the activity of CTLs through their interaction with CTLA-4 and programmed death receptor 1 (PD-1), respectively [88] (Fig. 13.2). Studies have demonstrated an association between overexpression of these immunosuppressive costimulatory signals and poor clinical outcomes in a variety of advanced stage solid tumors [87–90]. In addition, anti-CTLA-4 and anti-PD-L1 therapies have been shown to be effective in clinical trials [91]. These strategies build a resistance to cytotoxic killing by T cells and hinder the activity of immune stimulatory chemotherapeutic agents.

There are a number of other known alterations that cancer cells and CSCs utilize to evade the host immune system. Some of the changes in expression, such as overexpression of CD133, CD44, and CD47, are linked to increased metastatic potential of cancer cells and CSCs by evading the immune cell-mediated attacks and by mobilizing the protumorigenic immune responses. Cancers of varying origins may also have specific markers that they utilize to enhance CSC proliferation, survival, metastasis as well as immune evasion. For instance, CD47 interacts with and induces ligation of signal-regulatory protein alpha (SIRP-α) on macrophages to prevent phagocytosis [92]; expression of CD47 marks CSCs in many tumor types by inhibiting macrophage-mediated phagocytosis, and it has been shown to be effective targets of neutralizing monoclonal antibodies used to treat non-Hodgkin's lymphoma and other tumors [93]. Furthermore, another study showed the involvement of toll-like receptor 2 (TLR2) in maintenance of stem/progenitor cells in the intestinal and mammary epithelia [94,95]; TLR2 is a membrane receptor that plays an important role in pathogen recognition and activation of the innate immune response via NF-κB pathway [96]. In contrast to the hypothesis that TLR2 is an important receptor in pathogenesis of cancer, Ey et al. showed that deletion of *TLR2* and/or its downstream effector or coreceptors along with *MDR1A* multidrug resistance gene, can exacerbate colitis, increasing the risk of colitis-associated colon cancer [97]. Thus, further research is needed to understand the dual nature of this signaling pathway. In addition to the cellular alterations that give cancer cells and CSCs advantage to evade the immune responses, evidence points to the morphologic changes associated with EMT phenotype of the cancer cells and CSCs. EMT is associated with $CD24^-$/$CD44^+$/$ALDH^+$ stem cell population and enhances therapeutic resistance as well as escape from the immune surveillance [71]. EMT also involves the activation of several important growth regulatory pathways that help tumors survive and evolve into highly invasive and metastatic cancer cells. Recent investigations have revealed that Zinc Finger E-box Binding Homeobox 1 (ZEB1) suppresses microRNA-200 (miR-200) to promote EMT. MiR-200 is a known suppressor of PD-L1; suppression of miR-200, therefore, results in upregulation of PD-L1 on tumor cells and $CD8^+$ T cell immunosuppression [98]. Further research is warranted to acquire insights on how cancer cells, CSCs, and MCSCs alter their morphologies to evade the immune system and take advantage of the proinflammatory responses to promote tumor growth, survival, and metastasis. Identification of the specific targets may ultimately result in development of effective therapeutic and tumor-preventive agents.

Modulation of Tumor Microenvironment

A key player in the innate immune response and the generation of inflammatory cytokines are the tumor-associated macrophages (TAMs) (Fig. 13.2). Tumors infiltrated by activated macrophages are typically classified as aggressive and highly metastatic. Macrophages release several growth factors, such as epidermal growth factor (EGF), platelet-derived growth factor (PDGF), and hepatocyte growth factor (HGF) that enhance growth and survival of the tumor cells [99]. TAMs are recruited to the areas of hypoxia and necrosis (tumor microenvironment) and secrete vasoactive factors, such as IL-8, prostaglandin E2 (PGE_2), and vascular endothelial growth factor (VEGF) that induce angiogenesis in the tumor

[99]. Moreover, these macrophages also secrete various proteases, such as matrix metallopeptidase-9 (MMP-9), that enhances the bioactivity of the vasofactors [60,99]. Together, macrophages act to enhance proliferation, survival, and the metastatic potential of cancer cells. TAMs affect the growth and metastatic potential of cancer cells as well as CSCs. Jinushi et al. showed that TAMs and their downstream factor, milk-fat globule epidermal growth factor-VIII (MGF-E8), in synergy with IL-6, conferred CSCs with the ability to promote tumorigenesis and anticancer drug resistance in MC38-CSC colon and 3LL-CSC lung cancer cells [99]. IL-6 mediated activation of STAT3 transcription pathway is necessary for transformation of mutated cells, and in turn, activates the NF-κB pathway, completing a positive feedback loop [100]. Similar results by Zhou et al. reported that glioma stem cells secrete periostin to recruit TAMs enriched in glioblastoma multiformes [101]. Higher levels of periostin enhance tumor growth, and in turn, enhance TAM density in the tumor microenvironment through integrin $\alpha_v\beta_3$ [101]. Conversely, defects in macrophage lineage or a failure to activate macrophages have been associated with fewer metastases. For instance, mice with CSF-1 mutations have defects in macrophage development and, thus, rarely develop lung metastases [102]. In contrast, evidence demonstrated that a subset of macrophages or nonclassical patrolling monocytes with specific markers (CX3CR1^{high}Ly6C$^-$ in mouse; CX3CR1^{high}CD14^{dim}CD16$^+$ in humans) may inhibit metastasis by expressing the orphan nuclear receptor Nr4a1 [103].

Immunosuppression by Utilizing Regulatory Immune Responses

Suppression of the host immune response by using the host immune regulatory signaling is a relatively new area of research. Little is known about the factors that play a part in immunosuppression and their associated mechanisms. Immunosuppression requires the recruitment of immune regulatory signals that normally function to suppress the immune responses caused by a failure of self/nonself discrimination. Myeloid-derived suppressor cells (MDSCs) and regulatory T cells (T_{regs}) are pivotal components of immunosuppression in the tumor microenvironment. Presence of both types of suppressors in the tumor microenvironment is generally associated with poor clinical outcomes. Moreover, high level of MDSCs along with T_{reg} infiltration has been seen in a variety of cancers. T_{regs} are a subpopulation of CD8$^+$ T cells and CD4$^+$ (CD4$^+$/CD25^{high}/Foxp3$^+$) cells that have the ability to modulate immune response and maintain tolerance to self-antigens (Fig. 13.2). T_{regs} generally act to suppress immune and inflammatory responses by targeting the proliferation of effector (cytotoxic and helper) T cells [88]. Cancer cells, possibly CSCs as well, recruit T_{regs} in response to infiltration by immune and inflammatory cells of the tumor microenvironment to control the immune responses. Blocking the immunosuppression responses has implications for the application of immune stimulatory approaches for cancer treatment [88]. Recruitment of immunosuppressive T_{regs} by the cancer cells weakens the immune and inflammatory responses during the progression of neoplasms. Blocking immune-inhibitory signals by using agonist antibodies against PD-1/PD-L1 [104] and CTLA-4 and via antagonistic OX40/4-1BB [105,106] stimulation allow for increased T cell activation and show therapeutic promise in targeting tumor resistance to immunotherapies. However, only a subset of individuals with a preexisting T cell response derive clinical benefit from CTLA-4 or PD-1/PD-L1. Sprangler et al. study identified a melanoma-cell-intrinsic oncogenic Wnt/β-catenin signaling pathway that deters T cell infiltration in melanoma [107].

MDSCs have the characteristic ability to suppress both innate and adaptive immune responses and exert their effects through antigen-specific and antigen nonspecific mechanisms. In addition, MDSCs can induce the development of T_{regs}. Two major subtypes of MDSCs have been demonstrated to play an important role in tumor promotion—$CD11b^{+}/Gr1^{high}/Ly\text{-}6G^{+}/CD49d^{-}$ granulocytic MDSCs (G-MDSCs) and $CD11b^{+}/Gr1^{high}/Ly\text{-}6G^{high}/CD49d^{+}$ monocytic MDSCs (M-MDSCs) [108]. Although the impact of MDSC infiltration of the tumor microenvironment is well known, the underlying mechanisms and the cross talk between MDSCs and tumor cells are less well understood. The critical role of MDSCs in cancer progression through immunosuppression needs to be evaluated. Identification and association of subsets of MDSCs with types of cancer may provide insight into the immune regulation by MDSCs within the tumor microenvironment.

A more recently recognized class of B lymphocytes, known as B regulatory cells (B_{regs}), have been shown to play an immunosuppressive role in cancers. B_{regs} have been shown to inhibit other immune cells (primarily, T cells) through cytokine secretion and antigen presentation [109,110]. He et al. showed that B_{regs} secrete cytokines, such as IL-10 and TGF-β, that suppress apoptotic immune response [109,111]. Due to the recent emergence of the role of B cells in cancers, little is known about their modulatory effect on tumor microenvironment and their effect on cancer cells and CSCs. From the current perspective, B cells appear to play a dual role during tumor immunity. Since B cells are responsible, in part, for the induction of $CD4^{+}$ and $CD8^{+}$ T cells, depletion of $CD20^{+}$ B cells increases tumor burden in the lungs of mice injected with melanoma cells. In contrast, deficiencies of B cells have been shown to enhance $CD4^{+}$ T cells priming and the activity of $CD8^{+}$ T cells, which is necessary for cytotoxic killing of tumor cells [111].

Intravasation, Survival During Circulation, and Beyond Extravasation

Once cancer cells have infiltrated the adjacent tissue, cancer cells begin the process to enter the bloodstream. This process can be broken down into two subprocesses, namely [1] intravasation or endothelial transmigration of cancer cells into vasculature and [2] survival of the cancer cells in the bloodstream and lymphatic system. Together, these are termed hematogenous and lymphatic dissemination, respectively. Several studies suggest a role of TAMs involved in a paracrine loop with cancer cells, promoting metastasis and hematogenous survival. The recruitment of macrophages to the tumor site, first documented in breast cancer context, allows for metastasis using EGF/CSF-1 chemoattractant gradients [112,113]. CSF-1 is a strong chemoattractant for EGF-secreting macrophages, which in turn, promote chemotaxis of the cancer cells, resulting in a positive feedback loop [59,114]. Evidence for tumor–macrophage association comes from direct microscopic visualization of how perivascular macrophages in mammary tumors are responsible for intravasation of tumor cells [113]. In vitro experiments with CT26 colon cancer cells and RAW 264.7 macrophages have also showed that RAW 264.7 macrophages localize specifically to the tumor periphery and enhanced tumor growth, microvascular density, and metastasis [112]. In addition, infiltrating macrophages upregulate SDF-1α and VEGF, cytokines that contribute to cell migration [112]. Further, other factors, such as TGF-β, enhance intravasation through increasing penetrance through microvessel walls [115]. Together, these data suggest a cross talk between immune/inflammatory signaling and cancer cell metastasis to enhance the survival and metastatic potential of carcinoma.

Cancer cells that successfully intravasize and enter the bloodstream face two new challenges: (1) surviving the physical shearing forces of the bloodstream and (2) evading the immune surveillance throughout the circulation until they reach a new organ site. A variety of different immune cells within the bloodstream—platelets, lymphocytes, macrophages, mast cells—contribute to the metastatic process [116]. Extensive evidence shows that within the circulatory system, platelet activation and coagulation systems guard tumor cells from physical pressures, shield cells from immune elimination, and promote their arrest at the endothelium for the establishment of a secondary lesion. The link between platelets and cancer was first observed by Trousseau, in the mid-19th century, who observed an association between thrombosis and cancer, suggesting that the host hemostatic system supports the progression of malignancy [116]. Moreover, patients with cancer who show concomitant signs of thrombosis show poor clinical outcomes; thromboembolism is an early diagnostic characteristic of malignant tumors. Cancer cells upregulate fibrin on their cell surfaces; the interaction of fibrin with integrins results in the formation of tumor cell-fibrin-platelet aggregates, protecting the tumor cells at the core of this embolus from shearing stresses [117,118]. These embolus aggregates also protect tumor cells from immune-mediated lysis as well as from NK cells [118]. Thrombin formation through activation of platelets and coagulation cascade, in essence, allows CTCs to evade both physical forces and immune surveillance within the bloodstream.

Finally, CTCs or MCSCs that survive the circulation need to extravasate and survive at the distant metastatic site. The platelet-fibrin clot formation helps circulating cancer cells establish a site for proliferation, while TAMs and T_{regs} provide survival and proliferation signals at the new site. The initial retention of cancer cells at the metastatic site is also promoted through direct interaction with intercellular adhesion molecule 1 (ICAM1) with tumor-associated neutrophils that recruited in response to CXCL8 chemokine, secreted by the metastatic cancer cells [119]. Secreted chemokines attract $CD11b^+$ macrophages, called metastasis-associated macrophages (MAMs), which establish vasculature (via VEGF signaling) for the emigrated cells and regulate immune trafficking [119,120].

CONCLUSIONS AND FUTURE DIRECTION

Metastasis is a multistage process and is the leading cause of cancer-related deaths. Although metastasis of cancer cells has been studied extensively, the role of CSCs in metastasis and their dissemination has recently gained interest due to their characteristics in enhancing the metastatic potential of the cancer cells as well for the mechanisms they may employ in evading the host immune responses. This chapter provides a brief overview of the CSCs and evaluates the role of CSCs in cancer metastasis. Moreover, we reviewed recent findings that suggest a role of CSCs in evading the immune system as well as targeting immune and inflammatory responses to support tumor progression. In addition, this chapter provides a summary of the mechanisms that cancer cells and CSCs may utilize to evade the host innate and adaptive immune responses. Current therapies that bolster or enhance host immune response are not completely effective at eliminating cancer cells, as CSCs may use the immune responses to support tumor progression. Recruitment of immune regulatory cells, such as T_{regs} and B_{regs}, can reduce the effectiveness of immune-modulatory therapies. In addition, the effect of current therapies, such as anti-CTLA-4 and anti-PD-L1 antibodies that suppress the immune responses, on MCSCs is not well understood. Thus, understanding these roles may provide new insights into novel pathways for therapeutic targeting of CSCs with the goal of completely halting the progression of tumor cells and eliminating them. From a therapeutic perspective, analysis of CSCs and MCSCs can provide information on associated tumor type, grade, stage, and clinical outcomes. Combined therapies that take advantage of the "Achilles' heel" of CSCs and MCSCs by targeting CSC-specific markers and signaling pathways and enhancing immune surveillance may potentially prove useful in selective and effective elimination of the cancer lesions and mortality-related metastasis.

Glossary

Immunosuppression Suppression of the immune response by reducing the immune cell infiltration to the site of infection and/or by reducing the activity of immune system components.

Inflammation Localized, innate response to tissue damage or injury, characterized by redness, swollenness, heat, and pain.

Metastasis The process by which cancer cells break away from the site of the primary tumor and move to a distant location in the body to result in tumor at a distant site.

List of Acronyms and Abbreviations

ALDH Aldehyde dehydrogenase
APC Antigen-presenting cell
B_{regs} B regulatory cells
BMDC Bone marrow-derived cell
BMPR2 Bone morphogenetic protein receptor type 2
CD Cluster of differentiation
cDNA Complementary DNA
CML Chronic myeloid leukemia
COX Cyclooxygenase
CSC Cancer stem cell
CSF Colony stimulating factor
CTC Circulating tumor cell
CTL Cytotoxic T lymphocyte
CTLA Cytotoxic T lymphocyte antigen
CXCL Chemokine (C-X-C motif) ligand
CXCR Chemokines (C-X-C motif) receptor
DC-TIC Delayed contributing tumor-initiating cell
DNA Deoxyribonucleic acid
DTC Disseminated cancer cell
EGF Epithelial growth factor
EMT Epithelial-to-mesenchymal
EPCAM Epithelial cell adhesion molecule
ERK Extracellular signal-regulated kinase
HGF Hepatocyte growth factor
IL Interleukin
LSC Leukemia stem cell
LT-TIC Long-term tumor-initiating cell
MAMs Metastasis-associated macrophages
MCS Migrating cancer stem cell
MCSC Metastatic cancer stem cell
MDSC Myeloid-derived suppressor cell
MHC Major histocompatibility complex
MIC Metastasis-initiating cell
MMP Matrix metallopeptidase
NF-κB Nuclear factor kappa-light-chain-enhancer of activated B cells
NK cell Natural killer cell
PD-1 Programmed death [receptor] 1
PDGF Platelet-derived growth factor
PD-L1 Programmed death ligand 1
PSOTN Periostin
PVN Perivascular niche
STAT3 Signal transducer and activator of transcription 3
T_{regs} T regulatory cells
TAA Tumor-associated antigen
TAM Tumor-associated macrophage
TGF-β Transforming growth factor beta
TNF Tumor necrosis factor

TNFR TNF receptor
TLR Toll-like receptor
TRAIL TNF-related apoptosis inducing ligand
TSP-1 Endothelial-derived thrombospondin 1
T-TAC Tumor transient amplifying cell
VEGF Vascular endothelial growth factor
ZEB1 Zinc Finger E-Box Binding Homeobox 1

References

[1] Valastyan S, Weinberg RA. Tumor metastasis: molecular insights and evolving paradigms. Cell 2011;147(2):275–92.

[2] Lobo NA, Shimono Y, Qian D, Clarke MF. The biology of cancer stem cells. Annu Rev Cell Dev Biol 2007;23:675–99.

[3] Kreso A, Dick JE. Evolution of the cancer stem cell model. Cell Stem Cell 2014;14(3):275–91.

[4] Geng SQ, Alexandrou AT, Li JJ. Breast cancer stem cells: multiple capacities in tumor metastasis. Cancer Lett 2014;349(1):1–7.

[5] Liu H, Patel MR, Prescher JA, Patsialou A, Qian D, Lin J, et al. Cancer stem cells from human breast tumors are involved in spontaneous metastases in orthotopic mouse models. Proc Natl Acad Sci USA 2010;107(42):18115–20.

[6] Shiozawa Y, Nie B, Pienta KJ, Morgan TM, Taichman RS. Cancer stem cells and their role in metastasis. Pharmacol Ther 2013;138(2):285–93.

[7] Lapidot T, Sirard C, Vormoor J, Murdoch B, Hoang T, Caceres-Cortes J, et al. A cell initiating human acute myeloid leukaemia after transplantation into SCID mice. Nature 1994;367(6464):645–8.

[8] Hou YC, Chao YJ, Tung HL, Wang HC, Shan YS. Co-expression of CD44-positive/CD133-positive cancer stem cells and CD204-positive tumor-associated macrophages is a predictor of survival in pancreatic ductal adenocarcinoma. Cancer 2014;120(17):2766–77.

[9] Horst D, Kriegl L, Engel J, Kirchner T, Jung A. Prognostic significance of the cancer stem cell markers CD133, CD44, and CD166 in colorectal cancer. Cancer Invest 2009;27(8):844–50.

[10] Poste G, Fidler IJ. The pathogenesis of cancer metastasis. Nature 1980;283(5743):139–46.

[11] Zitvogel L, Kroemer G. Cancer: antibodies regulate antitumour immunity. Nature 2015;521(7550):35–7.

[12] Carmi Y, Spitzer MH, Linde IL, Burt BM, Prestwood TR, Perlman N, et al. Allogeneic IgG combined with dendritic cell stimuli induce antitumour T-cell immunity. Nature 2015;521(7550):99–104.

[13] Brabletz T, Jung A, Spaderna S, Hlubek F, Kirchner T. Opinion: migrating cancer stem cells – an integrated concept of malignant tumour progression. Nat Rev Cancer 2005;5(9):744–9.

[14] Hermann PC, Huber SL, Herrler T, Aicher A, Ellwart JW, Guba M, et al. Distinct populations of cancer stem cells determine tumor growth and metastatic activity in human pancreatic cancer. Cell Stem Cell 2007;1(3):313–23.

[15] Balic M, Lin H, Young L, Hawes D, Giuliano A, McNamara G, et al. Most early disseminated cancer cells detected in bone marrow of breast cancer patients have a putative breast cancer stem cell phenotype. Clin Cancer Res 2006;12(19):5615–21.

[16] Baccelli I, Schneeweiss A, Riethdorf S, Stenzinger A, Schillert A, Vogel V, et al. Identification of a population of blood circulating tumor cells from breast cancer patients that initiates metastasis in a xenograft assay. Nat Biotechnol 2013;31(6):539–44.

[17] Oskarsson T, Batlle E, Massague J. Metastatic stem cells: sources, niches, and vital pathways. Cell Stem Cell 2014;14(3):306–21.

[18] Ye X, Tam WL, Shibue T, Kaygusuz Y, Reinhardt F, Ng Eaton E, et al. Distinct EMT programs control normal mammary stem cells and tumour-initiating cells. Nature 2015;525(7568):256–60.

[19] Mani SA, Guo W, Liao MJ, Eaton EN, Ayyanan A, Zhou AY, et al. The epithelial-mesenchymal transition generates cells with properties of stem cells. Cell 2008;133(4):704–15.

[20] Fischer KR, Durrans A, Lee S, Sheng J, Li F, Wong ST, et al. Epithelial-to-mesenchymal transition is not required for lung metastasis but contributes to chemoresistance. Nature 2015;527(7579):472–6.

[21] Zheng X, Carstens JL, Kim J, Scheible M, Kaye J, Sugimoto H, et al. Epithelial-to-mesenchymal transition is dispensable for metastasis but induces chemoresistance in pancreatic cancer. Nature 2015;527(7579):525–30.

[22] Baccelli I, Trumpp A. The evolving concept of cancer and metastasis stem cells. J Cell Biol 2012;198(3):281–93.

[23] Pang R, Law WL, Chu AC, Poon JT, Lam CS, Chow AK, et al. A subpopulation of $CD26^+$ cancer stem cells with metastatic capacity in human colorectal cancer. Cell Stem Cell 2010;6(6):603–15.

[24] Dieter SM, Ball CR, Hoffmann CM, Nowrouzi A, Herbst F, Zavidij O, et al. Distinct types of tumor-initiating cells form human colon cancer tumors and metastases. Cell Stem Cell 2011;9(4):357–65.

[25] Nagler C, Zanker KS, Dittmar T. Cell fusion, drug resistance and recurrence CSCs. Adv Exp Med Biol 2011;714:173–82.

[26] Dittmar T, Nagler C, Schwitalla S, Reith G, Niggemann B, Zanker KS. Recurrence cancer stem cells – made by cell fusion? Med Hypotheses 2009;73(4):542–7.

[27] Clawson GA. Cancer. Fusion for moving. Science 2013;342(6159):699–700.

[28] Pawelek JM, Chakraborty AK. Fusion of tumour cells with bone marrow-derived cells: a unifying explanation for metastasis. Nat Rev Cancer 2008;8(5):377–86.

[29] Ding J, Jin W, Chen C, Shao Z, Wu J. Tumor associated macrophage × cancer cell hybrids may acquire cancer stem cell properties in breast cancer. PLoS One 2012;7(7):e41942.

[30] Al-Nedawi K, Meehan B, Micallef J, Lhotak V, May L, Guha A, et al. Intercellular transfer of the oncogenic receptor EGFRvIII by microvesicles derived from tumour cells. Nat Cell Biol 2008;10(5):619–24.

[31] Charafe-Jauffret E, Ginestier C, Iovino F, Tarpin C, Diebel M, Esterni B, et al. Aldehyde dehydrogenase 1-positive cancer stem cells mediate metastasis and poor clinical outcome in inflammatory breast cancer. Clin Cancer Res 2010;16(1):45–55.

[32] Croker AK, Goodale D, Chu J, Postenka C, Hedley BD, Hess DA, et al. High aldehyde dehydrogenase and expression of cancer stem cell markers selects for breast cancer cells with enhanced malignant and metastatic ability. J Cell Mol Med 2009;13(8B):2236–52.

[33] Todaro M, Gaggianesi M, Catalano V, Benfante A, Iovino F, Biffoni M, et al. CD44v6 is a marker of constitutive and reprogrammed cancer stem cells driving colon cancer metastasis. Cell Stem Cell 2014;14(3):342–56.

[34] Butler TP, Gullino PM. Quantitation of cell shedding into efferent blood of mammary adenocarcinoma. Cancer Res 1975;35(3):512–6.

[35] Yu M, Bardia A, Wittner BS, Stott SL, Smas ME, Ting DT, et al. Circulating breast tumor cells exhibit dynamic changes in epithelial and mesenchymal composition. Science 2013;339(6119):580–4.

[36] Theodoropoulos PA, Polioudaki H, Agelaki S, Kallergi G, Saridaki Z, Mavroudis D, et al. Circulating tumor cells with a putative stem cell phenotype in peripheral blood of patients with breast cancer. Cancer Lett 2010;288(1):99–106.

[37] Yang ZF, Ngai P, Ho DW, Yu WC, Ng MN, Lau CK, et al. Identification of local and circulating cancer stem cells in human liver cancer. Hepatology 2008;47(3):919–28.

[38] Cho EH, Wendel M, Luttgen M, Yoshioka C, Marrinucci D, Lazar D, et al. Characterization of circulating tumor cell aggregates identified in patients with epithelial tumors. Phys Biol 2012;9(1):016001.

[39] Aceto N, Bardia A, Miyamoto DT, Donaldson MC, Wittner BS, Spencer JA, et al. Circulating tumor cell clusters are oligoclonal precursors of breast cancer metastasis. Cell 2014;158(5):1110–22.

[40] Liao WT, Ye YP, Deng YJ, Bian XW, Ding YQ. Metastatic cancer stem cells: from the concept to therapeutics. Am J Stem Cells 2014;3(2):46–62.

[41] Bragado P, Estrada Y, Parikh F, Krause S, Capobianco C, Farina HG, et al. TGF-β2 dictates disseminated tumour cell fate in target organs through TGF-β-RIII and p38α/β signalling. Nat Cell Biol 2013;15(11):1351–61.

[42] Aguirre-Ghiso JA, Bragado P, Sosa MS. Metastasis awakening: targeting dormant cancer. Nat Med 2013;19(3):276–7.

[43] Ghajar CM. Metastasis prevention by targeting the dormant niche. Nat Rev Cancer 2015;15(4):238–47.

[44] Prost S, Relouzat F, Spentchian M, Ouzegdouh Y, Saliba J, Massonnet G, et al. Erosion of the chronic myeloid leukaemia stem cell pool by PPARγ agonists. Nature 2015.

[45] Sosa MS, Bragado P, Aguirre-Ghiso JA. Mechanisms of disseminated cancer cell dormancy: an awakening field. Nat Rev Cancer 2014;14(9):611–22.

[46] Gao H, Chakraborty G, Lee-Lim AP, Mo Q, Decker M, Vonica A, et al. The BMP inhibitor Coco reactivates breast cancer cells at lung metastatic sites. Cell 2012;150(4):764–79.

[47] Zhang XH, Wang Q, Gerald W, Hudis CA, Norton L, Smid M, et al. Latent bone metastasis in breast cancer tied to Src-dependent survival signals. Cancer Cell 2009;16(1):67–78.

[48] Becker PS. Dependence of acute myeloid leukemia on adhesion within the bone marrow microenvironment. Sci World J 2012;2012:856467.

[49] Ghajar CM, Peinado H, Mori H, Matei IR, Evason KJ, Brazier H, et al. The perivascular niche regulates breast tumour dormancy. Nat Cell Biol 2013;15(7):807–17.

[50] Malanchi I, Santamaria-Martinez A, Susanto E, Peng H, Lehr HA, Delaloye JF, et al. Interactions between cancer stem cells and their niche govern metastatic colonization. Nature 2012;481(7379):85–9.

[51] Callahan MK, Postow MA, Wolchok JD. CTLA-4 and PD-1 pathway blockade: combinations in the clinic. Front Oncol 2014;4:385.

[52] Wiemann B, Starnes CO. Coley's toxins, tumor necrosis factor and cancer research: a historical perspective. Pharmacol Ther 1994;64(3):529–64.

[53] Starnes CO. Coley's toxins in perspective. Nature 1992;357(6373):11–2.

[54] Correale P, Cusi MG, Micheli L, Nencini C, Del Vecchio MT, Torino F, et al. Chemo-immunotherapy of colorectal carcinoma: preclinical rationale and clinical experience. Invest New Drugs 2006;24(2):99–110.

[55] Correale P, Cusi MG, Tsang KY, Del Vecchio MT, Marsili S, Placa ML, et al. Chemo-immunotherapy of metastatic colorectal carcinoma with gemcitabine plus FOLFOX 4 followed by subcutaneous granulocyte macrophage colony-stimulating factor and interleukin-2 induces strong immunologic and antitumor activity in metastatic colon cancer patients. J Clin Oncol 2005;23(35):8950–8.

[56] Grivennikov SI, Greten FR, Karin M. Immunity, inflammation, and cancer. Cell 2010;140(6):883–99.

[57] Gupta GP, Massague J. Cancer metastasis: building a framework. Cell 2006;127(4):679–95.

[58] Montel AH, Bochan MR, Hobbs JA, Lynch DH, Brahmi Z. Fas involvement in cytotoxicity mediated by human NK cells. Cell Immunol 1995;166(2):236–46.

[59] Wyckoff J, Wang W, Lin EY, Wang Y, Pixley F, Stanley ER, et al. A paracrine loop between tumor cells and macrophages is required for tumor cell migration in mammary tumors. Cancer Res 2004;64(19):7022–9.

[60] Riabov V, Gudima A, Wang N, Mickley A, Orekhov A, Kzhyshkowska J. Role of tumor associated macrophages in tumor angiogenesis and lymphangiogenesis. Front Physiol 2014;5:75.

[61] Lal I, Dittus K, Holmes CE. Platelets, coagulation and fibrinolysis in breast cancer progression. Breast Cancer Res 2013;15(4):207.

[62] Gasic GJ, Gasic TB, Stewart CC. Antimetastatic effects associated with platelet reduction. Proc Natl Acad Sci USA 1968;61(1):46–52.

[63] Gajewski TF, Schreiber H, Fu YX. Innate and adaptive immune cells in the tumor microenvironment. Nat Immunol 2013;14(10):1014–22.

[64] Funada Y, Noguchi T, Kikuchi R, Takeno S, Uchida Y, Gabbert HE. Prognostic significance of $CD8^+$ T cell and macrophage peritumoral infiltration in colorectal cancer. Oncol Rep 2003;10(2):309–13.

[65] Holcombe RF, Jacobson J, Dakhil SR, Stewart RM, Betzing KS, Kannan K, et al. Association of immune parameters with clinical outcome in stage III colon cancer: results of Southwest Oncology Group Protocol 9009. Cancer Immunol Immunother 1999;48(9):533–9.

[66] Prall F, Duhrkop T, Weirich V, Ostwald C, Lenz P, Nizze H, et al. Prognostic role of $CD8^+$ tumor-infiltrating lymphocytes in stage III colorectal cancer with and without microsatellite instability. Hum Pathol 2004;35(7):808–16.

[67] Coca S, Perez-Piqueras J, Martinez D, Colmenarejo A, Saez MA, Vallejo C, et al. The prognostic significance of intratumoral natural killer cells in patients with colorectal carcinoma. Cancer 1997;79(12):2320–8.

[68] Dupaul-Chicoine J, Arabzadeh A, Daenais M, Douglas T, Champagne C, Morizot A. The Nlrp3 inflammasome suppresses colorectal cancer metastatic growth in the liver by promoting natural killer cell tumoricidal activity. Immunity 2015.

[69] Matsui S, Ahlers JD, Vortmeyer AO, Terabe M, Tsukui T, Carbone DP, et al. A model for $CD8^+$ CTL tumor immunosurveillance and regulation of tumor escape by CD4 T cells through an effect on quality of CTL. J Immunol 1999;163(1):184–93.

[70] Tanaka Y, Koido S, Xia J, Ohana M, Liu C, Cote GM, et al. Development of antigen-specific $CD8^+$ CTL in MHC class I-deficient mice through CD4 to CD8 conversion. J Immunol 2004;172(12):7848–58.

[71] Akalay I, Janji B, Hasmim M, Noman MZ, Andre F, De Cremoux P, et al. Epithelial-to-mesenchymal transition and autophagy induction in breast carcinoma promote escape from T-cell-mediated lysis. Cancer Res 2013;73(8):2418–27.

[72] Oettgen HF, Carswell EA, Kassel RL, Fiore N, Williamson B, Hoffmann MK, et al. Endotoxin-induced tumor necrosis factor. Recent Results Cancer Res 1980;75:207–12.

[73] Green S, Dobrjansky A, Carswell EA, Kassel RL, Old LJ, Fiore N, et al. Partial purification of a serum factor that causes necrosis of tumors. Proc Natl Acad Sci USA 1976;73(2):381–5.

[74] Carswell EA, Old LJ, Kassel RL, Green S, Fiore N, Williamson B. An endotoxin-induced serum factor that causes necrosis of tumors. Proc Natl Acad Sci USA 1975;72(9):3666–70.

[75] Hanahan D, Weinberg RA. Hallmarks of cancer: the next generation. Cell 2011;144(5): 646–74.

[76] van Horssen R, Ten Hagen TL, Eggermont AM. TNF-α in cancer treatment: molecular insights, antitumor effects, and clinical utility. Oncologist 2006;11(4):397–408.

[77] Culig Z. Proinflammatory cytokine interleukin-6 in prostate carcinogenesis. Am J Clin Exp Urol 2014;2(3):231–8.

[78] Chopra A, Anderson A, Giardina C. Novel piperazine-based compounds inhibit microtubule dynamics and sensitize colon cancer cells to tumor necrosis factor-induced apoptosis. J Biol Chem 2014;289(5):2978–91.

[79] Rivas MA, Carnevale RP, Proietti CJ, Rosemblit C, Beguelin W, Salatino M, et al. TNFα acting on TNFR1 promotes breast cancer growth via p42/P44 MAPK, JNK, Akt and NF-κ B-dependent pathways. Exp Cell Res 2008;314(3):509–29.

[80] Piguet PF, Kan CD, Vesin C. Role of the tumor necrosis factor receptor 2 (TNFR2) in cerebral malaria in mice. Lab Invest 2002;82(9):1155–66.

[81] Hamilton KE, Simmons JG, Ding S, Van Landeghem L, Lund PK. Cytokine induction of tumor necrosis factor receptor 2 is mediated by STAT3 in colon cancer cells. Mol Cancer Res 2011;9(12):1718–31.

[82] Karavitis J, Zhang M. COX2 regulation of breast cancer bone metastasis. Oncoimmunology 2013;2(3):e23129.

[83] Kurtova AV, Xiao J, Mo Q, Pazhanisamy S, Krasnow R, Lerner SP, et al. Blocking PGE2-induced tumour repopulation abrogates bladder cancer chemoresistance. Nature 2015;517(7533):209–13.

[84] Kratky W, Reis e Sousa C, Oxenius A, Sporri R. Direct activation of antigen-presenting cells is required for $CD8^+$ T-cell priming and tumor vaccination. Proc Natl Acad Sci USA 2011;108(42):17414–9.

[85] Waldhauer I, Steinle A. NK cells and cancer immunosurveillance. Oncogene 2008;27(45):5932–43.

[86] Sanjabi S, Zenewicz LA, Kamanaka M, Flavell RA. Anti-inflammatory and pro-inflammatory roles of TGF-β, IL-10, and IL-22 in immunity and autoimmunity. Curr Opin Pharmacol 2009;9(4):447–53.

[87] Driessens G, Kline J, Gajewski TF. Costimulatory and coinhibitory receptors in anti-tumor immunity. Immunol Rev 2009;229(1):126–44.

[88] Smyth MJ, Ngiow SF, Teng MW. Targeting regulatory T cells in tumor immunotherapy. Immunol Cell Biol 2014;92(6):473–4.

[89] Taube JM, Young GD, McMiller TL, Chen S, Salas JT, Pritchard TS, et al. Differential expression of immune-regulatory genes associated with PD-L1 display in melanoma: implications for PD-1 pathway blockade. Clin Cancer Res 2015;21(17):3969–76.

[90] Tumeh PC, Harview CL, Yearley JH, Shintaku IP, Taylor EJ, Robert L, et al. PD-1 blockade induces responses by inhibiting adaptive immune resistance. Nature 2014;515(7528):568–71.

[91] Herbst RS, Soria JC, Kowanetz M, Fine GD, Hamid O, Gordon MS, et al. Predictive correlates of response to the anti-PD-L1 antibody MPDL3280A in cancer patients. Nature 2014;515(7528):563–7.

[92] Takizawa H, Manz MG. Macrophage tolerance: CD47-SIRP-α-mediated signals matter. Nat Immunol 2007;8(12):1287–9.

[93] Chao MP, Alizadeh AA, Tang C, Myklebust JH, Varghese B, Gill S, et al. Anti-CD47 antibody synergizes with rituximab to promote phagocytosis and eradicate non-Hodgkin lymphoma. Cell 2010;142(5):699–713.

[94] Scheeren FA, Kuo AH, van Weele LJ, Cai S, Glykofridis I, Sikandar SS, et al. A cell-intrinsic role for TLR2-MYD88 in intestinal and breast epithelia and oncogenesis. Nat Cell Biol 2014;16(12):1238–48.

[95] Chefetz I, Alvero AB, Holmberg JC, Lebowitz N, Craveiro V, Yang-Hartwich Y, et al. TLR2 enhances ovarian cancer stem cell self-renewal and promotes tumor repair and recurrence. Cell Cycle 2013;12(3):511–21.

[96] Arbibe L, Mira JP, Teusch N, Kline L, Guha M, Mackman N, et al. Toll-like receptor 2-mediated NF-κB activation requires a Rac1-dependent pathway. Nat Immunol 2000;1(6):533–40.

[97] Ey B, Eyking A, Klepak M, Salzman NH, Gothert JR, Runzi M, et al. Loss of TLR2 worsens spontaneous colitis in MDR1A deficiency through commensally induced pyroptosis. J Immunol 2013;190(11):5676–88.

[98] Chen L, Gibbons DL, Goswami S, Cortez MA, Ahn YH, Byers LA, et al. Metastasis is regulated via microRNA-200/ZEB1 axis control of tumour cell PD-L1 expression and intratumoral immunosuppression. Nat Commun 2014;5:5241.

[99] Jinushi M, Chiba S, Yoshiyama H, Masutomi K, Kinoshita I, Dosaka-Akita H, et al. Tumor-associated macrophages regulate tumorigenicity and anticancer drug responses of cancer stem/initiating cells. Proc Natl Acad Sci USA 2011;108(30):12425–30.

[100] Karin M. NF-κB as a critical link between inflammation and cancer. Cold Spring Harb Perspect Biol 2009;1(5):a000141.

[101] Zhou W, Ke SQ, Huang Z, Flavahan W, Fang X, Paul J, et al. Periostin secreted by glioblastoma stem cells recruits M2 tumour-associated macrophages and promotes malignant growth. Nat Cell Biol 2015;17(2):170–82.

[102] Squadrito ML, De Palma M. A niche role for periostin and macrophages in glioblastoma. Nat Cell Biol 2015;17(2):107–9.

[103] Hanna RN, Cekic C, Sag D, Tacke R, Thomas GD, Nowyhed H, et al. Patrolling monocytes control tumor metastasis to the lung. Science 2015.

[104] Francisco LM, Salinas VH, Brown KE, Vanguri VK, Freeman GJ, Kuchroo VK, et al. PD-L1 regulates the development, maintenance, and function of induced regulatory T cells. J Exp Med 2009;206(13):3015–29.

[105] Smith SE, Hoelzinger DB, Dominguez AL, Van Snick J, Lustgarten J. Signals through 4-1BB inhibit T regulatory cells by blocking IL-9 production enhancing antitumor responses. Cancer Immunol Immunother 2011;60(12):1775–87.

[106] Hamano R, Huang J, Yoshimura T, Oppenheim JJ, Chen X. TNF optimally activatives regulatory T cells by inducing TNF receptor superfamily members TNFR2, 4-1BB and OX40. Eur J Immunol 2011;41(7):2010–20.

[107] Spranger S, Bao R, Gajewski TF. Melanoma-intrinsic β-catenin signalling prevents anti-tumour immunity. Nature 2015;523(7559):231–5.

[108] Gabrilovich DI, Ostrand-Rosenberg S, Bronte V. Coordinated regulation of myeloid cells by tumours. Nat Rev Immunol 2012;12(4):253–68.

[109] He Y, Qian H, Liu Y, Duan L, Li Y, Shi G. The roles of regulatory B cells in cancer. J Immunol Res 2014;2014:215471.

[110] Olkhanud PB, Damdinsuren B, Bodogai M, Gress RE, Sen R, Wejksza K, et al. Tumor-evoked regulatory B cells promote breast cancer metastasis by converting resting $CD4^+$ T cells to T-regulatory cells. Cancer Res 2011;71(10):3505–15.

[111] Balkwill F, Montfort A, Capasso M. B regulatory cells in cancer. Trends Immunol 2013;34(4):169–73.

[112] Wyckoff JB, Wang Y, Lin EY, Li JF, Goswami S, Stanley ER, et al. Direct visualization of macrophage-assisted tumor cell intravasation in mammary tumors. Cancer Res 2007;67(6): 2649–56.

[113] Condeelis J, Segall JE. Intravital imaging of cell movement in tumours. Nat Rev Cancer 2003;3(12):921–30.

[114] Goswami S, Sahai E, Wyckoff JB, Cammer M, Cox D, Pixley FJ, et al. Macrophages promote the invasion of breast carcinoma cells via a colony-stimulating factor-1/epidermal growth factor paracrine loop. Cancer Res 2005;65(12):5278–83.

[115] Giampieri S, Manning C, Hooper S, Jones L, Hill CS, Sahai E. Localized and reversible TGFβ signalling switches breast cancer cells from cohesive to single cell motility. Nat Cell Biol 2009;11(11):1287–96.

[116] Gay LJ, Felding-Habermann B. Contribution of platelets to tumour metastasis. Nat Rev 2011;11(2):123–34.

[117] Palumbo JS, Talmage KE, Massari JV, La Jeunesse CM, Flick MJ, Kombrinck KW, et al. Tumor cell-associated tissue factor and circulating hemostatic factors cooperate to increase metastatic potential through natural killer cell-dependent and-independent mechanisms. Blood 2007;110(1):133–41.

[118] Im JH, Fu W, Wang H, Bhatia SK, Hammer DA, Kowalska MA, et al. Coagulation facilitates tumor cell spreading in the pulmonary vasculature during early metastatic colony formation. Cancer Res 2004;64(23):8613–9.

[119] Kitamura T, Qian BZ, Pollard JW. Immune cell promotion of metastasis. Nat Rev Immunol 2015;15(2):73–86.

[120] Qian B, Deng Y, Im JH, Muschel RJ, Zou Y, Li J, et al. A distinct macrophage population mediates metastatic breast cancer cell extravasation, establishment and growth. PLoS One 2009;4(8):e6562.

CHAPTER 14

Cancer Stem Cells and Tumor-Associated Macrophages

W. Zhou, S. Bao
Lerner Research Institute, Cleveland, OH, United States

CONTENTS

INTRODUCTION

Cancer Stem Cells and Tumor Microenvironment

Decades of intensive study on cancer have led to substantial advances in tumor treatment, but the outcome is still far from satisfactory for most cancers. The failure of cancer therapy can be largely ascribed to the cellular heterogeneity and the differentiation hierarchy of cancer cells within tumors. Solid tumors contain a collection of heterogeneous neoplastic and nonneoplastic cell types as well as a variety of secreted factors and extracellular elements in the tumor microenvironment. Neoplastic tumor cells display remarkable hierarchy in their differentiation states. At the apex of the hierarchy are cancer stem cells (CSCs) or tumor-initiating cells (TICs) are a population of poorly differentiated cancer cells with the capacity to self-renew, differentiate, and repopulate the whole tumor bulk. Moreover, CSCs are more resistant to conventional therapies relative to matched nonstem tumor cells (NSTCs) [1–3] likely making them the main cause of tumor relapse. Thus targeting CSCs may effectively improve treatment in most tumor types [1].

To function as key players in tumor malignancy, CSCs are often localized in certain tumor microenvironments composed of heterogeneous cells and noncellular elements. Within the tumor microenvironment, cells can nourish each other and protect each other from adverse stimuli. Since different types of cells in the tumor microenvironment have distinct responses to therapeutic interventions, antitumor agents aiming at certain tumor components may not effectively target the tumor environment. Therefore the microenvironment can serve as both the soil and shelter for CSCs. Several lines of evidence have demonstrated that CSCs show preferences for and tend to be enriched in certain microenvironments called niches. The enrichment of CSCs in the proximity of vessels and the positive correlation between the vessel density and the CSC frequency strongly suggest the perivascular niche for CSCs [4]. CSCs in the perivascular niche may be dependent on endothelial cell-derived stimulatory signals. Likewise, the

Cancer Stem Cells. http://dx.doi.org/10.1016/B978-0-12-803892-5.00014-0

observation that other CSCs seem to concentrate in necrotic regions, invasive edges, and avascular areas implicates the existence of a hypoxic niche for CSCs in glioblastoma and breast cancer [5]. It has been reported that hypoxia-induced signaling pathways are critical for the maintenance and propagation of CSCs. Finally, accumulating evidence suggests a close interaction between CSCs and the tumor-infiltrating immune cells, particularly tumor-associated macrophages (TAMs), indicating that the immune microenvironment may play an important role in CSC maintenance and propagation. However, the tumor immune microenvironment and the perivascular or hypoxic niches are not mutually exclusive. Instead, the enrichment of CSCs and tumor-infiltrating immune cells is frequently observed in perivascular and hypoxic niches. Whereas researchers tend to focus on a specific type of CSC niche in their studies, one should always keep in mind the numerous environmental factors that may simultaneously regulate CSCs in tumors. Thus, deciphering the interaction between CSCs and their niches, including the immune microenvironment, is extremely important for the development of new antitumor therapies.

Tumor Immune Microenvironment and Tumor-Associated Macrophages

Many types of immune cells, including macrophages, T cells, B cells, dendritic cells, neutrophil granulocytes, and mast cells have been detected in tumor tissues [6–8]. B lymphocytes are generally found to favor disease-free survival of tumor patients [9,10]. The antitumor activity of B cells may be ascribed to their production of autoantibodies to target cancer cells in vivo [11]. Likewise, natural kill cells, in spite of their low density in tumors, demonstrate a capacity to lyse cancer cells [12]. T cells, the predominant tumor-infiltrating lymphocytes, are composed of several different subsets with distinct functions [13]. Whereas some subsets of T helper $CD4^+$ cells show protumor activities [14,15], cytotoxic $CD8^+$ cells are believed to attack cancer cells through adaptive immune response [15]. Moreover, $CD25^+Foxp3^+$ T regulatory cells have been found to inhibit the antitumor activities of T cells [16]. In general, even though some types of tumor-infiltrating immune cells may manifest antitumor phenotypes under certain circumstances, the tumor immune environment as a whole supports tumor progression.

At the early stage of tumor development, immune cell infiltration is regarded as an event of host immune response against tumor cells, which often fails during tumor progression. Intensive efforts have been made to understand the escape of tumor cells from immune surveillance thought to be the intrinsic function of tumor-infiltrating immune cells. However, contrary to this traditional concept, immune cell infiltration, such as TAMs, is closely related to tumor malignancy. Furthermore, infiltration of TAMs seems to be a prerequisite for active tumor growth and metastasis in a variety of tumors. Finally, abundant TAM infiltration in some cases predicts a poor prognosis for tumor patients with glioblastoma, hepatocytocellular carcinoma, and pancreatic tumor [17,18]. In fact,

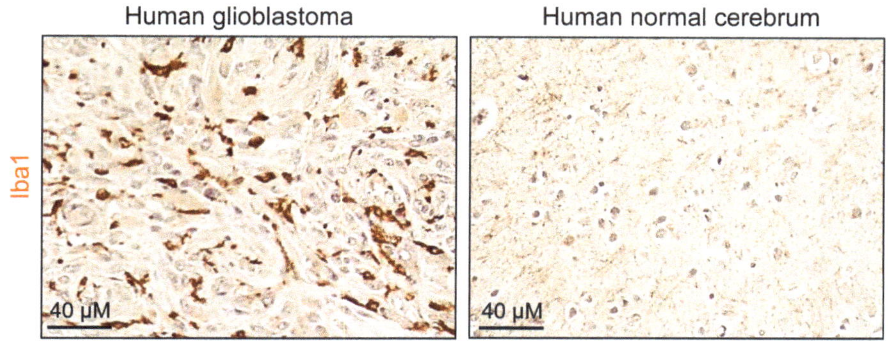

FIGURE 14.1 Massive macrophage (tumor-associated macrophage (TAM)) infiltration in human glioblastoma.
Immunohistochemical staining of the TAM marker Iba1 in a human primary glioblastoma specimen. A massive TAM infiltration was detected in the glioblastoma. Scale bar represents 40 μm.

discoveries underscore the tumor-supportive role of some infiltrating immune cells, especially TAMs, whereas other types of tumor-infiltrating immune cells, such as natural killer cells and $CD8^+$ cytotoxic T cells, can bear antitumor potentials [12,19]. However, the functions of the antitumor T cells and NK cells are frequently inhibited in the immune-suppressive tumor microenvironments [12,19]. Considering their enormous numbers and essential functions in tumor progression, the protumor immune cells constitute a supportive microenvironment for tumor growth and malignant development.

Among the many types of tumor-infiltrating immune cells, TAMs represent one of the most active protumorigenic players in tumor progression. TAMs are the dominant infiltrating immune cells that can make up to 50% of tumor mass in some solid tumors such as breast, pancreatic, and brain tumors [20]. Massive TAM infiltration is a common feature of high-grade malignant tumors such as glioblastoma (Fig. 14.1). TAMs have been found to be involved in critical processes in tumor growth, metastasis, neovascularization, and resistance to therapies. The density of TAMs is closely correlated to tumor grades and patient survival. Other immune cells, eg, lymphatic T-helper cells, plasmacytoid dendritic cells, and neutrophils also have tumor-supportive activities. These protumorigenic immune cells may follow similar signaling pathways utilized by TAMs. We will focus on TAMs to summarize our understanding and new advances on the tumor immune microenvironment.

THE ORIGIN OF TUMOR-ASSOCIATED MACROPHAGES

Peripheral Blood-Derived Monocytes

Classically, macrophages are derived from differentiation of bone marrow-derived monocytes. The same situation applies to TAMs in most solid tumors. In a 3,5-di-t-butyl-4-hydroxytoluene (BHT)-initiated murine lung tumor model, macrophage-specific Foxm1 deletion reduced the number and size of lung tumors in macFoxm1(−/−) mice. Importantly, transfer of wild-type monocytes into macFoxm1(−/−) mice restored the BHT-induced pulmonary inflammation and tumor initiation, confirming monocytes as the origin of TAMs [21]. Likewise, in search of TAM progenitors in mouse mammary tumors, the TAM pool within tumors was found to almost exclusively consist of $Ly6C^{hi}CX3CR1^{low}$ monocytes that continuously infiltrated into tumors and renewed TAMs [22]. Furthermore, during tumor metastasis, Gr1 positive inflammatory monocytes were found to be preferentially recruited to pulmonary metastases derived from mammary tumors in a CCL2-dependent manner, which in turn promote the extravasation of tumor cells [23]. These studies strongly suggest that monocytes as precursors of TAMs are involved in different stages of tumor progression.

Monocytes are originally produced by bone marrow. However, there is a splenic reservoir of undifferentiated monocytes that may serve as the intermediate supply of TAMs. In accordance with this hypothesis, the physical relocation of TAM precursors from spleen to tumor site was observed in a genetic mouse model of lung adenocarcinoma driven by K-ras activation plus p53 inactivation. Removal of the spleen before or after tumor initiation significantly reduced TAM infiltration and tumor growth, suggesting that monocytes from the spleen are an important source of TAMs [7]. In addition, it was found that hormone Angiotensin II (AngII) amplified macrophage progenitors in the spleen but not the bone marrow, which allowed the spleen in tumor patients to supply new TAMs throughout tumor progression [24]. In spite of these reports emphasizing the spleen as a critical supply of monocytes and TAMs, controversial results have been reported using the fluorescent tracking of the redistribution of monocytes from the bone marrow and spleen during tumor progression. Although late-stage tumors recruited monocytes from both the spleen and bone marrow, monocytes from bone marrow were younger than those from spleen and displayed better motility. As a result of their distinct characteristics, spleen monocytes made only a minor contribution (ie, 2%) to TAM precursors, whereas 27% of tumor-infiltrating monocytes were derived from the bone marrow [25]. Moreover, the same study showed that monocytes in spleen were rapidly replaced by monocytes from bone marrow at a rate of 9% per hour, indicating that the spleen is only a transition site rather than a major source of TAMs. However, more data from different cancer models are required to further address the role of the splenic monocyte reservoir in TAM production.

Peripheral blood monocytes in patients may promote tumorigenesis more actively than TAM progenitors. Circulating monocytes in tumor patients are different from those in healthy individuals in many aspects. Monocytes from the blood of glioma patients expressed much higher levels of epidermal growth factor (EGF) than those from healthy people, suggesting the existence of tumor-related paracrine signaling from these monocytes [26]. In addition, $CD11b^+$ monocytes isolated from peripheral blood of glioblastoma patients markedly suppressed T-cell function in coculture, indicative of monocyte-mediated immunosuppression [27]. However, it is not clear whether these circulating protumor monocytes end up as tumor-infiltrating macrophages or if they may stay in the circulation and other benign sites to function as an independent population of tumor-supportive immune cells.

Microglia in Brain Tumors

The possible origins of TAMs in brain tumors are more controversial due to the presence of the brain-specific immune cells named microglia. As an abundant cell population in the normal brain, microglia make up 5–20% of the total glial cell population. Microglia precursors derived from mesodermal hematopoietic cells may enter the developing brain before birth and maturate into microglial cells in adults [28]. Microglia bear cell surface markers are similar to those expressed in monocyte-derived macrophages, which makes it difficult to distinguish microglia from macrophages in brain tumors. TAMs in brain tumors may come from the resident microglia, and many reports use "microglia/macrophages" to denote TAMs in brain tumors. However, studies of the monocyte subset trafficking using the CX3CR1(+/GFP)CCR2(+/RFP) knock-in fluorescent protein reporter mice suggested that infiltrating macrophages and activated microglia in brains can be distinguished by their specific cell surface markers CCR2 and CX3CR3, respectively [29]. By application of these specific cell surface markers along with the pan-macrophage markers Iba1 and CD11b, we demonstrated in an orthotopic glioblastoma mouse model that TAMs in glioblastoma were $CX3CR1^-/CCR2^+$ macrophages derived from monocytes but not $CX3CR1^+/CCR2^-$ microglia [30]. Therefore, at least in high-grade brain tumors, TAMs infiltrating into the tumor mass mainly originate from peripheral blood but not resident microglia.

However, the microglia may still be a key player in brain tumors. It has been reported that macrophages and microglia are differentially localized in human glioblastoma specimens. Whereas macrophages infiltrated into and diffused within tumors, activated microglia tended to cluster at the tumor edge [31], indicating the different functions of the two populations in brain tumors. In fact, depletion of CX3CR1, which is likely to impact microglia, but not the monocyte-derived macrophages, resulted in an increase of tumor incidence and a shorter survival time in glioma-bearing mice [32]. Interestingly, loss of CX3CR1 enhanced the accumulation of monocytes in the tumor mass, suggesting an antagonistic role of CX3CR1 positive microglia in the recruitment of monocytes and TAMs [32]. It is possible that microglia may interact with monocytes and TAMs or else they may even independently function as antitumor immune cells to affect brain tumor progression.

GENERAL PROPERTIES OF TUMOR-ASSOCIATED MACROPHAGES

Impaired Immune Surveillance of Tumor-Associated Macrophages

Mobilized as classic immune cells, macrophages ingest foreign particles through phagocytosis and present antigens to T and B lymphocytes to trigger the elimination of invading materials. It is expected that macrophages can recognize and destroy tumor cells as alien cells to achieve immune surveillance in tumor patients. However, only some of the TAMs in tumors were positively stained with phagocyte markers whereas most TAMs showed no phagocytic activity in vivo. Similarly, TAMs isolated from mice bearing mammary tumors display alterations in their ability to lyse tumor cells, which may be associated with the reduced production of nitrogen oxides by TAMs relative to inflammatory macrophages [33]. Furthermore, whereas activated leukocytes outside brain tumors express class II major histocompatibility antigens, TAMs usually lack these antigens and demonstrate no antigen-presenting activity, indicating the impaired immune surveillance of TAMs as antigen-presenting cells [34]. Of note, phagocytic markers may be elevated in TAMs under some circumstances, but such phagocytic activity may still be regulated by tumor cells to facilitate tumor growth. For example, when treated with oncolytic viruses, TAMs in glioblastoma expressed both the phagocytic marker CD68 and the immunosuppressive macrophage marker CD163. However, it is likely that the elevation of $CD68^+$ TAMs was triggered by the increased number of apoptotic cancer cells infected by the oncolytic virus. By clearing these apoptotic cancer cells, the $CD68^+$ population sabotaged virus infection and antagonized the antitumor effect of the oncolytic viruses [35]. Thus immune surveillance mediated by macrophages is suppressed in tumors in general to provide a supportive environment for tumor growth.

M1 and M2 Subtypes of Tumor-Associated Macrophages

Monocytes can be activated to differentiate into different lineages of macrophages. Grossly, based on the secretion of IL-2 and IL-10 and the subsequent activation of Th1 and Th2 cells, macrophages can be divided into M1 and M2 subtypes, although a continuum of intermediates has been reported [36]. However, in tumors, M1 and M2 subtypes of macrophages are basically related to their tumor-suppressive or tumor-supportive functions, ie, M1 for antitumorigenic and M2 for protumorigenic macrophages. A variety of cell surface markers, eg, CD163, Fizz1, Arg1, MHCII, etc. [37], have been raised to identify M1 and M2 subtypes of macrophages in tumors. By using these markers, it has been found that M2 subtype but not M1 subtype macrophages are inclined to be enriched in high-grade tumors. For example, the M1 marker MHCII is significantly reduced, whereas the M2 marker CD163 is increased in glioblastomas relative to low-grade astrocytomas [38]. So far, most studies focused on the protumorigenic TAMs, ie, M2 subtype of macrophages, although antitumorigenic M1 subtype macrophages also exist in the proximity of tumors (Fig. 14.2).

Complicated mechanisms are found to be involved in the regulation of M1 or M2 activation of macrophages in tumors. NF-kappa B signaling may have disparate functions in TAM activation. It was reported that lack of p50/p65 NF-kappa B activation in murine and human TAMs is associated with the reduced IL-12 expression and increased IL-10 expression from TAMs, indicating the inactivated NF-kappa B signaling in M2 subtype TAMs [39]. However, Ikappa B kinase beta-mediated NF-kappa B activation was required for IL-1R and MyD88-mediated M2 subtype activation of macrophages. In this case, inhibition of the canonical NF-kappa B signaling in TAMs induced M1 subtype activation and macrophage cytotoxicity to tumor cells [40]. It is possible that NF-kappa B activity may vary in different subsets of M2 subtype macrophages. On the other hand, M2 TAMs in mouse tumors were found to have a lower level of Notch pathway activation. Forced activation of canonical Notch signaling mediated by RBPJ and SOCS3 increased M1 macrophages, resulting in elevated IL-12 production and antitumor capacity of tumor-infiltrating macrophages. Meanwhile, blockage of Notch signaling induced an M2 at the expense of an M1 response [41]. These data highlight the suppression of canonical Notch pathways as a molecular switch of monocyte to M2 macrophage differentiation. Moreover, TAMs infiltrating into subcutaneous lung cancer xenografts showed an elevated expression of interleukin receptor-associated kinase-M (IRAK-M), which may function to negatively regulate toll-like receptors (TLRs) and suppress antitumor innate immune responses. Tumors implanted in IRAK-M(−/−)

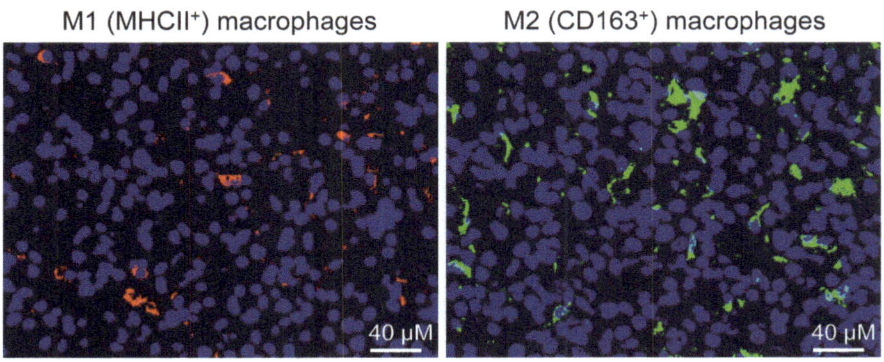

FIGURE 14.2 M1 and M2 subtypes of tumor-associated macrophages (TAMs) in glioblastoma. Immunofluorescent staining of the M1 subtype TAM marker MHCII (in red) and the M2 subtype TAM marker CD163 (in green) in human primary glioblastoma specimens. Both M1 and M2 TAMs reside in the tumor mass. Scale bar represents 40 μm.

mice had a five-fold reduction in growth compared to tumors in wild-type animals along with the M1 subtype activation of macrophages in tumors [42], indicating the requirement of IRAK-M activity by M2 subtype macrophages. Furthermore, the mammalian target of the rapamycin (mTOR) pathway may also be involved in M2 subtype activation. Inhibition of the mTOR pathway by rapamycin induced differentiation of human peripheral monocytes into M1 macrophages, whereas knockdown of the mTOR repressor tuberous sclerosis 2 (TSC2) promoted differentiation of monocytes into M2 macrophages. Importantly, infusion of TSC2-deficient or TSC2-overexpression monocytes into mice bearing tumor xenografts promoted or suppressed tumor angiogenesis, respectively [43], indicating the participation of the mTOR pathway in the differentiation of monocytes into M2 subtype macrophages rather than the maintenance of M2 phenotypes. In conclusion, M1 or M2 subtype activation of monocytes is determined by a network of multiple signaling pathways.

CANCER STEM CELLS AND TUMOR-ASSOCIATED MACROPHAGES

Tumor-Associated Macrophages Support Cancer Stem Cells

As two distinct pivotal populations in tumors, TAMs and CSCs are both enriched in perivascular niches as well as hypoxic niches [44]. In addition, TAMs and CSCs simultaneously increase in tumors after radiotherapy [45]. Furthermore, the number of TAMs is positively correlated with the density of CSCs in tumors [46]. These connections between TAMs and CSCs strongly suggest the functional association between the two cell populations.

Increasing evidence has indicated that TAMs may function as supporters of CSCs. In pancreatic tumors, targeting TAMs by blocking the myeloid cell receptors, colony-stimulating factor-1 receptor (CSF1R) or chemokine (C-C motif) receptor 2 (CCR2) decreased the number of CSCs [47]. In breast cancers, TAMs correlated with the CSC-enriched side population. TAMs may induce SOX2 expression and stem cell-like phenotypes through activating EGF receptor (EGFR) and signal transducers and activators of transcription 3 (STAT3) in breast tumor cells [48]. Consistently, in vitro coculture of M2 macrophages with breast cancer cell lines promoted the formation of mammospheres, which are widely regarded as surrogates of enrichment of CSCs [49]. Furthermore, $CD34^-$ melanoma CSCs were found to depend on M2 macrophages for survival and proliferation. In this case, TAM-derived transforming growth factor-beta (TGF-beta) and polyamines, products of the elevated M2 macrophage-specific arginase pathway, functioned to stimulate CSC growth [50]. Moreover, in xenografts derived from $CD44^+ALDH1^+$ colon cancer cells and $CD133^+ALDH1^+$ lung cancer cells, TAMs were found to secrete milk-fat globule-EGF-VIII (MFG-E8) to confer tumorigenicity and drug resistance on CSCs by activation of STAT3 and Sonic Hedgehog pathways in CSCs [51]. All these studies demonstrated that TAMs support CSCs through paracrine signaling involving multiple factors. In addition, TAMs may indirectly facilitate CSCs by affecting tumor vascularization, as described in detail in later paragraphs.

Cancer Stem Cells Recruit and Maintain Tumor-Associated Macrophages

The functional interplay between TAMs and CSCs appears to be mutual. Enrichment of TAMs in the proximity of CSCs is not spontaneous but is a deliberately regulated process. CSCs actively recruit TAMs into tumors. In glioblastomas, we found that peripheral blood monocytes infiltrate into tumors in response to glioma stem cell (GSC)-secreted Periostin (POSTN) that mediates through an integrin $\alpha_v\beta_3$ signaling pathway. Depletion of POSTN in GSCs reduced the infiltration of monocyte-derived TAMs, suppressed orthotopic tumor growth, and extended the survival of mice bearing tumors [30], illustrating a crucial role of the GSC-mediated TAM recruitment in tumor growth and progression. In addition, M2 subtype activation of TAMs can be regulated by CSCs. Disruption of POSTN in GSCs altered the tumor-infiltrating TAMs from M2 subtype to M1 subtype macrophages in vivo, whereas POSTN treatment suppressed expression of M1 subtype macrophage markers but elevated expression of M2 subtype macrophage markers in vitro [30]. Similarly, coculture of macrophages with ovarian CSCs increased the expression of M2 TAM markers but decreased the expression of M1 TAM markers [52]. Interestingly, when human monocytes were cultured together with breast cancer cell lines, the triple negative breast cancer cells that are enriched with CSCs promoted monocyte differentiation into M2 macrophages. In contrast, the less malignant ER^+ breast cancer cell lines induced M1-like macrophage differentiation, featured by an acute inflammatory response and increased antigen presentation [53]. Mechanistically, CSC-derived interferon-regulatory factor-5 (IRF5) and CSF1 are believed to be involved in the M2 subtype activation of TAMs [54]. Finally, CSCs may be required for TAM survival. In teratomas, the CSC-like cells derived from embryonic stem cells (ESCs) prevented TAMs from macrophage colony-stimulating factor (M-CSF/CSF1) withdrawal-induced apoptosis and prolonged TAM lifespan. IL-34 secreted by the ESC-derived cells may activate the ERK1/2 and PI3K/Akt pathways in TAMs and thus promotes TAM survival [55]. Collectively, CSCs recruit TAMs, stimulate M2 subtype activation, and promote TAM survival, resulting in the coenrichment of TAMs and CSCs. Thus the cellular and molecular interactions between CSCs and TAMs play critical roles in promoting tumor malignant progression.

PROTUMOR ACTIVITIES OF TUMOR-ASSOCIATED MACROPHAGES

Tumor-Associated Macrophages and Tumor Growth

TAMs serve as important sources of key cytokines essential for tumor cell proliferation and survival. In colon cancer, TAM-secreted IL-1beta induces phosphorylation of GSK3beta, stabilizes beta-catenin, and enhances the expression of Wnt target genes to promote tumor cell growth [56]. In melanoma, osteopontin-positive TAMs produced a high level of prostaglandin E2 (PGE2) in a cyclooxygenase-2 (COX-2)–dependent fashion. PGE2 in turn stimulated melanoma cancer cell growth [57]. In addition, in breast cancers, CCL18 derived from TAMs induced cancer cell epithelial to mesenchymal transition (EMT) and accelerated tumor progression [58]. Likewise, expression of IL-10 was found to be restricted in TAMs in human gliomas. IL-10 level was correlated with the glioma malignancy, and TAM-derived IL-10 increased cell proliferation of cultured glioma cells [59]. Expression of another protumorigenic cytokine, IL-6, was also detected in TAMs in gliomas [60,45]. Taken together, through production and secretion of multiple tumor-supportive factors, TAMs promote tumor growth and malignant behavior via paracrine signaling.

Tumor-Associated Macrophages and Neovascularization

Studies on different human tumors have demonstrated a positive correlation between TAM concentration and the vascular density which are both closely linked to patient prognosis. In fact, TAMs produce multiple soluble factors to promote tumor angiogenesis. For instance, TAMs in primary breast cancers had a high expression of macrophage chemoattractant protein-1 (MCP-1/CCL2), whose concentration was correlated with the levels of a series of potent angiogenic factors including vascular endothelial growth factor (VEGF), TNF-alpha, and IL-8 that synergized to accelerate neovascularization [61]. TAMs were also found to be resources of VEGF in colon and breast carcinomas [62]. In addition, in the transgenic PyMT mouse model susceptible to mammary cancer, the formation of a high-density vessel network in tumors was regulated by TAMs. This process relies on Wnt7b produced by TAMs as was observed in the deletion of Wnt7b in TAMs resulting in an impaired vessel network. During the formation of the vessel network, vascular endothelial cells, but not cancer cells, responded to TAM-derived Wnt7b through activation of the Wnt/beta-catenin pathway [63]. Moreover, platelet-derived endothelial cell growth factor (PD-ECGF) was detected in TAMs in glioblastoma along with a positive correlation between PD-ECGF and the degree of stromal vascularity, suggesting an upregulation of vascularization in glioblastoma by TAM-derived PD-ECGF [64]. Therefore TAMs produce a variety of angiogenic factors to stimulate endothelial cell growth and promote tumor neovascularization.

Interestingly, some metabolic enzymes in TAMs also seem to be important players in TAM-induced tumor neovascularization. For example, heme oxygenase-1 (HO-1), a rate-limiting enzyme in heme catabolism, was predominantly observed in TAMs. The mRNA level of HO-1 was correlated with vascular density in human glioma samples, suggesting a role for HO-1 in tumor vascularization [65,34]. Likewise, glioblastomas showed a high level of thymidine phosphorylase (TP) whose expression was restricted in TAMs and correlated with the microvessel density [66]. It is likely that intermediate metabolites generated by certain enzymes in TAMs may function as angiogenic signals in the tumor microenvironments.

Tumor-Associated Macrophages and Cancer Invasion

TAMs also actively participate in regulating cancer invasion and tumor metastasis. Coculture with macrophages remarkably elevated both invasive potential and matrix-degrading activities of lung cancer cell lines [67]. TAM-mediated tumor cell migration is at least partly related to degradation of the extracellular matrix. The stimulation of gastric and colorectal cancer cell motility by human macrophages relied on the matrix metalloproteinase (MMP) activity [68]. However, a study on macrophage infiltration into breast cancer SUM159PT spheroids indicated the existence of both MMP-dependent and MMP-independent tumor cell migration induced by macrophages [69]. Although TAM-mediated MMP-independent migration of tumor cells was not blocked by protease inhibitors [69], macrophage-induced migration of gastric and colorectal cancer cells was impaired by immune depletion of EGF [68]. Further studies demonstrated that activation of the EGF receptor, phospholipase C-gamma (PLC-gamma), and the GRB2-associated binding protein-1 (Gab1) in cancer cells was required for macrophage-induced migration [68]. In addition, TAMs were found to secrete Wnt5a in breast cancer, which triggered both canonical and noncanonical Wnt signaling in tumor cells. While the activation of the canonical Wnt pathway in tumor cells was a necessary prerequisite, experiments with agonists and antagonists revealed that noncanonical signaling via the Jun-N-terminal kinase (JNK) pathway was critical for cancer cell invasiveness [70]. Likewise, coculture of macrophages with ovarian or breast cancer cells led to TNF-alpha-dependent activation of JNK and the canonical NF-kappa B pathways in tumor cells to promote cancer cell invasion [71]. Several proteases may also play a role in the TAM-mediated tumor cell invasion. For example, high levels of cathepsin protease activity were detected in TAMs in pancreatic islet cancers, mammary tumors, and lung metastases, resulting in the release of cathepsins B and S that in turn enhanced the invasiveness of cancer cells [72]. Taken together, TAMs promote tumor cell motility both by remodeling the extracellular matrix in the tumor microenvironment and by secretion of soluble factors to activate the migration-related signaling pathways in tumor cells.

Tumor-Associated Macrophages and Immunosuppression

Most TAMs have lost traditional immune cell properties. Moreover, TAMs can impair the functions of other immune cells to help tumor cells escape from immune surveillance. In a colon cancer mouse model, inoculation of colon carcinoma cells rapidly elicited a population of $CD11b^{+}IL\text{-}4Ralpha^{+}$ monocytes as TAM progenitors. These $CD11b^{+}IL\text{-}4Ralpha^{+}$ cells produced IL-13 and IFN-gamma to suppress the antigen-activated $CD8^{+}$ T lymphocytes [73]. In addition, in murine mammary tumors, hypoxia induced the expression of transcription factor hypoxia-inducible factor-1alpha (HIF-1alpha) in TAMs and augmented TAM-mediated suppression of T-cells. HIF-1alpha-induced expression of iNOS and Arg-1 was supposed to be responsible for T-cell suppression by production of nitric oxide and depletion of L-arginine [74]. Furthermore, TAM-derived IL-1 suppressed the expression of cell surface antigen HLA class II in glioblastoma cells by antagonizing IFN-gamma signals [75], which may help glioma cells to escape from being targeted by T cells. Consistent with these observations, depletion of TAMs in mammary tumors restored tumor-infiltrating cytotoxic T cell responses and suppressed tumor growth [76]. Collectively, these studies indicated that TAMs participate in the establishment of an immunosuppressive environment of tumors by interfering with the normal function of other types of immune cells.

Tumor-Associated Macrophages and Therapeutic Resistance

TAMs are also involved in the resistance to therapies. In murine sarcomas and carcinomas receiving radiotherapy, TAM density is inversely correlated with tumor response [77]. Moreover, TAMs promote tumor metastasis and relapse after surgery. In a mouse model that develops spontaneous metastatic melanoma from the eyes, surgical treatment increased tumor cell proliferation and distal metastases, both correlated with the increased TAM density [78]. Depletion of TAMs significantly reduced postoperative tumor recurrence and inhibited metastatic outgrowth [78]. Furthermore, TAMs inhibited the effectiveness of several chemotherapeutic agents. In the hepatocellular carcinoma xenografts resistant to treatment with sorafenib, an inhibitor of multiple kinases, an increase of TAM recruitment was observed along with the elevation of angiogenic factors VEGF and stromal-derived factor 1 (SDF-1). Depletion of macrophages by clodrolip or zoledronic acid (ZA) significantly inhibited tumor angiogenesis, lung metastasis, and tumor progression during sorafenib treatment, indicating the substantial contribution of TAMs to tumor resistance against sorafenib [79]. Likewise, treatment of human gastrointestinal stromal tumors with the KIT oncoprotein inhibitor imatinib polarized TAMs to become tumor-supportive M2-like macrophages [80]. Moreover, antiandrogen therapy of prostate cancer enhanced the recruitment of TAMs that contributed to cancer recurrence [81]. Antiandrogen-induced expression of CSF1 in tumor cells was found to be responsible for the increased TAM infiltration, whereas blocking TAM influx with a CSF1R inhibitor disrupted tumor promotion [81]. TAMs may even promote tumor resistance to cutting-edge new therapies. Suppression of neovascularization by targeting VEGFR has been proposed as a new strategy to treat cancers. Despite the decreased vessel density, VEGFR inhibition increased TAM infiltration probably due to the creation of hypoxic regions within tumors. TAM infiltration was found to be inversely correlated with the overall survival of patients with recurrent gliomas after antiangiogenic therapy [82]. These studies emphasize the importance of prioritizing TAM inhibition as a critical adjuvant therapy to reduce tumor resistance and recurrence after antitumor treatments.

RECRUITMENT OF TUMOR-ASSOCIATED MACROPHAGES BY CANCER CELLS

The correlation between tumor-derived chemoattractants and macrophage density in neoplastic tissues has been recognized for a while. Several studies showed that tumor cells secrete multiple factors to recruit TAMs. Disruption of glial cell-derived neurotrophic factor (GDNF) in mouse glioma cells attenuated TAM infiltration in vivo [83], indicating GDNF as a chemoattractant to TAMs. The angiogenic factor VEGF may also promote TAM recruitment. Antibody blockade of VEGFR1 diminished TAM accumulation in gliomas [84]. Selective inhibition of VEGFR2 also reduced TAM infiltration in orthotopic pancreatic tumors [85]. Moreover, in prostate cancer, cancer cell-derived neuroblastoma overexpressed (NOV/CCN3) was correlated with TAM infiltration, which could be inhibited by an anti-CCN3 neutralizing antibody [86]. Besides soluble factors, some enzymes such as heparanase, in tumor cells are involved in the recruitment of TAMs. For example overexpression of heparanase, the predominant enzyme degrading heparin sulfate, was associated with increased TAM infiltration in pancreatic ductal adenocarcinoma. The macrophages from heparanase-rich tumors displayed a strong M2 phenotype that accelerates tumor growth [86]. Therefore cancer cells recruit TAMs through the production and secretion of chemoattractants.

Some pathophysiological conditions can augment the recruitment of TAMs by cancer cells. Several lines of evidence suggest that hypoxia increases the capacity of tumor cells to recruit TAMs. In Lewis lung carcinomas, hypoxia induced tumor cells to express Semaphorin 3A (Sema3A) that acts as a chemoattractant for TAMs. Sema3A triggers TAM infiltration by phosphorylation of VEGFR1 in TAMs through the associated holoreceptor composed of Neuropilin-1 (Nrp-1) and PlexinA1/PlexinA4. Interestingly, in hypoxic regions where Nrp1 was downregulated, Sema3A elicited PlexinA1/PlexinA4-mediated stop signals, which retained TAMs inside the hypoxic niche [87]. Furthermore, hypoxia increased expression of SDF1 in glioma cells, which in turn recruited TAM progenitor myeloid cells. SDF1 was shown to be partially controlled by HIF1alpha, as seen in SDF1 levels decreasing in the absence of HIF1alpha accompanied by reduced TAM infiltration [88]. In addition, metabolic enzymes responsible for TAM recruitment may be regulated by hypoxia. 5-Lipoxygenase (5-LOX), a key enzyme assisting conversion of arachidonic acid to 5-heme and leukotrienes, was found to be highly expressed in hypoxic areas of human ovarian tumor tissues. Through increased production of 5-LOX metabolites, ovarian cancer cells in hypoxic areas promoted TAM infiltration [89]. Furthermore, hypoxia-induced recruitment of TAMs by cancer cells appears to be a complex process involving multiple cell components. It has been shown that hypoxia enhanced the interaction between breast cancer cells and mesenchymal stem cells, resulting in the production of CSF1 from the former and chemokine (C-C motif) ligand 5 (CCL5) from the latter, both functioning as TAM chemoattractants in breast tumors [90]. Similar to hypoxia, irradiation can induce HIF1alpha expression in tumors. Irradiation caused an influx of bone marrow-derived macrophage precursors that function to restore tumor growth in gliomas. Inhibition of HIF1alpha or SDF1 prevented this irradiation-induced influx and attenuated tumor recurrence after irradiation [91]. Taken together, cancer cell-mediated TAM recruitment is dynamically regulated by pathophysiological conditions, which frequently occur during tumor progression and tumor treatment.

Interestingly, some tumor-derived factors seem to prevent TAM infiltration. Glioma cells showed high expression of the soluble decoy receptor 3 (DcR3). However, ectopic overexpression of DcR3 in glioma cells substantially decreased TAM infiltration in tumors [92]. On the other hand, free docosahexaenoic acid in the glioma microenvironment was found to induce macrophage apoptosis [93]. It is possible that tumors develop an inhibitory mechanism to deny the entry of specific types of macrophages. However, the clinical implication of this selectivity in TAM recruitment remains to be elucidated.

EDUCATION OF TUMOR-ASSOCIATED MACROPHAGES BY CANCER CELLS

TAMs gain the tumor supportive features in response to cancer cell-derived stimuli. Conditioned medium from glioma cells suppressed phagocytic activity of macrophages. Instead, macrophages were turned into M2 protumorigenic subtypes with the secretion of immunosuppressive cytokine IL-10 and the capacity to inhibit T-cell proliferation [94]. However, direct contact between cancer cells and macrophages may be required for the TAM polarization, as evidenced by the upregulation of the M2 markers, the mannose and scavenger receptors in macrophages, by coculture with ovarian cancer cells but not the conditioned medium [95].

The induction of the M2 phenotype in TAMs by tumor cells may be a rapid process that happens in the early stage of monocyte to macrophage differentiation. Coculture of healthy human $CD14^{+}$ monocytes with human glioblastoma cells elicited immunosuppressive properties in monocytes, including reduced CD14 (but not CD11b) expression, increased interleukin-10 secretion, decreased phagocytic ability, and increased toxicity to activated lymphocytes. However, normal human astrocytes could not induce such responses in cocultured monocytes [96]. This indicated that glioblastoma cells have the capacity to transform monocytes/macrophages into M2 immunosuppressive subtypes. Moreover, in chemically induced primary murine lung tumors, the polarization of TAMs into $Arg1^{high}iNOS^{low}$ M2 subtype occurred within 1 week after carcinogen treatment, followed by the similar activation of TAM precursor monocytes a few days later [97]. Furthermore, in K-ras and FGF10-driven models of lung cancers, silencing of the transgenes caused rapid tumor regression accompanied by shifting of TAMs from $Arg1^{high}iNOS^{low}$ polarized status to $Arg1^{low}iNOS^{low}$ inactive status [97]. Therefore polarization of TAMs by cancer cells is a dynamic and reversible process.

Cancer cells may educate TAMs through secretion of soluble factors. CSF1 is a well-documented factor with a strong capacity to induce M2 polarization of macrophages. Expression of CSF1 significantly correlated with the M2 TAM infiltration and histological malignancy in human brain tumors [38]. In addition, a soluble form of vacuolar ATPase composed of its cleaved N-terminal domain of alpha2 isoform (alpha2NTD) was detected in breast and skin cancers. Alpha2NTD from cancer cells stimulated TAMs in cancers to elevate the expression of M2 markers including mannose receptor, Arg1, IL-10, and TGF-beta [98].

Cancer cells may need to be directly or indirectly stimulated by TAMs in order to gain the capacity to educate TAMs. During coculture of human primary macrophages with breast cancer cell MCF-7, apoptosis of cancer cells conferred macrophages with M2 phenotypes featuring increased IL-10 and decreased IL-12 expression [99]. As a control, coculture with the apoptosis-resistant Bcl-2-overexpressing MCF-7 cells failed to induce macrophage polarization. Sphingosine-1-phosphate (S1P) produced by cancer cells was found to contribute to the polarization of TAMs by apoptotic cancer cells [99]. Further studies showed that S1P triggered the Src-dependent shuttling of cytosolic tyrosine kinase receptor A (TRKA) to the TAM plasma membrane via S1P receptor signaling. TRKA was then activated by the autocrine-secreted nerve growth factor, which in turn activated the PI3K/Akt and p38 MAPK pathways to induce IL-10 expression by TAMs [100]. It is likely that other kinds of mediators besides proapoptotic signals on cancer cells may be required for the cancer cell-mediated TAM education.

Noticeably, education of TAMs by cancer cells may be associated with the low expression of proinflammatory factors. For instance, granulocyte-colony-stimulating factor (G-CSF) that induces the classic activation of macrophages was barely detectable in malignant glioblastomas [101]. The activation status of TAMs may rely on the balance of proinflammatory and protumorigenic signals produced by different cell components within the tumor microenvironment.

TUMOR-ASSOCIATED MACROPHAGES AND CANCER THERAPIES

Targeting Tumor-Associated Macrophages

Considering the critical role of TAMs in promoting tumor malignant progression, elimination of TAMs might provide a new strategy to improve cancer treatment. Several attempts have been made to target TAMs and the preliminary outcomes in preclinical models are encouraging. Application of a recombinant immunotoxin to Folate receptor beta (FR-beta) to mice bearing C6 glioma xenografts significantly depleted FR-beta positive TAMs and reduced tumor growth [102]. In addition, an atypical methylxanthine propentofylline (PPF) remarkably inhibited rat glioblastoma growth by targeting TAMs but not tumor cells [103,88]. Another chemotherapeutic agent, trabectedin, selectively induced caspase-8-dependent apoptosis of monocytes and macrophages in blood, spleens, and tumors, resulting in reduced angiogenesis and attenuated tumor growth [104]. Likewise, the bisphosphonate zoledronic acid disrupted multiple kinds of macrophages including human macrophages, murine bone marrow-derived macrophages, and microglia but not cancer cells. Zoledronic acid treatment not only inhibited tumor growth but also reduced brain tumor cell invasion into normal tissues [105]. Furthermore, introduction of an RNA aptamer targeting IL-4Ralpha into tumor-bearing mice preferentially disrupted TAMs and inhibited tumor growth [106].

Of note, education of TAMs by cancer cells may be a rapid process along with a fast continuing infiltration of monocytes into tumors. Thus this may narrow the time window for the therapeutic efficiency of drugs targeting TAMs. Moreover, many studies have pointed out that monocytes in tumor patients may bear tumor-supportive activities even before infiltrating into tumors. Therefore depletion of TAMs in tumors may not be enough to achieve a promising antitumor effect. In the future, researchers may want to find a way to durably eliminate TAMs and TAM precursor cells with the minimum collateral damage of normal monocytes and macrophages.

Converting M2 Macrophages Into M1 Macrophages

Due to the enormous number of TAMs in tumors and the continuing recruitment of TAMs by cancer cells including CSCs from peripheral blood, effective elimination of TAMs is quite challenging. A more attractive idea is to convert the protumorigenic M2 macrophages to antitumorigenic M1 macrophages that could attack cancer cells in situ. Preclinical studies have supported the feasibility of this strategy. IL-12 is a proinflammatory cytokine recognized as a marker of M1 subtype activation of macrophages. Interestingly, treatment of TAMs isolated from tumors with IL-12 induced a rapid reduction of protumor cytokines and a concomitant increase of proinflammatory cytokines, suggesting that the polarized tumor-infiltrating macrophages retained the ability to shift into classically activated macrophages [107].

The most encouraging results came from the preclinical studies focusing on CSF1, the critical cytokine in inducing M2 polarization. In a pioneering study, oligonucleotides against CSF1 suppressed breast tumor growth in mice by up to 50% reduction of tumor mass [108]. Further studies showed that blocking CSF1R activation by a monoclonal antibody killed CSF1-induced differentiated macrophages in vitro. Moreover, administration of the anti-CSF1R antibody to the diffuse-type giant cell tumor in patients led to a striking reduction of $CD163^+$ M2 macrophages in tumors along with an increase of proinflammatory $CD8^+CD4^+$ T cells [109]. Alternatively, treatment with a selective pharmacologic CSF1R inhibitor GW2580 reduced the expression of proangiogenic and immunosuppressive genes in TAM precursor monocytes and suppressed tumor growth of orthotopic prostate cancers [110]. Importantly, blockade of CSF1R by its inhibitor BLZ945 significantly slowed intracranial growth of patient-derived glioma xenografts [111]. Rather than depletion of TAMs, inhibition of CSF1R signaling decreased M2 markers in TAMs. Survival of TAMs after BLZ945 treatment may be facilitated by glioma-derived GM-CSF and IFN-gamma that were reported to favor M1 activation of macrophages [111]. Similarly, inhibition of CSF1R in a mouse model of pancreatic ductal adenocarcinoma led to functional reprogramming of TAMs that enhanced antigen presentation and antitumor T-cell responses [112]. Therefore disruption of the CSF1/CSF1R signaling in TAMs alters TAM phenotypes, reduces production of protumor cytokines, enhances the macrophage-mediated antigen presentation, relieves TAM-induced immunosuppression of T-cells, and thus overall provides a treatment benefit.

Multiple signaling pathways have been reported in the regulation of M2 polarization of TAMs. Manipulation of these polarization-associated signaling pathways also may have the potential to reprogram TAMs. For example, in ovarian tumor xenografts, targeting NF-kappa B in TAMs promoted tumor regression by induction of macrophage tumoricidal activity [40]. Likewise, inhibition of the STAT3 pathway reversed phenotypes of TAMs isolated from glioblastoma patients with the induction of proinflammatory cytokines IL-2 and IL-12 [113]. Some soluble factors demonstrated the potential to skew TAM activation away from the M2 to an M1-like phenotype. For instance, GM-CSF treatment of murine mammary tumors resulted in inhibition of angiogenesis, metastasis, and tumor growth, accompanied by more M1-like macrophages in tumors [114]. In addition, forced expression of the histidine-rich glycoprotein (HRG) in different cancer cells improved chemotherapy treatment of xenografts. HRG was found to convert M2 TAMs into an M1-like status through downregulation of placental growth factor in these tumors [17]. It is expected that additional approaches will be developed in the future to reprogram TAMs in situ and effectively improve cancer treatment.

CONCLUDING REMARKS

The existence of a tumor immune microenvironment has been recognized as an inevitable issue in tumor treatment. So far, our knowledge on the tumor immune microenvironments is still limited. However, studies on protumorigenic tumor-infiltrating immune cells, tumor-supportive TAMs in particular, can be summarized as follows: first, TAMs have impaired immune surveillance, as demonstrated by a lack of phagocytosis and antigen-presenting activity. In fact, TAMs synthesize and secrete stimulatory factors to function as tumor supportive cells. Second, TAMs are closely associated with critical events during malignant progression, such as tumor growth, neovascularization, metastasis, and tumor relapse after treatment (Fig. 14.3). Third, TAMs gain the tumor-supportive activities in response to signals from tumor cells, particularly CSCs. Thus through reciprocal supportive interplay between TAMs and tumor cells, tumors display potent ability to manipulate different components in the microenvironment to achieve a maximal malignant growth.

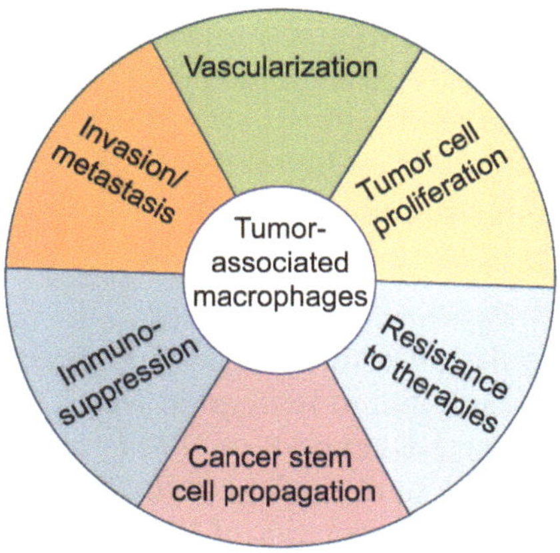

FIGURE 14.3 Schematic representation of tumor malignant processes affected by tumor-associated macrophages (TAMs).
TAMs promote multiple malignant processes including vascularization, invasion, immunosuppression, tumor cell proliferation, therapeutic resistance, and propagation of cancer stem cells.

Several lines of evidence suggest a strong connection between CSCs and the tumor immune microenvironments. Enrichment of CSCs and TAMs are detected in similar niches, including perivascular areas and hypoxic regions. In tumor samples treated with chemo- or radiotherapies, increased numbers of CSCs as well as tumor-infiltrating TAMs have been observed. Recent studies have shown that the immune microenvironment could provide another critical CSC niche in addition to the well-known perivascular and hypoxic niches. Similar to the situations between tumor cells and stromal cells, the interaction between CSCs and TAMs is bidirectional. On one hand, CSCs recruit TAMs and tune them into a tumor-supportive fashion. On the other hand, TAMs promote CSC propagation and survival (Fig. 14.4). Given the importance of both populations in tumor progression, disruption of this positive feedback loop between CSCs and TAMs may provide a crucial therapeutic strategy in cancer treatment.

The investigation on molecular interaction between CSCs and TAMs is greatly hampered by the difficulty in isolating these cells from tumors and maintaining their bona fide characteristics in vitro. CSCs, once sorted from tumor samples, are prone to differentiation. To make matters worse, it is unlikely that tumor-infiltrating TAMs can be maintained under current culture conditions without undergoing massive death. Normal human monocytes and macrophages, as the subject of mechanistic studies, have the same problem with respect to their maintenance in vitro. By using in vitro priming of immortalized monocytes, people have tried to mimic M2 polarization of macrophages. However, these

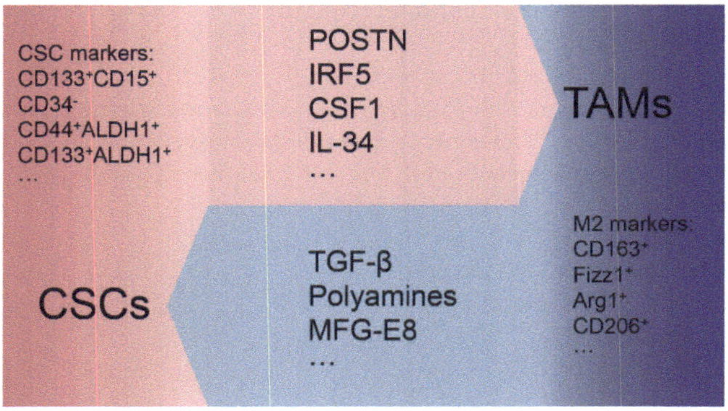

FIGURE 14.4 Schematic representation of the reciprocal supportive signaling between cancer stem cells (CSCs) and tumor-associated macrophages (TAMs).

CSCs enriched with different cell surface markers in different cancers secrete multiple soluble factors to recruit TAMs and promote protumorigenic phenotypes of TAMs. Protumorigenic TAMs with different cell surface markers in turn secrete multiple cytokines to promote CSC maintenance and propagation. Of note, whereas many subtypes of TAMs with various surface markers have been reported to participate in cancer progression, only those involved in CSC maintenance are included in this schematic representation.

immortalized monocytes themselves may have tumorigenic potential, which significantly limits in-depth studies of the properties of nonneoplastic TAMs. Despite the high cost and long duration, implantation of tumor xenografts into genetically engineered rodents may still be the best current method to explore key pathways in TAMs. Until now, there are no reports on the combination of in vivo CSC tracing in the mice with genetically manipulated monocytes/macrophages. Such studies will help to clarify the effects of TAMs on CSCs. Further studies are required to fully elucidate the critical roles of the cellular and molecular interplay between CSCs and the tumor immune microenvironment in malignant progression. The new knowledge from this exciting area will greatly help improve cancer immunotherapy to benefit patients.

List of Acronyms and Abbreviations

5-LOX 5-Lipoxygenase
ALDH1 Aldehyde dehydrogenase 1
AngII Angiotensin II
Arg-1 Arginase 1
Bcl-2 B-cell CLL/lymphoma 2
BHT 3,5-Di-t-butyl-4-hydroxytoluene
CCL5 Chemokine (C-C motif) ligand 5
CCR2 Chemokine (C-C motif) receptor 2
CD11b CD11 antigen-like family member B (macrophage antigen alpha)
CD133 Prominin 1 (Antigen AC133)
CD14 Myeloid cell-specific leucine-rich glycoprotein (CD14 antigen)
CD163 Hemoglobin scavenger receptor (CD163 antigen)
CD34 CD34 cell surface antigen
CD4 T-cell surface antigen T4/Leu-3 (CD4 antigen)
CD44 Cell surface glycoprotein CD44 (hematopoietic cell E– and L-selectin ligand)
CD68 Scavenger receptor class D, member 1 (macrophage antigen CD68)
CD8 T-lymphocyte differentiation antigen T8/Leu-2 (CD8a antigen)
COX2 Cyclooxygenase-2
CSC Cancer stem cell
CSF1R Colony-stimulating factor-1 receptor
CX3CR1 Chemokine (C-X3-C motif) receptor 1
DcR3 Soluble decoy receptor 3
EFG Epidermal growth factor
EGFR Epidermal growth factor receptor
EMT Epithelial to mesenchymal transition
ERK1/2 Extracellular signal-regulated kinase 1/2
ESC Embryonic stem cell
FGF10 Fibroblast growth factor 10
Fizz1 Cysteine-rich secreted protein A12-alpha-like protein 1 (Resistin-like beta)
FR-beta Folate receptor beta TIC: Tumor-initiating cell
Gab1 The GRB2-associated binding protein-1
GBM Glioblastoma

G-CSF Granulocyte-colony-stimulating factor
GDNF Glial cell-derived neurotrophic factor
GSC Glioma stem cell
GSK3beta Glycogen synthase 3 beta
HIF-1alpha Hypoxia-inducible factor-1alpha
HO-1 Heme oxygenase-1
HRG Histidine-rich glycoprotein
Iba1 Allograft inflammatory factor 1 (AIF-1)
IFN-gamma Interferon gamma
IL-1 Interleukin 1
IL-1beta Interleukin 1 beta
IL-1R Interleukin 1 receptor
IL-2 Interleukin 2
IL-4Ralpha Interleukin 4 receptor alpha
IL-6 Interleukin 6
IL-10 Interleukin 10
IL-11 Interleukin 11
IL-12 Interleukin 12
IL-13 Interleukin 13
iNOS Inducible nitric oxide synthase
IRAK-M Interleukin receptor-associated kinase-M
IRF5 Interferon-regulatory factor-5
JNK Jun-N-terminal kinase
K-Ras Kirsten rat sarcoma viral oncogene homolog
M-CSF/CSF1 Macrophage colony-stimulating factor
MCP-1/CCL2 Macrophage chemoattractant protein-1
MFG-E8 Milk-fat globule-epidermal growth factor-VIII
MHCII Major histocompatibility complex, class II
MMP Matrix metalloproteinase
mTOR Mammalian target of rapamycin
NF-kappa B Nuclear factor kappa B
NOV/CCN3 Neuroblastoma overexpressed
Nrp-1 Neuropilin-1
NSTC Nonstem tumor cell
p38 MAPK Mitogen-activated protein kinase p38
p53 Tumor suppressor protein 53
PD-ECGF Platelet-derived endothelial cell growth factor
PGE2 Prostaglandin E2
PI3K/Akt Proto-oncogene C-Akt (Protein kinase B alpha)
PLC-gamma Phospholipase C-gamma
PlexinA1/PlexinA4 Semaphorin receptor A1/A4
POSTN Periostin, osteoblast-specific factor (OSF2)
PPF Propentofylline
RBPJ Recombination signal binding protein for immunoglobulin kappa J region
RFP Red fluorescent protein
S1P Sphingosine-1-phosphate
SDF-1 Stromal-derived factor 1
Sema3A Semaphorin 3A

SOCS3 Suppressor of cytokine signaling 3
SOX2 SRY-related HMG-Box gene 2 (sex-determining region Y-box 2)
STAT3 Signal transducers and activators of transcription 3
TAM Tumor-associated macrophage
TGF-beta Transforming growth factor beta
TLR Toll-like receptor
TNF-alpha Tumor necrosis factor alpha
TP Thymidine phosphorylase
TRKA Tyrosine kinase receptor A
TSC2 Tuberous sclerosis 2
VEGF Vascular endothelial growth factor
VEGFR1/2 Vascular endothelial growth factor Receptor ½
Wnt Wingless-type MMTV integration site family
Wnt7b Wingless-type MMTV integration site family, member 7B
ZA Zoledronic acid

References

[1] Nguyen LV, Vanner R, Dirks P, Eaves CJ. Cancer stem cells: an evolving concept. Nat Rev Cancer 2012;12(2):133–43.

[2] Chen J, Li Y, Yu TS, McKay RM, Burns DK, Kernie SG, et al. A restricted cell population propagates glioblastoma growth after chemotherapy. Nature 2012;488(7412):522–6.

[3] Bao S, Wu Q, McLendon RE, Hao Y, Shi Q, Hjelmeland AB, et al. Glioma stem cells promote radioresistance by preferential activation of the DNA damage response. Nature 2006;444(7120):756–60.

[4] Calabrese C, Poppleton H, Kocak M, Hogg TL, Fuller C, Hamner B, et al. A perivascular niche for brain tumor stem cells. Cancer Cell 2007;11(1):69–82.

[5] Xing F, Okuda H, Watabe M, Kobayashi A, Pai SK, Liu W, et al. Hypoxia-induced jagged2 promotes breast cancer metastasis and self-renewal of cancer stem-like cells. Oncogene 2011;30(39):4075–86.

[6] Gajewski TF, Schreiber H, Fu YX. Innate and adaptive immune cells in the tumor microenvironment. Nat Immunol 2013;14(10):1014–22.

[7] Cortez-Retamozo V, Etzrodt M, Newton A, Rauch PJ, Chudnovskiy A, Berger C, et al. Origins of tumor-associated macrophages and neutrophils. Proc Natl Acad Sci USA 2012;109(7):2491–6.

[8] Chanmee T, Ontong P, Konno K, Itano N. Tumor-associated macrophages as major players in the tumor microenvironment. Cancers (Basel) 2014;6(3):1670–90.

[9] Bindea G, Mlecnik B, Tosolini M, Kirilovsky A, Waldner M, Obenauf AC, et al. Spatiotemporal dynamics of intratumoral immune cells reveal the immune landscape in human cancer. Immunity 2013;39(4):782–95.

[10] Schmidt M, Bohm D, von Torne C, Steiner E, Puhl A, Pilch H, et al. The humoral immune system has a key prognostic impact in node-negative breast cancer. Cancer Res 2008;68(13):5405–13.

[11] Preuss KD, Zwick C, Bormann C, Neumann F, Pfreundschuh M. Analysis of the B-cell repertoire against antigens expressed by human neoplasms. Immunol Rev 2002;188:43–50.

[12] Biroccio A, Cherfils-Vicini J, Augereau A, Pinte S, Bauwens S, Ye J, et al. TRF2 inhibits a cell-extrinsic pathway through which natural killer cells eliminate cancer cells. Nat Cell Biol 2013;15(7):818–28.

[13] Savage PA, Leventhal DS, Malchow S. Shaping the repertoire of tumor-infiltrating effector and regulatory T cells. Immunol Rev 2014;259(1):245–58.

[14] Curiel TJ, Coukos G, Zou L, Alvarez X, Cheng P, Mottram P, et al. Specific recruitment of regulatory T cells in ovarian carcinoma fosters immune privilege and predicts reduced survival. Nat Med 2004;10(9):942–9.

[15] Hansen W, Hutzler M, Abdel S, Alter C, Stockmann C, Kliche S, et al. Neuropilin 1 deficiency on $CD4^{+}Foxp3^{+}$ regulatory T cells impairs mouse melanoma growth. J Exp Med 2012;209(11):2001–16.

[16] Rech AJ, Mick R, Martin S, Recio A, Aqui NA, Powell Jr DJ, et al. CD25 blockade depletes and selectively reprograms regulatory T cells in concert with immunotherapy in cancer patients. Sci Transl Med 2012;4(134):134ra62.

[17] Rolny C, Mazzone M, Tugues S, Laoui D, Johansson I, Coulon C, et al. HRG inhibits tumor growth and metastasis by inducing macrophage polarization and vessel normalization through downregulation of PlGF. Cancer Cell 2011;19(1):31–44.

[18] Abou-Ghazal M, Yang DS, Qiao W, Reina-Ortiz C, Wei J, Kong LY, et al. The incidence, correlation with tumor-infiltrating inflammation, and prognosis of phosphorylated STAT3 expression in human gliomas. Clin Cancer Res 2008;14(24):8228–35.

[19] Johnston RJ, Comps-Agrar L, Hackney J, Yu X, Huseni M, Yang Y, et al. The immunoreceptor TIGIT regulates antitumor and antiviral CD8(+) T cell effector function. Cancer Cell 2014;26(6):923–37.

[20] Solinas G, Germano G, Mantovani A, Allavena P. Tumor-associated macrophages (TAM) as major players of the cancer-related inflammation. J Leukoc Biol 2009;86(5):1065–73.

[21] Balli D, Ren X, Chou FS, Cross E, Zhang Y, Kalinichenko VV, et al. Foxm1 transcription factor is required for macrophage migration during lung inflammation and tumor formation. Oncogene 2012;31(34):3875–88.

[22] Movahedi K, Laoui D, Gysemans C, Baeten M, Stange G, Van den Bossche J, et al. Different tumor microenvironments contain functionally distinct subsets of macrophages derived from Ly6C(high) monocytes. Cancer Res 2010;70(14):5728–39.

[23] Qian BZ, Li J, Zhang H, Kitamura T, Zhang J, Campion LR, et al. CCL2 recruits inflammatory monocytes to facilitate breast-tumour metastasis. Nature 2011;475(7355):222–5.

[24] Cortez-Retamozo V, Etzrodt M, Newton A, Ryan R, Pucci F, Sio SW, et al. Angiotensin II drives the production of tumor-promoting macrophages. Immunity 2013;38(2):296–308.

[25] Shand FH, Ueha S, Otsuji M, Koid SS, Shichino S, Tsukui T, et al. Tracking of intertissue migration reveals the origins of tumor-infiltrating monocytes. Proc Natl Acad Sci USA 2014;111(21):7771–6.

[26] Fries G, Perneczky A, Kempski O. Glioblastoma-associated circulating monocytes and the release of epidermal growth factor. J Neurosurg 1996;85(4):642–7.

[27] Sippel TR, White J, Nag K, Tsvankin V, Klaassen M, Kleinschmidt-DeMasters BK, et al. Neutrophil degranulation and immunosuppression in patients with GBM: restoration of cellular immune function by targeting arginase I. Clin Cancer Res 2011;17(22):6992–7002.

[28] Ling EA, Wong WC. The origin and nature of ramified and amoeboid microglia: a historical review and current concepts. Glia 1993;7(1):9–18.

[29] Mizutani M, Pino PA, Saederup N, Charo IF, Ransohoff RM, Cardona AE. The fractalkine receptor but not CCR2 is present on microglia from embryonic development throughout adulthood. J Immunol 2012;188(1):29–36.

[30] Zhou W, Ke SQ, Huang Z, Flavahan W, Fang X, Paul J, et al. Periostin secreted by glioblastoma stem cells recruits M2 tumour-associated macrophages and promotes malignant growth. Nat Cell Biol 2015;17(2):170–82.

[31] Roggendorf W, Strupp S, Paulus W. Distribution and characterization of microglia/macrophages in human brain tumors. Acta Neuropathol 1996;92(3):288–93.

[32] Feng X, Szulzewsky F, Yerevanian A, Chen Z, Heinzmann D, Rasmussen RD, et al. Loss of CX3CR1 increases accumulation of inflammatory monocytes and promotes gliomagenesis. Oncotarget 2015;6(17):15077–94.

[33] Dinapoli MR, Calderon CL, Lopez DM. The altered tumoricidal capacity of macrophages isolated from tumor-bearing mice is related to reduce expression of the inducible nitric oxide synthase gene. J Exp Med 1996;183(4):1323–9.

[34] Black KL, Chen K, Becker DP, Merrill JE. Inflammatory leukocytes associated with increased immunosuppression by glioblastoma. J Neurosurg 1992;77(1):120–6.

[35] Fulci G, Dmitrieva N, Gianni D, Fontana EJ, Pan X, Lu Y, et al. Depletion of peripheral macrophages and brain microglia increases brain tumor titers of oncolytic viruses. Cancer Res 2007;67(19):9398–406.

[36] Gordon S, Taylor PR. Monocyte and macrophage heterogeneity. Nat Rev Immunol 2005;5(12):953–64.

[37] Qian BZ, Pollard JW. Macrophage diversity enhances tumor progression and metastasis. Cell 2010;141(1):39–51.

[38] Komohara Y, Ohnishi K, Kuratsu J, Takeya M. Possible involvement of the M2 anti-inflammatory macrophage phenotype in growth of human gliomas. J Pathol 2008;216(1):15–24.

[39] Sica A, Saccani A, Bottazzi B, Polentarutti N, Vecchi A, van Damme J, et al. Autocrine production of IL-10 mediates defective IL-12 production and NF-kappa B activation in tumor-associated macrophages. J Immunol 2000;164(2):762–7.

[40] Hagemann T, Lawrence T, McNeish I, Charles KA, Kulbe H, Thompson RG, et al. "Re-educating" tumor-associated macrophages by targeting NF-kappaB. J Exp Med 2008;205(6):1261–8.

[41] Wang YC, He F, Feng F, Liu XW, Dong GY, Qin HY, et al. Notch signaling determines the M1 versus M2 polarization of macrophages in antitumor immune responses. Cancer Res 2010;70(12):4840–9.

[42] Standiford TJ, Kuick R, Bhan U, Chen J, Newstead M, Keshamouni VG. TGF-beta-induced IRAK-M expression in tumor-associated macrophages regulates lung tumor growth. Oncogene 2011;30(21):2475–84.

[43] Chen W, Ma T, Shen XN, Xia XF, Xu GD, Bai XL, et al. Macrophage-induced tumor angiogenesis is regulated by the TSC2-mTOR pathway. Cancer Res 2012;72(6):1363–72.

[44] Laoui D, Van Overmeire E, Di Conza G, Aldeni C, Keirsse J, Morias Y, et al. Tumor hypoxia does not drive differentiation of tumor-associated macrophages but rather fine-tunes the M2-like macrophage population. Cancer Res 2014;74(1):24–30.

[45] Deininger MH, Pater S, Strik H, Meyermann R. Macrophage/microglial cell subpopulations in glioblastoma multiforme relapses are differentially altered by radiochemotherapy. J Neurooncol 2001;55(3):141–7.

[46] Fan QM, Jing YY, Yu GF, Kou XR, Ye F, Gao L, et al. Tumor-associated macrophages promote cancer stem cell-like properties via transforming growth factor-beta1-induced epithelial-mesenchymal transition in hepatocellular carcinoma. Cancer Lett 2014;352(2):160–8.

[47] Mitchem JB, Brennan DJ, Knolhoff BL, Belt BA, Zhu Y, Sanford DE, et al. Targeting tumor-infiltrating macrophages decreases tumor-initiating cells, relieves immunosuppression, and improves chemotherapeutic responses. Cancer Res 2013;73(3):1128–41.

[48] Yang J, Liao D, Chen C, Liu Y, Chuang YH, Xiang R, et al. Tumor-associated macrophages regulate murine breast cancer stem cells through a novel paracrine EGFR/Stat3/Sox-2 signaling pathway. Stem Cells 2013;31(2):248–58.

[49] Ward R, Sims AH, Lee A, Lo C, Wynne L, Yusuf H, et al. Monocytes and macrophages, implications for breast cancer migration and stem cell-like activity and treatment. Oncotarget 2015;6(16):14687–99.

[50] Tham M, Tan KW, Keeble J, Wang X, Hubert S, Barron L, et al. Melanoma-initiating cells exploit M2 macrophage TGFbeta and arginase pathway for survival and proliferation. Oncotarget 2014;5(23):12027–42.

[51] Jinushi M, Chiba S, Yoshiyama H, Masutomi K, Kinoshita I, Dosaka-Akita H, et al. Tumor-associated macrophages regulate tumorigenicity and anticancer drug responses of cancer stem/initiating cells. Proc Natl Acad Sci USA 2011;108(30):12425–30.

[52] Deng X, Zhang P, Liang T, Deng S, Chen X, Zhu L. Ovarian cancer stem cells induce the M2 polarization of macrophages through the PPARgamma and NF-kappaB pathways. Int J Mol Med 2015;36(2):449–54.

[53] Hollmen M, Roudnicky F, Karaman S, Detmar M. Characterization of macrophage–cancer cell crosstalk in estrogen receptor positive and triple-negative breast cancer. Sci Rep 2015;5:9188.

[54] Jinushi M. Role of cancer stem cell-associated inflammation in creating pro-inflammatory tumorigenic microenvironments. Oncoimmunology 2014;3:e28862.

[55] Chen T, Wang X, Guo L, Wu M, Duan Z, Lv J, et al. Embryonic stem cells promoting macrophage survival and function are crucial for teratoma development. Front Immunol 2014;5:275.

[56] Kaler P, Augenlicht L, Klampfer L. Macrophage-derived IL-1beta stimulates Wnt signaling and growth of colon cancer cells: a crosstalk interrupted by vitamin D3. Oncogene 2009;28(44):3892–902.

[57] Kale S, Raja R, Thorat D, Soundararajan G, Patil TV, Kundu GC. Osteopontin signaling upregulates cyclooxygenase-2 expression in tumor-associated macrophages leading to enhanced angiogenesis and melanoma growth via alpha9beta1 integrin. Oncogene 2014;33(18):2295–306.

[58] Su S, Liu Q, Chen J, Chen J, Chen F, He C, et al. A positive feedback loop between mesenchymal-like cancer cells and macrophages is essential to breast cancer metastasis. Cancer Cell 2014;25(5):605–20.

[59] Wagner S, Czub S, Greif M, Vince GH, Suss N, Kerkau S, et al. Microglial/macrophage expression of interleukin 10 in human glioblastomas. Int J Cancer 1999;82(1):12–6.

[60] Samaras V, Piperi C, Korkolopoulou P, Zisakis A, Levidou G, Themistocleous MS, et al. Application of the ELISPOT method for comparative analysis of interleukin (IL)-6 and IL-10 secretion in peripheral blood of patients with astroglial tumors. Mol Cell Biochem 2007;304(1–2):343–51.

[61] Ueno T, Toi M, Saji H, Muta M, Bando H, Kuroi K, et al. Significance of macrophage chemoattractant protein-1 in macrophage recruitment, angiogenesis, and survival in human breast cancer. Clin Cancer Res 2000;6(8):3282–9.

[62] Barbera-Guillem E, Nyhus JK, Wolford CC, Friece CR, Sampsel JW. Vascular endothelial growth factor secretion by tumor-infiltrating macrophages essentially supports tumor angiogenesis, and IgG immune complexes potentiate the process. Cancer Res 2002;62(23):7042–9.

[63] Yeo EJ, Cassetta L, Qian BZ, Lewkowich I, Li JF, Stefater 3rd JA, et al. Myeloid WNT7b mediates the angiogenic switch and metastasis in breast cancer. Cancer Res 2014;74(11):2962–73.

[64] Nakayama Y, Sueishi K, Oka K, Kono S, Tomonaga M. Stromal angiogenesis in human glioma: a role of platelet-derived endothelial cell growth factor. Surg Neurol 1998;49(2):181–7. discussion 187–8.

[65] Nishie A, Ono M, Shono T, Fukushi J, Otsubo M, Onoue H, et al. Macrophage infiltration and heme oxygenase-1 expression correlate with angiogenesis in human gliomas. Clin Cancer Res 1999;5(5):1107–13.

[66] Hirano H, Tanioka K, Yokoyama S, Akiyama S, Kuratsu J. Angiogenic effect of thymidine phosphorylase on macrophages in glioblastoma multiforme. J Neurosurg 2001;95(1):89–95.

[67] Chen JJ, Lin YC, Yao PL, Yuan A, Chen HY, Shun CT, et al. Tumor-associated macrophages: the double-edged sword in cancer progression. J Clin Oncol 2005;23(5):953–64.

[68] Cardoso AP, Pinto ML, Pinto AT, Oliveira MI, Pinto MT, Goncalves R, et al. Macrophages stimulate gastric and colorectal cancer invasion through EGFR Y(1086), c-Src, Erk1/2 and Akt phosphorylation and smallGTPase activity. Oncogene 2014;33(16):2123–33.

[69] Guiet R, Van Goethem E, Cougoule C, Balor S, Valette A, Al Saati T, et al. The process of macrophage migration promotes matrix metalloproteinase-independent invasion by tumor cells. J Immunol 2011;187(7):3806–14.

[70] Pukrop T, Klemm F, Hagemann T, Gradl D, Schulz M, Siemes S, et al. Wnt 5a signaling is critical for macrophage-induced invasion of breast cancer cell lines. Proc Natl Acad Sci USA 2006;103(14):5454–9.

[71] Hagemann T, Wilson J, Kulbe H, Li NF, Leinster DA, Charles K, et al. Macrophages induce invasiveness of epithelial cancer cells via NF-kappa B and JNK. J Immunol 2005;175(2):1197–205.

[72] Gocheva V, Wang HW, Gadea BB, Shree T, Hunter KE, Garfall AL, et al. IL-4 induces cathepsin protease activity in tumor-associated macrophages to promote cancer growth and invasion. Genes Dev 2010;24(3):241–55.

[73] Gallina G, Dolcetti L, Serafini P, De Santo C, Marigo I, Colombo MP, et al. Tumors induce a subset of inflammatory monocytes with immunosuppressive activity on $CD8^+$ T cells. J Clin Invest 2006;116(10):2777–90.

[74] Doedens AL, Stockmann C, Rubinstein MP, Liao D, Zhang N, DeNardo DG, et al. Macrophage expression of hypoxia-inducible factor-1 alpha suppresses T-cell function and promotes tumor progression. Cancer Res 2010;70(19):7465–75.

[75] Cowan EP, Pierce ML, Dhib-Jalbut S. Interleukin-1 beta decreases HLA class II expression on a glioblastoma multiforme cell line. J Neuroimmunol 1991;33(1):17–28.

[76] Franklin RA, Liao W, Sarkar A, Kim MV, Bivona MR, Liu K, et al. The cellular and molecular origin of tumor-associated macrophages. Science 2014;344(6186):921–5.

[77] Milas L, Wilke J, Hunter N, Volpe J, Basic I. Macrophage content of murine sarcomas and carcinomas: associations with tumor growth parameters and tumor radiocurability. Cancer Res 1987;47(4):1069–75.

[78] Tham M, Khoo K, Yeo KP, Kato M, Prevost-Blondel A, Angeli V, et al. Macrophage depletion reduces postsurgical tumor recurrence and metastatic growth in a spontaneous murine model of melanoma. Oncotarget 2015;6(26):22857–68.

[79] Zhang W, Zhu XD, Sun HC, Xiong YQ, Zhuang PY, Xu HX, et al. Depletion of tumor-associated macrophages enhances the effect of sorafenib in metastatic liver cancer models by antimetastatic and antiangiogenic effects. Clin Cancer Res 2010;16(13):3420–30.

[80] Cavnar MJ, Zeng S, Kim TS, Sorenson EC, Ocuin LM, Balachandran VP, et al. KIT oncogene inhibition drives intratumoral macrophage M2 polarization. J Exp Med 2013;210(13):2873–86.

[81] Escamilla J, Schokrpur S, Liu C, Priceman SJ, Moughon D, Jiang Z, et al. CSF1 receptor targeting in prostate cancer reverses macrophage-mediated resistance to androgen blockade therapy. Cancer Res 2015;75(6):950–62.

[82] Lu-Emerson C, Snuderl M, Kirkpatrick ND, Goveia J, Davidson C, Huang Y, et al. Increase in tumor-associated macrophages after antiangiogenic therapy is associated with poor survival among patients with recurrent glioblastoma. Neuro Oncol 2013;15(8):1079–87.

[83] Ku MC, Wolf SA, Respondek D, Matyash V, Pohlmann A, Waiczies S, et al. GDNF mediates glioblastoma-induced microglia attraction but not astrogliosis. Acta Neuropathol 2013;125(4):609–20.

[84] Stefanik DF, Fellows WK, Rizkalla LR, Rizkalla WM, Stefanik PP, Deleo AB, et al. Monoclonal antibodies to vascular endothelial growth factor (VEGF) and the VEGF receptor, FLT-1, inhibit the growth of C6 glioma in a mouse xenograft. J Neurooncol 2001;55(2):91–100.

[85] Dineen SP, Lynn KD, Holloway SE, Miller AF, Sullivan JP, Shames DS, et al. Vascular endothelial growth factor receptor 2 mediates macrophage infiltration into orthotopic pancreatic tumors in mice. Cancer Res 2008;68(11):4340–6.

[86] Chen PC, Cheng HC, Wang J, Wang SW, Tai HC, Lin CW, et al. Prostate cancer-derived CCN3 induces M2 macrophage infiltration and contributes to angiogenesis in prostate cancer microenvironment. Oncotarget 2014;5(6):1595–608.

[87] Casazza A, Laoui D, Wenes M, Rizzolio S, Bassani N, Mambretti M, et al. Impeding macrophage entry into hypoxic tumor areas by Sema3A/Nrp1 signaling blockade inhibits angiogenesis and restores antitumor immunity. Cancer Cell 2013;24(6):695–709.

[88] Du R, Lu KV, Petritsch C, Liu P, Gnass R, Passeque E, et al. HIF1alpha induces the recruitment of bone marrow-derived vascular modulatory cells to regulate tumor angiogenesis and invasion. Cancer Cell 2008;13(3):206–20.

[89] Wen Z, Liu H, Li M, Li B, Gao W, Shao Q, et al. Increased metabolites of 5-lipoxygenase from hypoxic ovarian cancer cells promote tumor-associated macrophage infiltration. Oncogene 2015;34(10):1241–52.

[90] Chaturvedi P, Gilkes DM, Takano N, Semenza GL. Hypoxia-inducible factor-dependent signaling between triple-negative breast cancer cells and mesenchymal stem cells promotes macrophage recruitment. Proc Natl Acad Sci USA 2014;111(20):E2120–9.

[91] Kioi M, Vogel H, Schultz G, Hoffman RM, Harsh GR, Brown JM. Inhibition of vasculogenesis, but not angiogenesis, prevents the recurrence of glioblastoma after irradiation in mice. J Clin Invest 2010;120(3):694–705.

[92] Roth W, Isenmann S, Nakamura M, Platten M, Wick W, Kleihues P, et al. Soluble decoy receptor 3 is expressed by malignant gliomas and suppresses CD95 ligand-induced apoptosis and chemotaxis. Cancer Res 2001;61(6):2759–65.

[93] Hirano H, Takeshima H, Niiro M, Nagayama T, Oyoshi T, Kuratsu J. The environment of increased concentration of docosahexaenoic acid in glioblastoma may suppress the anti-tumor effect of macrophages. Oncol Rep 2005;13(6):1185–91.

[94] Wu A, Wei J, Kong LY, Wang Y, Priebe W, Qiao W, et al. Glioma cancer stem cells induce immunosuppressive macrophages/microglia. Neuro Oncol 2010;12(11):1113–25.

[95] Hagemann T, Wilson J, Burke F, Kulbe H, Li NF, Pluddemann A, et al. Ovarian cancer cells polarize macrophages toward a tumor-associated phenotype. J Immunol 2006;176(8):5023–32.

[96] Rodrigues JC, Gonzalez GC, Zhang L, Ibrahim G, Kelly JJ, Gustafson MP, et al. Normal human monocytes exposed to glioma cells acquire myeloid-derived suppressor cell-like properties. Neuro Oncol 2010;12(4):351–65.

[97] Redente EF, Dwyer-Nield LD, Merrick DT, Raina K, Agarwal R, Pao W, et al. Tumor progression stage and anatomical site regulate tumor-associated macrophage and bone marrow-derived monocyte polarization. Am J Pathol 2010;176(6):2972–85.

[98] Katara GK, Jaiswal MK, Kulshrestha A, Kolli B, Gilman-Sachs A, Beaman KD. Tumor-associated vacuolar ATPase subunit promotes tumorigenic characteristics in macrophages. Oncogene 2014;33(49):5649–54.

[99] Weigert A, Tzieply N, von Knethen A, Johann AM, Schmidt H, Geisslinger G, et al. Tumor cell apoptosis polarizes macrophages role of sphingosine-1-phosphate. Mol Biol Cell 2007;18(10):3810–9.

[100] Ley S, Weigert A, Weichand B, Henke N, Mille-Baker B, Janssen RA, et al. The role of TRKA signaling in IL-10 production by apoptotic tumor cell-activated macrophages. Oncogene 2013;32(5):631–40.

[101] Stan AC, Walter GF, Welte K, Pietsch T. Immunolocalization of granulocyte-colony-stimulating factor in human glial and primitive neuroectodermal tumors. Int J Cancer 1994;57(3):306–12.

[102] Nagai T, Tanaka M, Tsuneyoshi Y, Xu B, Michie SA, Hasui K, et al. Targeting tumor-associated macrophages in an experimental glioma model with a recombinant immunotoxin to folate receptor beta. Cancer Immunol Immunother 2009;58(10):1577–86.

[103] Jacobs VL, Landry RP, Liu Y, Romero-Sandoval EA, De Leo JA. Propentofylline decreases tumor growth in a rodent model of glioblastoma multiforme by a direct mechanism on microglia. Neuro Oncol 2012;14(2):119–31.

[104] Germano G, Frapolli R, Belgiovine C, Anselmo A, Pesce S, Liquori M, et al. Role of macrophage targeting in the antitumor activity of trabectedin. Cancer Cell 2013;23(2):249–62.

[105] Rietkotter E, Menck K, Bleckmann A, Farhat K, Schaffrinski M, Schultz M, et al. Zoledronic acid inhibits macrophage/microglia-assisted breast cancer cell invasion. Oncotarget 2013;4(9):1449–60.

[106] Roth F, De La Fuente AC, Vella JL, Zoso A, Inverardi L, Serafini P. Aptamer-mediated blockade of IL4Ralpha triggers apoptosis of MDSCs and limits tumor progression. Cancer Res 2012;72(6):1373–83.

[107] Watkins SK, Egilmez NK, Suttles J, Stout RD. IL-12 rapidly alters the functional profile of tumor-associated and tumor-infiltrating macrophages in vitro and in vivo. J Immunol 2007;178(3):1357–62.

[108] Aharinejad S, Paulus P, Sioud M, Hofmann M, Zins K, Schafer R, et al. Colony-stimulating factor-1 blockade by antisense oligonucleotides and small interfering RNAs suppresses growth of human mammary tumor xenografts in mice. Cancer Res 2004;64(15):5378–84.

[109] Ries CH, Cannarile MA, Hoves S, Benz J, Wartha K, Runza V, et al. Targeting tumor-associated macrophages with anti-CSF-1R antibody reveals a strategy for cancer therapy. Cancer Cell 2014;25(6):846–59.

[110] Priceman SJ, Sung JL, Shaposhnik Z, Burton JB, Torres-Collado AX, Moughon DL, et al. Targeting distinct tumor-infiltrating myeloid cells by inhibiting CSF-1 receptor: combating tumor evasion of antiangiogenic therapy. Blood 2010;115(7):1461–71.

[111] Pyonteck SM, Akkari L, Schuhmacher AJ, Bowman RL, Sevenich L, Quail DF, et al. CSF-1R inhibition alters macrophage polarization and blocks glioma progression. Nat Med 2013;19(10):1264–72.

[112] Zhu Y, Knolhoff BL, Meyer MA, Nywening TM, West BL, Luo J, et al. CSF1/CSF1R blockade reprograms tumor-infiltrating macrophages and improves response to T-cell checkpoint immunotherapy in pancreatic cancer models. Cancer Res 2014;74(18):5057–69.

[113] Hussain SF, Kong LY, Jordan J, Conrad C, Madden T, Fokt I, et al. A novel small molecule inhibitor of signal transducers and activators of transcription 3 reverses immune tolerance in malignant glioma patients. Cancer Res 2007;67(20):9630–6.

[114] Eubank TD, Roberts RD, Khan M, Curry JM, Nuovo GJ, Kuppusamy P, et al. Granulocyte macrophage colony-stimulating factor inhibits breast cancer growth and metastasis by invoking an anti-angiogenic program in tumor-educated macrophages. Cancer Res 2009;69(5):2133–40.

CHAPTER 15

The Mechanisms of Therapy Resistance in Cancer Stem Cells

C.-H. Chang, J.M. Rosen
Baylor College of Medicine, Houston, TX, United States

CONTENTS

EVIDENCE OF CANCER STEM CELLS

In 1994, John Dick et al. identified cancer stem cells (CSCs) from acute myeloid leukemia (AML), the most common malignant myeloid disorder in adults, using the cell surface markers CD34 and CD38 [1]. This study became the paradigm for later CSC research in both leukemia and solid tumors. The first solid tumor CSC population was identified in breast cancer in 2003 based on cell surface markers CD44 and CD24 [2]. During the following several years, CSCs were identified in several other human solid tumors, including glioblastoma (GBM) [3], colon cancer [4,5], pancreatic cancer [6], and melanomas [7]. In addition, during the past decade, there have been a growing number of studies identifying CSCs in genetically engineered mouse models [8–14]. With the appropriate microenvironment for tumor transplantation and an intact immune system, these findings in syngeneic mouse models provided additional support for the existence of CSCs.

Technical advances in DNA sequencing and RNA profiling as well as studies characterizing CSCs have shown that cancers exhibit extremely complex intertumoral as well as intratumoral heterogeneity. Hence, it has been very difficult to target all of the different subpopulations in a given tumor with conventional therapies. Moreover, increasing evidence has suggested that CSCs contribute to therapy resistance and tumor recurrence.

EVIDENCE OF THERAPY RESISTANCE

Conventional therapies for cancers include, but are not limited to, ionizing radiation (IR), chemotherapy, targeted therapy, and hormonal therapy. While radiation and chemotherapy cause DNA damage in most of the tumor cells or rapidly dividing cells [15,16], targeted and hormonal therapies, such as imatinib, trastuzumab, and tamoxifen, inhibit the specific functions of certain types

Cancer Stem Cells. http://dx.doi.org/10.1016/B978-0-12-803892-5.00015-2

of tumor cells. In many cases, however, tumors still relapse after treatment, and the recurrent tumors often exhibit resistance to original therapies. More detailed treatment methods and evidence of resistance of each cancer type, including leukemia, breast cancer, and GBM, are described below.

There are four main types of leukemia, including AML, acute lymphoblastic leukemia (ALL), chronic lymphocytic leukemia (CLL), and chronic myeloid leukemia (CML). Different types of treatments are applied in different leukemias. Cytosine arabinoside (Ara-C), a standard therapy used to treat AML, is a chemotherapeutic agent, which inhibits DNA synthesis. Using a primary AML xenograft model [17], human AML stem cells were shown to be more resistant to Ara-C-induced apoptosis in vivo than the other leukemic cells. In CML, imatinib (Gleevec), a tyrosine-kinase inhibitor, is the most prevalent and effective treatment [18]. However, relapses are still observed in part due to therapy resistance of CML stem cells. Residual malignant stem cells, which are *BCR/ABL*-containing $CD34^+$ cells, have been detected even in patients who achieved a complete cytogenetic response after imatinib treatment [19]. In addition, patient-derived primary CML stem cells are also more resistant to imatinib in vitro [20–22]. Moreover, in a mouse model of CML, CSCs are more resistant to both imatinib and radiation treatments [23].

Breast cancer is a heterogeneous disease based on receptor classification or molecular subtypes. Using global gene expression analyses, breast cancer can be classified into five molecular subtypes, including Luminal A, Luminal B, HER2-enriched, Claudin-low, and Basal-like [24,25]. These subtypes are treated with different therapeutic regimens. Based upon their receptor status, breast cancer also can be categorized as estrogen receptor-positive (ER+), progesterone receptor-positive (PR+), HER2-positive, and triple negative, ie, lacking all three receptors [26]. These classifications are critical for determining targeted adjuvant therapies for patients. For example, ER^+ breast cancer can be treated with tamoxifen, an estrogen receptor antagonist; and $HER2^+$ patients can be treated with trastuzumab (Herceptin), a monoclonal antibody that blocks the HER2 receptor. In spite of a wide variety of treatment choices, tumor recurrence is still observed.

Several preclinical animal as well as clinical studies have demonstrated that breast CSCs exhibit increased therapy resistance as compared to the bulk tumor cells. In a phase 2 clinical trial performed from 2000 to 2006, biopsies were collected before and at several times after treatment from HER2-negative patients treated with conventional chemotherapy (docetaxel or doxorubicin and cyclophosphamide). In these biopsies, Li et al. observed a significant increase in the tumorigenic $CD44^+CD24^{low}$ subpopulation in patients after 12 weeks of therapy, suggesting that the CSC subpopulation was more resistant to chemotherapy [27]. An extension of this clinical study defined a common gene expression signature in both sorted $CD44^+CD24^{low}$ cells and cultured mammospheres (MS) from breast cancer patients, and showed that this $CD44^+CD24^{low}$-MS gene signature is enriched in the Claudin-low subtype, characterized by the expression of epithelial–mesenchymal transition (EMT)–associated genes. Both the $CD44^+CD24^{low}$-MS and Claudin-low signatures also were enriched in luminal breast cancers after endocrine therapy (letrozole) or chemotherapy (docetaxel), indicating the residual tumor cells are more stemlike and exhibit EMT properties [28]. Simultaneously, in vitro cell culture studies also showed that stemlike cells, which were isolated as a side population by Hoechst dye staining and flow cytometry, in the luminal MCF-7 cell line were more resistant to IR [29]. Furthermore, the Wnt/β-catenin pathway was shown to mediate radiation resistance in primary mouse mammary epithelial progenitors. Correspondingly, subsequent mouse studies using a p53-null syngeneic tumor model, which mimics several different subtypes of human breast cancer [30], demonstrated that CSCs from these mouse mammary tumors also were more radioresistant possibility due to increased Wnt signaling, increased DNA repair activity, and decreased apoptosis [12,31,32]. The mechanisms involved will be discussed in detail later in this chapter. Additional studies in cell culture and mouse models have also provided evidence to support that breast CSCs are more resistant to either IR or chemotherapy [33–35].

GBM is the most common and lethal form of brain cancer. The conventional therapy for GBM employs surgical resection followed by radiation plus temozolomide (TMZ) chemotherapy [36]. However, the median survival time after diagnosis of GBM is less than a year likely due to therapy resistance [37]. Studies in both xenografts and in vitro cell culture models have shown that $CD133^+$ glioma stem cells exhibit increased radioresistance by preferential activation of the DNA damage checkpoint, more efficient DNA repair, and lower apoptosis [38]. Also, in a GBM mouse model, a subset of cells with properties similar to CSCs is more resistant to TMZ treatment, and contributes to tumor regrowth [39]. CSCs in other solid cancers, including colon cancer and melanomas, are also more resistant to conventional therapies in xenograft or cell culture models [40,41].

RESISTANCE MECHANISMS AND STRATEGIES TO SENSITIZE CANCER STEM CELLS

As illustrated in Fig. 15.1, CSCs from different cancers may utilize different mechanisms to defend themselves from therapy toxicity eventually resulting in tumor recurrence. Understanding these resistance mechanisms, which will be discussed in this section, is critical for designing new strategies to target CSCs.

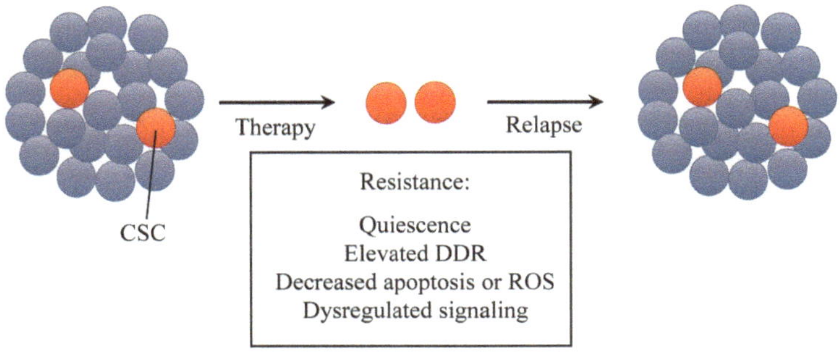

FIGURE 15.1 The mechanisms of therapy resistance in cancer stem cells (CSCs). CSCs within the tumor are more resistant to therapies resulting in tumor recurrence. CSCs may differentially respond to treatments. Generally, the mechanisms of the resistance include, but are not limited to quiescence, elevated activity of DNA damage response (DDR), decreased apoptosis or reactive oxygen species (ROS), and dysregulated signaling pathways, such as Notch and Wnt.

Dormancy

Most chemotherapies target rapidly dividing cells, so quiescent cells may escape treatment and contribute to tumor recurrence. Indeed, slow-cycling CSCs have been identified in multiple cancers and contribute to therapy resistance. Using a xenograft model for studying human primary leukemia stem cells (LSCs), Ishikawa et al. [17] observed that human AML stem cells are more resistant to Ara-C chemotherapy-induced apoptosis, and the majority (around 85%) of these human $CD34^{+}CD38^{-}$ cells resides in the G0 phase of cell cycle. In contrast only around 35% of the other differentiated cells were quiescent. These results suggested that the quiescence of these AML stem cells contributes to therapy resistance [17]. In a subsequent study, Ishikawa et al. showed that granulocyte colony–stimulating factor not only induced AML stem cells to enter the cell cycle from the G0 phase, but also sensitized the LSCs to chemotherapy-induced apoptosis in xenograft models [42].

In glioma, it has been shown that slow-cycling label-retaining primary human GMB cells exhibit CSC features [43]. Furthermore, quiescent stem cells in a GBM mouse model can escape from TMZ treatment and reinitiate tumor growth [39]. Targeting these quiescent stem cells by genetic elimination prevents development of high-grade gliomas. In addition, a dramatic inhibition of GMB progression was observed when TMZ was administered in combination with stem cell ablation [39]. In other tumor studies, using a human colorectal cancer xenograft model following lentivirus transduction to label different tumor subclones, Dick et al. demonstrated that resting or slow-proliferating colorectal cancer subclones can tolerate oxaliplatin, a commonly used chemotherapy treatment, and reinitiate tumor growth [40]. In melanoma, a subpopulation of $JARID1B^{high}$ slow-cycling cells was shown to be more resistant to multiple chemotherapy treatments, including cisplatin and vemurafenib [41,44]. Although these JARID1B^{high} cells do not necessarily follow a hierarchical CSC model, these studies still indicated the importance of targeting slow-cycling cells in tumors. In addition, knockdown of JARID1B increased cell proliferation in vitro and sensitized the xenograft tumors to drug treatment. Similarly, a small fraction of drug resistant, largely quiescent cells, has been detected in several different human cancer–derived cell lines, including those derived from non-small cell lung cancer, melanoma, colorectal cancer, and breast cancer. These drug-resistant cells exhibit CSC properties, including surface marker expression, and the ability to reestablish the heterogeneity of parental cells [45].

DNA Damage Response

Most chemotherapy agents and radiation elicit their cytotoxic effects by inducing DNA damage. Hence, an increasing number of studies have focused on the DNA damage response (DDR) in CSCs and therapy resistance. The DDR involves cell cycle arrest, DNA repair, gene expression, and alternatively, apoptosis or senescence [46]. It has been demonstrated that CSCs in different cancers exhibit increased cell cycle arrest, DNA repair, or decreased levels of apoptosis, to ensure their resistance to treatment. In this section, we will discuss each aspect of DDR separately.

With respect to cell cycle arrest, human glioma stem cells in xenografts are more resistant to IR due to increased activation of cell cycle checkpoints, as shown by increased levels of phosphorylated ataxia-telangiectasia-mutated (ATM), Chk1, and Chk2. In addition, chemical inhibition of Chk1 and Chk2 by debromohymenialdisine (DBH) sensitized CSCs to IR treatment in vivo and in vitro [38]. Conversely, a subsequent study indicated that loss of Chk2 actually increased the radioresistance in glioma in vivo. However, these studies also showed that loss of Chk2 indeed disrupted cell cycle checkpoints in neurospheres after IR [47]. Similarly, in vitro studies suggested that stem-like cells in human head and neck, prostate, and breast cancers are resistant to UV- or drug-induced apoptosis and exhibit a prolonged G2-phase of the cell cycle perhaps as a result of increased levels of G2 checkpoint proteins. In these studies, inhibiting Chk1 and Chk2 using DBH increased the level of apoptosis of stemlike cells after treatment [48].

With respect to DNA repair, the two major pathways utilized to repair DNA double-strand breaks (DSB), which are mainly induced by IR treatment, are homologous recombination (HR) and nonhomologous end joining (NHEJ). In adult stem cells in the hematopoietic system and hair follicles, it has been shown that stem cells are more radioresistant and exhibit a more efficient DSB repair accompanied by increased activity of NHEJ [49,50]. In breast cancer and glioma, using phospho-Histone H2AX (Ser139) (γH2AX) staining and comet assays, CSCs also have been shown to exhibit more efficient DSB repair as compared to the bulk of the tumor cells [31,38]. However, little is known about the repair mechanisms utilized in CSCs. Using a high-throughput NHEJ assay in primary mammary epithelial cells and tumor cells from mouse models, a recent study showed that both normal mammary stem cells (MaSCs) and CSCs exhibit increased NHEJ activity, which is accompanied by increased expression of 53BP1, an important regulator of the NHEJ pathway [32]. These studies suggest that it may be possible to sensitize CSCs to IR treatment by inhibiting the NHEJ activity. In other DDR studies, glioma stem cells were shown to have an increased basal level of single-strand break repair activity, which contributed to their higher stress tolerance. By inhibiting poly-ADP-ribose polymerase (PARP), an important mediator of single-strand break repair, Rich et al. sensitized glioma stem cells to IR treatment, and inhibited growth, self-renewal, and DNA damage repair [51].

Another important aspect of the DDR is damage-induced apoptosis. It has been demonstrated that mouse hair follicle bulge stem cells are more resistance to DNA damage–induced apoptosis, due to higher expression of Bcl-2, an antiapoptotic protein [49]. Similarly, mouse hematopoietic stem cells are also more resistant to IR-induced apoptosis, and they have increased expression of prosurvival genes, such as Mcl-1 [50]. Studies in CSCs have used apoptosis as a biomarker of therapy resistance and showed relatively lower levels of apoptosis in CSCs following treatment [17,38]. However, once again there are only limited studies investigating the mechanisms of apoptosis resistance in CSCs. In recent studies in the mouse mammary gland and a p53-null mammary tumor model, both MaSCs and CSCs were shown to be more resistant to apoptosis in vivo after IR as determined by Annexin V, cleaved Caspase-3, or TUNEL staining. Decreased apoptosis was correlated with downregulation of proapoptotic proteins, such as BAX and Caspase-3, in MaSCs, and upregulation of stress-response proteins, such as p-AKT and p-p38, in CSCs [32]. Furthermore, inhibition of Bcl-2 family members has become an emerging strategy for sensitizing CSCs. For example, inhibition of Mcl-1 overcomes drug resistance in AML and potently kills all the different subpopulations, including AML stem cells, but not normal hematopoietic stem cells [52]. In gliomas, dual inhibition of Bcl-2 and the phosphatidylinositol 3-kinase (PI3K) signaling pathway decreases cellular viability in glioma cells and abrogates neurosphere forming efficiency [53].

Role of Reactive Oxygen Species

Genome integrity can also be impacted by the intracellular level of reactive oxygen species (ROS), which is tightly regulated through cell metabolism and induces DNA single-strand breaks. ROS has been shown to be lower in tissue stem cells, including mouse hematopoietic stem cells, rat glial precursor cells, and mouse MaSCs [33,54,55]. Similarly, the ROS level is lower in CSCs derived from primary human breast and mouse MMTV-Wnt1 mammary tumors [33]. In addition, these mouse CSCs developed less DNA damage and were enriched after IR treatment as compared to non-CSCs. Importantly, CSCs can be sensitized to IR through pharmacological manipulation of ROS levels in vitro. In a later study, RNA-Seq and metabolic analyses revealed that CSCs from MMTV-Wnt1 mouse mammary tumors prefer glycolysis over oxidative phosphorylation as expected from the Warburg hypothesis, and this, therefore, might be one of the reasons they exhibit lower ROS. Importantly, pharmacological promotion of oxidative phosphorylation can preferentially target CSCs in vitro and in vivo [56]. Other in vitro studies using cell line models have also indicated that $CD44^{+}CD24^{low}$ breast CSCs are more resistant to IR possibly due to their lower ROS levels [34,57]. Interestingly, higher ROS levels were found in cultured CSCs derived from human GBM as compared to non-CSCs, possibly contributing to the increased activity of single-strand break repair in CSCs [51].

Notch, Wnt, and Other Signaling Pathways

CSCs share many similar characteristics with normal stem cells, including the process of self-renewal. Signaling pathways regulating normal stem cell functions, such as the Notch and Wnt pathways, and Ezh2, the catalytic subunit of the Polycomb repressive complex 2 (PRC2) also affect CSC fate and contribute to therapy resistance. Notch signaling is critically important for maintaining adult stem cells by promoting self-renewal and inhibiting differentiation [58]. It has been shown that pharmacologically inhibiting Notch signaling using γ-secretase inhibitors (GSI) sensitized glioma stem cells to radiation through enhancing apoptosis in vitro. Additionally, Notch increases radioresistance possibly through regulating the PI3K/Akt pathway, as well as Mcl-1 [59]. In preclinical and clinical studies of breast cancer, GSI treatment reduced CSCs in xenografts, and GSI plus chemotherapy decreased CSCs and MS-forming efficiency in tumor cells from patients [60]. Similarly, using a Notch2/3 targeting antibody in multiple xenograft models, including pancreatic, breast, lung, and ovarian cancers, it has been shown that CSCs frequency and tumor recurrence was decreased in combination with chemotherapy [61].

The Wnt signaling pathway is well known for its roles in embryonic development, stem cell regulation, and carcinogenesis [62]. Progenitor cells from mammary epithelium are more resistant to IR in vivo and in vitro due to upregulation of canonical Wnt signaling [29,63]. A subsequent study using a p53-null mammary tumor model demonstrated that CSCs from these tumors are also more radioresistant in vivo likely due to increased activation of Akt and canonical Wnt signaling pathways, as shown at both the protein level as well as by Wnt reporter expression [31]. Additionally in this study, using perifosine, an Akt inhibitor, to decrease Akt as well as canonical Wnt activities, the proportion of CSCs was reduced in the tumor, and more importantly, CSCs were sensitized to IR treatment. In other animal models, for example, Wnt inhibition via monoclonal antibody binding to Frizzled receptors inhibits tumor growth and reduces CSC frequency in multiple types of xenograft tumors, including breast and pancreatic cancers [64]. In addition, inhibition of canonical Wnt signaling has been used as a combination treatment to target drug-resistant CSCs in CML [65]. Other studies have focused on the self-renewal ability, the key stemness property, of CSCs. For example, CSCs from breast cancer cell lines exhibit increased self-renewal ability after IR [66]. Reducing self-renewal through Bmi1 inhibition in human colorectal tumor xenograft models may provide a therapeutic approach for targeting CSCs [67].

EZH2 is the functional enzymatic component of the PRC2 complex, which is important for stem cell function. Overexpression of EZH2 has been observed in a variety of cancers and linked to tumor malignancy [68,69] as well as CSC maintenance [70,71]. Furthermore, EZH2 is essential for radioresistance in CSCs from GBM [72]. Other signaling pathways, including NF-κB and TGF-β, are also involved in therapy resistance in glioma stem cells and breast CSCs [73,74].

RESISTANCE AND TUMOR EVOLUTION

From a genetic perspective, therapy resistance may be due to (1) tumor cells accumulating mutations during disease progression, resulting in heterogeneous subclones that respond differentially to treatment (Darwinian selection) and (2) genetic alterations caused by the therapies which contribute to resistance (acquired drug resistance). Therefore, studying the genetic evolution of tumor cells during tumor growth and after treatment is critical for understanding tumor heterogeneity and the cause of therapy resistance. An increasing number of studies have emerged in the field of tumor evolution due to the recent technical advances in DNA sequencing. Using deep sequencing or DNA copy number analysis, extensive genomic heterogeneity has been observed in multiple types of cancers, including leukemia [75,76], breast cancer [77–80], and pancreatic cancer [81,82].

For example, using whole-genome sequencing to identify the mutations in primary tumor genomes and after relapse from AML patients, it has been shown that additional acquired mutations emerge after therapy, some of which may contribute to clonal selection and chemotherapy resistance [75]. Other studies in leukemia also tracked subclonal heterogeneity before and after treatment indicating that genetic alterations selected during treatment contributed to relapse [76,83–85]. Moreover, it has been shown that genetic diversity also occurs in LSCs in ALL using xenograft models [84,86], again supporting the hypothesis that CSCs undergo clonal evolution. In breast cancer, extensive clonal diversity has been observed in the different ER+, HER2-positive, and triple-negative breast cancer (TNBC) subtypes [77,78]. In addition, TNBC cells exhibit an increased mutation rate as compared to ER+ cells [79]. Furthermore, sequencing of circulating cell-free tumor DNA has identified mutations associated with acquired resistance in advanced breast cancer [87].

These results from breast cancer and leukemia, however, did not take into account the CSC paradigm, so the connection between CSC evolution after treatment and tumor relapse has not yet been directly established. Interestingly, therapy or microenvironment-induced alterations in gene expression have been reported to promote CSC activity or maintenance [28,88,89]. Hence, it is likely that there is a relationship among therapy resistance, CSCs, and genetic plasticity. In addition, since resistance of drug-tolerant subpopulations in cell lines can be reversed through histone and DNA modifications, epigenetic mechanisms may also regulate therapy resistance [45]. For example, studies have shown the effects of microenvironmentally driven epigenetic diversity on CSC progression [90,91]. Together these results suggest that genetic and epigenetic changes, CSC maintenance, and therapy resistance are interrelated (Fig. 15.2).

CONCLUSION

During the past two decades, targeted adjuvant therapy in combination with radiation and/or chemotherapy has strongly improved the prognosis for several cancers, including leukemias and breast cancer. However, the enormous heterogeneity and continuing evolution of cancer cells are two main reasons preventing more successful treatment outcomes. As a result, in the current era of cancer research, the mission of finding the magic bullet to defeat all cancer cells has shifted to targeting the Achilles heel of therapy-resistant cells, which in many cases, may be CSCs. CSCs in most types of cancers have been reported

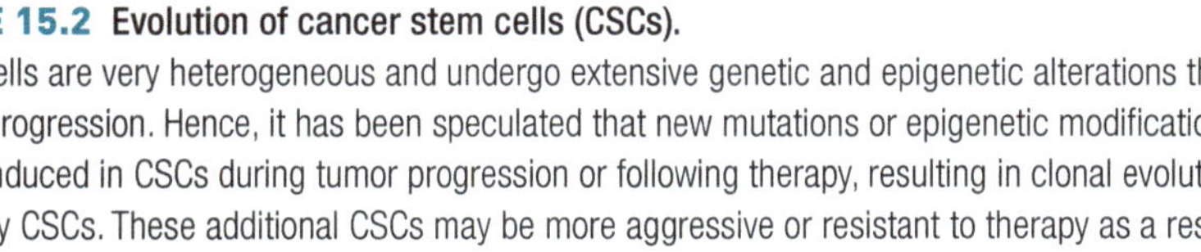

FIGURE 15.2 Evolution of cancer stem cells (CSCs).

Cancer cells are very heterogeneous and undergo extensive genetic and epigenetic alterations throughout disease progression. Hence, it has been speculated that new mutations or epigenetic modifications also may be induced in CSCs during tumor progression or following therapy, resulting in clonal evolution of secondary CSCs. These additional CSCs may be more aggressive or resistant to therapy as a result of Darwinian selection.

to be more resistant to therapies. The mechanisms of resistance may be contributed by (1) quiescence, since most chemotherapy treatments target rapidly dividing cells; (2) elevated DDR activity, including increased activation of cell cycle arrest and NHEJ activity; (3) increased resistance to apoptotic cell death or cellular stresses, such as ROS; and (4) dysregulated signaling pathways, including Notch and Wnt signaling. Over the past decade, many studies have focused on sensitizing CSCs based upon targeting these mechanisms. For example, inducing cell proliferation of quiescent LSCs can sensitize them to chemotherapy. Promising results also have been observed when treating CSCs with inhibitors of Chk1/2, as well as Notch and Wnt signaling. Hopefully, additional therapeutic targets will be identified in the near future as more CSC-resistance mechanisms are elucidated. However, because of heterogeneity and continued evolution due to both genetic and epigenetic mechanisms upon disease progression and therapy it is unlikely that single agents will be effective on these constantly moving targets. More than 50 years ago early treatment successes in leukemias involved combination chemotherapy using agents such as methotrexate, vincristine, and prednisone, each targeting quite different pathways yet this type of combinatorial approach is only just being applied clinically in solid cancers. CSCs are also genetically heterogeneous and may exhibit plasticity, adding additional layers of complexity for developing effective treatments. To fully understand the resistance mechanisms of CSCs, ongoing studies will need to focus on genetic and epigenetic alterations in CSCs both prior to and after treatment especially in the metastatic setting. This will require improved methods of CSC identification perhaps using pathway reporters and other biomarkers instead of only surface markers and fluorescence-activated cell sorting isolation.

List of Abbreviations

ALL Acute lymphoblastic leukemia
AML Acute myeloid leukemia
Ara-C Cytosine arabinoside
ATM Ataxia-telangiectasia-mutated
CLL Chronic lymphocytic leukemia
CML Chronic myeloid leukemia
CSCs Cancer stem cells
DBH Debromohymenialdisine
DDR DNA damage response
DSB Double-strand breaks
ER^+ Estrogen receptor-positive
GBM Glioblastoma
GSI γ-secretase inhibitors
HR Homologous recombination
IR Ionizing radiation

LSCs Leukemia stem cells
MaSCs Mammary stem cells
MS Mammospheres
NHEJ Non-homologous end joining
PARP Poly-ADP-ribose polymerase
PI3K Phosphatidylinositol 3-kinase
PR⁺ progesterone receptor-positive
PRC2 Polycomb repressive complex 2
ROS Reactive oxygen species
TMZ Temozolomide
TNBC Triple-negative breast cancers
γH2AX Phospho-Histone H2AX

Acknowledgments

We apologize to those authors whose work we were unable to cite due to space limitations. These studies were supported by grant CA148761.

References

[1] Lapidot T, Sirard C, Vormoor J, Murdoch B, Hoang T, Caceres-Cortes J, et al. A cell initiating human acute myeloid leukaemia after transplantation into SCID mice. Nature 1994;367(6464):645–8.

[2] Al-Hajj M, Wicha MS, Benito-Hernandez A, Morrison SJ, Clarke MF. Prospective identification of tumorigenic breast cancer cells. Proc Natl Acad Sci USA 2003;100(7):3983–8.

[3] Singh SK, Hawkins C, Clarke ID, Squire JA, Bayani J, Hide T, et al. Identification of human brain tumour initiating cells. Nature 2004;432(7015):396–401.

[4] O'Brien CA, Pollett A, Gallinger S, Dick JE. A human colon cancer cell capable of initiating tumour growth in immunodeficient mice. Nature 2007;445(7123):106–10.

[5] Ricci-Vitiani L, Lombardi DG, Pilozzi E, Biffoni M, Todaro M, Peschle C, et al. Identification and expansion of human colon-cancer-initiating cells. Nature 2007;445(7123):111–5.

[6] Li C, Heidt DG, Dalerba P, Burant CF, Zhang L, Adsay V, et al. Identification of pancreatic cancer stem cells. Cancer Res 2007;67(3):1030–7.

[7] Schatton T, Murphy GF, Frank NY, Yamaura K, Waaga-Gasser AM, Gasser M, et al. Identification of cells initiating human melanomas. Nature 2008;451(7176):345–9.

[8] So CW, Karsunky H, Passegue E, Cozzio A, Weissman IL, Cleary ML. MLL-GAS7 transforms multipotent hematopoietic progenitors and induces mixed lineage leukemias in mice. Cancer Cell 2003;3(2):161–71.

[9] Huntly BJ, Shigematsu H, Deguchi K, Lee BH, Mizuno S, Duclos N, et al. MOZ-TIF2, but not BCR-ABL, confers properties of leukemic stem cells to committed murine hematopoietic progenitors. Cancer Cell 2004;6(6):587–96.

[10] Liu JC, Deng T, Lehal RS, Kim J, Zacksenhaus E. Identification of tumorsphere- and tumor-initiating cells in HER2/Neu-induced mammary tumors. Cancer Res 2007;67(18):8671–81.

[11] Cho RW, Wang X, Diehn M, Shedden K, Chen GY, Sherlock G, et al. Isolation and molecular characterization of cancer stem cells in MMTV-Wnt-1 murine breast tumors. Stem Cells (Dayton, Ohio) 2008;26(2):364–71.

[12] Zhang M, Behbod F, Atkinson RL, Landis MD, Kittrell F, Edwards D, et al. Identification of tumor-initiating cells in a p53-null mouse model of breast cancer. Cancer Res 2008;68(12):4674–82.

[13] Read TA, Fogarty MP, Markant SL, McLendon RE, Wei Z, Ellison DW, et al. Identification of CD15 as a marker for tumor-propagating cells in a mouse model of medulloblastoma. Cancer Cell 2009;15(2):135–47.

[14] Barker N, Ridgway RA, van Es JH, van de Wetering M, Begthel H, van den Born M, et al. Crypt stem cells as the cells-of-origin of intestinal cancer. Nature 2009;457(7229):608–11.

[15] Bernier J, Hall EJ, Giaccia A. Radiation oncology: a century of achievements. Nat Rev Cancer 2004;4(9):737–47.

[16] Chabner BA, Roberts Jr TG. Timeline: chemotherapy and the war on cancer. Nat Rev Cancer 2005;5(1):65–72.

[17] Ishikawa F, Yoshida S, Saito Y, Hijikata A, Kitamura H, Tanaka S, et al. Chemotherapy-resistant human AML stem cells home to and engraft within the bone-marrow endosteal region. Nat Biotechnol 2007;25(11):1315–21.

[18] Druker BJ, Guilhot F, O'Brien SG, Gathmann I, Kantarjian H, Gattermann N, et al. Five-year follow-up of patients receiving imatinib for chronic myeloid leukemia. N Engl J Med 2006;355(23):2408–17.

[19] Bhatia R, Holtz M, Niu N, Gray R, Snyder DS, Sawyers CL, et al. Persistence of malignant hematopoietic progenitors in chronic myelogenous leukemia patients in complete cytogenetic remission following imatinib mesylate treatment. Blood 2003;101(12):4701–7.

[20] Jiang X, Zhao Y, Smith C, Gasparetto M, Turhan A, Eaves A, et al. Chronic myeloid leukemia stem cells possess multiple unique features of resistance to BCR-ABL targeted therapies. Leukemia 2007;21(5):926–35.

[21] Graham SM, Jorgensen HG, Allan E, Pearson C, Alcorn MJ, Richmond L, et al. Primitive, quiescent, Philadelphia-positive stem cells from patients with chronic myeloid leukemia are insensitive to STI571 in vitro. Blood 2002;99(1):319–25.

[22] Corbin AS, Agarwal A, Loriaux M, Cortes J, Deininger MW, Druker BJ. Human chronic myeloid leukemia stem cells are insensitive to imatinib despite inhibition of BCR-ABL activity. J Clin Investig 2011;121(1):396–409.

[23] Neering SJ, Bushnell T, Sozer S, Ashton J, Rossi RM, Wang PY, et al. Leukemia stem cells in a genetically defined murine model of blast-crisis CML. Blood 2007;110(7):2578–85.

[24] Perou CM, Sorlie T, Eisen MB, van de Rijn M, Jeffrey SS, Rees CA, et al. Molecular portraits of human breast tumours. Nature 2000;406(6797):747–52.

[25] Prat A, Perou CM. Deconstructing the molecular portraits of breast cancer. Mol Oncol 2011;5(1):5–23.

[26] Sotiriou C, Pusztai L. Gene-expression signatures in breast cancer. N Engl J Med 2009;360(8):790–800.

[27] Li X, Lewis MT, Huang J, Gutierrez C, Osborne CK, Wu MF, et al. Intrinsic resistance of tumorigenic breast cancer cells to chemotherapy. J Natl Cancer Inst 2008;100(9):672–9.

[28] Creighton CJ, Li X, Landis M, Dixon JM, Neumeister VM, Sjolund A, et al. Residual breast cancers after conventional therapy display mesenchymal as well as tumor-initiating features. Proc Natl Acad Sci USA 2009;106(33):13820–5.

[29] Woodward WA, Chen MS, Behbod F, Alfaro MP, Buchholz TA, Rosen JM. WNT/beta-catenin mediates radiation resistance of mouse mammary progenitor cells. Proc Natl Acad Sci USA 2007;104(2):618–23.

[30] Herschkowitz JI, Zhao W, Zhang M, Usary J, Murrow G, Edwards D, et al. Comparative oncogenomics identifies breast tumors enriched in functional tumor-initiating cells. Proc Natl Acad Sci USA 2012;109(8):2778–83.

[31] Zhang M, Atkinson RL, Rosen JM. Selective targeting of radiation-resistant tumor-initiating cells. Proc Natl Acad Sci USA 2010;107(8):3522–7.

[32] Chang CH, Zhang M, Rajapakshe K, Coarfa C, Edwards D, Huang S, et al. Mammary stem cells and tumor-initiating cells are more resistant to apoptosis and exhibit increased DNA repair activity in response to DNA damage. Stem Cell Rep 2015;5(3):378–91.

[33] Diehn M, Cho RW, Lobo NA, Kalisky T, Dorie MJ, Kulp AN, et al. Association of reactive oxygen species levels and radioresistance in cancer stem cells. Nature 2009;458(7239):780–3.

[34] Phillips TM, McBride WH, Pajonk F. The response of CD24(−/low)/CD44$^+$ breast cancer-initiating cells to radiation. J Natl Cancer Inst 2006;98(24):1777–85.

[35] Shafee N, Smith CR, Wei S, Kim Y, Mills GB, Hortobagyi GN, et al. Cancer stem cells contribute to cisplatin resistance in Brca1/p53-mediated mouse mammary tumors. Cancer Res 2008;68(9):3243–50.

[36] Johnson DR, O'Neill BP. Glioblastoma survival in the United States before and during the temozolomide era. J Neuro Oncol 2012;107(2):359–64.

[37] Ramirez YP, Weatherbee JL, Wheelhouse RT, Ross AH. Glioblastoma multiforme therapy and mechanisms of resistance. Pharmaceuticals (Basel, Switzerland) 2013;6(12):1475–506.

[38] Bao S, Wu Q, McLendon RE, Hao Y, Shi Q, Hjelmeland AB, et al. Glioma stem cells promote radioresistance by preferential activation of the DNA damage response. Nature 2006;444(7120):756–60.

[39] Chen J, Li Y, Yu TS, McKay RM, Burns DK, Kernie SG, et al. A restricted cell population propagates glioblastoma growth after chemotherapy. Nature 2012;488(7412):522–6.

[40] Kreso A, O'Brien CA, van Galen P, Gan OI, Notta F, Brown AM, et al. Variable clonal repopulation dynamics influence chemotherapy response in colorectal cancer. Science (New York, NY) 2013;339(6119):543–8.

[41] Roesch A, Vultur A, Bogeski I, Wang H, Zimmermann KM, Speicher D, et al. Overcoming intrinsic multidrug resistance in melanoma by blocking the mitochondrial respiratory chain of slow-cycling JARID1B(high) cells. Cancer Cell 2013;23(6):811–25.

[42] Saito Y, Uchida N, Tanaka S, Suzuki N, Tomizawa-Murasawa M, Sone A, et al. Induction of cell cycle entry eliminates human leukemia stem cells in a mouse model of AML. Nat Biotechnol 2010;28(3):275–80.

[43] Deleyrolle LP, Harding A, Cato K, Siebzehnrubl FA, Rahman M, Azari H, Olson S, et al. Evidence for label-retaining tumour-initiating cells in human glioblastoma. Brain 2011;134:1331–43.

[44] Roesch A, Fukunaga-Kalabis M, Schmidt EC, Zabierowski SE, Brafford PA, Vultur A, et al. A temporarily distinct subpopulation of slow-cycling melanoma cells is required for continuous tumor growth. Cell 2010;141(4):583–94.

[45] Sharma SV, Lee DY, Li B, Quinlan MP, Takahashi F, Maheswaran S, et al. A chromatin-mediated reversible drug-tolerant state in cancer cell subpopulations. Cell 2010;141(1):69–80.

[46] Harper JW, Elledge SJ. The DNA damage response: ten years after. Mol Cell 2007;28(5):739–45.

[47] Squatrito M, Brennan CW, Helmy K, Huse JT, Petrini JH, Holland EC. Loss of ATM/Chk2/p53 pathway components accelerates tumor development and contributes to radiation resistance in gliomas. Cancer Cell 2010;18(6):619–29.

[48] Harper LJ, Costea DE, Gammon L, Fazil B, Biddle A, Mackenzie IC. Normal and malignant epithelial cells with stem-like properties have an extended G2 cell cycle phase that is associated with apoptotic resistance. BMC Cancer 2010;10:166.

[49] Sotiropoulou PA, Candi A, Mascre G, De Clercq S, Youssef KK, Lapouge G, et al. Bcl-2 and accelerated DNA repair mediates resistance of hair follicle bulge stem cells to DNA-damage-induced cell death. Nat Cell Biol 2010;12(6):572–82.

[50] Mohrin M, Bourke E, Alexander D, Warr MR, Barry-Holson K, Le Beau MM, et al. Hematopoietic stem cell quiescence promotes error-prone DNA repair and mutagenesis. Cell Stem Cell 2010;7(2):174–85.

[51] Venere M, Hamerlik P, Wu Q, Rasmussen RD, Song LA, Vasanji A, et al. Therapeutic targeting of constitutive PARP activation compromises stem cell phenotype and survival of glioblastoma-initiating cells. Cell Death Differ 2014;21(2):258–69.

[52] Pan R, Ruvolo VR, Wei J, Konopleva M, Reed JC, Pellecchia M, et al. Inhibition of Mcl-1 with the pan-Bcl-2 family inhibitor (-)BI97D6 overcomes ABT-737 resistance in acute myeloid leukemia. Blood 2015;126(3):363–72.

[53] Pareja F, Macleod D, Shu C, Crary JF, Canoll PD, Ross AH, et al. PI3K and Bcl-2 inhibition primes glioblastoma cells to apoptosis through downregulation of Mcl-1 and Phospho-BAD. Mol Cancer Res 2014;12(7):987–1001.

[54] Ito K, Hirao A, Arai F, Matsuoka S, Takubo K, Hamaguchi I, et al. Regulation of oxidative stress by ATM is required for self-renewal of haematopoietic stem cells. Nature 2004;431(7011):997–1002.

[55] Smith J, Ladi E, Mayer-Proschel M, Noble M. Redox state is a central modulator of the balance between self-renewal and differentiation in a dividing glial precursor cell. Proc Natl Acad Sci USA 2000;97(18):10032–7.

[56] Feng W, Gentles A, Nair RV, Huang M, Lin Y, Lee CY, et al. Targeting unique metabolic properties of breast tumor initiating cells. Stem Cells (Dayton, Ohio) 2014;32(7):1734–45.

[57] Karimi-Busheri F, Rasouli-Nia A, Mackey JR, Weinfeld M. Senescence evasion by MCF-7 human breast tumor-initiating cells. Breast Cancer Res 2010;12(3):R31.

[58] Liu J, Sato C, Cerletti M, Wagers A. Notch signaling in the regulation of stem cell self-renewal and differentiation. Curr Top Dev Biol 2010;92:367–409.

[59] Wang J, Wakeman TP, Lathia JD, Hjelmeland AB, Wang XF, White RR, et al. Notch promotes radioresistance of glioma stem cells. Stem Cells (Dayton, Ohio) 2010;28(1):17–28.

[60] Schott AF, Landis MD, Dontu G, Griffith KA, Layman RM, Krop I, et al. Preclinical and clinical studies of gamma secretase inhibitors with docetaxel on human breast tumors. Clin Cancer Res 2013;19(6):1512–24.

[61] Yen WC, Fischer MM, Axelrod F, Bond C, Cain J, Cancilla B, et al. Targeting Notch signaling with a Notch2/Notch3 antagonist (tarextumab) inhibits tumor growth and decreases tumor-initiating cell frequency. Clin Cancer Res 2015;21(9):2084–95.

[62] MacDonald BT, Tamai K, He X. Wnt/beta-catenin signaling: components, mechanisms, and diseases. Dev Cell 2009;17(1):9–26.

[63] Chen MS, Woodward WA, Behbod F, Peddibhotla S, Alfaro MP, Buchholz TA, et al. Wnt/beta-catenin mediates radiation resistance of $Sca1^+$ progenitors in an immortalized mammary gland cell line. J Cell Sci 2007;120(Pt 3):468–77.

[64] Gurney A, Axelrod F, Bond CJ, Cain J, Chartier C, Donigan L, et al. Wnt pathway inhibition via the targeting of Frizzled receptors results in decreased growth and tumorigenicity of human tumors. Proc Natl Acad Sci USA 2012;109(29):11717–22.

[65] Heidel FH, Bullinger L, Feng Z, Wang Z, Neff TA, Stein L, et al. Genetic and pharmacologic inhibition of beta-catenin targets imatinib-resistant leukemia stem cells in CML. Cell Stem Cell 2012;10(4):412–24.

[66] Lagadec C, Vlashi E, Della Donna L, Meng Y, Dekmezian C, Kim K, et al. Survival and self-renewing capacity of breast cancer initiating cells during fractionated radiation treatment. Breast Cancer Res 2010;12(1):R13.

[67] Kreso A, van Galen P, Pedley NM, Lima-Fernandes E, Frelin C, Davis T, et al. Self-renewal as a therapeutic target in human colorectal cancer. Nat Med 2014;20(1):29–36.

[68] Radulovic V, de Haan G, Klauke K. Polycomb-group proteins in hematopoietic stem cell regulation and hematopoietic neoplasms. Leukemia 2013;27(3):523–33.

[69] Sauvageau M, Sauvageau G. Polycomb group proteins: multi-faceted regulators of somatic stem cells and cancer. Cell Stem Cell 2010;7(3):299–313.

[70] Kim E, Kim M, Woo DH, Shin Y, Shin J, Chang N, et al. Phosphorylation of EZH2 activates STAT3 signaling via STAT3 methylation and promotes tumorigenicity of glioblastoma stem-like cells. Cancer Cell 2013;23(6):839–52.

[71] van Vlerken LE, Kiefer CM, Morehouse C, Li Y, Groves C, Wilson SD, et al. EZH2 is required for breast and pancreatic cancer stem cell maintenance and can be used as a functional cancer stem cell reporter. Stem Cells Transl Med 2013;2(1):43–52.

[72] Kim SH, Joshi K, Ezhilarasan R, Myers TR, Siu J, Gu C, et al. EZH2 protects glioma stem cells from radiation-induced cell death in a MELK/FOXM1-dependent manner. Stem Cell Rep 2015;4(2):226–38.

[73] Bhat KP, Balasubramaniyan V, Vaillant B, Ezhilarasan R, Hummelink K, Hollingsworth F, et al. Mesenchymal differentiation mediated by NF-kappaB promotes radiation resistance in glioblastoma. Cancer Cell 2013;24(3):331–46.

[74] Bhola NE, Balko JM, Dugger TC, Kuba MG, Sanchez V, Sanders M, et al. TGF-beta inhibition enhances chemotherapy action against triple-negative breast cancer. J Clin Investig 2013;123(3):1348–58.

[75] Ding L, Ley TJ, Larson DE, Miller CA, Koboldt DC, Welch JS, et al. Clonal evolution in relapsed acute myeloid leukaemia revealed by whole-genome sequencing. Nature 2012;481(7382):506–10.

[76] Schuh A, Becq J, Humphray S, Alexa A, Burns A, Clifford R, et al. Monitoring chronic lymphocytic leukemia progression by whole genome sequencing reveals heterogeneous clonal evolution patterns. Blood 2012;120(20):4191–6.

[77] Shah SP, Roth A, Goya R, Oloumi A, Ha G, Zhao Y, et al. The clonal and mutational evolution spectrum of primary triple-negative breast cancers. Nature 2012;486(7403):395–9.

[78] Nik-Zainal S, Van Loo P, Wedge DC, Alexandrov LB, Greenman CD, Lau KW, et al. The life history of 21 breast cancers. Cell 2012;149(5):994–1007.

[79] Wang Y, Waters J, Leung ML, Unruh A, Roh W, Shi X, et al. Clonal evolution in breast cancer revealed by single nucleus genome sequencing. Nature 2014;512(7513):155–60.

[80] Navin N, Kendall J, Troge J, Andrews P, Rodgers L, McIndoo J, et al. Tumour evolution inferred by single-cell sequencing. Nature 2011;472(7341):90–4.

[81] Campbell PJ, Yachida S, Mudie LJ, Stephens PJ, Pleasance ED, Stebbings LA, et al. The patterns and dynamics of genomic instability in metastatic pancreatic cancer. Nature 2010;467(7319):1109–13.

[82] Yachida S, Jones S, Bozic I, Antal T, Leary R, Fu B, et al. Distant metastasis occurs late during the genetic evolution of pancreatic cancer. Nature 2010;467(7319):1114–7.

[83] Mullighan CG, Phillips LA, Su X, Ma J, Miller CB, Shurtleff SA, et al. Genomic analysis of the clonal origins of relapsed acute lymphoblastic leukemia. Science (New York, NY) 2008;322(5906):1377–80.

[84] Anderson K, Lutz C, van Delft FW, Bateman CM, Guo Y, Colman SM, et al. Genetic variegation of clonal architecture and propagating cells in leukaemia. Nature 2011;469(7330):356–61.

[85] Landau DA, Carter SL, Stojanov P, McKenna A, Stevenson K, Lawrence MS, et al. Evolution and impact of subclonal mutations in chronic lymphocytic leukemia. Cell 2013;152(4):714–26.

[86] Notta F, Mullighan CG, Wang JC, Poeppl A, Doulatov S, Phillips LA, et al. Evolution of human BCR-ABL1 lymphoblastic leukaemia-initiating cells. Nature 2011;469(7330):362–7.

[87] Murtaza M, Dawson SJ, Tsui DW, Gale D, Forshew T, Piskorz AM, et al. Non-invasive analysis of acquired resistance to cancer therapy by sequencing of plasma DNA. Nature 2013;497(7447):108–12.

[88] Lagadec C, Vlashi E, Della Donna L, Dekmezian C, Pajonk F. Radiation-induced reprogramming of breast cancer cells. Stem Cells (Dayton, Ohio) 2012;30(5):833–44.

[89] Heddleston JM, Li Z, McLendon RE, Hjelmeland AB, Rich JN. The hypoxic microenvironment maintains glioblastoma stem cells and promotes reprogramming towards a cancer stem cell phenotype. Cell Cycle (Georgetown, Tex) 2009;8(20):3274–84.

[90] Fan Y, Potdar AA, Gong Y, Eswarappa SM, Donnola S, Lathia JD, et al. Profilin-1 phosphorylation directs angiocrine expression and glioblastoma progression through HIF-1alpha accumulation. Nat Cell Biol 2014;16(5):445–56.

[91] van den Beucken T, Koch E, Chu K, Rupaimoole R, Prickaerts P, Adriaens M, et al. Hypoxia promotes stem cell phenotypes and poor prognosis through epigenetic regulation of DICER. Nat Commun 2014;5:5203.

CHAPTER 16

Adipose Tissue-Derived Stem Cells in Regenerative Medicine and Impact on Cancer

P.S. Thiagarajan[1], O. Reizes[1,2,3]

[1]Cleveland Clinic, Cleveland, OH, United States; [2]Cleveland Clinic Lerner College of Medicine of Case Western Reserve University, Cleveland, OH, United States; [3]Case Western Reserve University, Cleveland, OH, United States

ADULT SOMATIC STEM CELLS

Adult stem cells were first identified by Till and McCulloch in the hematopoietic tissue by observing clonal expansion of colonies in the spleen (CFU-S) of irradiated mice after transplantation of the bone marrow [1]. Maintenance of homeostasis in different tissues of the body is a critical physiological process and shown to occur in a tissue-specific manner in adult somatic stem cells. Tissue-committed adult stem cells are characterized by their potential to self-renew and differentiate. Based upon their capacity to differentiate, embryonic stem cells can differentiate into a multilineage of cells while adult tissues-committed stem cells have limited differentiation potential [2,3]. Elimination of the aged or injured terminally differentiated cells by programmed cell death and rigorous replacement with freshly differentiated mature cells is an essential physiological process. This replacement is carried out by the adult stem cells maintaining at least a basal number of stem cells by self-renewal to maintain tissue homeostasis [4,5].

Adult Stem Cells and Cancer

Adult stem cells are postulated to be the cells of origin for epithelial cancers [6]. Several earlier studies support that functional dysregulation of stem cells can initiate tumor and potentiate tumor growth and progression supporting the cancer stem cell (CSC) hypothesis [7,8]. In 1994, the isolation of leukemic stem cells by Lapidot et al. confirmed the CSC hypothesis which states that the hierarchical organization of tumor cells contributes to tumorigenesis and cellular heterogeneity [9]. It postulates that a small number of CSCs in cancers at the top of the hierarchy promote tumor growth and regeneration after injury and metastasis [9,10]. There has been a surge of interest in the CSC model

CONTENTS

Cancer Stem Cells. http://dx.doi.org/10.1016/B978-0-12-803892-5.00016-4

after this landmark study. A decade later, CSCs in solid cancers were defined and enriched based on the profiles of their markers [11,12]. The fine balance between self-renewal and terminal differentiation is regulated by evolutionarily conserved signaling pathways including Wnt, Notch, and Hedgehog pathways, small molecules such as prostaglandins and retinoic acid, developmental factors such as bone morphogenetic protein (BMP), epigenetic and cell cycle regulators, and transcription factors [13]. Any dysregulation of this physiological balance by intrinsic factors such as mutations and cell cycle state and extrinsic factors such as inflammation and tumor microenvironment enhances the tumorigenic capacity of stem cells [6]. Understanding the deciding factors and identifying the cellular or secreted components that influence the stem cell fate is an area of intense research [14].

Mesenchymal Stem Cells

Adult stem cell populations have been isolated from many tissues in the human body. Adult stem cells are tissue-specific and are of intrinsic importance in the repair and regeneration of many tissues and organs [15]. Most stem cells are tissue-specific except for mesenchymal stem cells (MSCs), also known as the multipotent adult progenitor cells, which were originally derived from the stromal adherent fraction of the bone marrow [16]. Bone marrow transplantation is the earliest known cell-based therapy. MSCs were first identified by the studies conducted by Friedenstein and Petrakova [16a] in which bone-forming progenitor cells were isolated from the rat marrow [17]. Identification of MSCs is a crucial discovery in the field of stem cell biology. Despite more than 30 years of extensive research, many fundamental questions still remain unanswered such as characterization of a distinct set of markers for the reliable isolation of pure MSCs from other stem cells. Also, the ex vivo MSC population has been extensively studied and characterized, but the in vivo MSC biology is still in the early stages requiring further understanding [18]. The poor understanding of MSC biology makes the phenotypic definition and the physiological roles of MSCs undefined and unclear. MSCs can be broadly defined as mesenchymal cells with fibroblastic morphology, which possess the robust ability to self-renew and are capable of differentiation into mesenchymal lineages such as adipocytes, chondrocytes, and osteoblasts in adipose and connective tissue, cartilage, marrow, muscle, tendon ligaments and bone, and also into neurons, liver, lung, and intestinal tissues [19,20]. Stem cells derived from amniotic fluid and placenta have been shown to self-renew and differentiate into distinct cell types as well. They have also been characterized as having stem cell properties hierarchically between embryonic stem cells and adult stem cells [21,22]. Though MSCs were initially identified and are commonly isolated from the bone marrow, they were subsequently isolated in diverse tissues such as the subcutaneous adipose tissue, bone, muscle synovial fluid, amniotic fluid, umbilical cord blood, and fetal tissues in humans and other mammals [23,24]. Tissue-specific multipotent MSCs are well studied for their essential role in normal adult tissue homeostasis, tissue maintenance, and repair in response to injury and wound healing [25].

Recent evidence suggests that MSCs from bone marrow can be recruited into the sites of injury and inflammation to serve as regenerative stem cells [26]. MSCs are easily isolated and expanded from adult human tissue collected from healthy volunteers. The majority of the self-renewing and multipotent precursor MSC populations can be isolated and expanded in vitro in cell culture [27]. MSCs are shown to express CD73, CD90, CD105 (endoglin), and CD106 (VCAM-1). Though they are classified based on their common mesenchymal markers, MSC-specific markers have yet to be definitively characterized [28,29]. Different groups opt for different methods of isolation and culture conditions to expand MSCs in tissue culture plates with high efficiency and induced to differentiate to multiple lineages under defined culture conditions [20]. However, further in vivo studies defining the physiological functions of MSCs in the maintenance and regeneration of different tissues are necessary. In more recent studies, ease of isolation and expansion of in vitro multipotent mesenchymal stromal cell cultures are attractive agents for sustaining tissue homeostasis. There is increased interest in understanding MSC biology as a model for studying the molecular basis of differentiation and because of their therapeutic potential for tissue repair and immune modulation [3]. Several studies have demonstrated that MSCs can be selectively mobilized into tumors [30–33]. Nevertheless, their therapeutic potential has been consistently demonstrated in numerous credible studies. Apart from the previously mentioned tissues, adult stem cells have shown to be extensively isolated from the adipose tissue [34].

Dual Role of Mesenchymal Stem Cells in Cancer

The involvement of MSCs is still controversial in carcinogenesis. MSCs have been postulated to play a dual role in tumor biology both by promoting tumorigenesis by regulating immune surveillance, growth and angiogenesis, and also possess tumor suppressive effects [35,36]. By direct or indirect interaction with tumor cells, MSCs exert stimulatory effect on cancer cell growth by engrafting into sites of tumors and possess the ability to migrate and invade. MSCs have been shown to induce metastases and epithelial to mesenchymal transition [35,36]. On the contrary, MSCs have shown to be tumor suppressive by modulating Wnt and Akt signaling pathways [37]. MSCs have been shown to be tumor suppressive in other tumor types such as in gliomas and when coadministered, show a decrease in vascularization and tumor volume [37–39]. These contrasting results may be due to differences in donors, variable experimental factors such as different cellular sources of tissues such as bone marrow or fat tissue, and variations in timing of administration of MSCs.

ADIPOSE TISSUE

Adipose tissue has long been considered to be primarily a storage organ for excess energy from food intake. However, evidence challenges this notion and indicates that adipose tissue possesses immunological functions and through secretions of key cytokines can communicate with other organs including the brain tissue [40]. In addition, there is an interest in application of adipose tissue in cell transfer–based therapies that offers a unique therapeutic utility for adipose cells [41].

Adipose tissue is widely distributed in the body. The size of the tissue is dynamic and may increase or decrease in response to the balance between the nutritional intake and energy usage. Adding to the complexity of fat storage, gender difference also impacts the mass and distribution of adipose tissue [42]. The two widely known distinct types of adipose tissues are namely the white adipose tissue (WAT) and the brown adipose tissue (BAT) [43]. A third type of adipose tissue called the beige adipocytes has been shown to develop from the WAT due to various activators [44]. The interconversion between WAT and BAT is still an ongoing debate [45,46]. Adipose tissue is comprised of adipocytes, preadipocytes, fibroblasts, stromal vascular cells, and mast cells. Importantly, anatomic location of the adipose depot determines the properties exhibited by the tissue. The primary distinction of adipose tissue is based on its subcutaneous and omental localization. Importantly, blood supply to the adipose tissue in the mesenteric region is the most highly innervated site with the highest blood circulation in the body indicating increased cardiac output to this region [47].

Adipose Tissue as an Endocrine Organ

Adipose tissue is now appreciated to function as a dynamic robust endocrine organ, dysfunction of which leads to numerous pathologies [48]. Understanding the properties of some of the secreted factors can provide an idea of the scope of the systemic and local effects of adipose tissue. Adipose tissue is extremely metabolically active, as documented by its capacity to secrete hormones, growth factors, and cytokines by both mature adipocytes and adipose tissue-derived stem cells (ASCs). The different cells that comprise adipose tissue can secrete different biological molecules. The major group of factors secreted by both mature adipocytes and ASCs are the hormones, growth factors, and adipokines or adipocytokines [49,50]. In response to adipose tissue expansion, which is characteristic of obesity, chronic low-grade inflammation, and metabolic disorders, adipokines are secreted. The major adipokines that have been discovered are leptin, adiponectin, fibroblast growth factor 21 (FGF21), retinol-binding protein 4 (RBP4), dipeptidyl peptidase 4 (DPP-4), BMP-4, BMP-7, vaspin, apelin, and progranulin, plasminogen activator inhibitor-1, angiotensinogen, serum amyloid A, resistin, visfatin, zinc-α2-glycoprotein, omentin, and chemerin. Obesity-specific adipokine profile and obesity-induced release of proinflammatory cytokines play an important role in obesity-related induction of breast cancer. However, the role of some adipokines still remains unclear [51].

Adipose Tissue-Derived Stem Cells

Adipose tissue, which is of mesodermal origin, has been a recent focal point in the stem cell field of research. Zuk et al. [41] was the first group to document that ASCs can be isolated from human lipoaspirates. The group identified a subpopulation of cells with stem cell properties including multipotency and ability to differentiate into distinct types of cells (Fig. 16.1), dissociated from collagenase-treated WAT [41,52,53]. ASCs also contribute to the cellular turnover of the adipose tissue [54,55]. Initially adipose-derived stromal cells had been termed "preadipocytes," and recent studies by multiple independent investigators have demonstrated that they exhibit multipotency, with chondrogenic, neuronal-like, and osteogenic differentiation capabilities [56]. Hence, they are now referred to as adipose-derived stromal/stem cells or ASCs [57–60].

Isolation of Adipose Tissue-Derived Stem Cells

Adipose tissue is mainly composed of adipocytes and a stromal vascular fraction (SVF) consisting of pericytes, endothelial cells, monocytes, macrophages, and ASCs [61]. ASCs are dissociated from the vascular fraction in the stroma of lipoaspirates. Freshly dissociated ASCs are also usually associated with different types of cells such as endothelial cells, fibroblasts, preadipocytes, mast cells, and cells from the smooth muscle. After a few early passages of cells, the

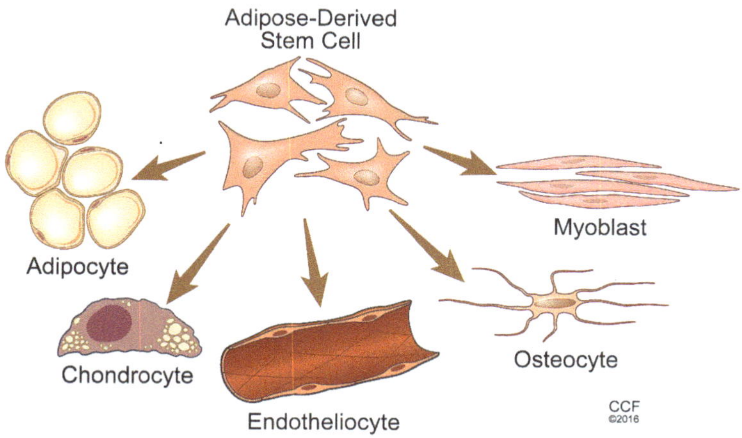

FIGURE 16.1 Multilineage differentiation potential of adipose tissue-derived stem cells (ASCs). ASCs have the ability to differentiate into different cell types under appropriate conditions such as the adipocytes, chondrocytes, endotheliocytes, osteocytes, and myoblasts.

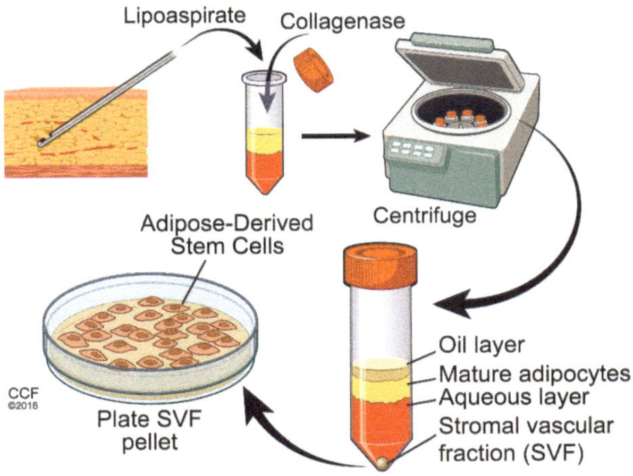

FIGURE 16.2 Isolation of adipose tissue-derived stem cells (ASCs) from lipoaspirates. Schematic outline shows the isolation of ASCs from lipoaspirates followed by the enzymatic digestion of the aspirate with collagenase. The SVF is isolated by centrifugation. The SVF pellet is then plated and expanded ex vivo.

ASCs in culture survive leading to homogeneity of the population. To isolate ASCs, adipose tissue derived from liposuction involves mincing of the adipose tissue sample, digestion with collagenase, and extensive washing of the tissue with phosphate-buffered saline followed by centrifugation to separate the stromal population in the bottom layer of the pellet from the adipocytes in the top layer as shown in Fig. 16.2 [53,62,63]. In short, ASCs are isolated from fat tissues by digestion of collagen and differential centrifugation followed by separation of ASCs from SVF by adhesion on plastic dishes [53,63]. The extraction of ASCs is being done using liposuction, and studies have utilized these cells in xenotransplantation experiments involving rats and mice [64,65].

ASCs can be maintained in well-defined media conditions with and without serum. The cellular components in the isolation of ASCs vary based on the influence of multiple factors such as age, tissue type, isolation procedures, culture and experimental conditions, passage number of the dissociated cells, and storage of cells. ASCs can be purified based on the expression profile of surface marker antigens like bone marrow MSCs. The lack of reliable cell surface marker antigen for the enrichment of ASCs is the problem in isolating a pure ASC population. While magnetic fractionation by immunologic markers or cell sorting by flow cytometry has been attempted, this approach has led to limited unconvincing results [53,66].

OBESITY AND CANCER

Obesity refers to an excess accumulation of hypertrophic and hyperplastic body fat cells while overweight refers to an excess amount of body weight that may come from muscles, bone, fat, and water [67]. Obesity is a health crisis on a global scale with an alarming proportion of more than two-thirds of American adults being either overweight or obese, compared with the 1990s, when less than 15% of the population were obese in most states in the United States [68]. Being overweight or obese is the culmination of a multitude of genetic, hormonal, environmental, emotional, and cultural factors. Obesity is also associated with the risk of many adverse health conditions such as heart diseases, hypertension, type II diabetes, and cancer [69].

Of relevance to this chapter, epidemiological evidence strongly associates obesity with an increased risk of cancer initiation and malignant progression. Obesity is a worldwide epidemic with serious implications in cancer-related incidence and mortality. Current data show that about 40% of the population with one in two men and one in three women will be diagnosed with cancer, and the trends strongly indicate that obesity is a major risk factor for the incidence of multiple types of cancer [70–74]. Apart from increasing the risk of developing cancer, obesity also impacts survival of cancer patients. Obesity is associated with a 52% and 88% increase in cancer mortality rate among men and women, respectively. Being either overweight or obese has been shown to correlate with 14% and 20% of all cancer deaths in men and women, respectively. It is important to note that obesity-related disorders, such as stroke, cardiovascular disease, renal disease, metabolic syndrome, also compound the complexity of the link between obesity and cancer mortality [70]. Though the profound negative effects of obesity on tumorigenesis are well recognized, detailed insights into the mechanism of the relationship remain poorly understood. Investigation into the molecular mechanisms has begun only in the recent past for better understanding of the obese state with cancer etiology.

Obesity is considered to be a promoter of tumor progression rather than as an etiology of cancer. The tumorigenic impact of obesity has a disproportionately serious consequence among women compared with men. Many tumors both grow in close proximity to adipose tissue such as breast cancer and also metastasize to adipose tissue-rich environment such as the abdominal cavity, particularly gastric and ovarian cancers [75]. Obesity has been associated with increased incidence and mortality of breast cancer. A large number of epidemiological studies on cancer incidence correlate obesity with the propensity to develop aggressive cancers like renal, esophageal, gastrointestinal, colon, prostate, thyroid, liver, lung, blood, and reproductive cancers in men and women [70–74].

Increased adipose tissue promotes the development of tumor by virtue of its anatomical proximity to different tumor types and its state of low-grade inflammation. Obesity-related chronic low-grade inflammation is an important link between obesity and cancer. In this activated state, adipose tissue-related cells and inflammatory cells secrete hormones, adipokines, and cytokines which stimulate the adhesion, migration, and tumor cell invasion.

The relative abundance of adipocytes in the microenvironment also supports tumor progression and growth [76,77]. All these characteristics of hypertrophied adipose tissue have been attributed to the increased morbidity and mortality of obesity-associated cancers [78].

Obesity, Adipose Tissue-Derived Stem Cells, and Cancer Triad

Tumors arising from the oncogenic transformation of epithelial cells represent the most common cancer type among humans. Adult stem cells have been shown to be the cancer cells of origin for epithelial cancers [79]. Obesity is a well-recognized risk factor for numerous epithelial cancer subtypes with some cancers occurring more commonly in women. Though adipose tissue has been widely observed to play a crucial role in the development and mortality in a subset of cancers, detailed mechanistic insights underlying this relationship are still being addressed. The altered systemic endocrine state in obesity coupled with the unique adipose tissue microenvironment favor tumor initiation and progression [78]. Understanding ASCs is important to understand the impact of obesity on the incidence of many cancers. For instance, the obese setting provides a unique microenvironment of the adipose tissue with concomitant ASCs alterations that favor both tumor initiation and progression. Altered ASC state is intimately linked to endocrine and paracrine dysregulation of adipose tissue in obesity-related cancers. As an abundant source of ASCs, adipose tissue has received intense attention for its role in cancer pathology. Significant advances have been made to identify cells of origin of epithelial cancers, and evidence points toward ASCs as the first cells to initiate a tumor. ASCs, by virtue of their properties, make an attractive target of tumorigenesis, which benefit malignancies in more ways than expected [80]. ASCs have a long life suggesting that there are higher chances of accumulation of DNA damage over time [81,82]. The multipotent differentiation potential of ASCs suggests that ASCs can generate a number of different types of cells which explains the heterogeneity found within most tumors. The role of adipose tissue, and more specifically ASCs, in tumor initiation, growth, and metastasis, is a relatively new area of investigation [83].

An increased number of ASCs are observed with increased accumulation of the adipose tissue. ASCs have been found to be within the stromal fraction of the adipose tissue. Biological properties in the intratumoral and tumor-adipose-stromal fractions are distinct from the adjacent healthy tissues. Major differences in the interaction between ASCs and the neighboring tissues have been observed within tumors that distinguish them from nontransformed healthy ASCs and tissues. Importantly, considerable tumorigenic changes are induced by ASCs in the SVF that surrounds them. Obesity can induce characteristic transformation of the ASC biology enhancing tumorigenesis [83]. Mutual interactions between ASCs and cancer cells may account for the strong correlation between obesity and cancer. ASCs in the tumor stroma impact cancer progression by secreting adipokines and growth factors thereby recruiting more ASCs to the tumor site [84]. ASCs can also drive the development and progression of tumors, influenced by the inflammation, the formation of new blood vessels, and by the altered microenvironment in obese states as shown in Fig. 16.3 [85]. Signals between ASCs, adipocytes, and other cells in the niche provide a conducive environment for the growth and metastasis of tumors. Stromal environment promotes tumor growth by inducing epithelial–mesenchymal transition, genetic instability, and angiogenesis [86]. Apart from the above-mentioned reasons, adipose tissue is the major source of circulating blood estrogen, and the levels correlate with BMI [75]. The biology of ASCs is strongly influenced by estrogen-dependent signaling with potent ability to induce oncogenic transformation of cells and neighboring tissues [87]. The surrounding microenvironment and the paracrine effect of ASCs on

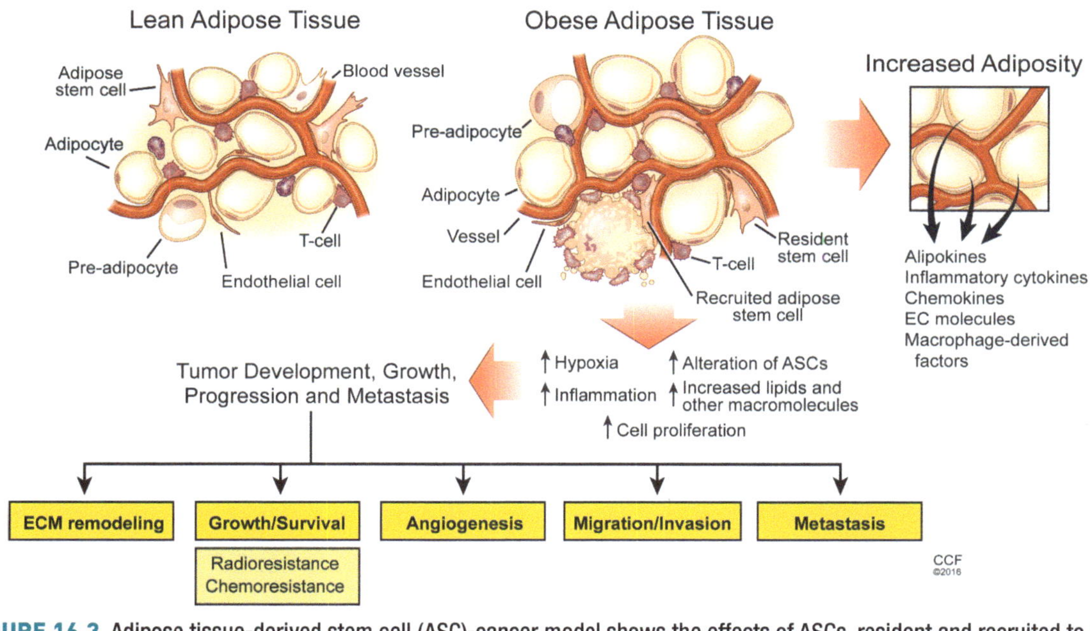

FIGURE 16.3 Adipose tissue-derived stem cell (ASC)-cancer model shows the effects of ASCs, resident and recruited to the tumor site establishing the link between obesity and breast cancer.

ASCs are crucial for normal tissue homeostasis, but by promoting proliferation, growth, survival, secretion of local factors and extracellular matrix (ECM) components, angiogenesis, and invasion, they have also been shown to support tumorigenesis. Many factors such as secretion of adipokines, increased local estrogen concentrations, increased collagen and fibronectin deposition, secretion of cytokines and other growth factors and hypoxic conditions in the tumor microenvironment attract ASCs to the site of the tumor. Expansion of the adipose tissue during weight gain and its contribution to signals and factors induce trafficking of ASCs. The model shows how obesity enhances the recruitment of ASCs and induces both local and systemic changes implicated in breast cancer progression. Bidirectional signaling between tissue-specific cells and ASCs are crucial for the creation of a niche that is permissive for cancer initiation, growth, and metastasis. ASCs stimulate the acquisition of cancer-associated functions in cells which support tumor growth and survival indicating the crucial role of environmental stimuli in tumorigenesis.

the epithelial cells are important for normal tissue homeostasis. Malignant transformation of these cells is supported by the stromal tissue around the tumor that shapes the microenvironment. Evidence shows that a homeostatic microenvironment can inhibit the growth of the tumor in an effective manner. Surrounding tumor stroma influences the interchange between various states of differentiation depicting the dynamic state of cancer stemness [50].

Inflammatory microenvironments, such as those of tumors, also attract regenerative cells including ASCs. Studies strongly suggest that ASC is a key player in the tumor microenvironment by promoting several mechanisms of pathogenic processes such as angiogenesis and invasion. Cancer cell phenotypes have shown to be regulated primarily by the surrounding cells in the stroma such as local invasion and proliferation. Stromal signals are exerted in the form of cell–cell communication, locally secreted factors, and extracellular matrix (ECM) components. Highly expressed interleukins IL-6 and IL-8 in breast cancer cells have been shown to induce significant recruitment of ASCs. Signals from the tumor such as TGF-β induce stem cells to form cancer-associated fibroblasts (CAF) and smooth muscle actin [88,89]. Recent findings show that CAFs secrete ECM protein tenascin C (TNC). Stromal-derived and CAF-derived TNC make the microenvironment conducive for the engraftment of metastatic cancer cells by enhancing the response to embryonic self-renewal pathways such as canonical Wnt and Notch signaling pathways. Cells that secrete TNC have also been observed to prevent apoptosis in CSCs and stimulate neoangiogenesis in the growing metastatic niche [90]. Studies have showed that coculture of human ASCs with breast cancer cells significantly increases secretion of stromal cell-derived factor-1α. CXCR-4, the chemokine receptor for SDF-1α, is overexpressed in highly invasive breast tumor lines and CSC-enriched spheres. The fact that the SDF-1α-CXCR4 axis is tightly regulated in a tumor microenvironment underscores the critical role of stromal cells in supporting tumor progression. This signaling pathway has been studied to contribute to resistance to chemotherapy in cancer cells by regulating the survival pathways such as Notch, AKT, and the Bcl-2 family. Tumors with high levels of SDF-1α within tumors recruit ASCs to the tumor stroma [57]. The tumor–stromal components sustain inflammation and induce stem cell pathways for sustaining tumor homeostasis. All these factors act in concert to establish self-promoting and self-sustaining autocrine feedback loops in tumor and tumor-associated stroma [91].

Though ASCs have been explored for clinical applications, their relation to incidence of breast cancer has not been extensively studied. Since the adipose tissue is abundantly endowed with blood vessels, every adipocyte is encased by a rich network of capillaries. Interestingly, studies have shown that a number of ASCs correlate with the density of capillaries in the adipose tissue. ASCs have been shown to migrate to the location of the tumor. ASCs combined with 4T1 cells have been observed to promote the growth of the tumor by migrating into the vessels within the tumor thereby inducing tumorigenesis [92,93]. Breast tissue-derived ASCs located in close proximity to micro vessels have been shown to promote tumorigenesis. As soon as the circulating CSCs home to the future metastatic niche, the novel microenvironment ASC interactions might change the CSC phenotype depending on the location. Intraabdominal tumors, such as ovarian cancer, have the highest metastatic propensity to the tissue consisting primarily of adipocytes, the omentum. Omental metastases in ovarian cancer reflect the largest tumors in the abdominal cavities [94]. Interestingly, the distant the CSC gets from the origin, more is the influence of the destination site and its interaction with the new environment. ASC-derived factors have also been shown to contribute to the initial quiescence of the metastatic seed early on in the formation of the metastasis. Recently developed concept is that circulating or metastatic CSCs could be the driver themselves and activate the formation of secondary metastases. All the previously described potential events are the result of the high plasticity displayed by CSCs under the influence of ASCs. The feedback loop between obesity and ASCs as well as the mechanisms by which ASCs from obese patients alter the biology of cancer cells and enhance tumorigenesis will need to be addressed [49,92].

Cancer cells induce resident stem cells to play tumor-supportive roles. The critical questions to be addressed are how does the primary tumor attract stem cells and if metastases are formed by disseminated cancer cells or cancer cells accompanied by stem cells. Interesting findings such as formation of metastases several weeks after resection of in situ breast tumors indicate that CSCs can disseminate. This dissemination could be triggered during either surgical procedure leading to activation of quiescent metastatic CSCs influencing stem cells at different distant sites supporting the development of future metastases [92].

Wicha et al. have demonstrated that resident stem cells by secreting cytokines such as IL-6, IL-8, and CXCL7 contribute to breast cancer tumorigenesis [95]. As shown earlier, TNC and also periostin have been observed to be an important ECM molecule found in stem cell niches, mainly the breast cancer. Periostin was found to be expressed at the end buds of mammary glands and upregulated in CAFs and lung metastases of breast cancer. Also, it induces the homing of canonical Wnt ligands in the metastatic site for potential circulating, metastasis-initiating CSCs [90]. Periostin also induced tumorsphere formation and periostin-knockout mice showed significantly reduced tumorsphere formation which was further rescued partly by extrinsic periostin [96]. By recruiting TAM, periostin has been shown to play a role in the progression of other cancers such as colon and glioblastoma [97,98]. It has also been shown to promote the invasiveness of cancer cells and enhances their survival under conditions of hypoxia. Taken together, these results corroborate the principle that ASCs in the tumor microenvironment enhance the production of tumor-promoting factors and niche that attract colonization of CSCs and metastatic cancer cells [99].

BROAD UTILITY OF ADIPOSE TISSUE-DERIVED STEM CELLS IN REGENERATIVE MEDICINE

ASCs were identified primarily based on their quiescent nature and their ability to regenerate and reconstitute by differentiating into different cell types in their resident tissues as shown in Fig. 16.4 [100]. Major adult organs comprise ASCs where they maintain tissue homeostasis and regeneration in response to injury [6]. The simplicity of the autologous fat grafting procedure coupled with the absence of a postsurgical scar has sparked an interest in the applicability of this technique. The characteristics of ASCs being one of the most abundant stem cells and available in parts of the body with easy access make them an attractive target for application in regenerative medicine [101]. The high degree of plasticity of ASCs isolated from the SVF of adipose tissue combined with its potential for differentiation and inducing angiogenesis and vascular remodeling are important features for its application in regenerative medicine and surgery. ASCs, upon natural turnover or in response to injury can give rise to restricted progenitors of different lineages called transit amplifying cells (TACs). TACs divide and terminally differentiate into resident cells of the tissue

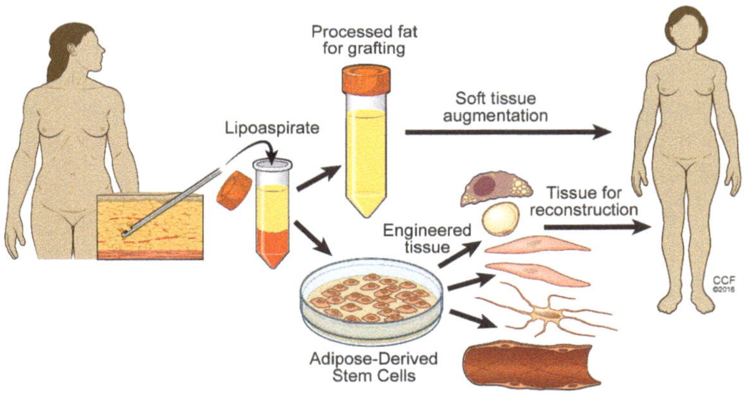

FIGURE 16.4 Therapeutic applications of adipose tissue-derived stem cell (ASCs) derived from lipoaspirates.

ASCs are isolated from lipoaspirates and biopsy tissues. The isolated ASCs can be utilized for soft tissue augmentation after processing the fat for grating. The stromal vascular fraction can be plated and expanded followed by eventual selection of the population of ASCs. Induction of ASCs under defined conditions can result in their differentiation to multiple cell types such as adipocytes, chondrocytes, endotheliocytes, osteocytes, and myoblasts. The differentiation potential of ASCs facilitates the repair and regeneration of damaged or diseased tissues. Furthermore, it could also be used for the cell tissue reconstruction.

or organ in a rapid fashion to generate the bulk of the tissue cells required for turnover or regeneration. Intriguingly, chemotherapeutic drugs exert their effect on rapidly dividing cells including TACs [102–104].

Bone marrow stroma cell (BMSC) in bone marrow is approximately 1 in 25,000 to 1 in 100,000 nucleated cells, and the average frequency of ASC in lipoaspirates is approximately 2% of nucleated cells. Another advantage of ASCs over BMSCs is the safety of the liposuction procedures accounting for no deaths between 1994 and 2000 [105]. Though the therapeutic potentials of ASCs and BMSCs are almost identical, there are a number of advantages in utilizing ASCs over other stem cells for various clinical applications [106]. Lee et al. [107] observed ASCs to be superior to bone marrow stromal cells with respect to their proliferation potential. Interestingly, microarray gene expression analysis comparing ASCs and bone marrow reveals differentially expressed genes, but their phenotypes and the gene expression profiles are similar. A disadvantage with the use of ASCs is that they have a potential to induce transformation of cells into cancerous cells by virtue of their ability to lose the expression of embryonic markers such as Oct-4, UTF-1, and Nodal even in short-term culture [107].

Approximately 2% of adipose tissue population compared with a low 0.002% in the bone marrow consists of cells with multilineage differentiation properties [108]. ASCs differentiate into a broad spectrum of cell lineages. ASCs have the highest plasticity potential compared with all adult tissues. Adipose tissue is a highly vascular tissue enforcing the concept that it is an excellent source for harvesting ASCs in large numbers. Interestingly, ASCs can be retrieved in higher numbers and also expanded in vitro with ease unlike bone marrow stem cells. A higher number of ASCs are present mainly in the WAT of the subcutaneous tissues and abdominal viscera. These cells possess enhanced proliferation and plasticity in comparison with the stem cells in the BAT found especially in the newborns. When multilineage differentiation is aimed for an adipose tissue-derived cell population in cell culture, special stimulants are used. Stimulants like insulin, dexamethasone, and indomethacin are used to obtain adipogenic cells while dexamethasone, ascorbate, and glycerophosphate are used to obtain osteogenic cells. ASCs display properties similar to that observed in BMSCs [109]. Upon induction, ASCs undergo myogenic, cardiogenic, vascular, endocrine, hepatic, hematopoietic, chondrogenic, adipogenic, and neurogenic differentiation in vitro for further potential clinical applications. For osteoarthritis treatment, ASCs have a high propensity for cartilage and skeletal joint repair [104].

ASCs have been shown to differentiate into a variety of lineages including adipogenic, osteogenic, myogenic, and chondrogenic lineages as shown in Fig. 16.1. Besides differentiating into these lineages, ASCs have been shown to differentiate toward other lineages to a lesser capacity. Factors that regulate their differentiation include soluble factors, small molecules, mechanical stimulation, factors of the ECM, and genetic manipulation. Using soluble factors, BMSCs and ASCs have a high tendency to differentiate into cells of mesenchymal lineage, including bone, adipose tissue, and cartilage [107,110]. Although greater than 90% similarity in the immunophenotypes of ASCs and bone marrow MSCs has been observed, there are also differences in surface antigen expression. CD14 and HLA-DR have been identified in early passage of human ASCs but were observed to be absent in bone marrow MSCs. The cell surface expression of ASCs has been shown to resemble BMSC closely with the expression of CD34 on the surface. For example, the nature of the ASCs is so dynamic that the expression profile changes from adherence during culture or just the passage number. Freshly dissociated ASCs express CD34 but lose the expression of CD34 with an increase in cell passage number. Tissues of origin have shown CD34 positivity and over time, the expression is lost during culture. This consistent finding is characteristic of all MSCs including ASCs. Recent findings have also shown that CD34 might be a marker of differentiation from progenitor state into a specific lineage of differentiation [111].

Determining whether ASCs are capable of maintaining a chondrogenic phenotype remains controversial. While culturing ASCs is observed to be unfavorable to retain chondrogenic potential, culturing these cells in elastin-like polypeptide enhanced chondrogenesis. Also, enhanced chondrogenesis was observed with exposure of these cells in chondrogenic cocktail including TGF-β, expression of collagen type II, aggrecan, sulfated proteoglycans and upon treatment with BMP-6. Chondrogenic potential was observed up to 12 weeks in in vivo studies using ASCs. Adipogenic differentiation is induced by 1-methyl-3-isobutylxanthine, dexamethasone, insulin, and indomethacin. Differentiation into the osteogenic lineage can be induced by culturing the cells in the presence of dexamethasone, ascorbic acid, and β-glycerophosphate. Growth factors for the chondrogenic differentiation include transforming growth factor-β3 (TGF-β3) and bone morphogenetic protein-6 (BMP-6) [112].

Other lineage differentiation include induction of smooth muscle phenotype by adding platelet-derived growth factor (PDGF) and TGF-β3. 5-Azacytidine treatment induced the expression of cardiac markers β-myosin heavy chain, desmin, and α-cardiac actin leading to cardiac cells formation. Mechanical stimulation is also a potent cell phenotype regulator like restructuring of the bone by compression from gravitational forces and fluid shear stress on the blood vessels. Hence, the role of mechanical factors on ASCs has become an interesting area of research. ECMs primarily provide matrix rigidity, matrix patterning, and matrix composition which vary from soft tissues like the brain to hard tissues of the bone. Apart from providing structural integrity, ECMs induce differentiation by providing molecular cues regulating the cellular phenotype. Aside from soluble, mechanical, and ECM factors in the microenvironment, intracellular factors such as microRNAs have also been observed to induce differentiation by regulating posttranscriptional gene expression. Overexpression of miR-21-induced adipogenesis of human ASCs is achieved by decreasing TGF-β signaling pathway member, TGFBR2 both at the mRNA and protein levels. Taken together, these results suggest that ASC differentiation is regulated by both intracellular signals and multiple microenvironmental factors [113].

During early passages, freshly isolated ASCs elicit a T-cell proliferative response whereas later passage cells do not elicit this response. Due to the absence of major histocompatibility complex (MHC) Class II antigens on their surface, ASCs have been found to be immunosuppressive. Furthermore, ASCs exhibit an inhibitory effect on allogenic lymphocytes in vitro. Adult stem cells derived from different tissue origins with beneficial immunomodulatory and regenerative properties have recently received attention for their potential in developing as effective therapeutic agents. The selection of a small subpopulation of pure stem cells from an isolated tissue among a heterogeneous mixture of progenitor and other cell types is absolutely critical in both basic and clinical stem cell research. For tissue engineering and regenerative medicine applications, the major strategy is to differentiate the cells in vitro into lineages of interest followed by in vivo delivery for therapeutic treatment [114].

Evolving Role of Adipose Tissue-Derived Stem Cells in Reconstructive Surgery and in Tumorigenesis

Grafting of adipose tissue with ASCs after surgical resection of breast tumors to restore breast defects and improve the regenerative ability of the graft and retention has gained immense attention recently. Recent finding that the expression of ASCs in the adipose tissue and its regulation in the cellular turnover in the adipose tissue requires further consideration. This also calls for caution in understanding the potential implications in the application of enriched ASCs in postsurgery breast cancer reconstruction [115]. It is crucial to understand the interactions between resident or grafted ASCs, residual active or dormant breast cancer cells, cellular components of the adipose tissue, and the adjacent microenvironment [116]. Studies have also shown that ASCs have been widely proposed to improve graft retention and the restorative capacity of the graft, and supplementation with ASCs for breast reconstruction did not impact breast tumor recurrence rates in a number of case reports and small studies after short-term follow-up. Despite the potential of ASC supplementation in reconstructive surgeries, studies addressing the safety of this procedure have not been performed and followed up for a long term; hence, ASCs supplementation in breast and other reconstructive surgeries still remain debatable [80,117]. Long-term safety for ASCs grafting is unknown as the interactions between ASCs and residual breast cancer cells are not entirely understood [84]. In vivo and in vitro studies have been conducted to verify and compare the influence of ASCs on active and dormant tumor cells. Since the tumor cells in active state require growth factors while the dormant tumor cells are autonomous and their growth is slower, grafted ASCs favor the growth of active tumor cells [115,118,119]. By incorporating into the vessels of the tumor, ASCs have the propensity to differentiate into endothelial cells contributing to angiogenesis. These results highlight the similarities between formation of tumor stroma and wound-healing process. ASC-induced angiogenesis has been observed raising a concern regarding using enriched ASCs for grafting. Mixed results have been observed in laboratory studies studying the effects of ASCs on breast cancer growth and metastasis [84].

ASCs are of crucial importance in breast cancer since the breast tissue is abundant in adipose tissue. The tumor-promoting effect of ASCs was validated using a human breast cancer line MDA-MB-231 cells. Also, when ASCs were coinjected with 4T1 cells, a faster growth of tumors was induced. ASCs induce the secretion of SDF-1α which has a paracrine effect on tumor cells enhancing motility, invasion, and metastasis. This effect on tumor enhancement was compromised by CXCR4 knockdown in the tumor cells. Other studies have also demonstrated that the paracrine effect of ASCs stimulated human triple-negative breast cancer cell line MDA-MB-231 migration by factors independent of the differentiation status of ASCs. ASCs home to the site of the tumor, promoting growth of the tumor by both local coinjection and intravenous injection. ASCs, adjacent to primary breast cancer cells, may interact with the tumor cells directly [84,99].

MDA-MB-231 cells when cocultured with ASCs showed increased migration accompanied by increased spindle-shaped morphology. Differentiation of ASCs involves epidermal growth factor receptor (EGFR), fibroblast growth factor receptor (FGFR), ERBB2, and Akt signaling pathways that have been shown to be dysregulated in breast cancer initiation and progression. These findings are consistent with recent reports that ASCs enhanced cancer cell migration and invasion capacity. Breast cancer cells possessing epithelial characteristics undergo mesenchymal transition by displaying features such as loss of polarity, spindle shape that favors motility, invasion, and survival [120]. Enhancement of early stages of metastasis from primary breast tumor is induced by ASC promoting the pattern of tumor cell dissemination to first pass organs such as lung, liver, kidney, and spleen. ASCs induce EMT and angiogenesis and promotes prometastatic paracrine factors leading to an increase in cancer cell growth, migration, invasion, and metastasis of MDA-MB-231 tumor cells both in vitro and in vivo [117]. Both in vivo and in vitro experiments report that ASCs increase the deposition of ECM and angiogenesis, strongly suggesting that ASCs contribute to desmoplastic reaction in breast cancer [54]. In this context, ASCs may interact with breast cancer cells through the formation of gap junctions that allows intercellular communication and the exchange of low-molecular weight compounds [121]. Gap junction intercellular communication has been shown to correlate with increased tumor progression regulating the metastatic capacity of breast cancer cells [122]. It is postulated that gap junction inhibition could block the stem cell-mediated induction of growth of breast cancer cells [120]. In summary, pros and cons should be weighed before applying ASCs in graft procedures.

Studies using breast cancer cell lines cocultured with ASCs and coinjection of ASCs with breast cancer cells into mice promoted growth and invasion of breast cancer cells in vitro and promoted growth of primary tumor xenografts. Coinjection of murine 4T1 cancer cell line with murine ASCs increases primary tumor volume coupled with increase in lung metastases. Study using human ASCs derived from reduction mammoplasty showed inhibition of primary breast cancer xenograft growth and metastasis. Combined together, ASCs have been shown to promote the breast cancer metastases especially during the initial stages. It is still not clear if the risk of de novo initiation of cancer or recurrence occurs with resident or grafted ASCs. The current evidence presents a contrasting picture concerning the role of ASCs on breast cancer initiation and progression [92].

MDA-MB-231 was employed to study the effect of ASCs on early micro metastases. Tumors extracted from one of the three donors showed elevated angiogenesis accompanied by changes in tumor stroma along with lung and liver metastases. ASCs stimulated the invasive and metastatic spread of MDA-MB-231 breast tumor xenografts to different tissues in the mice [84]. Strong et al. observed that abdominal ASCs derived from obese patients increased MCF-7 ER-positive breast cancer cell proliferation in vitro and tumor xenograft growth in vivo [83]. The same group also studied the effect of ASCs on MDA-MB-231 cell proliferation in vitro and found that ASCs increased MDA-MB-231 growth approximately twofold after 7 days coculture regardless of ASC depot site or donor BMI [117].

Role of Adipose Tissue-Derived Stem Cells in Anticancer Stem Cell Treatment Strategies

Reconstructive and regenerative therapies are not the only benefit of studying stem cell biology, but a better understanding of stem cell biology will help distinguish them from nonstem cells which will likely contribute to fine-tune the development of targeted drug treatment of cancers. Progenitor cells and stem cells are recruited by many ways to a tumor microenvironment. Andreeff et al. was the first group to address the potential antitumor benefits of this interaction by demonstrating that in many cancers, systemic injection of modified stem cells homed within primary and secondary tumor sites with pronounced proapoptotic effects. More importantly, homing of stem cells toward tumor sites is extremely useful in undetectable tumors or surgically inaccessible tumors. Development of novel antitumor agents is aimed at exploiting the tumor-homing properties of stem cells [123]. Stem cells have beneficial effects in brainstem gliomas, HeLa cells, and in CSC subpopulations. Modulation of the critical self-renewal Wnt pathway in this manner suggests that it may provide additional benefits by targeting cancer stemness traits [124].

A model for tumor therapy was developed based on ASCs as cellular vehicles for ganciclovir-mediated bystander killing of tumor cells, and noninvasive bioluminescent imaging was employed to monitor both the tumor cells and ASCs in real time. The therapy used to deliver ASCs sustained a longer time within tumors, and the therapeutic effect was assessed continuously. Treatment of mice bearing prostate tumors with therapeutic ASCs reduced the number of tumor cells compared with the control tumors. ASCs could be effective vehicles to deliver therapeutics efficiently, with a tremendous potential for clinical utility in tumors that cannot be operated on and surgical borders after tumor resection [125]. The identification of the phenotypic changes and the mechanism of induction of chemosensitivity by the interaction between stromal and cancer cells are yet to be well defined. Interestingly, ASCs have been shown to function as a viable vehicle for delivery in vivo. Introduction of a potential attachment surface to adhere circulating tumor cells in gastric cancer significantly promoted the chemosensitivity of the nonadherent gastric cancer cells [126]. While the functional ability of ASCs to support or suppress tumor cell proliferation is unclear, this therapeutic model approach should improvise the selection of cell types, evaluation of dosages, effective therapeutics, and treatment strategies for cell-based therapies of tumors.

High-grade gliomas always recur even after standard treatment owing to the high infiltrating nature of these tumors. The infiltrating parts of gliomas resist and respond poorly to radiation and chemotherapy treatment and remain after surgery. A study by Lamfers et al. has shown that ASCs demonstrated a similar tropism as the neural stem cells to home to different types of intracerebral pathology [127]. The main drawback in employing neural stem cells is the major limitation of harvesting and expansion of these cells ex vivo. Interestingly, irrespective of the side of injection in brains of glioma-bearing mice, ASCs extensively homed to the tumor. This study strongly suggests that ASCs are an excellent choice for further development of cell-based therapy of gliomas [127].

ASCs are considered to be a promising tool for cell and gene therapy for a variety of diseases. ASCs derived from the adipose tissue, an endocrine organ, possess a strong secretory potential. Disrupting the complex autocrine and paracrine mechanisms by which the cells of the tumor interact with the environment results in dysregulation of cell proliferation, adhesion, function, and migration inducing tumor progression. ASCs provide functional cell support and modulating immune and inflammatory functions [128]. ASCs have been shown to alter the progression of cell cycle by interfering with the tumor cell proliferation [125,127]. A multitude effect of ASCs on cancer cells has been hypothesized. A study aimed at investigating the potential of ASCs to treat breast cancer in an animal model of breast cancer [129]. Cellular adoptive immunotherapy using ASCs injected into the tumors did not affect tumor growth or progression. While injection of ASCs into the stroma encompassing the tumor showed decreased tumor growth and complete recovery was sustained throughout a 6-month study period [129]. Hence, the ASCs secretory potential is a promising utility against breast cancer cells.

CONCLUSIONS AND PERSPECTIVES

Regenerative stem cells maintain homeostasis of normal tissue by transdifferentiation to replace mature cell types upon injury or apoptosis. Tumorigenesis and repair of tissue involve similar processes, and interactions within the aberrant tumor–stromal environment can induce stem cells to support tumor growth. Similarities between healing wounds and tumors can be drawn in the tissue-specific resident cell subpopulations. Microenvironmental effects are crucial in inducing the different fates of stem cells in both wound healing and tumor progression. Current evidence strongly supports a model in which ASCs-induced reprogrammed cells gain the capacity to grow into a tumor [130]. These results point to the dynamic interaction between resident ASCs and tumor cells influencing the nature of the tumor. Accumulating evidence supports the postulate that ASCs affect the growth and progression of distant tumors but when applied therapeutically to wounds suggest that the wound microenvironment can retain ASCs, preventing their homing and stromal contribution to distant neoplastic foci. Further studies are needed to better clarify this in vivo interaction and define how to selectively stimulate ASCs regenerative function without influencing tumorigenesis.

Through a better understanding of the tumor biology and their microenvironment, targeted delivery of therapeutic agents can be carried out by stem cells. The utility of modified stem cells to target particularly insidious small subpopulations of CSCs is very attractive. Finally, stem cells are increasingly being implemented for wound closure and applications of regeneration of stroma after surgical removal of solid tumors and eradication of any residual CSCs. To summarize, stem cells are perceived as double-edged swords because while they support tumor initiation and progression, they also have immense regenerative properties in tissue injury. The possibility of the involvement of ASCs in cancers, other diseases, and disorders could explain the pathophysiological processes involved in those conditions. A clear understanding of stem cells will certainly lead to new therapies in the context of tumors with the aim of undermining key aspects of the intrinsic qualities sustaining tumors.

The prospect of translating the potential utility of stem cells into clinical application remains exciting, motivating the efforts toward improved understanding of the mechanisms driving their behavior. ASCs are highly abundant and can be readily extracted by minimally invasive clinical procedures without affecting the physiological functions providing a feasible, autologous, and an ideal translatable utility. The major advantage of easy access to ASCs has led to the development of automated devices designed for automated isolation of vascular stromal fraction which are commercially available now. In this chapter, we have discussed the varied roles and the potential implications of ASCs in cancer and regenerative medicine.

Glossary

Autologous fat grafting It is the process of using the patient's own fat to increase the fat tissue in another area of their body.

Endocrine Of, relating to, or denoting glands that secrete hormones or other products directly into the blood.

Lipoaspirate The material removed by lipoaspiration.

Mesoderm The middle layer of an embryo in early development, between the endoderm and the ectoderm.

Paracrine Of, relating to, or denoting a hormone that has effect only in the vicinity of the gland secreting it.

Plasticity The capacity of tissue-derived stem cells to exhibit a phenotypic potential that extends beyond the differentiated cell phenotypes of their resident tissue.

Stem cells An undifferentiated cell of a multicellular organism that is capable of giving rise to indefinitely more cells of the same type and give rise to other types of cells by differentiation.

Tumor microenvironment The milieu surrounding the tumor consisting of cells, vessels, soluble factors, and molecules that can influence and be influenced by the tumor's growth.

Xenograft A tissue graft or organ transplant from a donor of a different species than the recipient.

List of Acronyms and Abbreviations

ASC Adipose tissue-derived stem cells
BAT Brown adipose tissue
BMI Body mass index
BMP Bone morphogenic protein
BMSC Bone marrow stroma cell
CAF Cancer-associated fibroblasts
CD Cluster of differentiation
CFU-S Colony forming unit—spleen
CSC Cancer stem cell
DPP Dipeptidyl peptidase
ECM Extracellular matrix
EGFR Epidermal growth factor receptor
EMT Epithelial–mesenchymal transition
ER Estrogen receptor
FGF Fibroblast growth factor

FGFR Fibroblast growth factor receptor
HLA Human leukocyte antigen
IL Interleukin
MHC Major histocompatibility complex
MSC Mesenchymal stem cell
PDGF Platelet-derived growth factor
SVF Stromal vascular fraction
TAC Transit amplifying cell
TGF-β Transforming growth factor-β
TGFBR Transforming growth factor beta receptor
TNC Tenascin C
WAT White adipose tissue

References

[1] McCulloch EA, Till JE. Proliferation of hemopoietic colony-forming cells transplanted into irradiated mice. Radiat Res June 1964;22:383–97.

[2] Goodell MA, Jackson KA, Majka SM, Mi T, Wang H, Pocius J, et al. Stem cell plasticity in muscle and bone marrow. Ann N Y Acad Sci June 2001;938:208–18. discussion 18–20.

[3] Herzog EL, Chai L, Krause DS. Plasticity of marrow-derived stem cells. Blood November 15, 2003;102(10):3483–93.

[4] Siminovitch L, McCulloch EA, Till JE. The distribution of colony-forming cells among spleen colonies. J Cell Physiol December 1963;62:327–36.

[5] Till JE, McCulloch EA, Siminovitch L. A stochastic model of stem cell proliferation, based on the growth of spleen colony-forming cells. Proc Natl Acad Sci USA January 1964;51:29–36.

[6] White AC, Lowry WE. Refining the role for adult stem cells as cancer cells of origin. Trends Cell Biol January 2015;25(1):11–20.

[7] Hamburger AW, Salmon SE. Primary bioassay of human tumor stem cells. Science July 29, 1977;197(4302):461–3.

[8] Stenning SP, Parkinson MC, Fisher C, Mead GM, Cook PA, Fossa SD, et al. Postchemotherapy residual masses in germ cell tumor patients: content, clinical features, and prognosis. Medical Research Council Testicular Tumour Working Party. Cancer October 1, 1998;83(7):1409–19.

[9] Lapidot T, Sirard C, Vormoor J, Murdoch B, Hoang T, Caceres-Cortes J, et al. A cell initiating human acute myeloid leukaemia after transplantation into SCID mice. Nature February 17, 1994;367(6464):645–8.

[10] Visvader JE, Lindeman GJ. Cancer stem cells: current status and evolving complexities. Cell Stem Cell June 14, 2012;10(6):717–28.

[11] Hemmati HD, Nakano I, Lazareff JA, Masterman-Smith M, Geschwind DH, Bronner-Fraser M, et al. Cancerous stem cells can arise from pediatric brain tumors. Proc Natl Acad Sci USA December 9, 2003;100(25):15178–83.

[12] Singh SK, Hawkins C, Clarke ID, Squire JA, Bayani J, Hide T, et al. Identification of human brain tumour initiating cells. Nature November 18, 2004;432(7015):396–401.

[13] Liu N, Lu M, Tian X, Han Z. Molecular mechanisms involved in self-renewal and pluripotency of embryonic stem cells. J Cell Physiol May 2007;211(2):279–86.

[14] Romano G. Artificial reprogramming of human somatic cells to generate pluripotent stem cells: a possible alternative to the controversial use of human embryonic stem cells. Drug News Perspect October 2008;21(8):440–5.

[15] Young HE, Duplaa C, Katz R, Thompson T, Hawkins KC, Boev AN, et al. Adult-derived stem cells and their potential for use in tissue repair and molecular medicine. J Cell Mol Med 2005 July–September;9(3):753–69.

[16] Colter DC, Sekiya I, Prockop DJ. Identification of a subpopulation of rapidly self-renewing and multipotential adult stem cells in colonies of human marrow stromal cells. Proc Natl Acad Sci USA July 3, 2001;98(14):7841–7845.

[16a] Friedenstein AJ, Piatetzky S, II, Petrakova KV. Osteogenesis in transplants of bone marrow cells. J Embryol Exp Morphol 1966;16(3):381–90.

[17] Friedenstein AJ, Petrakova KV, Kurolesova AI, Frolova GP. Heterotopic of bone marrow. Analysis of precursor cells for osteogenic and hematopoietic tissues. Transplantation March 1968;6(2):230–47.

[18] Wagers AJ, Weissman IL. Plasticity of adult stem cells. Cell March 5, 2004;116(5):639–48.

[19] Friedenstein AJ, Deriglasova UF, Kulagina NN, Panasuk AF, Rudakowa SF, Luria EA, et al. Precursors for fibroblasts in different populations of hematopoietic cells as detected by the in vitro colony assay method. Exp Hematol 1974;2(2):83–92.

[20] Pittenger MF, Mackay AM, Beck SC, Jaiswal RK, Douglas R, Mosca JD, et al. Multilineage potential of adult human mesenchymal stem cells. Science April 2, 1999;284(5411):143–7.

[21] Bhartiya D, Shaikh A, Nagvenkar P, Kasiviswanathan S, Pethe P, Pawani H, et al. Very small embryonic-like stem cells with maximum regenerative potential get discarded during cord blood banking and bone marrow processing for autologous stem cell therapy. Stem Cells Dev January 2012;21(1):1–6.

[22] Dziadosz M, Basch RS, Young BK. Human amniotic fluid: a source of stem cells for possible therapeutic use. Am J Obstet Gynecol January 6, 2016.

[23] Campagnoli C, Roberts IA, Kumar S, Bennett PR, Bellantuono I, Fisk NM. Identification of mesenchymal stem/progenitor cells in human first-trimester fetal blood, liver, and bone marrow. Blood October 15, 2001;98(8):2396–402.

[24] Phinney DG, Prockop DJ. Concise review: mesenchymal stem/multipotent stromal cells: the state of transdifferentiation and modes of tissue repair – current views. Stem Cells November 2007;25(11):2896–902.

[25] Sasaki M, Abe R, Fujita Y, Ando S, Inokuma D, Shimizu H. Mesenchymal stem cells are recruited into wounded skin and contribute to wound repair by transdifferentiation into multiple skin cell type. J Immunol February 15, 2008;180(4):2581–7.

[26] Waterman RS, Tomchuck SL, Henkle SL, Betancourt AM. A new mesenchymal stem cell (MSC) paradigm: polarization into a pro-inflammatory MSC1 or an Immunosuppressive MSC2 phenotype. PLoS One 2010;5(4):e10088.

[27] Beyer Nardi N, da Silva Meirelles L. Mesenchymal stem cells: isolation, in vitro expansion and characterization. Handb Exp Pharmacol 2006;174:249–82.

[28] Maleki M, Ghanbarvand F, Reza Behvarz M, Ejtemaei M, Ghadirkhomi E. Comparison of mesenchymal stem cell markers in multiple human adult stem cells. Int J Stem Cells November 2014;7(2):118–26.

[29] Zimmerlin L, Donnenberg VS, Rubin JP, Donnenberg AD. Mesenchymal markers on human adipose stem/progenitor cells. Cytometry A January 2013;83(1):134–40.

[30] Gao Y, Yao A, Zhang W, Lu S, Yu Y, Deng L, et al. Human mesenchymal stem cells overexpressing pigment epithelium-derived factor inhibit hepatocellular carcinoma in nude mice. Oncogene May 13, 2010;29(19):2784–94.

[31] Khakoo AY, Pati S, Anderson SA, Reid W, Elshal MF, Rovira II, et al. Human mesenchymal stem cells exert potent antitumorigenic effects in a model of Kaposi's sarcoma. J Exp Med May 15, 2006;203(5):1235–47.

[32] Nakamizo A, Marini F, Amano T, Khan A, Studeny M, Gumin J, et al. Human bone marrow-derived mesenchymal stem cells in the treatment of gliomas. Cancer Res April 15, 2005;65(8):3307–18.

[33] Kidd S, Spaeth E, Klopp A, Andreeff M, Hall B, Marini FC. The (in) auspicious role of mesenchymal stromal cells in cancer: be it friend or foe. Cytotherapy 2008;10(7):657–67.

[34] Zuk PA, Zhu M, Ashjian P, De Ugarte DA, Huang JI, Mizuno H, et al. Human adipose tissue is a source of multipotent stem cells. Mol Biol Cell December 2002;13(12):4279–95.

[35] Karnoub AE, Dash AB, Vo AP, Sullivan A, Brooks MW, Bell GW, et al. Mesenchymal stem cells within tumour stroma promote breast cancer metastasis. Nature October 4, 2007;449(7162):557–63.

[36] Kabashima-Niibe A, Higuchi H, Takaishi H, Masugi Y, Matsuzaki Y, Mabuchi Y, et al. Mesenchymal stem cells regulate epithelial-mesenchymal transition and tumor progression of pancreatic cancer cells. Cancer Sci February 2013;104(2):157–64.

[37] Ho IA, Toh HC, Ng WH, Teo YL, Guo CM, Hui KM, et al. Human bone marrow-derived mesenchymal stem cells suppress human glioma growth through inhibition of angiogenesis. Stem Cells January 2013;31(1):146–55.

[38] Dasari VR, Kaur K, Velpula KK, Gujrati M, Fassett D, Klopfenstein JD, et al. Upregulation of PTEN in glioma cells by cord blood mesenchymal stem cells inhibits migration via downregulation of the PI3K/Akt pathway. PLoS One 2010;5(4):e10350.

[39] Otsu K, Das S, Houser SD, Quadri SK, Bhattacharya S, Bhattacharya J. Concentration-dependent inhibition of angiogenesis by mesenchymal stem cells. Blood April 30, 2009;113(18):4197–205.

[40] Trayhurn P, Beattie JH. Physiological role of adipose tissue: white adipose tissue as an endocrine and secretory organ. Proc Nutr Soc August 2001;60(3):329–39.

[41] Zuk PA, Zhu M, Mizuno H, Huang J, Futrell JW, Katz AJ, et al. Multilineage cells from human adipose tissue: implications for cell-based therapies. Tissue Eng April 2001;7(2):211–28.

[42] Sanchez-Gurmaches J, Guertin DA. Adipocyte lineages: tracing back the origins of fat. Biochim Biophys Acta March 2014;1842(3):340–51.

[43] Chabowska-Kita A, Kozak LP. The critical period for brown adipocyte development: genetic and environmental influences. Obesity (Silver Spring) February 2016;24(2):283–90.

[44] Harms M, Seale P. Brown and beige fat: development, function and therapeutic potential. Nat Med October 2013;19(10):1252–63.

[45] Kim SH, Plutzky J. Brown fat and browning for the treatment of obesity and related metabolic disorders. Diabetes Metab J February 2016;40(1):12–21.

[46] Giralt M, Villarroya F. White, brown, beige/brite: different adipose cells for different functions? Endocrinology September 2013;154(9):2992–3000.

[47] Bjorndal B, Burri L, Staalesen V, Skorve J, Berge RK. Different adipose depots: their role in the development of metabolic syndrome and mitochondrial response to hypolipidemic agents. J Obes 2011;2011:490650.

[48] Kershaw EE, Flier JS. Adipose tissue as an endocrine organ. J Clin Endocrinol Metab June 2004;89(6):2548–56.

[49] Kilroy GE, Foster SJ, Wu X, Ruiz J, Sherwood S, Heifetz A, et al. Cytokine profile of human adipose-derived stem cells: expression of angiogenic, hematopoietic, and pro-inflammatory factors. J Cell Physiol September 2007;212(3):702–9.

[50] Fruhbeck G, Gomez-Ambrosi J, Muruzabal FJ, Burrell MA. The adipocyte: a model for integration of endocrine and metabolic signaling in energy metabolism regulation. Am J Physiol Endocrinol Metab June 2001;280(6):E827–47.

[51] Fasshauer M, Bluher M. Adipokines in health and disease. Trends Pharmacol Sci July 2015;36(7):461–70.

[52] Guilak F, Lott KE, Awad HA, Cao Q, Hicok KC, Fermor B, et al. Clonal analysis of the differentiation potential of human adipose-derived adult stem cells. J Cell Physiol January 2006;206(1):229–37.

[53] Gimble JM, Katz AJ, Bunnell BA. Adipose-derived stem cells for regenerative medicine. Circ Res May 11, 2007;100(9):1249–60.

[54] Wang YY, Lehuede C, Laurent V, Dirat B, Dauvillier S, Bochet L, et al. Adipose tissue and breast epithelial cells: a dangerous dynamic duo in breast cancer. Cancer Lett November 28, 2012;324(2):142–51.

[55] Strawford A, Antelo F, Christiansen M, Hellerstein MK. Adipose tissue triglyceride turnover, de novo lipogenesis, and cell proliferation in humans measured with $2H_2O$. Am J Physiol Endocrinol Metab April 2004;286(4):E577–88.

[56] Im GI, Shin YW, Lee KB. Do adipose tissue-derived mesenchymal stem cells have the same osteogenic and chondrogenic potential as bone marrow-derived cells? Osteoarthritis Cartilage October 2005;13(10):845–53.

[57] Zhao M, Dumur CI, Holt SE, Beckman MJ, Elmore LW. Multipotent adipose stromal cells and breast cancer development: think globally, act locally. Mol Carcinog November 2010;49(11):923–7.

[58] Mitchell JB, McIntosh K, Zvonic S, Garrett S, Floyd ZE, Kloster A, et al. Immunophenotype of human adipose-derived cells: temporal changes in stromal-associated and stem cell-associated markers. Stem Cells February 2006;24(2):376–85.

[59] Deslex S, Negrel R, Vannier C, Etienne J, Ailhaud G. Differentiation of human adipocyte precursors in a chemically defined serum-free medium. Int J Obes 1987;11(1):19–27.

[60] Hauner H, Entenmann G, Wabitsch M, Gaillard D, Ailhaud G, Negrel R, et al. Promoting effect of glucocorticoids on the differentiation of human adipocyte precursor cells cultured in a chemically defined medium. J Clin Invest November 1989;84(5):1663–70.

[61] Bourin P, Bunnell BA, Casteilla L, Dominici M, Katz AJ, March KL, et al. Stromal cells from the adipose tissue-derived stromal vascular fraction and culture expanded adipose tissue-derived stromal/stem cells: a joint statement of the International Federation for Adipose Therapeutics and Science (IFATS) and the International Society for Cellular Therapy (ISCT). Cytotherapy June 2013;15(6):641–8.

[62] Yu G, Floyd ZE, Wu X, Halvorsen YD, Gimble JM. Isolation of human adipose-derived stem cells from lipoaspirates. Methods Mol Biol 2011;702:17–27.

[63] Cervelli V, Gentile P, Scioli MG, Grimaldi M, Casciani CU, Spagnoli LG, et al. Application of platelet-rich plasma in plastic surgery: clinical and in vitro evaluation. Tissue Eng Part C Methods December 2009;15(4):625–34.

[64] Bunnell BA, Flaat M, Gagliardi C, Patel B, Ripoll C. Adipose-derived stem cells: isolation, expansion and differentiation. Methods June 2008;45(2):115–20.

[65] Dubois SG, Floyd EZ, Zvonic S, Kilroy G, Wu X, Carling S, et al. Isolation of human adipose-derived stem cells from biopsies and liposuction specimens. Methods Mol Biol 2008;449:69–79.

[66] Rajala K, Lindroos B, Hussein SM, Lappalainen RS, Pekkanen-Mattila M, Inzunza J, et al. A defined and xeno-free culture method enabling the establishment of clinical-grade human embryonic, induced pluripotent and adipose stem cells. PLoS One 2010;5(4):e10246.

[67] Jo J, Gavrilova O, Pack S, Jou W, Mullen S, Sumner AE, et al. Hypertrophy and/or hyperplasia: dynamics of adipose tissue growth. PLoS Comput Biol March 2009;5(3):e1000324.

[68] Ogden CL, Carroll MD, Fryar CD, Flegal KM. Prevalence of obesity among adults and youth: United States, 2011–2014. NCHS Data Brief November 2015;219:1–8.

[69] Wang Y, Beydoun MA. The obesity epidemic in the United States – gender, age, socioeconomic, racial/ethnic, and geographic characteristics: a systematic review and meta-regression analysis. Epidemiol Rev 2007;29:6–28.

[70] Calle EE, Rodriguez C, Walker-Thurmond K, Thun MJ. Overweight, obesity, and mortality from cancer in a prospectively studied cohort of U.S. adults. N Engl J Med April 24, 2003;348(17):1625–38.

[71] Bianchini F, Kaaks R, Vainio H. Overweight, obesity, and cancer risk. Lancet Oncol September 2002;3(9):565–74.

[72] Riondino S, Roselli M, Palmirotta R, Della-Morte D, Ferroni P, Guadagni F. Obesity and colorectal cancer: role of adipokines in tumor initiation and progression. World J Gastroenterol May 14, 2014;20(18):5177–90.

[73] Ntikoudi E, Kiagia M, Boura P, Syrigos KN. Hormones of adipose tissue and their biologic role in lung cancer. Cancer Treat Rev February 2014;40(1):22–30.

[74] Duan XF, Tang P, Li Q, Yu ZT. Obesity, adipokines and hepatocellular carcinoma. Int J Cancer October 15, 2013;133(8):1776–83.

[75] Maccio A, Madeddu C. Obesity, inflammation, and postmenopausal breast cancer: therapeutic implications. Sci World J 2011;11:2020–36.

[76] Nieman KM, Romero IL, Van Houten B, Lengyel E. Adipose tissue and adipocytes support tumorigenesis and metastasis. Biochim Biophys Acta October 2013;1831(10):1533–41.

[77] Schwartz B, Yehuda-Shnaidman E. Putative role of adipose tissue in growth and metabolism of colon cancer cells. Front Oncol 2014;4:164.

[78] Booth A, Magnuson A, Fouts J, Foster M. Adipose tissue, obesity and adipokines: role in cancer promotion. Horm Mol Biol Clin Investig January 2015;21(1):57–74.

[79] Aguilar-Gallardo C, Simon C. Cells, stem cells, and cancer stem cells. Semin Reprod Med January 2013;31(1):5–13.

[80] Rowan BG, Lacayo EA, Sheng M, Anbalagan M, Gimble JM, Jones RK, et al. Human adipose tissue-derived stromal/stem cells promote migration and early metastasis of head and neck cancer xenografts. Aesthet Surg J January 2016;36(1):93–104.

[81] Sharpless NE, DePinho RA. How stem cells age and why this makes us grow old. Nat Rev Mol Cell Biol September 2007;8(9):703–13.

[82] Josephson R, Sykes G, Liu Y, Ording C, Xu W, Zeng X, et al. A molecular scheme for improved characterization of human embryonic stem cell lines. BMC Biol 2006;4:28.

[83] Strong AL, Strong TA, Rhodes LV, Semon JA, Zhang X, Shi Z, et al. Obesity associated alterations in the biology of adipose stem cells mediate enhanced tumorigenesis by estrogen dependent pathways. Breast Cancer Res 2013;15(5):R102.

[84] Bielli A, Scioli MG, Gentile P, Agostinelli S, Tarquini C, Cervelli V, et al. Adult adipose-derived stem cells and breast cancer: a controversial relationship. Springerplus 2014;3:345.

[85] Zhang Y, Bellows CF, Kolonin MG. Adipose tissue-derived progenitor cells and cancer. World J Stem Cells October 26, 2010;2(5):103–13.

[86] Heneidi S, Simerman AA, Keller E, Singh P, Li X, Dumesic DA, et al. Awakened by cellular stress: isolation and characterization of a novel population of pluripotent stem cells derived from human adipose tissue. PLoS One 2013;8(6):e64752.

[87] Ghosh S, Dean A, Walter M, Bao Y, Hu Y, Ruan J, et al. Cell density-dependent transcriptional activation of endocrine-related genes in human adipose tissue-derived stem cells. Exp Cell Res August 1, 2010;316(13):2087–98.

[88] Razmkhah M, Jaberipour M, Erfani N, Habibagahi M, Talei AR, Ghaderi A. Adipose derived stem cells (ASCs) isolated from breast cancer tissue express IL-4, IL-10 and TGF-beta1 and upregulate expression of regulatory molecules on T cells: do they protect breast cancer cells from the immune response? Cell Immunol 2011;266(2):116–22.

[89] Razmkhah M, Jaberipour M, Hosseini A, Safaei A, Khalatbari B, Ghaderi A. Expression profile of IL-8 and growth factors in breast cancer cells and adipose-derived stem cells (ASCs) isolated from breast carcinoma. Cell Immunol 2010;265(1):80–5.

[90] Jotzu C, Alt E, Welte G, Li J, Hennessy BT, Devarajan E, et al. Adipose tissue derived stem cells differentiate into carcinoma-associated fibroblast-like cells under the influence of tumor derived factors. Cell Oncol (Dordr) February 2011;34(1):55–67.

[91] Jotzu C, Alt E, Welte G, Li J, Hennessy BT, Devarajan E, et al. Adipose tissue-derived stem cells differentiate into carcinoma-associated fibroblast-like cells under the influence of tumor-derived factors. Anal Cell Pathol (Amst) 2010;33(2):61–79.

[92] Muehlberg FL, Song YH, Krohn A, Pinilla SP, Droll LH, Leng X, et al. Tissue-resident stem cells promote breast cancer growth and metastasis. Carcinogenesis April 2009;30(4):589–97.

[93] Maumus M, Peyrafitte JA, D'Angelo R, Fournier-Wirth C, Bouloumie A, Casteilla L, et al. Native human adipose stromal cells: localization, morphology and phenotype. Int J Obes (Lond) September 2011;35(9):1141–53.

[94] Nieman KM, Kenny HA, Penicka CV, Ladanyi A, Buell-Gutbrod R, Zillhardt MR, et al. Adipocytes promote ovarian cancer metastasis and provide energy for rapid tumor growth. Nat Med 2011;17(11):1498–503.

[95] Korkaya H, Liu S, Wicha MS. Breast cancer stem cells, cytokine networks, and the tumor microenvironment. J Clin Invest October 2011;121(10):3804–9.

[96] Heo SC, Lee KO, Shin SH, Kwon YW, Kim YM, Lee CH, et al. Periostin mediates human adipose tissue-derived mesenchymal stem cell-stimulated tumor growth in a xenograft lung adenocarcinoma model. Biochim Biophys Acta December 2011;1813(12):2061–70.

[97] Bao S, Ouyang G, Bai X, Huang Z, Ma C, Liu M, et al. Periostin potently promotes metastatic growth of colon cancer by augmenting cell survival via the Akt/PKB pathway. Cancer Cell April 2004;5(4):329–39.

[98] Zhou W, Ke SQ, Huang Z, Flavahan W, Fang X, Paul J, et al. Periostin secreted by glioblastoma stem cells recruits M2 tumour-associated macrophages and promotes malignant growth. Nat Cell Biol February 2015;17(2):170–82.

[99] Chandler EM, Seo BR, Califano JP, Andresen Eguiluz RC, Lee JS, Yoon CJ, et al. Implanted adipose progenitor cells as physicochemical regulators of breast cancer. Proc Natl Acad Sci USA June 19, 2012;109(25):9786–91.

[100] Rumman M, Dhawan J, Kassem M. Concise review: quiescence in adult stem cells: biological significance and relevance to tissue regeneration. Stem Cells October 2015;33(10):2903–12.

[101] Kolaparthy LK, Sanivarapu S, Moogla S, Kutcham RS. Adipose tissue – adequate, accessible regenerative material. Int J Stem Cells November 2015;8(2):121–7.

[102] Mizuno H. Adipose-derived stem cells for tissue repair and regeneration: ten years of research and a literature review. J Nippon Med Sch April 2009;76(2):56–66.

[103] Mizuno H, Tobita M, Uysal AC. Concise review: adipose-derived stem cells as a novel tool for future regenerative medicine. Stem Cells May 2012;30(5):804–10.

[104] Mizuno H. Adipose-derived stem and stromal cells for cell-based therapy: current status of preclinical studies and clinical trials. Curr Opin Mol Ther August 2010;12(4):442–9.

[105] Housman TS, Lawrence N, Mellen BG, George MN, Filippo JS, Cerveny KA, et al. The safety of liposuction: results of a national survey. Dermatol Surg November 2002;28(11):971–8.

[106] Strioga M, Viswanathan S, Darinskas A, Slaby O, Michalek J. Same or not the same? Comparison of adipose tissue-derived versus bone marrow-derived mesenchymal stem and stromal cells. Stem Cells Dev September 20, 2012;21(14):2724–52.

[107] Lee RH, Kim B, Choi I, Kim H, Choi HS, Suh K, et al. Characterization and expression analysis of mesenchymal stem cells from human bone marrow and adipose tissue. Cell Physiol Biochem 2004;14(4–6):311–24.

[108] Li CY, Wu XY, Tong JB, Yang XX, Zhao JL, Zheng QF, et al. Comparative analysis of human mesenchymal stem cells from bone marrow and adipose tissue under xeno-free conditions for cell therapy. Stem Cell Res Ther 2015;6:55.

[109] De Girolamo L, Stanco D, Salvatori L, Coroniti G, Arrigoni E, Silecchia G, et al. Stemness and osteogenic and adipogenic potential are differently impaired in subcutaneous and visceral adipose derived stem cells (ASCs) isolated from obese donors. Int J Immunopathol Pharmacol 2013 Jan-Mar;26(1 Suppl.):11–21.

[110] Naderi N, Combellack EJ, Griffin M, Sedaghati T, Javed M, Findlay MW, et al. The regenerative role of adipose-derived stem cells (ADSC) in plastic and reconstructive surgery. Int Wound J February 1, 2016.

[111] Traktuev DO, Merfeld-Clauss S, Li J, Kolonin M, Arap W, Pasqualini R, et al. A population of multipotent CD34-positive adipose stromal cells share pericyte and mesenchymal surface markers, reside in a periendothelial location, and stabilize endothelial networks. Circ Res January 4, 2008;102(1):77–85.

[112] Awad HA, Wickham MQ, Leddy HA, Gimble JM, Guilak F. Chondrogenic differentiation of adipose-derived adult stem cells in agarose, alginate, and gelatin scaffolds. Biomaterials July 2004;25(16):3211–22.

[113] Tobita M, Orbay H, Mizuno H. Adipose-derived stem cells: current findings and future perspectives. Discov Med February 2011;11(57):160–70.

[114] Kim I, Bang SI, Lee SK, Park SY, Kim M, Ha H. Clinical implication of allogenic implantation of adipogenic differentiated adipose-derived stem cells. Stem Cells Transl Med November 2014;3(11):1312–21.

[115] Donnenberg VS, Zimmerlin L, Rubin JP, Donnenberg AD. Regenerative therapy after cancer: what are the risks? Tissue Eng Part B Rev December 2010;16(6):567–75.

[116] Zimmerlin L, Park TS, Zambidis ET, Donnenberg VS, Donnenberg AD. Mesenchymal stem cell secretome and regenerative therapy after cancer. Biochimie December 2013;95(12):2235–45.

[117] Rowan BG, Gimble JM, Sheng M, Anbalagan M, Jones RK, Frazier TP, et al. Human adipose tissue-derived stromal/stem cells promote migration and early metastasis of triple negative breast cancer xenografts. PLoS One 2014;9(2):e89595.

[118] Sun B, Roh KH, Park JR, Lee SR, Park SB, Jung JW, et al. Therapeutic potential of mesenchymal stromal cells in a mouse breast cancer metastasis model. Cytotherapy 2009;11(3):289–98. 1 p. following 98.

[119] Zimmerlin L, Donnenberg AD, Rubin JP, Basse P, Landreneau RJ, Donnenberg VS. Regenerative therapy and cancer: in vitro and in vivo studies of the interaction between adipose-derived stem cells and breast cancer cells from clinical isolates. Tissue Eng Part A January 2011;17(1–2):93–106.

[120] Hass R, Otte A. Mesenchymal stem cells as all-round supporters in a normal and neoplastic microenvironment. Cell Commun Signal 2012;10(1):26.

[121] Donahue HJ, Saunders MM, Li Z, Mastro AM, Gay CV, Welch DR. A potential role for gap junctions in breast cancer metastasis to bone. J Musculoskelet Neuronal Interact June 2003;3(2):156–61.

[122] Mandel K, Yang Y, Schambach A, Glage S, Otte A, Hass R. Mesenchymal stem cells directly interact with breast cancer cells and promote tumor cell growth in vitro and in vivo. Stem Cells Dev December 1, 2013;22(23):3114–27.

[123] Studeny M, Marini FC, Dembinski JL, Zompetta C, Cabreira-Hansen M, Bekele BN, et al. Mesenchymal stem cells: potential precursors for tumor stroma and targeted-delivery vehicles for anticancer agents. J Natl Cancer Inst November 3, 2004;96(21):1593–603.

[124] Ilmer M, Vykoukal J, Recio Boiles A, Coleman M, Alt E. Two sides of the same coin: stem cells in cancer and regenerative medicine. FASEB J July 2014;28(7):2748–61.

[125] Vilalta M, Degano IR, Bago J, Aguilar E, Gambhir SS, Rubio N, et al. Human adipose tissue-derived mesenchymal stromal cells as vehicles for tumor bystander effect: a model based on bioluminescence imaging. Gene Ther April 2009;16(4):547–57.

[126] Liu G, Neumeister M, Reichensperger J, Yang RD. Therapeutic potential of human adipose stem cells in a cancer stem cell-like gastric cancer cell model. Int J Oncol October 2013;43(4):1301–9.

[127] Lamfers M, Idema S, van Milligen F, Schouten T, van der Valk P, Vandertop P, et al. Homing properties of adipose-derived stem cells to intracerebral glioma and the effects of adenovirus infection. Cancer Lett February 8, 2009;274(1):78–87.

[128] Casteilla L, Planat-Benard V, Laharrague P, Cousin B. Adipose-derived stromal cells: their identity and uses in clinical trials, an update. World J Stem Cells April 26, 2011;3(4):25–33.

[129] Illouz YG. Breast cancer treatment by adipose-derived stem cells: an experimental study. J Stem Cells 2014;9(4):211–7.

[130] Abd Elmageed ZY, Yang Y, Thomas R, Ranjan M, Mondal D, Moroz K, et al. Neoplastic reprogramming of patient-derived adipose stem cells by prostate cancer cell-associated exosomes. Stem Cells April 2014;32(4):983–97.

4

SECTION

Clinical Applications

CHAPTER 17

From Research to the Clinic: Targeting Stem Cell Pathways in Cancer

A. Gurney, T. Hoey
OncoMed Pharmaceuticals, Inc., Redwood City, CA, United States

CONTENTS

INTRODUCTION

The hypothesis that solid tumors comprise heterogeneous cell types, including a subpopulation of cells termed cancer stem cells (CSCs) that underlie the growth, proliferation, and metastasis of the tumor, has been supported by numerous lines of evidence [1–3]. In the CSC model for tumor growth, tumors can be viewed as having a hierarchical organization of cells with a tumorigenic cell population that can self-renew and proliferate indefinitely at the top of this hierarchy. CSCs also possess the capacity to differentiate into other tumor cells, typically comprising the majority of the tumor, which may be either proliferating or postmitotic, but have reduced capacity to self-renew and are therefore nontumorigenic or less tumorigenic [4]. Extensive evidence supports the concept that CSCs are preferentially resistant to many conventional therapies and thus mediate tumor recurrence [5,6]. Experimental data demonstrating the presence of CSCs has been derived for many distinct tumor types, including breast, colon, lung, pancreas, prostate, ovarian, melanoma, and glioblastoma [7]. Consequently, therapeutic strategies aimed at reducing CSC frequency within solid tumors offer the promise of limiting disease progression and, in some instances, possibly providing durable clinical benefit. The Notch and Wnt pathways are two key signaling cascades that govern cell fate during development and have been shown to be involved in CSCs in a variety of disease settings [8]. This review will focus on drug discovery and development efforts targeting these pathways with an emphasis on agents that have reached the stage of clinical testing.

WNT PATHWAY ANTAGONISTS

The Wnt pathway is known to regulate determination of stem cell fate in a variety of settings both in normal development and various diseases including

Cancer Stem Cells. http://dx.doi.org/10.1016/B978-0-12-803892-5.00017-6

cancer [9]. The pathway was initially discovered through the *wingless* mutation in Drosophila that regulates embryonic pattern formation; elucidation of Wnt signaling mechanisms has been facilitated through genetic studies [10]. The Wnt1 gene was identified in mammals in a screen of mouse mammary tumor virus (MMTV) integration sites that resulted in mammary tumors in mice [11]. In subsequent years, it has become widely appreciated that the Wnt pathway is frequently altered in human cancer through a variety of genetic and epigenetic mechanisms, and targets in the Wnt pathway are being extensively pursued for the development of new anticancer therapies [12].

Signaling by the Wnt pathway is exceedingly complex in mammalian cells [13]. There are 19 members of the Wnt gene family in the human genome. These Wnt proteins are secreted ligands that bind to one of ten frizzled (FZD) receptors. FZD proteins are seven transmembrane domain receptors and are characterized by having a relatively large extracellular cysteine-rich domain that binds the Wnt ligands. In addition to FZD, there are two coreceptors, LRP5 and LRP6 that interact with FZD on the membrane. In the absence of Wnt, beta-catenin is phosphorylated and highly unstable. Key negative regulators of Wnt-FZD signaling include the axin and adenomatous polyposis coli (APC) proteins which are part of the "destruction complex" together with glycogen synthase kinase 3 (GSK3) that is known to participate in degradation of beta-catenin. Loss-of-function mutations in APC, activating Wnt signaling, are the most common mutations in colorectal cancer, strongly implicating the Wnt pathway in this disease. Binding of Wnt to FZD initiates signal transduction events through the Disheveled protein that inhibit the destruction complex and thereby stabilize beta-catenin. Subsequently, beta-catenin protein levels accumulate and the protein functions as a transcriptional coactivator by interacting with TCF/LEF DNA binding proteins to regulate downstream target genes. Wnt-FZD signaling through beta-catenin is known as the "canonical" pathway and this is thought to be the key pathway deregulated in cancer [14]. Wnt signaling also plays other diverse roles in the regulation of cell fate that are likely to contribute to its oncogenic activity including mediating "noncanonical" signaling, and the regulation of cell adhesion, cell polarity, and mitosis.

Several specific inhibitors of the Wnt pathway have been developed that disrupt ligand-dependent signaling. These included a multi-FZD targeting antibody (OMP-18R5 or vantictumab; [15]), a decoy receptor (OMP-54F28 or ipafricept; [16]), and a small molecule inhibitor of Wnt signaling (LGK974 or Wnt-974; [17]). A current review of various Wnt signaling inhibitors that act at the level of ligand–receptor interaction has recently been published [18]. Wnt signaling plays a key role in bone homeostasis [19]. In the clinic, impacts on bone biology have emerged as an important dose-limiting toxicity for the Wnt antagonists. Fortunately, it has been possible to monitor patients using serum-based biomarkers, such beta-CTX (beta-isomerized C-terminal telopeptide of collagen type I), for early signs of bone turnover and intervene with bone protecting agents, such as zoledronic acid, to mitigate the effects of Wnt blockade on bone metabolism [20].

Vantictumab

A key hurdle to developing a biologic agent capable of antagonizing the Wnt pathway is the complexity and redundancy of Wnt signaling through multiple Wnt ligands and FZD receptors. For example, antibodies selective for individual FZD receptors with antagonist activity have not been reported. Therefore, an effort was made to find a cross-reactive antibody capable of targeting multiple FZDs that would be broadly active in blocking canonical Wnt signal transduction. This approach yielded OMP-18R5 (vantictumab) which was originally identified by binding to FZD7 and subsequently found to cross-react with FZD1, 2, 5, and 8 [15]. Epitope mapping indicated that vantictumab binds to a conserved epitope on the FZD extracellular domain which was later shown through structural studies to be part of the Wnt-binding region on FZD and the region that interacts with the lipid moiety on Wnt [21]. Vantictumab has been shown to be broadly active in reducing growth and CSC frequency in patient-derived xenografts from many different tumor types including breast, pancreatic, and lung cancers [15]. Vantictumab has completed Phase 1a clinical testing and is currently in Phase 1b trials in combination with standard of care chemotherapeutic agents in breast, lung, and pancreatic cancers (www.clinicaltrials.gov).

Ipafricept

An alternative strategy to targeting FZDs for biologic drug development is to target the Wnt ligands. A soluble, decoy receptor comprised of the extracellular domain of FZD8 fused to an IgG Fc domain has been developed that is capable of binding and sequestering Wnts and thereby blocking signaling [16]. This compound has completed Phase 1a clinical testing and is currently in Phase 1b trials (www.clinicaltrials.gov). A hallmark of Wnt pathway blockade in tumors is that this sensitizes tumor cells to the cytotoxic effect of chemotherapeutic agents. Wnt inhibition has been found to be particularly effective in combination with taxane treatment [22].

Porcupine Inhibitor (LGK974)

All Wnt proteins are posttranslationally modified by addition of palmitoleic acid moiety. This lipid modification is required for both the secretion of Wnts from the ligand-expressing cell and the binding of Wnt to FZD on the signal-receiving cell. The enzyme involved in lipid modification of Wnts is known as Porcupine (Porcn), based on its initial genetic characterization in Drosophila. A highly potent and selective Porcn inhibitor has been developed by Novartis [17] and is currently in Phase 1b clinical testing in combination with an EGFR inhibitor and a B-Raf inhibitor in a subset of colorectal cancer with Wnt pathway mutations (RNF3 mutations or R-spondins (Rspo) fusions; www.clinicaltrials.gov). Porcn inhibitors have also been shown to have selective cytotoxicity in head and neck squamous cancer lines particularly those with loss-of-function mutation in Notch [17,23].

Tankyrase Inhibitors

Several groups have identified small molecule tankyrase inhibitors as selective blockers of Wnt signaling based on high-throughput cell-based assays including researchers at Novartis and Genentech [24,25]. Tankyrase is an enzyme that posttranslationally modifies various proteins including axin, a component of the destruction complex that regulates beta-catenin stability downstream of FZD. Notably tankyrase inhibitors have been shown to be able to block the growth of colorectal tumors with mutations in the APC tumor suppressor gene, a property which has not been reported for agents that work at the level of Wnt–FZD interaction. This class of compounds is known to have on-target toxicity in the GI tract through inhibiting Wnt-dependent maturation of the intestinal epithelium [25]. Despite significant efforts to optimize tankyrase inhibitors, none of these compounds have been reported to have reached the stage of clinical development. Tankyrase has also been shown to play a role in other signaling pathways distinct from Wnt, for example, FGFR2 signaling [26]; therefore tankyrase inhibition would be expected to have more pleiotropic on-target effects relative to more selective Wnt inhibitors.

PRI-724

Another mechanism that has been reported for targeting Wnt signaling is to inhibit the transcriptional activity of beta-catenin in the nucleus. ICG-001 is a compound that was identified by cell-based screening and works by blocking the interaction of beta-catenin and the transcriptional coactivator CREB-binding protein (cAMP response element binding protein) [27]. PRI-724 is a derivative of ICG-001 that is currently in Phase 1 clinical testing. Evidence for the lack of selectivity of ICG-001/PRI-724 for Wnt pathway signaling is highlighted by a recent publication showing that cytotoxicity of this compound can be Wnt-independent [28].

Anti-Rspo3

The Rspo is a family of four secreted proteins that serve to amplify Wnt signaling. The mechanisms for Rspo signaling and how they interact with the Wnt pathway are now well characterized [29]. Rspo proteins bind with high affinity to LGR4, 5, or 6 along with a coreceptor, RNF43 (or the related protein ZNFR3). RNF43 and ZNFR3 are membrane-bound ubiquitin ligases and, in the absence of Rspo proteins, they act to ubiquitinate and thereby destabilize FZD. Rspo binding sequesters the RNF43 or ZNFR3 away from FZD, thus enhancing Wnt signal strength. Several lines of evidence indicate that Rspo proteins can be upregulated in cancer. Notably, expression of Rspo2 or Rspo3 can be elevated by genomic translocation in certain colorectal tumors [30] leading to activation of Wnt pathway signaling. OncoMed Pharmaceuticals has developed an antagonist of Rspo3, OMP-131R10, that has been shown to inhibit tumor growth and CSC frequency of Rspo3-overexpressing tumors of various tumor types including colon, lung, and ovarian cancers [31].

NOTCH PATHWAY ANTAGONISTS

The Notch pathway is an ancient and evolutionarily conserved signaling system that plays diverse roles in normal tissue development and homeostasis [32,33]. It serves to transduce signals between adjacent cells by means of transmembrane ligands and receptors. In humans the Notch pathway comprises four receptors (Notch1–4) and five ligands, JAG1, JAG2, DLL1, DLL3, and DLL4. In addition to critical roles in the specification of cell fate during development, the Notch pathway mediates intercellular communication and regulates cell fate decisions in adult organs and cell lineages. Important examples include roles for Notch in intestinal homeostasis [34–36], hematopoiesis [37], and angiogenesis [38]. Not surprisingly, the Notch pathway has been implicated in cancer. Cancer can be considered as an aberrant and mutant recapitulation of normal tissue development and homeostasis. Cancer cells inappropriately activate self-renewal pathways and this enables their ability to grow indefinitely. Numerous lines of evidence link the Notch pathway to cancer, for example, activating mutations in Notch1 are detected in T-cell leukemias and components of the pathway have been linked to the progression and metastasis of solid tumors. These observations suggest that therapeutic strategies directed at inhibition of the Notch pathway may offer promise in the treatment of solid tumors [39,40].

We undertook a concerted effort to generate blocking antibodies to the ligands and receptors of this pathway. These antibodies were screened for antitumor activity using xenograft models with minimally passaged human tumors. As a result of these efforts, several antibodies have been moved forward into clinical development, including an anti-DLL4 antibody (OMP-21M18, now termed demcizumab, which was the first therapeutic entity that selectively targeted the Notch pathway to enter human clinical trials), an anti-Notch2 and 3 antibody (OMP-59R5, now termed tarextumab) and an anti-Notch1 antibody (OMP-52M51, now termed brontictuzumab). Recently, a second generation bispecific agent targeting both DLL4 and VEGF (OMP-305B83) has also entered clinical testing. Each of these agents is currently in active clinical development. In addition to the agents advanced by OncoMed, two other anti-DLL4 antibodies have now entered clinical trials, MEDI0639, from Medimmune, and Enoticumab from Regeneron. Gamma-secretase inhibitors (GSIs) that inhibit the ligand-dependent cleavage of Notch receptors have also been entered into development as anticancer therapies [41]. GSIs have been found to result in severe gastrointestinal toxicity, which may limit their therapeutic utility, due to the combined inhibition of both Notch1 and Notch2 within the stem-progenitor compartment of the intestinal crypt [42,43]. Gamma-secretase cleaves many membrane proteins, in addition to processing Notch proteins, and is therefore involved in signaling pathways apart from Notch, including such diverse proteins as CD44, HLA, syndicans, and b-amyloid precursor protein. These pleiotropic effects are also likely to contribute to the toxicity of GSIs [44,45]. The GSI agents will not be further discussed in this review. Among the various monoclonal antibody agents targeting specific Notch pathway components, demcizumab (anti-DLL4) and tarextumab (anti-Notch2 and 3), are the most advanced in clinical testing and are currently in multiple randomized Phase 2 trials.

Demcizumab

DLL4 is in many respects a remarkable therapeutic target. Our understanding of the mechanism of action of anti-DLL4 now appreciates the existence of three distinct therapeutic impacts: inhibition of tumorigenicity/CSC frequency; inhibition of functional tumor angiogenesis; and, most recently, a growing recognition of the impact of anti-DLL4 on antitumor immune response. These three mechanisms are highlighted in Fig. 17.1 and will be discussed later.

Anti-DLL4 was initially flagged as an active anticancer agent at OncoMed through tumor xenograft studies utilizing minimally passaged human tumors. The parent anti-DLL4 antibody of demcizumab, 21M18, was a murine hybridoma–derived mAb that blocked DLL4 interaction with Notch receptors. It is therefore a functional antagonist antibody. The antibody is specific for human DLL4 and does not bind to murine DLL4 [6]. This is notable because it means that it does not disrupt murine DLL4 in the host and its effects in this xenograft system are mediated through DLL4 expressed on the human tumor cells. Among the findings of these studies were the observations that anti-DLL4, both as a single agent and in combination with chemotherapy reduced the tumor-initiating frequency of treated tumors, as assessed by serial transplant of cells from previously treated tumors to new mice in the absence of further therapy [6]. These studies thus highlighted the first mechanism of anti-DLL4 activity: reduction of CSC activity.

DLL4 expression has been shown to be expressed in many contexts. In developing mouse embryos expression was noted in the vascular system, the nervous system, the gastrointestinal system, the glomeruli of the kidney, and the thymus [46]. Notably, deletion of a single allele of DLL4 results in embryonic lethality

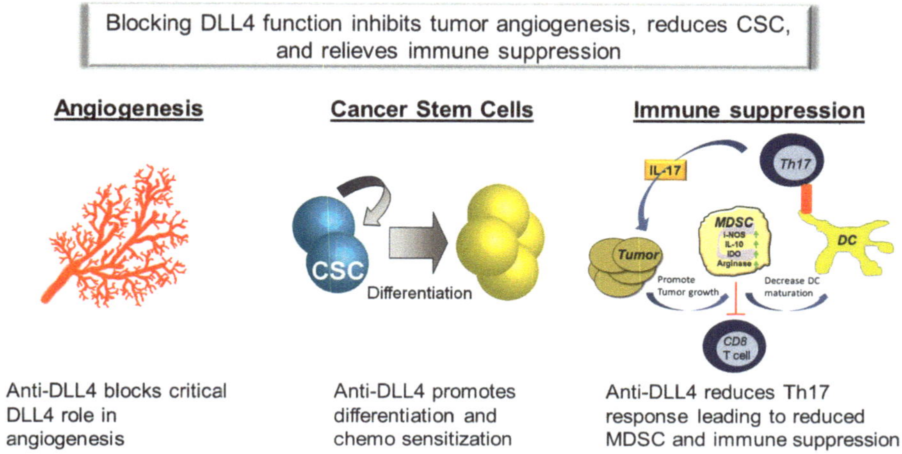

FIGURE 17.1

Schematic illustration of three distinct mechanisms by which inhibition of the Notch ligand DLL4 impacts tumor growth. The mechanisms highlight the diverse biological roles of key stem cell pathways such as Notch, and the potential for therapeutic agents targeting these pathways to impact multiple aspects of tumor biology. *CSC*, cancer stem cell; *MDSC*, myeloid-derived suppressor cells.

due to defects in development of the vasculature [47,48]. DLL4 has been reported to be overexpressed in tumor vasculature and tumor cells, and has the potential to activate Notch signaling at these sites [49–55]. Further studies have shown that antibody-mediated inhibition of DLL4 causes a hyperproliferation of endothelial cells coupled to a dysregulated, nonproductive angiogenesis; thus highlighting a potential role for anti-DLL4 as a novel antiangiogenic agent [56–58]. Consistent with these results when we performed xenograft studies with an anti-DLL4 antibody that was specific for murine DLL4, but did not bind human DLL4, we also observed broad antitumor activity. In these models, performed with immunocompromised mice and human tumors, the effect of anti-DLL4 is not likely immune-mediated nor is it impacting DLL4 expressed by the human tumor [6]. Histology of the treated tumors revealed extensive nonperfused vasculature. These studies thus highlight the second mechanism of anti-DLL4 activity: promotion of dysfunctional angiogenesis. These first two mechanisms, anti-CSC activity and promotion of dysfunctional angiogenesis are complementary. Xenograft studies that examined antihuman, antimurine, or the combination of anti-DLL4 mAbs found additive activity. Interestingly, when tumor-initiating frequency was examined, decrease in tumor-initiating frequency was mediated solely by the antihuman DLL4 mAb, despite the significant impact the antimurine DLL4 mAb had on tumor volume [51].

Based on the compelling efficacy demonstrated in preclinical xenograft models, OMP-21M18/demcizumab, a humanized IgG2 mAb derived from the original 21M18 murine mAb, was advanced into clinical trials. An initial Phase 1a study examined demcizumab as a single agent in a dose escalation study in patients with advanced malignancies. In this 55-patient study a formal maximum tolerated dose was not reached [59]. Hypertension was the most commonly observed adverse event that was clearly related to treatment. Prolonged administration was associated with an increased risk of congestive heart failure, particularly at the highest dose. Encouragingly, there was evidence of clinical benefit, and 16 of 25 patients at the highest dose tested had evidence of stable disease or tumor shrinkage. Additional phase 1b studies were conducted to explore demcizumab in select indications in combination with standard of care chemotherapy. These studies enabled optimization of dose and the development of a dosing schedule to minimize cardiovascular toxicity. A key aspect of the dosing strategy has been to provide the agent for a fixed duration of treatment that terminates before the occurrence of cardiovascular toxicities associated with chronic administration. These studies have in turn provided the foundation for two ongoing Phase 2 trials. The YOSEMITE trial is evaluating demcizumab and gemcitabine plus Abraxane in patients with first-line metastatic pancreatic cancers. The DENALI trial is evaluating the efficacy of demcizumab with carboplatin and pemetrexed in first-line nonsquamous nonsmall cell lung cancer, NSCLC. One of several striking observations reported during the Phase 1b experience was the observation of a notable "tail" on the OS Kaplan–Meier curve in the nonsquamous NSCLC trial [60]. This suggests that a subset of patients may experience prolonged overall survival. This marked result is reminiscent of the impact of the "checkpoint inhibitor" agents that target CLTA4 or PD-1. An intriguing question posed by these results is whether anti-DLL4 may also promote an immune response against the tumor. DLL4 is expressed by myeloid cells including macrophage and dendritic cells and plays a role in modulating adaptive immune responses [61–64]. Conditional gene disruption studies have also demonstrated DLL4 plays an indispensable role in thymic T-cell development [65,66].

To examine the potential role of anti-DLL4 in modulating antitumor immune response, tumor-graft studies were performed with murine tumors in syngeneic immune competent mice. Single agent anti-DLL4 inhibited tumor growth, a result consistent with its previously defined mechanisms of reduction in CSC frequency and inhibition of functional angiogenesis. Interestingly however, ELISpot analysis revealed decreased IL-17 production suggesting anti-DLL4 had also altered the character of the antitumor immune response [67]. This result is in agreement with reports that DLL4 can promote IL-17 production [68–70]. Moreover when anti-DLL4 was tested in combination with anti-PD-1 increased antitumor efficacy was noted compared to either agent alone and a greater percentage of mice were cleared of tumor. Again IL-17 was reduced in the anti-DLL4+ anti-PD-1 combo treatment, an effect not seen with anti-PD-1 alone. Following anti-DLL4 treatment and also in combination with anti-PD-1 there was an increase in the percentage of splenic central and effector memory CD8 T-cells. In follow-up experiments, it was noted that mice that had previously been cleared of tumor by combined anti-DLL4+ anti-PD-1 treatment were better protected from subsequent rechallenge with fresh tumor cells in the absence of treatment than were mice that had been cleared of tumor by treatment with anti-PD-1 alone. These results suggest that the presence of anti-DLL4 may help to promote a more robust antitumor immune response and more substantial immune memory. These studies thus highlight the third mechanism of anti-DLL4 activity: modulation of antitumor immune response. Further investigation of the consequence of anti-DLL4 treatment has uncovered a potential mechanism for the enhanced antitumor immune response. Treatment with anti-DLL4 greatly reduces the presence of myeloid-derived suppressor cells (MDSC) [67]. MDSC are a potent immunosuppressive cell population. It is tempting to speculate that by reducing Th17 response, which serves to mobilize the recruitment and generation of myeloid cells during normal inflammation, anti-DLL4 is able to reduce the generation and accumulation of MDSC within tumors and thereby promote a more effective immune response.

Tarextumab

Dysregulated Notch2 and/or Notch3 activity has been associated with several human tumor types including lung cancer [71–76], pancreatic cancer [77–79], breast cancer [80–82], ovarian cancer [83,84], colorectal cancer [82,85,86], and hepatocellular carcinoma [87–90]. Notch2 has been found to be amplified in brain tumors [91]; whereas Notch3 has been demonstrated to exhibit frequent gene amplification in ovarian cancer [92,93] and to play a functional role in tumor growth [94]. OMP-59R5, tarextumab, is a fully human mAb that inhibits ligand-dependent signaling through both Notch2 and Notch3 [95]. Tarextumab was tested in a range of xenograft models and found to be active in breast, ovarian, lung, and pancreatic tumors. Administration of tarextumab delayed tumor recurrence after chemotherapeutic treatment and reduced tumor-initiating cell frequency, particularly in combination with chemotherapy [95]. Mechanism of action studies indicated that tarextumab blocked Notch signaling in both tumor and stromal cells and induced differentiation of tumor cells. In the stromal cells, tarextumab inhibited gene expression required for the function of pericytes, vascular support cells essential for tumor angiogenesis [96].

Tarextumab is currently in two Phase 2 trials. The ALPINE trial is assessing tarextumab with Abraxane plus gemcitabine in first-line advanced pancreatic cancer patients. The PINNACLE trial is testing tarextumab in combination with etoposide and cisplatin or etoposide and carboplatin in first-line extensive stage small cell lung cancer (SCLC) patients. In the Phase 1b portions of the ALPINE and

PINNACLE clinical trials, tarextumab was well tolerated in combination with standard of care chemotherapies [97,98]. In SCLC, tarextumab demonstrated dose-dependent and biomarker-driven activity. In pancreatic cancer, antitumor responses and prolonged survival rates versus standard of care were observed, particularly in patients whose tumors were identified as high in Notch3 gene expression. Randomized Phase 2 clinical trials of tarextumab are ongoing (www.clinicaltrials.gov).

Brontictuzumab

Brontictuzumab (anti-Notch1, OMP-52M51) is an antibody that binds the Notch1 receptor and blocks ligand-dependent signaling. In preclinical models, blocking Notch1 has been shown to inhibit tumor growth, decrease CSC frequency, promote tumor cell differentiation, and disrupt tumor angiogenesis [99,100]. Brontictuzumab is being studied as a single agent in two ongoing Phase 1a clinical trials among patients with advanced solid tumors or hematologic malignancies. The Phase 1a solid tumor trial is currently enrolling patients whose tumors demonstrate overexpression of the activated form of Notch1 (www.clinicaltrials.gov).

Hedgehog Pathway

In addition to the Notch and Wnt pathways, Hedgehog (Hh) signaling is another important pathway in regulating cell fate in embryonic development that can be deregulated in cancer [101]. Hh is a secreted factor that signals through its receptor, a seven transmembrane G protein-coupled receptor known as Smoothened (Smo) which is part of a complex with another transmembrane protein known as Patched (Ptch). Mutations in Smo or Ptch have been implicated in two types of cancer: basal cell carcinoma (BCC) of the skin and medulloblastoma. Several Smo small molecule antagonists have been developed that inhibit signaling. The most clinically advanced of these is GDC-0449 (vismodegib) which has been approved for the treatment of BCC [101]. Vismodegib has also been shown to have activity in a Phase 2 trial in medulloblastoma, particularly in the context of patients with oncogenic mutations in the Hh pathway [102].

CONCLUSIONS

There has been recognition, for the past several decades, of the involvement of the Wnt and Notch stem cell pathways in cancer. Indeed both mammalian Wnt and mammalian Notch proteins were first identified in the context of cancer. The growing recognition that there is functional significance to tumor heterogeneity, as well as accumulating evidence supportive of the stem cell hypothesis, has provided a framework with which to appreciate the relationship of these stem cell pathways to cancer. Despite extensive genetic and functional analysis implicating these pathways, effective therapeutic targeting of both Notch and

Table 17.1 Summary of Stem Cell Pathway Agents in Clinical Development

Agent	Target	Pathway	Preclinical Efficacy	Trial Status
Demcizumab	DLL4	Notch	Diverse solid tumors	Phase 2 in NSCLC, pancreatic
MEDI0639	DLL4	Notch		Phase 1
Enotictumab	DLL4	Notch	Ovarian, RCC	Phase 1
OMP-305B83	DLL4 + VEGF	Notch, VEGF	Diverse solid tumors	Phase 1
Tarextumab	Notch 2,3	Notch	SCLC, breast, lung, ovarian, pancreatic	Phase 2 in SCLC, pancreatic
Brontictuzumab	Notch1	Notch	Diverse solid tumors, T-All	Phase 1b colorectal carcinoma
Vantictumab	Frizzleds 1,2,5,7,8	Wnt	Diverse solid tumors	Phase 1b in lung, breast, pancreatic
Ipafracept	Wnts	Wnt	Diverse solid tumors	Phase 1b, hepatocellular, ovarian, pancreatic
LGK974	PORCN	Wnt	Breast, head and neck, glioblastoma	Phase 1
Vismodegib	Smoothened	Hedgehog	Basal cell carcinoma	Approved for BCC
OMP-131R10	RSPO3	RSPO/LGR	Lung, colon, ovarian	Phase 1

Agents described in this chapter are highlighted. The molecular targets, tumor-type efficacy observed in preclinical studies, and current stage of clinical development are indicated.
BCC, *basal cell carcinoma;* LGR, *leucine-rich repeat-containing G-protein-coupled receptor;* NSCLC, *nonsmall cell lung cancer;* RCC, *renal cell carcinoma;* RSPO, *R-spondins;* SCLC, *small cell lung cancer.*

Wnt has been difficult to achieve. This is in large measure due to the complexities of these pathways which possess multiple receptors and ligands and the potential for redundancy. In recent years, however, several strategies have been developed to generate functional antagonists of these pathways. Several agents are now progressing in clinical trials (see Table 17.1) and the coming several years will likely provide significant answers to the therapeutic benefit that may be obtained by targeting of Notch and Wnt. Key questions related to safety and the ability to manage undesired toxicities that are likely inherent when modulating these powerful signaling axes may best be addressed by clinical experience. Future directions for analysis of the role of these pathways in cancer will also likely include more extensive focus on the ways in which these pathways contribute to the establishment of the tumor microenvironment. This microenvironment includes not only mesenchymal stroma and endothelial vascularization, but a complex interplay of immune cells that participate in the creation of an immunosuppressive environment. It is likely that dysregulation of the stem cell pathways contributes to the establishment of a fertile environment for tumor growth. Finally, these key stem cell pathways play central roles in the regulation of differentiation and self-renewal and thus these ongoing clinical trials may also, hopefully, help to provide clinical evaluation of the CSC hypothesis.

List of Abbreviations

APC Adenomatous polyposis coli
BCC Basal cell carcinoma
beta-CTX Beta-isomerized C-terminal telopeptide of collagen type I
CSC Cancer stem cell
FZD Frizzled receptor
GSK3 Glycogen synthase kinase 3
MMTV Mouse mammary tumor virus
Rspo R-spondin

References

[1] Reya T, Morrison SJ, Clarke MF, Weissman IL. Stem cells, cancer, and cancer stem cells. Nature 2001;414(6859):105–11.

[2] Clarke MF, Dick JE, Dirks PB, Eaves CJ, Kamieson CH, Jones DL, et al. Cancer stem cells–perspectives on current status and future directions: AACR Workshop on cancer stem cells. Cancer Res 2006;66(19):9339–44.

[3] Clevers H. The cancer stem cell: premises, promises and challenges. Nat Med 2011;17(3): 313–9.

[4] Clarke MF, Fuller M. Stem cells and cancer: two faces of eve. Cell 2006;124(6):1111–5.

[5] Dean M, Fojo T, Bates S. Tumour stem cells and drug resistance. Nat Rev Cancer 2005;5(4):275–84.

[6] Hoey T, Yen WC, Axelrod F, Basi J, Donigian L, Dylla S, et al. DLL4 blockade inhibits tumor growth and reduces tumor-initiating cell frequency. Cell Stem Cell 2009;5(2):168–77.

[7] Frank NY, Schatton T, Frank MH. The therapeutic promise of the cancer stem cell concept. J Clin Invest 2010;120(1):41–50.

[8] Reya T, Clevers H. Wnt signalling in stem cells and cancer. Nature 2005;434(7035):843–50.

[9] Clevers H, Loh KM, Nusse R. Stem cell signaling. An integral program for tissue renewal and regeneration: Wnt signaling and stem cell control. Science 2014;346(6205):1248012.

[10] Klingensmith J, Nusse R. Signaling by wingless in *Drosophila*. Dev Biol 1994;166(2): 396–414.

[11] Nusse R, Varmus HE. Many tumors induced by the mouse mammary tumor virus contain a provirus integrated in the same region of the host genome. Cell 1982;31(1):99–109.

[12] Takebe N, Harris PJ, Warren RQ, Ivy SP. Targeting cancer stem cells by inhibiting Wnt, Notch, and Hedgehog pathways. Nat Rev Clin Oncol 2011;8(2):97–106.

[13] Saito-Diaz K, Chen TW, Wang X, Thorne CA, Wallace HA, Page-McCaw A, et al. The way Wnt works: components and mechanism. Growth Factors 2013;31(1):1–31.

[14] Takebe N, Miele L, Harris PJ, Jeong W, Bando H, Kahn M, et al. Targeting Notch, Hedgehog, and Wnt pathways in cancer stem cells: clinical update. Nat Rev Clin Oncol 2015;12(8): 445–64.

[15] Gurney A, Axelrod F, Bond CJ, Cain J, Chartier C, Donigan L, et al. Wnt pathway inhibition via the targeting of Frizzled receptors results in decreased growth and tumorigenicity of human tumors. Proc Natl Acad Sci USA 2012;109(29):11717–22.

[16] Yeung P, Beviglia L, Cancilla B, Dee-Hoskins C, Evans JW, Fischer MM, et al. Abstract 1907: Wnt pathway antagonist OMP-54F28 (FZD8-Fc) inhibits tumor growth and reduces tumor-initiating cell frequency in patient-derived hepatocellular carcinoma and ovarian cancer xenograft models. In: AACR Annual Meeting. 2014.

[17] Liu J, Pan S, Hsieh MH, Ng N, Sun F, Wang T, et al. Targeting Wnt-driven cancer through the inhibition of Porcupine by LGK974. Proc Natl Acad Sci USA 2013;110(50):20224–9.

[18] Madan B, Virshup DM. Targeting Wnts at the source–new mechanisms, new biomarkers, new drugs. Mol Cancer Ther 2015;14(5):1087–94.

[19] Baron R, Kneissel M. WNT signaling in bone homeostasis and disease: from human mutations to treatments. Nat Med 2013;19(2):179–92.

[20] Smith DC, Rosen LS, Chugh R, Goldman JW, Xu L, Kapoun A, et al. Abstract 2540: first-in-human evaluation of the human monoclonal antibody vantictumab (OMP-18R5; anti-Frizzled) targeting the WNT pathway in a Phase I study for patients with advanced solid tumors. In: ASCO Annual Meeting. 2013.

[21] Janda CY, Waghray D, Levin AM, Thomas C, Garcia KC. Structural basis of Wnt recognition by frizzled. Science 2012;337(6090):59–64.

[22] Yen WC, Fischer M, Cancilla B, Cattaruzza F, Tang T, Yeung P, et al. Abstract 2576: enhanced anti-tumor efficacy by sequential application of Wnt pathway antagonists in combination with taxanes. In: AACR Annual Meeting. 2015.

[23] Andreu P, Colnot S, Godard C, Gad S, Chafey P, Niwa-Kawakita M, et al. Crypt-restricted proliferation and commitment to the Paneth cell lineage following Apc loss in the mouse intestine. Development 2005;132(6):1443–51.

[24] Huang SM, Mishina YM, Liu S, Cheung A, Stegmeier F, Michaud GA, et al. Tankyrase inhibition stabilizes axin and antagonizes Wnt signalling. Nature 2009;461(7264):614–20.

[25] Lau T, Chan E, Callow M, Waaler J, Boggs J, Blake RA, et al. A novel tankyrase small-molecule inhibitor suppresses APC mutation-driven colorectal tumor growth. Cancer Res 2013;73(10): 3132–44.

[26] Schoumacher M, Hurov KE, Lehar J, Yan-Neale Y, Mishina Y, Sonkin D, et al. Inhibiting Tankyrases sensitizes KRAS-mutant cancer cells to MEK inhibitors via FGFR2 feedback signaling. Cancer Res 2014;74(12):3294–305.

[27] Emami KH, Nguyen C, Ma H, Kim DH, Jeong KW, Eguchi M, et al. A small molecule inhibitor of beta-catenin/CREB-binding protein transcription [corrected]. Proc Natl Acad Sci USA 2004;101(34):12682–7.

[28] Grigson ER, Ozerova M, Pisklakova A, Liu H, Sullivan DM, Nefedova Y. Canonical Wnt pathway inhibitor ICG-001 induces cytotoxicity of multiple myeloma cells in Wnt-independent manner. PLoS One 2015;10(S):e0117693.

[29] de Lau W, Peng WC, Gros P, Clevers H. The R-spondin/Lgr5/Rnf43 module: regulator of Wnt signal strength. Genes Dev 2014;28(4):305–16.

[30] Seshagiri S, Stawiski EW, Durinck S, Modrusan Z, Storm EE, Conboy CB, et al. Recurrent R-spondin fusions in colon cancer. Nature 2012;488(7413):660–4.

[31] Chartier C, Raval J, Axelrod F, Bond C, Cain J, Dee-Hoskins C, et al. Therapeutic targeting of tumor-derived R-spondin attenuates beta-catenin signaling and tumorigenesis in multiple cancer types. Cancer Res 2016;76(3):713–23.

[32] Hori K, Sen A, Artavanis-Tsankonas S. Notch signaling at a glance. J Cell Sci 2013;126(Pt 10): 2135–40.

[33] Palermo R, Checquolo S, Bellavia D, Talora C, Screpanti I, et al. The molecular basis of notch signaling regulation: a complex simplicity. Curr Mol Med 2014;14(1):34–44.

[34] Fre S, Bardin A, Robine S, Louvard D. Notch signaling in intestinal homeostasis across species: the cases of Drosophila, Zebrafish and the mouse. Exp Cell Res 2011;317(19):2740–7.

[35] Koch U, Lehal R, Radtke F. Stem cells living with a Notch. Development 2013;140(4): 689–704.

[36] Vooijs M, Liu Z, Kopan R. Notch: architect, landscaper, and guardian of the intestine. Gastroenterology 2011;141(2):448–59.

[37] Suresh S, Irvine AE. The NOTCH signaling pathway in normal and malignant blood cell production. J Cell Commun Signal 2015;9(1):5–13.

[38] Kangsamaksin T, Tattersall IW, Kitajewski J. Notch functions in developmental and tumour angiogenesis by diverse mechanisms. Biochem Soc Trans 2014;42(6):1563–8.

[39] Grishina IB. Mini-review: does Notch promote or suppress cancer? New findings and old controversies. Am J Clin Exp Urol 2015;3(1):24–7.

[40] Alketbi A, Attoub S. Notch signaling in Cancer: rationale and strategies for targeting. Curr Cancer Drug Targets 2015;15(5):364–74.

[41] Takebe N, Nguyen D, Yang SX. Targeting notch signaling pathway in cancer: clinical development advances and challenges. Pharmacol Ther 2014;141(2):140–9.

[42] Milano J, McKay J, Dagenais C, Foster-Brown L, Pognan F, Gadient R, et al. Modulation of notch processing by gamma-secretase inhibitors causes intestinal goblet cell metaplasia and induction of genes known to specify gut secretory lineage differentiation. Toxicol Sci 2004;82(1):341–58.

[43] Imbimbo BP. Therapeutic potential of gamma-secretase inhibitors and modulators. Curr Top Med Chem 2008;8(1):54–61.

[44] Beel AJ, Sanders CR. Substrate specificity of gamma-secretase and other intramembrane proteases. Cell Mol Life Sci 2008;65(9):1311–34.

[45] Hemming ML, Elias JE, Gygi SP, Selkoe DJ. Proteomic profiling of gamma-secretase substrates and mapping of substrate requirements. PLoS Biol 2008;6(10):e257.

[46] Benedito R, Duarte A. Expression of Dll4 during mouse embryogenesis suggests multiple developmental roles. Gene Expr Patterns 2005;5(6):750–5.

[47] Gale NW, Dominguez MG, Noguera I, Pan L, Hughes V, Valenzuela DM, et al. Haploinsufficiency of delta-like 4 ligand results in embryonic lethality due to major defects in arterial and vascular development. Proc Natl Acad Sci USA 2004;101(45):15949–54.

[48] Krebs LT, Shutter JR, Tanigaki K, Honjo T, Stark KL, Gridley T. Haploinsufficient lethality and formation of arteriovenous malformations in Notch pathway mutants. Genes Dev 2004;18(20):2469–73.

[49] Patel NS, Li JL, Generali D, Poulsom R, Cranston DW, Harris AL. Up-regulation of delta-like 4 ligand in human tumor vasculature and the role of basal expression in endothelial cell function. Cancer Res 2005;65(19):8690–7.

[50] Ishigami S, Arigami T, Uenosono Y, Okumura H, Kurahara H, Uchikado Y, et al. Clinical implications of DLL4 expression in gastric cancer. J Exp Clin Cancer Res 2013;32:46.

[51] Yen WC, Fischer MM, Hynes M, Wu J, Kim E, Beviglia L, et al. Anti-DLL4 has broad spectrum activity in pancreatic cancer dependent on targeting DLL4-Notch signaling in both tumor and vasculature cells. Clin Cancer Res 2012;18(19):5374–86.

[52] Zhang YZ, Qin F, Han ZG, Liu Q, Zhou L, Wang YW. Prognostic significance of DLL4 expression in papillary thyroid cancer. Eur Rev Med Pharmacol Sci 2015;19(15):2901–5.

[53] Zheng Y, Hao Z, Ding Y, Wang Q, Li S, Xia G, et al. Expression of delta-like 4 (Drosophila) and vascular endothelial growth factor A in colon cancer and association with tumour angiogenesis. J Int Med Res 2015;43(4):535–43.

[54] Zhou L, Yu L, Ding G, Chen W, Zheng S, Cao L. Overexpressions of DLL4 and CD105 are associated with poor prognosis of patients with pancreatic ductal adenocarcinoma. Pathol Oncol Res 2015;21.

[55] Wang W, Yu Y, Wang Y, Li X, Bao J, Wu G, et al. Delta-like ligand 4: a predictor of poor prognosis in clear cell renal cell carcinoma. Oncol Lett 2014;8(6):2627–33.

[56] Noguera-Troise I, Daly C, Papadopoulos NJ, Coetzee S, Boland P, Gale NW, et al. Blockade of Dll4 inhibits tumour growth by promoting non-productive angiogenesis. Nature 2006;444(7122):1032–7.

[57] Ridgway J, Zhang G, Wu Y, Stawicki S, Liang WC, Chanthery Y, et al. Inhibition of Dll4 signalling inhibits tumour growth by deregulating angiogenesis. Nature 2006;444(7122):1083–7.

[58] Scehnet JS, Jiang W, Kumar SR, Krasnoperov V, Trindade A, Benedito R, et al. Inhibition of Dll4-mediated signaling induces proliferation of immature vessels and results in poor tissue perfusion. Blood 2007;109(11):4753–60.

[59] Smith DC, Eisenberg PD, Manikhas G, Chugh R, Gubens MA, Stagg RJ, et al. A phase I dose escalation and expansion study of the anticancer stem cell agent demcizumab (anti-DLL4) in patients with previously treated solid tumors. Clin Cancer Res 2014;20(24):6295–303.

[60] Kotasek D, et al. Abstract #8045: A Phase 1b study of the anti-cancer stem cell agent demcizumab (DEM), pemetrexed (PEM) & carboplatin (CARBO) in patients with 1st line non-squamous NSCLC. In: ASCO Annual Meeting. 2015.

[61] Ito T, Schaller M, Hogaboam CM, Standiford TJ, Sandor M, Lukacs NW, et al. TLR9 regulates the mycobacteria-elicited pulmonary granulomatous immune response in mice through DC-derived Notch ligand delta-like 4. J Clin Invest 2009;119(1):33–46.

[62] Kassner N, Krueger M, Yagita H, Szionek A, Hutloff A, Kroczek R, et al. Cutting edge: Plasmacytoid dendritic cells induce IL-10 production in T cells via the Delta-like-4/Notch axis. J Immunol 2010;184(2):550–4.

[63] Reynolds ND, Lukacs NW, Long N, Karpus WJ. Delta-like ligand 4 regulates CNS T cell accumulation during experimental autoimmune encephalomyelitis. J Immunol 2011;187.

[64] Laky K, Evans S, Perez-Diez A, Fowlkes BJ. Notch signaling regulates antigen sensitivity of naive CD4+ T cells by tuning co-stimulation. Immunity 2015;42(1):80–94.

[65] Koch U, Fiorini E, Benedito R, Besseyrias V, Schuster-Gossler K, Pierres M, et al. Delta-like 4 is the essential, nonredundant ligand for Notch1 during thymic T cell lineage commitment. J Exp Med 2008;205(11):2515–23.

[66] Hozumi K, Mailhos C, Negishi N, Hirano K, Yahata Y, Ando K, et al. Delta-like 4 is indispensable in thymic environment specific for T cell development. J Exp Med 2008;205(11):2507–13.

[67] Srivastava M, Murriel CL, Roda J, Jie HB, Axelrod F, Xie MH, et al. Dual targeting of delta-like ligand 4 (DLL4) and programmed death 1 (PD1) inhibits tumor growth and generates enhanced long-term immunological memory. In: AACR Annual Meeting. 2015.

[68] Mochizuki K, Xie F, He S, Tong Q, Liu Y, Mochizuki I, et al. Delta-like ligand 4 identifies a previously uncharacterized population of inflammatory dendritic cells that plays important roles in eliciting allogeneic T cell responses in mice. J Immunol 2013;190(7):3772–82.

[69] Eixarch H, Mansilla MJ, Costa C, Kunkel SL, Montalban X, Godessart N, et al. Inhibition of delta-like ligand 4 decreases Th1/Th17 response in a mouse model of multiple sclerosis. Neurosci Lett 2013;541:161–6.

[70] Mukherjee S, Schaller MA, Neupane R, Kunkel SL, Lukacs NW. Regulation of T cell activation by Notch ligand, DLL4, promotes IL-17 production and Rorc activation. J Immunol 2009;182(12):7381–8.

[71] Dang TP, Gazdar AF, Virmani AK, Sepetavec T, Hande KR, Minna JD, et al. Chromosome 19 translocation, overexpression of Notch3, and human lung cancer. J Natl Cancer Inst 2000;92(16):1355–7.

[72] Konishi J, Kawaguchi KS, Vo H, Haruki N, Gonzalez A, Carbone DP, et al. Gamma-secretase inhibitor prevents Notch3 activation and reduces proliferation in human lung cancers. Cancer Res 2007;67(17):8051–7.

[73] Arasada RR, Amann JM, Rahman MA, Huppert SS, Carbone DP. EGFR blockade enriches for lung cancer stem-like cells through Notch3-dependent signaling. Cancer Res 2014;74(19):5572–84.

[74] Ye YZ, Zhang ZH, Fan XY, Xu XL, Chen ML, Chang BW, et al. Notch3 overexpression associates with poor prognosis in human non-small-cell lung cancer. Med Oncol 2013;30(2):595.

[75] Liu L, Chen X, Wang Y, Qu Z, Lu Q, Zhao J, et al. Notch3 is important for TGF-beta-induced epithelial-mesenchymal transition in non-small cell lung cancer bone metastasis by regulating ZEB-1. Cancer Gene Ther 2014;21(9):364–72.

[76] Zheng Y, de la Cruz CC, Sayles LC, Alleyne-Chin C, Vaka D, Knaak TD, et al. A rare population of CD24(+)ITGB4(+)Notch(hi) cells drives tumor propagation in NSCLC and requires Notch3 for self-renewal. Cancer Cell 2013;24(1):59–74.

[77] Doucas H, Mann CD, Sutton CD, Garcea G, Neal CP, Bery DP, et al. Expression of nuclear Notch3 in pancreatic adenocarcinomas is associated with adverse clinical features, and correlates with the expression of STAT3 and phosphorylated Akt. J Surg Oncol 2008;97(1):63–8.

[78] Lee KM, Yasuda H, Hollingsworth MA, Ouellette MM. Notch 2-positive progenitors with the intrinsic ability to give rise to pancreatic ductal cells. Lab Invest 2005;85(8):1003–12.

[79] Zhou ZC, Dong QG, Fu DL, Gong YY, Ni QX. Characteristics of Notch2(+) pancreatic cancer stem-like cells and the relationship with centroacinar cells. Cell Biol Int 2013;37(8):805–11.

[80] Adelaide J, Finetti P, Bekhouche I, Repellini L, Geneix J, Sircoulomb F, et al. Integrated profiling of basal and luminal breast cancers. Cancer Res 2007;67(24):11565–75.

[81] Florena AM, Tripodo C, Guarnotta C, Ingrao S, Porcasi R, Martorana A, et al. Associations between Notch-2, Akt-1 and HER2/neu expression in invasive human breast cancer: a tissue microarray immunophenotypic analysis on 98 patients. Pathobiology 2007;74(6):317–22.

[82] Ozawa T, Kazama S, Akiyoshi T, Murono K, Yoneyama S, Tanaka T, et al. Nuclear Notch3 expression is associated with tumor recurrence in patients with stage II and III colorectal cancer. Ann Surg Oncol 2014;21(8):2650–8.

[83] Kang H, Jeong JY, Song JY, Kim TH, Kim G, Huh JH, et al. Notch3-specific inhibition using siRNA knockdown or GSI sensitizes paclitaxel-resistant ovarian cancer cells. Mol Carcinog 2015. http://dx.doi.org/10.1002/mc.22363. [Epub ahead of print].

[84] Brown CW, Brodsky AS, Freiman RN. Notch3 overexpression promotes anoikis resistance in epithelial ovarian cancer via upregulation of COL4A2. Mol Cancer Res 2015;13(1):78–85.

[85] Pasto A, Serafin V, Pilotto G, Lago C, Bellio C, Trusolino L, et al. NOTCH3 signaling regulates MUSASHI-1 expression in metastatic colorectal cancer cells. Cancer Res 2014;74(7): 2106–18.

[86] Wang XW, Xi XQ, Wu J, Wan YY, Hui HX, Cao XF. MicroRNA-206 attenuates tumor proliferation and migration involving the downregulation of NOTCH3 in colorectal cancer. Oncol Rep 2015;33(3):1402–10.

[87] Hu L, Xue F, Shao M, Deng A, Wei G. Aberrant expression of Notch3 predicts poor survival for hepatocellular carcinomas. Biosci Trends 2013;7(3):152–6.

[88] Gramantieri L, Giovannini C, Lanzi A, Chieco P, Ravaioli M, Venturi A, et al. Aberrant Notch3 and Notch4 expression in human hepatocellular carcinoma. Liver Int 2007;27(7):997–1007.

[89] Cardinale V, Carpino G, Reid LM, Gaudio E, Alvaro D. Notch2 signaling and undifferentiated liver cancers: evidence of hepatic stem/progenitor cell origin. Hepatology 2013;58(3):1188.

[90] Zhu P, Wang Y, Du Y, He L, Huang G, Zhang G, et al. C8orf4 negatively regulates self-renewal of liver cancer stem cells via suppression of NOTCH2 signalling. Nat Commun 2015;6:7122.

[91] Fan X, Mikolaenko I, Elhassan I, Ni X, Wang Y, Ball D, et al. Notch1 and notch2 have opposite effects on embryonal brain tumor growth. Cancer Res 2004;64(21):7787–93.

[92] Nakayama K, Nakayama N, Jinawath N, Salani R, Kurman RJ, Shih leM, et al. Amplicon profiles in ovarian serous carcinomas. Int J Cancer 2007;120(12):2613–7.

[93] Park JT, Li M, Nakayama K, Mao TL, Davidson B, Zhang Z, et al. Notch3 gene amplification in ovarian cancer. Cancer Res 2006;66(12):6312–8.

[94] Choi JH, Park JT, Davidson B, Morin PJ, Shih leM, Wang TL. Jagged-1 and Notch3 juxtacrine loop regulates ovarian tumor growth and adhesion. Cancer Res 2008;68(14):5716–23.

[95] Yen WC, Fischer MM, Axelrod F, Bond C, Cain J, Cancilla B, et al. Targeting Notch signaling with a Notch2/Notch3 antagonist (tarextumab) inhibits tumor growth and decreases tumor-initiating cell frequency. Clin Cancer Res 2015;21(9):2084–95.

[96] Armulik A, Abramsson A, Betsholtz C. Endothelial/pericyte interactions. Circ Res 2005;97(6):512–23.

[97] O'Reilly E, Smith L, Bendell JC, Strickler J, Zalupski M, Gluck W, et al. Final results of a phase 1b of tarextumab (TRXT, OMP-59R5, anti-notch2/3) in combination with Nab-paclitaxel and gemcitabine (Nab P+Gem) in patients (pts) with untreated metastatic pancreatic cancer (mPC). In: Alpine study. ASCO Gastrointestinal Cancers Symposium. 2015.

[98] Pietanza MC, et al. Abstract #7508: Results of phase 1b of tarextumab (TRXT, OMP-59R5, anti-notch 2/3) in combination with etoposide and platinum therapy (EP) in patients (pts) with untreated extensive-stage small-cell lung cancer (ED-SCLC). In: ASCO Annual Meeting. 2015.

[99] Agnusdei V, Minuzzo S, Frasson C, Grassi A, Axelrod F, Satyal S, et al. Therapeutic antibody targeting of Notch1 in T-acute lymphoblastic leukemia xenografts. Leukemia 2014;28(2):278–88.

[100] Wu Y, Cain-Hom C, Choy L, Hagenbeek TJ, de Leon GP, Chen Y, et al. Therapeutic antibody targeting of individual Notch receptors. Nature 2010;464(7291):1052–7.

[101] Beachy PA, Karhadkar SS, Berman DM. Tissue repair and stem cell renewal in carcinogenesis. Nature 2004;432(7015):324–31.

[102] Robinson GW, Orr BA, Wu G, Gururangan S, Lin T, Qaddoumi I, et al. Vismodegib exerts targeted efficacy against recurrent sonic hedgehog-subgroup medulloblastoma: results from phase II pediatric brain tumor consortium studies PBTC-025B and PBTC-032. J Clin Oncol 2015;33(24):2646–54.

CHAPTER 18

Targeted Therapies for Glioma Stem Cells

S.J. Sundar, J. Wright, J.K. Hseih, A.E. Sloan

University Hospitals-Case Medical Center & Case Comprehensive Cancer Center, Cleveland, OH, United States

CONTENTS

INTRODUCTION

Glioblastoma (GBM) is the most common primary malignant brain tumor as well as the most aggressive with a median survival of 15 months [1] and a 26–33% 2-year survival even after standard therapy which comprises surgery and concurrent radiation and chemotherapy [2–4]. The hallmarks of GBM are intratumoral heterogeneity, invasion within the brain, and treatment resistance. Recurrence is nearly universal [4]. While these phenomena are not well understood, they are thought to be due in part to cancer stem cells (CSCs)—or in the case of GBM, glioma stem cells (GSCs).

The CSC hypothesis proposes that there is a cellular hierarchy within individual tumors [5]. CSCs represent a fraction of the cells at the apex of the hierarchy and have the capability to self-renew as well as divide to give rise to the variety of cells populating a tumor. As tumor cells differentiate, their ability to self-renew is reduced, and they lose their "stemness." The hierarchy is dynamic with respect to cell types (CSCs, non-CSCs) and is maintained by the balance between self-renewal and differentiation [5–7]. Thus, tumors can be considered as aberrant organs comprised of heterogeneous cell types derived from CSCs, rather than simply an accumulation of diverse neoplastic clones. Recent studies have demonstrated that differentiation of CSCs is not unidirectional, however. Stimuli such as hypoxia and acidic stress as well as therapeutics such as temozolomide (TMZ) can induce some nonstem cancer cells to adopt a CSC phenotype [8–11]. Given the propensity of tumor cells to move in either direction along the tumor hierarchy—both toward differentiation and toward as dedifferentiation, the highly variable molecular characteristics of CSCs and their potential to change phenotype in response to internal and external signals–CSCs must be seen as dynamic entities shifting fluidly among different molecular and functional states. CSCs need not originate from aberrant

Cancer Stem Cells. http://dx.doi.org/10.1016/B978-0-12-803892-5.00018-8

stem cells, and some different tumors may arise from stem cells or restricted-lineage multipotent precursors, while others may arise from nonstem cells or more than one cell type. A single definition of CSCs is not universally agreed upon by researchers; however, a well-accepted functional definition of CSC includes the following traits: proliferation; self-renewal, the capacity to differentiate to give rise to heterogeneous cell populations; and tumorigenicity in the host or in immunosuppressed hosts [12,13].

In addition to meeting this functional definition, GSCs often share much in common with neural stem cells (NSCs) including gene expression profiles. NSCs possess the ability to self-renew and generate neurons, oligodendrocytes, and astrocytes [14]. These NSCs isolated from embryonic mice propagated in epidermal growth factor have demonstrated the ability to form "neurospheres," which contain neurons and glia [15–18]. There are many common pathways between GSCs and NSCs mutated to acquire tumorigenicity or if they originate from mature cells that dedifferentiated and then acquired the ability to self-renew.

There are multiple putative biomarkers for GSC. Among these, CD133 is the best-studied GSC biomarker and is often used experimentally to identify and enrich tumor-propagating and initiating cells. Also known as Prominin-1, CD133 is associated with normal NSCs and is expressed during embryonic development [19–21]. Tumor cells isolated from GBM that grew neurospheres in serum-free media (indicating self-renewal capabilities) and grew tumors phenotypically similar to GBM were found to be CD133+, while tumor cells lacking CD133 expression did not demonstrate self-renewal or tumorigenicity in xenotransplantation studies [22–25]. One report demonstrated that small hairpin RNA (ShRNA) knockdown of CD133 in putative CSCs resulted in loss of both these properties, and upon reexpression of CD133 in the same cells, the CSCs' maintenance ability and tumorigenicity returned [6,26,27]. However, despite the evidence outlining a crucial relationship between GSC and CD133, the latter is not a universal marker to identify CSCs. Several studies have demonstrated that CD133–GSC are also capable of tumor initiation and some GBM tumors do not contain any CD133+ cells [28–35]. It has also been proposed that tumor-initiating CD133-cells may, in fact, actually express CD133 at low levels below experimental thresholds. The subtleties of this relationship remain unclear, and the essential role of CD133 in CSC maintenance remains an area of investigation. Microenvironmental interactions, such as those mediated by CD15 and local growth factors, may complement the function of CD133 in GSC maintenance and preserve stemness in GSC populations expressing very low or cell cycle–dependent levels of CD133 [23,28].

GSCs also demonstrate increased expression of SOX2, a transcription factor associated with multipotency via the TGF-β signaling pathway, which also promotes self-renewal of GSCs [36,37]. In addition, GSCs also typically have increased expression of Nestin, an intermediate filament seen in NSCs, although this is a better marker in mouse tumors than in human GBM [37,38]. Integrin α6 is another biomarker that is highly expressed in GSCs and silencing it through shRNA knockdown is associated with failure of GSCs to self-renew or grow tumors [39]. Epidermal growth factor receptor (EGFR) is expressed in over half of GBM patients and may increase tumorigenicity and activate those characteristics of CSCs that promote treatment resistance [40,41].

Additional biomarkers that have been studied in GSC include, CD15, CD36, A2B5, L1CAM, CD44, and CXCR4 [32,39,42–45]. Although these markers are useful in furthering our understanding of GSC function and regulation and may be involved in targets for therapies against GSCs, no single marker can definitively identify nor define GSCs.

TUMOR NICHES/MICROENVIRONMENTS

Normal NSCs reside in particular anatomical regions known as niches, a microenvironment comprised of somatic cells and the extracellular matrix [46]. The relationship between stem cells and these niches are not passive. Rather, NSCs interact dynamically with their microenvironment [47]. They actively influence their microenvironments and, in turn, are regulated by signaling from that same microenvironment [47].

Similarly, GSCs also exist in specific niches that play a role in regulation of tumorigenicity. The microenvironment not only plays a role in helping to maintain GSCs and the tumor but can also impact response to therapy. The tumor microenvironment and the GSC niche is an active area of investigation, and GSCs are thought to occupy, among others, perivascular, hypoxic, and necrotic niches, as well as tumor border regions, impacting GBM's invasive properties [48–51]. The concept of CSCs is not unique to glioma, having been initially discovered in human acute myeloid leukemia progenitor cells in 1994. CSCs have also been discovered in breast, colon, lung, and brain tumors [25,52–54].

Perivascular Niche

Perhaps the best established tumor niche for GSC is the perivascular niche [48]. Many stem cells tend to be located close to the endothelial cells lining capillaries, particularly in the subventricular zone and the hippocampus [55–61]. These niches contain vascular factors that appear to regulate stem cells. These endothelial vascular factors have not precisely been elucidated; however, studies suggest the involvement of vascular endothelial growth factor C (VEGFC) and brain-derived neurotrophic factor (BDNF) [62,63]. NSCs not only influence the surrounding microenvironment, but they also actively regulate it by secreting VEGF and BDNF to promote angiogenesis and contribute to this dynamic, cyclical relationship [47,64]. Similarly, studies have demonstrated that GBM characteristically contains abnormal perivascular niches, and highly disorganized vasculature is characteristic of this tumor. Historically, hypervascularity was thought to be in response to the rapidly growing tumor; however, these aberrant vascular niches appear critical to maintaining the CSC population [65–67]. Some studies have suggested that vascular density of GBM correlates with the amount of GSCs and even patient prognosis [46,66]. Bevacizumab, an anti-VEGF antibody, is often used as part of salvage therapy for GBM patients [68]. The interaction of GSCs with endothelial cells promotes activity in critical stem pathways such as Notch signaling, which contributes to their self-renewal abilities [69]. The CD133+ GSCs are shown to express greater levels of VEGF, which lead to increased tumor vascularity, compared to CD133- cells [70]. GSCs may even be capable of differentiating into cells functionally resembling pericytes, supplying raw material necessary to continue supporting the perivascular niche [48].

Hypoxic/Necrotic Niche

In addition to aberrant vasculature, GBM is characterized by regions of intratumor necrosis surrounded by a rim of densely packed tumor cells, known as pseudopalisading necrosis [71]. These areas are believed to be another niche for GSCs to demonstrate increased self-renewal and differentiation due to the hypoxia in the environment [72]. Oxygen levels drop further from vessels because of rapid uptake [73], and hypoxia-inducible factors play an important role in embryonic stem cells and NSCs to promote proliferation [49,72]. Hypoxia has also been demonstrated to upregulate VEGF in GSCs and increase angiogenesis [74]. Hypoxia-induced activation of HIF-1α promotes self-renewal in CD133+ glioma-derived GSCs, resulting in expansion of the GSC population, while knockdown of HIF-1α, or inhibition of the phosphoinositol 3-kinase (PI3K)-Akt or ERK1/2 pathways reduced this effect [75]. HIF-2α and its target genes were also found to be preferentially expressed in glioblastoma GSCs, and HIF-2 has been found to colocalize with CSC markers [11]. Studies suggest that hypoxia may have the ability to induce CD133 expression and Notch signaling in GSCs, both of which are important for self-renewal [15]. The pseudopalisading regions have shown CD133+ immunoreactivity as well, further supporting that these hypoxic and necrotic niches are involved in supporting GSCs [76].

Invasion/Tumor Edge Niche

The outer edge of the tumor and its invasive properties are believed to constitute another niche for GSCs. Glioma invasion is another poorly understood but critical area of study. At the tumor boundary, TGF-β secreted from tumor macrophages may stimulate GSCs to expand the tumor by invasion of surrounding normal parenchyma [77]. Astrocytes also may play a role in GBM invasion through the activation of matrix metalloproteinases (MMPs) and initiating the sonic hedgehog (SHH) signaling pathway [78,79]. MMPs may help impair the integrity of the surrounding normal parenchyma and its matrix, while SHH is known to promote self-renewal of stem cells [78]. GBM invasion is also regulated by the chemokine receptor type 4 (CXCR4), which is increased in GSCs [80]. CXCR4 can help attract CSCs to nearby endothelial cells, reinitiating the cycle of invasion, tumor growth, and endothelium proliferation [74].

GLIOMA STEM CELL AS A THERAPEUTIC TARGET

The failure of conventional approaches to significantly improve the prognosis of patient with GBM over three decades has led to interest in novel approaches. Given their critical role in tumor initiation, propagation, and maintenance, CSCs offer an attractive therapeutic target. An ideal GBM treatment that targets CSCs would be able to distinguish between NSCs and CSCs and selectively eliminate only the CSCs. Targeted therapies have the potential for decreased morbidity after treatment by potentially decreasing damage to healthy tissue and potentially improved efficacy in targeting GSC. There are several aspects of GSCs that are potential targets of therapy including small molecule inhibitors of signaling pathways, CSC surface markers, microenvironment inhibitors, and immunotherapy. Nearly all the targets noted above have been targeted in numerous in vitro or preclinical xenograft studies. Clinical trials targeting GSC are outlined in Table 18.1.

Immunotherapy

Immunotherapy has been proposed for numerous types of cancer, including gliomas. In an animal study conducted by Xu et al., in 2009, a dendritic cell vaccine was created against tumor-associated antigens specific to CSCs [81]. The results showed that such a vaccination was able to induce cytotoxic T lymphocytes (CTLs) against CSC antigens, and in the 9L rat glioma model, survival was increased [81]. A Phase 1 vaccination study to treat GBM in humans was completed by Phuphanick et al., in 2012 using vaccines that targeted tumor antigens highly specific to GSCs [82]. The vaccine developed was found to be nontoxic, and for the 16 patients who received the vaccine, median progression-free survival (PFS) was 16.9 months and median overall survival (OS) was 38.4 months. A Phase 2-randomized clinical trial for this vaccine is currently ongoing (Clinicaltrials.gov #NCT01280552).

Stem Cell Signaling

Another approach to GBM therapy may be targeting signaling pathways critical to CSC renewal and proliferation (such as SHH or Notch) with small molecule inhibitors [25]. GDC-0449, known as vismodegib, can inhibit SHH signaling, and there is currently a recently completed Phase 2 clinical trial (Clinicaltrials.gov

Table 18.1 Selected Clinical Trials Targeting Cancer Stem Cells for Treatment of GBM

Authors & Year *or* Clinical Trial Number	Trial Design	Agent (Target)	Outcomes
Phuphanich et al. (2013)	Phase 1 single-arm	ICT-107, an autologous dendritic cell pulsed vaccine (tumor-associated antigens overexpressed on cancer stem cells)	21 total patients Nontoxic 33% immunological response rate Nonsignificant trend toward increased progression-free survival (PFS), but not overall survival (OS), for vaccine responders
Sloan et al. (2014)	Phase 0/2 randomized	Vismodegib (sonic hedgehog pathway)	40 total patients Well-tolerated No difference in PFS6 or overall survival as single agent Achieved therapeutic intratumoral concentration Decreased sonic hedgehog signaling Decreased cancer stem cell prolifera-tion and self-renewal
Clinicaltrials.gov (NCT01280552)	Phase 2 randomized	ICT-107 (tumor-associated antigens)	In progress
Clinicaltrials.gov (NCT01122901)	Phase 2 Nonrandomized	RO4929097 (gamma-secretase, Notch signaling pathway)	In progress
Clinicaltrials.gov (NCT01119599)	Phase 1 single-arm	R04929097 in combination with TMZ and radiotherapy (gamma-secretase, Notch signaling pathway)	In progress
Clinicaltrials.gov (NCT01189240)	Phase 1/2 randomized	R04929097 with bevacizumab (gamma-secretase, Notch signaling pathway)	In progress

#NCT00980343) demonstrated that the drug reached the tumor, inhibited stemness, and downregulated the SHH signaling pathway, though there was little improvement in PFS or OS with use as a single agent in recurrent GBM [83]. A new trial targeting both the SHH and another metabolic biomarker identified by this trial is currently in development for the Ohio Neuro-Oncology Collaborative (ONOC). It has also been demonstrated that the SHH and the PI3K pathways have a synergistic relationship in tumor proliferation, and there is interest in targeting both these pathways simultaneously with small molecular inhibitors. In vitro and in vivo tests have already demonstrated reduced GBM growth, so this may be a viable option for treating humans [84].

Similarly, Notch signaling is important for CSC self-renewal. Phase 1 and 2 clinical trials targeting the Notch pathway with small molecule inhibitors have recently been completed in GBM (Clinicaltrials.gov #NCT01122901, #NCT01119599, #NCT01189240).

Targeting the Perivascular Niche

The microenvironment of CSCs and their anatomical niche, which includes extracellular matrix and somatic cells, plays an important role in the growth and maintenance of tumor [46,47]. Interactions between CSCs are dynamic and may influence therapeutic responses. CSCs tend to be located in close proximity to capillaries and the molecules located in these abnormal perivascular niches may regulate CSCs [56–59]. Tumor growth depends on angiogenesis, which relies on the expression of factors such as VEGF, fibroblast growth factor (FGF), and hypoxia-inducible factor 1 alpha (HIF-1a) [85–88]. Numerous clinical trials with the goals of decreasing VEGF or interfering with VEGF signaling have been instituted, and several have been completed. These trials have primarily been designed to inhibit angiogenesis, but given the importance of the perivascular niche in GSC, these cells might also be considered a target of such trials. Early studies utilizing bevacizumab, an anti-VEGF monoclonal antibody with or without irinotecan demonstrated improvements in radiographic and increases in overall and PFS in patients with recurrent GBM [68,89]. However, a subsequent trial with first line use of bevacizumab did not demonstrate similar significant improvements in OS and PFS [90]. Another strategy utilized kinase inhibitors to interrupt VEGF signaling at the cellular level using various combinations of sunitinib, sorafenib, erlotinib with or without TMZ and or irinotecan with similarly underwhelming results [91–93]. Antiangiogenic targeted therapies remain an active area of investigation.

CONCLUSION

The treatment resistance of glioblastoma is likely due in part to CSCs and their inherent properties of self-renewal, proliferation, and ability to create heterogeneous tumor. Therapies that target these CSCs and their inherent niches have the potential to advance GBM treatment and improve outcomes by causing less damage to healthy tissue and more effective killing of malignant cells. Clinical trials utilizing small molecule inhibitors of signaling pathways, dendritic cell vaccines, and microenvironment inhibitors targeting GSC are all underway. The results of these clinical trials will serve to guide future treatment of these insidious tumors.

List of Acronyms and Abbreviations

BDNF Brain-derived neurotrophic factor
CSC Cancer stem cell
CTL Cytotoxic T lymphocyte
CXCR Chemokine receptor
EGFR Epidermal growth factor receptor
ERK Extracellular signal-regulated kinase

FGF Fibroblast growth factor
GBM Glioblastoma
GSC Gglioma stem cell
HIF Hypoxia-induced factor
MMP Matrix metalloproteinases
NSC Neural stem cell
OS Overall survival
PDGF Platelet-derived growth factor
PFS Progression-free survival
PI3K Phosphoinositol 3-kinase
SHH Sonic hedgehog
shRNA Small hairpin RNA
TGF Tumor growth factor
TMZ Temozolomide
VEGF Vascular endothelial growth factor

References

[1] Ostrom QT, Gittleman H, Fulop J, Liu M, Blanda R, Kromer C, et al. Cbtrus statistical report: primary brain and central nervous system tumors diagnosed in the United States in 2008–2012. Neuro Oncol 2015;17(Suppl. 4):iv1–62.

[2] Lapidot T, Sirard C, Vormoor J, Murdoch B, Hoang T, Caceres-Cortes J, et al. A cell initiating human acute myeloid leukaemia after transplantation into SCID mice. Nature 1994;367:645–8.

[3] Lawrence YR, Mishra MV, Werner-Wasik M, Andrews DW, Showalter TN, Glass J, et al. Improving prognosis of glioblastoma in the 21st century: who has benefited most? Cancer 2012;118:4228–34.

[4] Stupp R, Mason WP, van den Bent MJ, Weller M, Fisher B, Taphoorn MJB, et al. Radiotherapy plus concomitant and adjuvant temozolomide for glioblastoma. N Engl J Med 2005;352:987–96.

[5] Jackson EL, Alvarez-Buylla A. Characterization of adult neural stem cells and their relation to brain tumors. Cells Tissues Organs 2008;188:212–24.

[6] Godlewski J, Nowicki MO, Bronisz A, Williams S, Otsuki A, Nuovo G, et al. Targeting of the Bmi-1 oncogene/stem cell renewal factor by microRNA-128 inhibits glioma proliferation and self-renewal. Cancer Res 2008;68:9125–30.

[7] Yu J, Thomson JA. Pluripotent stem cell lines. Genes Dev 2008;22:1987–97.

[8] Auffinger B, Tobias AL, Han Y, Lee G, Guo D, Dey M, et al. Conversion of differentiated cancer cells into cancer stem-like cells in a glioblastoma model after primary chemotherapy. Cell Death Differ 2014;21:1119–31.

[9] Chen J, Li Y, Yu T-S, McKay RM, Burns DK, Kernie SG, et al. A restricted cell population propagates glioblastoma growth following chemotherapy. Nature 2012;488:522–6.

[10] Hjelmeland AB, Wu Q, Heddleston JM, Choudhary GS, MacSwords J, Lathia JD, et al. Acidic stress promotes a glioma stem cell phenotype. Cell Death Differ 2011;18:829–40.

[11] Li Z, Bao S, Wu Q, Wang H, Eyler C, Sathornsumetee S, et al. Hypoxia-inducible factors regulate tumorigenic capacity of glioma stem cells. Cancer Cell 2009;15:501–13.

[12] Antoniou A, Hébrant A, Dom G, Dumont JE, Maenhaut C. Cancer stem cells, a fuzzy evolving concept: a cell population or a cell property? Cell Cycle 2013;12:3743–8.

[13] Reya T, Morrison SJ, Clarke MF, Weissman IL. Stem cells, cancer, and cancer stem cells. Nature 2001;414:105–11.

[14] Leder K, Pitter K, Laplant Q, Hambardzumyan D, Ross BD, Chan TA, et al. Mathematical modeling of PDGF-driven glioblastoma reveals optimized radiation dosing schedules. Cell 2014;156:603–16.

[15] Huse JT, Holland EC. Genetically engineered mouse models of brain cancer and the promise of preclinical testing. Brain Pathol 2009;19:132–43.

[16] Jackson EL, Garcia-Verdugo JM, Gil-Perotin S, Roy M, Quinones-Hinojosa A, VandenBerg S, et al. PDGFR alpha-positive B cells are neural stem cells in the adult SVZ that form glioma-like growths in response to increased PDGF signaling. Neuron 2006;51:187–99.

[17] Reynolds BA, Tetzlaff W, Weiss S. A multipotent EGF-responsive striatal embryonic progenitor cell produces neurons and astrocytes. J Neurosci 1992;12:4565–74.

[18] Reynolds BA, Weiss S. Generation of neurons and astrocytes from isolated cells of the adult mammalian central nervous system. Science 1992;255:1707–10.

[19] Coskun V, Wu H, Blanchi B, Tsao S, Kim K, Zhao J, et al. CD133+ neural stem cells in the ependyma of mammalian postnatal forebrain. Proc Natl Acad Sci USA 2008;105:1026–31.

[20] Pfenninger CV, Roschupkina T, Hertwig F, Kottwitz D, Englund E, Bengzon J, et al. CD133 is not present on neurogenic astrocytes in the adult subventricular zone, but on embryonic neural stem cells, ependymal cells, and glioblastoma cells. Cancer Res 2007;67:5727–36.

[21] Zacchigna S, Oh H, Wilsch-Bräuninger M, Missol-Kolka E, Jászai J, Jansen S, et al. Loss of the cholesterol-binding protein prominin-1/CD133 causes disk dysmorphogenesis and photoreceptor degeneration. J Neurosci 2009;29:2297–308.

[22] Dirks PB. Brain tumor stem cells: bringing order to the chaos of brain cancer. J Clin Oncol 2008;26:2916–24.

[23] Lathia JD, Hitomi M, Gallagher J, Gadani SP, Adkins J, Vasanji A, et al. Distribution of CD133 reveals glioma stem cells self-renew through symmetric and asymmetric cell divisions. Cell Death Dis 2011;2:e200.

[24] Singh SK, Clarke ID, Terasaki M, Bonn VE, Hawkins C, Squire J, et al. Identification of a cancer stem cell in human brain tumors. Cancer Res 2003;63:5821–8.

[25] Singh SK, Hawkins C, Clarke ID, Squire JA, Bayani J, Hide T, et al. Identification of human brain tumour initiating cells. Nature 2004;432:396–401.

[26] Artavanis-Tsakonas S, Delidakis C, Fehon RG. The Notch locus and the cell biology of neuroblast segregation. Annu Rev Cell Biol 1991;7:427–52.

[27] Silber J, Lim DA, Petritsch C, Persson AI, Maunakea AK, Yu M, et al. miR-124 and miR-137 inhibit proliferation of glioblastoma multiforme cells and induce differentiation of brain tumor stem cells. BMC Med 2008;6:14.

[28] Beier D, Hau P, Proescholdt M, Lohmeier A, Wischhusen J, Oefner PJ, et al. CD133(+) and CD133(−) glioblastoma-derived cancer stem cells show differential growth characteristics and molecular profiles. Cancer Res 2007;67:4010–5.

[29] Campos B, Zeng L, Daotrong PH, Eckstein V, Unterberg A, Mairbäurl H, et al. Expression and regulation of AC133 and CD133 in glioblastoma. Glia 2011;59:1974–86.

[30] Gambelli F, Sasdelli F, Manini I, Gambarana C, Oliveri G, Miracco C, et al. Identification of cancer stem cells from human glioblastomas: growth and differentiation capabilities and CD133/prominin-1 expression. Cell Biol Int 2012;36:29–38.

[31] Lottaz C, Beier D, Meyer K, Kumar P, Hermann A, Schwarz J, et al. Transcriptional profiles of CD133+ and CD133− glioblastoma-derived cancer stem cell lines suggest different cells of origin. Cancer Res 2010;70:2030–40.

[32] Son MJ, Woolard K, Nam D-H, Lee J, Fine HA. SSEA-1 is an enrichment marker for tumor-initiating cells in human glioblastoma. Cell Stem Cell 2009;4:440–52.

[33] Wang J, Sakariassen PØ, Tsinkalovsky O, Immervoll H, Bøe SO, Svendsen A, et al. CD133 negative glioma cells form tumors in nude rats and give rise to CD133 positive cells. Int J Cancer 2008;122:761–8.

[34] Yan X, Ma L, Yi D, Yoon J, Diercks A, Foltz G, et al. A CD133-related gene expression signature identifies an aggressive glioblastoma subtype with excessive mutations. Proc Natl Acad Sci USA 2011;108:1591–6.

[35] Zarkoob H, Taube JH, Singh SK, Mani SA, Kohandel M. Investigating the link between molecular subtypes of glioblastoma, epithelial-mesenchymal transition, and CD133 cell surface protein. PLoS One 2013;8:e64169.

[36] Boumahdi S, Driessens G, Lapouge G, Rorive S, Nassar D, Le Mercier M, et al. SOX2 controls tumour initiation and cancer stem-cell functions in squamous-cell carcinoma. Nature 2014;511(7508):246–50. http://dx.doi.org/10.1038/nature13305.

[37] Ikushima H, Todo T, Ino Y, Takahashi M, Saito N, Miyazawa K, et al. Glioma-initiating cells retain their tumorigenicity through integration of the Sox axis and Oct4 protein. J Biol Chem 2011;286:41434–41.

[38] Ikushima H, Todo T, Ino Y, Takahashi M, Miyazawa K, Miyazono K. Autocrine TGF-beta signaling maintains tumorigenicity of glioma-initiating cells through Sry-related HMG-box factors. Cell Stem Cell 2009;5:504–14.

[39] Lathia JD, Gallagher J, Heddleston JM, Wang J, Eyler CE, Macswords J, et al. Integrin alpha 6 regulates glioblastoma stem cells. Cell Stem Cell 2010;6:421–32.

[40] Jin X, Yin J, Kim S-H, Sohn Y-W, Beck S, Lim YC, et al. EGFR-AKT-Smad signaling promotes formation of glioma stem-like cells and tumor angiogenesis by ID3-driven cytokine induction. Cancer Res 2011;71:7125–34.

[41] Mazzoleni S, Politi LS, Pala M, Cominelli M, Franzin A, Sergi Sergi L, et al. Epidermal growth factor receptor expression identifies functionally and molecularly distinct tumor-initiating cells in human glioblastoma multiforme and is required for gliomagenesis. Cancer Res 2010;70:7500–13.

[42] Anido J, Sáez-Borderías A, Gonzàlez-Juncà A, Rodón L, Folch G, Carmona MA, et al. TGF-β receptor inhibitors target the CD44(high)/Id1(high) glioma-initiating cell population in human glioblastoma. Cancer Cell 2010;18:655–68.

[43] Bao S, Wu Q, Li Z, Sathornsumetee S, Wang H, McLendon RE, et al. Targeting cancer stem cells through L1CAM suppresses glioma growth. Cancer Res 2008;68:6043–8.

[44] Ogden AT, Waziri AE, Lochhead RA, Fusco D, Lopez K, Ellis JA, et al. Identification of A2B5+CD133− tumor-initiating cells in adult human gliomas. Neurosurgery 2008;62:505–14. discussion 514–15.

[45] Ping Y, Yao X, Jiang J, Zhao L, Yu S, Jiang T, et al. The chemokine CXCL12 and its receptor CXCR4 promote glioma stem cell-mediated VEGF production and tumour angiogenesis via PI3K/AKT signalling. J Pathol 2011;224:344–54.

[46] Scheres B. Stem-cell niches: nursery rhymes across kingdoms. Nat Rev Mol Cell Biol 2007;8:345–54.

[47] Scadden DT. The stem-cell niche as an entity of action. Nature 2006;441:1075–9.

[48] Cheng L, Huang Z, Zhou W, Wu Q, Donnola S, Liu JK, et al. Glioblastoma stem cells generate vascular pericytes to support vessel function and tumor growth. Cell 2013;153:139–52.

[49] Gustafsson MV, Zheng X, Pereira T, Gradin K, Jin S, Lundkvist J, et al. Hypoxia requires notch signaling to maintain the undifferentiated cell state. Dev Cell 2005;9:617–28.

[50] Tavazoie M, Van der Veken L, Silva-Vargas V, Louissaint M, Colonna L, Zaidi B, et al. A specialized vascular niche for adult neural stem cells. Cell Stem Cell 2008;3:279–88.

[51] Wang R, Chadalavada K, Wilshire J, Kowalik U, Hovinga KE, Geber A, et al. Glioblastoma stem-like cells give rise to tumour endothelium. Nature 2010;468:829–33.

[52] Al-Hajj M, Wicha MS, Benito-Hernandez A, Morrison SJ, Clarke MF. Prospective identification of tumorigenic breast cancer cells. Proc Natl Acad Sci 2003;100:3983–8.

[53] O'Brien CA, Pollett A, Gallinger S, Dick JE. A human colon cancer cell capable of initiating tumour growth in immunodeficient mice. Nature 2007;445:106–10.

[54] Ricci-Vitiani L, Lombardi DG, Pilozzi E, Biffoni M, Todaro M, Peschle C, et al. Identification and expansion of human colon-cancer-initiating cells. Nature 2007;445:111–5.

[55] Gage FH. Mammalian neural stem cells. Science 2000;287:1433–8.

[56] Kuo CT, Mirzadeh Z, Soriano-Navarro M, Rasin M, Wang D, Shen J, et al. Postnatal deletion of Numb/Numblike reveals repair and remodeling capacity in the subventricular neurogenic niche. Cell 2006;127:1253–64.

[57] Merkle FT, Tramontin AD, García-Verdugo JM, Alvarez-Buylla A. Radial glia give rise to adult neural stem cells in the subventricular zone. Proc Natl Acad Sci USA 2004;101:17528–32.

[58] Palmer TD, Willhoite AR, Gage FH. Vascular niche for adult hippocampal neurogenesis. J Comp Neurol 2000;425:479–94.

[59] Riquelme PA, Drapeau E, Doetsch F. Brain micro-ecologies: neural stem cell niches in the adult mammalian brain. Philos Trans R Soc Lond B Biol Sci 2008;363:123–37.

[60] Sanai N, Tramontin AD, Quiñones-Hinojosa A, Barbaro NM, Gupta N, Kunwar S, et al. Unique astrocyte ribbon in adult human brain contains neural stem cells but lacks chain migration. Nature 2004;427:740–4.

[61] Shen Q, Goderie SK, Jin L, Karanth N, Sun Y, Abramova N, et al. Endothelial cells stimulate self-renewal and expand neurogenesis of neural stem cells. Science 2004;304:1338–40.

[62] Le Bras B, Barallobre M-J, Homman-Ludiye J, Ny A, Wyns S, Tammela T, et al. VEGF-C is a trophic factor for neural progenitors in the vertebrate embryonic brain. Nat Neurosci 2006;9:340–8.

[63] Leventhal C, Rafii S, Rafii D, Shahar A, Goldman SA. Endothelial trophic support of neuronal production and recruitment from the adult mammalian subependyma. Mol Cell Neurosci 1999;13:450–64.

[64] Li Q, Ford MC, Lavik EB, Madri JA. Modeling the neurovascular niche: VEGF- and BDNF-mediated cross-talk between neural stem cells and endothelial cells: an in vitro study. J Neurosci Res 2006;84:1656–68.

[65] Bao S, Wu Q, McLendon RE, Hao Y, Shi Q, Hjelmeland AB, et al. Glioma stem cells promote radioresistance by preferential activation of the DNA damage response. Nature 2006;444:756–60.

[66] Calabrese C, Poppleton H, Kocak M, Hogg TL, Fuller C, Hamner B, et al. A perivascular niche for brain tumor stem cells. Cancer Cell 2007;11:69–82.

[67] Folkins C, Man S, Xu P, Shaked Y, Hicklin DJ, Kerbel RS. Anticancer therapies combining antiangiogenic and tumor cell cytotoxic effects reduce the tumor stem-like cell fraction in glioma xenograft tumors. Cancer Res 2007;67:3560–4.

[68] Vredenburgh JJ, Desjardins A, Herndon JE, Dowell JM, Reardon DA, Quinn JA, et al. Phase II trial of bevacizumab and irinotecan in recurrent malignant glioma. Clin Cancer Res 2007;13:1253–9.

[69] Zhu TS, Costello MA, Talsma CE, Flack CG, Crowley JG, Hamm LL, et al. Endothelial cells create a stem cell niche in glioblastoma by providing NOTCH ligands that nurture self-renewal of cancer stem-like cells. Cancer Res 2011;71:6061–72.

[70] Bao S, Wu Q, Sathornsumetee S, Hao Y, Li Z, Hjelmeland AB, et al. Stem cell-like glioma cells promote tumor angiogenesis through vascular endothelial growth factor. Cancer Res 2006;66:7843–8.

[71] Rong Y, Durden DL, Van Meir EG, Brat DJ. "Pseudopalisading" necrosis in glioblastoma: a familiar morphologic feature that links vascular pathology, hypoxia, and angiogenesis. J Neuropathol Exp Neurol 2006;65:529–39.

[72] McCord AM, Jamal M, Shankavaram UT, Shankavarum UT, Lang FF, Camphausen K, et al. Physiologic oxygen concentration enhances the stem-like properties of CD133+ human glioblastoma cells in vitro. Mol Cancer Res 2009;7:489–97.

[73] Persano L, Rampazzo E, Della Puppa A, Pistollato F, Basso G. The three-layer concentric model of glioblastoma: cancer stem cells, microenvironmental regulation, and therapeutic implications. ScientificWorldJournal 2011;11:1829–41.

[74] Hardee ME, Zagzag D. Mechanisms of glioma-associated neovascularization. Am J Pathol 2012;181:1126–41.

[75] Soeda A, Park M, Lee D, Mintz A, Androutsellis-Theotokis A, McKay RD, et al. Hypoxia promotes expansion of the CD133-positive glioma stem cells through activation of HIF-1alpha. Oncogene 2009;28:3949–59.

[76] Heddleston JM, Li Z, McLendon RE, Hjelmeland AB, Rich JN. The hypoxic microenvironment maintains glioblastoma stem cells and promotes reprogramming towards a cancer stem cell phenotype. Cell Cycle 2009;8:3274–84.

[77] Ye XZ, Xu SL, Xin YH, Yu SC, Ping YF, Chen L, et al. Tumor-associated microglia/macrophages enhance the invasion of glioma stem-like cells via TGF-β1 signaling pathway. J Immunol 2012;189:444–53.

[78] Clement V, Sanchez P, de Tribolet N, Radovanovic I, Ruiz i Altaba A. HEDGEHOG-GLI1 signaling regulates human glioma growth, cancer stem cell self-renewal, and tumorigenicity. Curr Biol 2007;17:165–72.

[79] Le DM, Besson A, Fogg DK, Choi K-S, Waisman DM, Goodyer CG, et al. Exploitation of astrocytes by glioma cells to facilitate invasiveness: a mechanism involving matrix metalloproteinase-2 and the urokinase-type plasminogen activator-plasmin cascade. J Neurosci 2003;23:4034–43.

[80] Zagzag D, Lukyanov Y, Lan L, Ali MA, Esencay M, Mendez O, et al. Hypoxia-inducible factor 1 and VEGF upregulate CXCR4 in glioblastoma: implications for angiogenesis and glioma cell invasion. Lab Invest 2006;86:1221–32.

[81] Xu Q, Liu G, Yuan X, Xu M, Wang H, Ji J, et al. Antigen-specific T-cell response from dendritic cell vaccination using cancer stem-like cell-associated antigens. Stem Cells 2009;27:1734–40.

[82] Phuphanich S, Wheeler CJ, Rudnick JD, Mazer M, Wang H, Nuño MA, et al. Phase I trial of a multi-epitope-pulsed dendritic cell vaccine for patients with newly diagnosed glioblastoma. Cancer Immunol Immunother 2013;62:125–35.

[83] Sloan AE, Nock C, Supko J, Ye X, Rich J, Prados M, et al. Targeting glioma initiating cells in GBM: ABTC-0904, a randomized biomarker and phase 0/II study of vismodegib (GDC-0904) in patients with recurrent GBM. 2014.

[84] Gruber Filbin M, Dabral SK, Pazyra-Murphy MF, Ramkissoon S, Kung AL, Pak E, et al. Coordinate activation of Shh and PI3K signaling in PTEN-deficient glioblastoma: new therapeutic opportunities. Nat Med 2013;19:1518–23.

[85] Ferrara N. VEGF as a therapeutic target in cancer. Oncology 2005;69(Suppl. 3):11–6.

[86] Ferrara N, Gerber H-P, LeCouter J. The biology of VEGF and its receptors. Nat Med 2003;9:669–76.

[87] Folkman J. Tumor angiogenesis: therapeutic implications. N Engl J Med 1971;285:1182–6.

[88] Hanahan D, Folkman J. Patterns and emerging mechanisms of the angiogenic switch during tumorigenesis. Cell 1996;86:353–64.

[89] Reardon DA, Wen PY, Desjardins A, Batchelor TT, Vredenburgh JJ. Glioblastoma multiforme: an emerging paradigm of anti-VEGF therapy. Expert Opin Biol Ther 2008;8:541–53.

[90] Gilbert MR, Dignam JJ, Armstrong TS, Wefel JS, Blumenthal DT, Vogelbaum MA, et al. A randomized trial of bevacizumab for newly diagnosed glioblastoma. N Engl J Med 2014;370:699–708.

[91] Peereboom DM, Ahluwalia MS, Ye X, Supko JG, Hilderbrand SL, Phuphanich S, et al. NABTT 0502: a phase II and pharmacokinetic study of erlotinib and sorafenib for patients with progressive or recurrent glioblastoma multiforme. Neuro Oncol 2013;15:490–6.

[92] Reardon DA, Vredenburgh JJ, Desjardins A, Peters K, Gururangan S, Sampson JH, et al. Effect of CYP3A-inducing anti-epileptics on sorafenib exposure: results of a phase II study of sorafenib plus daily temozolomide in adults with recurrent glioblastoma. J Neurooncol 2011;101:57–66.

[93] Zustovich F, Landi L, Lombardi G, Porta C, Galli L, Fontana A, et al. Sorafenib plus daily low-dose temozolomide for relapsed glioblastoma: a phase II study. Anticancer Res 2013;33:3487–94.

CHAPTER 19

Circulating Tumor Cells, Cancer Stem Cells, and Emerging Microfluidic Detection Technologies With Clinical Applications

M. Kozminsky, S. Nagrath
University of Michigan, Ann Arbor, MI, United States

CONTENTS

INTRODUCTION

The responsibility for the vast majority of cancer-related deaths lies with metastasis, the multistep process by which cancer spreads [1]. Evidence of this progression taking place in the blood was first discovered in 1869 when Thomas Ashworth observed cells morphologically consistent with cancer in a vein in the leg of a cancer patient [2]. These cells have now been identified as circulating tumor cells (CTCs) or those cells shed from the primary tumor that are circulating in the bloodstream and lymphatic vessels of cancer patients [3]. The study of these unusual cells was hindered by surrounding normal blood cells, leading to initial separation attempts using filtration [4]. Additional filtration technologies were ultimately commercialized in 2000 [5], but a major turning point in this field was the introduction of CellSearch [6], which ultimately became the first FDA-approved CTC isolation technology in breast, prostate, and colon cancers. The second defining moment in the development of CTC technologies was the advent of the CTC chip in 2007 [7], marking the first use of microfluidics in CTC isolation and the inception of a new branch of this field. While there are other macroscale CTC isolation technologies still in development [8–10], a primary thrust going forward is the development of next generation microfluidic technologies with an emphasis on increased specificity, sensitivity, throughput, and ease of downstream analysis to maximize their utility in the clinic.

Clinical interest in CTCs lies in their implication in the metastatic cascade. In initiating metastasis, the cells must first leave the primary tumor, travel to a blood or lymphatic vessel, and intravasate, entering the lymphatic system or bloodstream, though those in the lymphatic system often ultimately reach blood circulation [11]. Once in the bloodstream, CTCs face numerous obstacles, such as challenges from the immune system and the possibility of

Cancer Stem Cells. http://dx.doi.org/10.1016/B978-0-12-803892-5.00019-X

apoptosis or anoikis. Not all remaining viable cells are able to extravasate from the bloodstream into a secondary location, and fewer still are able to proliferate. This sequence of events demonstrates the inherent inefficiency of metastasis while underscoring the need to identify those cells that are able to complete the process. CTCs represent a subpopulation of cells that have completed the initial steps of metastasis, and the ability to study them could enable clinicians, scientists, and engineers to identify the cells that pose the greatest threat to the patient. However, there is heterogeneity within the subset of cells that are able to enter circulation, in part because some CTCs may undergo the epithelial to mesenchymal transition (EMT) [12].

The underlying heterogeneity of cells within a patient's disease and within individual tumors is at the root of the cancer stem cell (CSC) hypothesis. In contrast with traditional views on cancer, the CSC hypothesis posits that certain cells within a tumor have the ability to ultimately proliferate and differentiate into multiple cell phenotypes [13]. These CSCs have stemness properties, as the name suggests, and may be able to regenerate a tumor following a therapy that might debulk the majority of the tumor [14]. CSCs are largely defined based on the presence or absence of molecular markers such as CD44, CD24, CD133, $\alpha 2\beta 1$ integrin, and aldehyde dehydrogenase (ALDH) [15–19]; it is worth noting that many of these markers are also associated with other types of stem cells. The most definitive method of identifying CSCs is through the injection of a cell population into a SCID or immune-deficient mouse to check for tumor-initiating capacity [20], though there is a strong interest in the development of alternative in vitro assays. Of particular concern is the interaction between CSCs and the tumor microenvironment with regards to the influence of chemical and mechanical cues from both proximate cells and various extracellular matrix components as well as their potential ability to travel in the blood [21].

There is evidence to support the presence of a subpopulation of CSCs within the population of cancer cells circulating in the bloodstream, prompting questions about the relationship between CTCs and CSCs [22]. Additionally the mutual association between CTCs along the EMT spectrum and similarities between the EMT and CSC phenotypes motivate further investigation. In light of the shared interests and potential overlaps, this chapter will describe the impetus for CTC isolation, current techniques for doing so, and the role of such techniques in the clinic and in the study of cancer and CSCs.

CLINICAL AND BIOLOGICAL SIGNIFICANCE OF CIRCULATING TUMOR CELLS

CTCs are highly relevant in the study of cancer both because of their identity and location. That CTC travel in the bloodstream presents an opportunity for clinicians and researchers as they are easily accessible through a simple blood draw. This "liquid biopsy," as opposed to a traditional biopsy, is relatively noninvasive, lessening the discomfort felt by the patient and allowing for repeated sampling. This positions CTCs to contribute to the initial diagnosis as well as to serve as a marker to track disease progression or response to treatment [3]. Additionally, as mentioned earlier, blood sampling selects for those cells that have already completed the preliminary steps in the metastatic cascade.

The presence of CTCs in the blood reflects their partial representation of a heterogeneous primary tumor. In that respect, CTCs are not only a subset of an initial population of tumor cells but a heterogeneous population of cells unto themselves. CTCs are often defined in terms of immunostaining, with CTCs staining positive for cytokeratin (CK, an epithelial intermediate filament) and 2-(4-amidinophenyl)-1H-indole-6-carboxamidine (DAPI, a stain for A-T rich regions of DNA used to denote the nucleus) and negative for CD45 (a leukocyte marker) [23]. However, this belies the spectrum of physical and molecular traits that may be embodied by CTCs. There is substantial evidence that CTCs may undergo EMT, which would affect the expression of epithelial markers such as the epithelial cell adhesion molecule (EpCAM), CK, and certain cadherins through downregulation, while leading to the upregulation of mesenchymal markers including vimentin and different cadherins [24]. Additionally, cells have been shown to migrate in different modes [25], ameboid and collective, each with different phenotypes, which could potentially lead to different phenotypes of those cells which are ultimately found in the blood. There are further differences among CTCs based on whether or not they express markers from the tissue of origin or if those are lost as they progress through the metastatic cascade [26]. Besides variations in molecular marker expression, CTCs have been shown to be of different sizes. While some cancers shed CTCs that are larger in size than a type mononuclear blood cell, there are reports of CTCs ranging from under 4 to over 30 μm [6]. Other distinctions between CTCs in the blood can be made based on their survival ability. Many CTCs succumb to anoikis, and those remaining may be targeted by the immune system [1]. Those able to survive have been known to coopt healthy cells as a cloaking mechanism, such as using platelets to protect against lysis mediated by natural killer cells [27]. Additionally, interactions with platelets can activate the TGFβ and NF-κB pathways to promote a more invasive and EMT-like phenotype [28]. These sources of heterogeneity show the largely underinterrogated breadth of CTCs of which we are aware, and this is without the introduction of the heterogeneity apparent once CTCs reach the secondary location. The different phenotypes present at the stages leading up to, during, and following the transport of CTCs are in line with the idea of a subpopulation that is able to survive and differentiate into different phenotypes as the disease progresses, often in parallel with the primary tumor; this is the CSC hypothesis (Fig. 19.1).

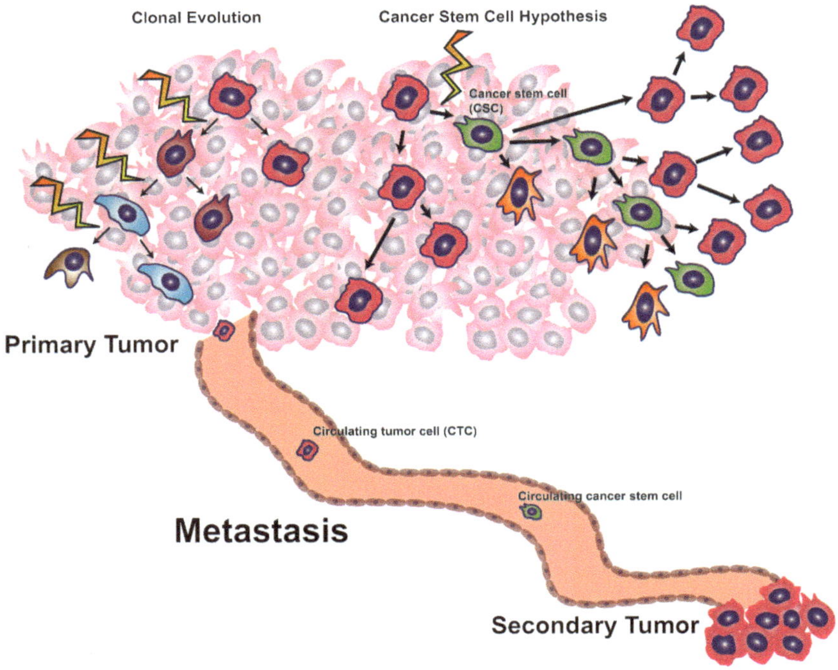

FIGURE 19.1
The biology of CTCs and CSCs. The cancer stem cell hypothesis contrasts with traditional evolution (mutations denoted by lightning bolts). Circulating tumor cells have left the primary tumor and travel to a secondary location.

CSCs are often invoked in the discussion of therapeutic resistance. While many conventional therapies place a focus on debulking the primary tumor, these therapies may fail to eliminate the CSCs that could lead to disease recurrence [29]. This may be in part due to an EMT or EMT-like phenotype embodied by CSCs [30]. Once more, there are parallels between the CSC model of disease progression and the natural selection undergone by CTCs as they progress through the metastatic cascade, potentially as a result of EMT. In that vein, CTCs present an opportunity both for the study of cells which have survived their time in the primary tumor as well as cells that could serve as potential targets en route to their metastatic location.

Beyond biological study, the medical relevance of CTCs has been shown through a number of clinical trials in which CTCs have been shown to have prognostic and predictive significance. The development of these cells as a highly informative biomarker is tantalizing as their accessibility suggests applications in disease profiling, the development and testing of therapeutics, patient stratification, and personalized medicine [31,32]. To this end, studies have been conducted in a number of different cancers, including the three in which the CTC isolation technology CellSearch has been FDA approved: breast, prostate, and colorectal. Generally, a cutoff of a certain number of CTCs per 1 or 7.5 mL blood is decided, after which the two groups established as CTC positive or negative are examined.

Using 5 CTCs/7.5 mL, CTC counts were shown to predict overall survival (OS) and progression-free survival (PFS) when patients were sampled over several time intervals [33,34]. CTCs have contributed information to the HER2 status of the patient, at times contrasting with the HER2 status of the primary tumor [35]. CTC staining results for various relevant markers including HER2, ER, BCL-2, and Ki67 have been incorporated into an index to determine the benefit of therapies such as endocrine therapy (being assessed in a clinical trial at the time of publication) [36]. In prostate cancer, CTC counts have been compared and contrasted with another blood-based biomarker—serum levels of the prostate-specific antigen (PSA). In one study, CTC counts relative to a cutoff were shown to better predict OS than PSA [37]. Baseline CTC counts have also been shown to predict survival, while higher CTC counts were associated with failure of chemotherapeutic regimens or metastasis to the bone [38]. Beyond enumeration, the correlation of ERG rearrangement as determined by fluorescence in situ hybridization (FISH) with response to abiraterone acetate suggests applications in the clinic [39]. In colorectal cancer, clinical utility of CTC counts can be inferred from the correlations found with OS and PFS [40,41] as well as disease stage [42].

The clinical data generated regarding CTCs support the use of information gleaned in the clinical setting. Evidence from CTC counts has been described by the American Society of Clinical Oncology (ASCO) as a three out of five, with one being the strongest evidence [43]. This evidence has been consequently cited in ASCO recommendations. Such data has also been contrasted with TNM staging, with the suggestion of the incorporation of CTC counts into this system [32]. However, there is still a clear need for further information and trials, and given the vast interest in CTCs in the literature, the field seems to be working to address this.

To gain such information and perform the analysis necessary to exploit the full utility of CTCs, a logistical obstacle must be addressed: CTCs exist surrounded by billions of normal blood cells, and they must be isolated to conduct further study. Such isolation is challenging for a number of reasons. As previously stated, there is great biological variability among the various subpopulations of CTCs, making them difficult to identify and separate from nontarget populations of cells such as white blood cells. CTCs present with different surface markers and diameters, complicating the development of technologies that seek to capture these cells with high yield, sensitivity, specificity, throughput, and viability ideally from whole blood. To meet these challenges, CTC isolation technologies have evolved as a field with an emphasis on microfluidic technologies. The development and application of these microfluidic technologies will be the focus of the remainder of this chapter.

EVOLUTION OF CIRCULATING TUMOR CELL ISOLATION TECHNOLOGIES

The general strategy for isolating CTCs from surrounding normal blood cells is to take advantage of the ways that they differ. Broadly, this has manifested in two categories of separation technologies based on differences in surface expression or in physical properties. While many early examples involved separation based on the oft-observed large size of CTCs relative to white blood cells [4], including the commercialized isolation by size of epithelial tumor cells (ISET) [5], a milestone in CTC isolation technologies separated cells of interest based on EpCAM expression. An early example of immunomagnetic separation, CellSearch isolated cells using antibodies against EpCAM conjugated to magnetic beads and is currently the only FDA-approved CTC isolation technology [6]. The CellSearch system is both commercially available and present in analysis laboratories, and allows the automated enumeration of CTCs based on a DAPI+/CK+/CD45-immunostained phenotype. However, the requirement for a large footprint system that fixes cells and allows limited flexibility in further analysis prompted the development of additional technologies (described later; performance metrics are included in Fig. 19.2 and Table 19.1 to facilitate comparison).

Microfluidics is a natural fit for use in CTC isolation. Established techniques can be used to fabricate systems that feature small footprint, low reagent usage, and operation in the laminar flow regime and on the length scale analogous to that experienced in the human body [44,45]. These "labs-on-a-chip" (LOC) or micro total analysis systems (μTAS) have been applied to a number of biological procedures, including PCR [46] and ELISA [47]. The cost of such technologies can vary based on material and fabrication considerations, but the commonly used elastomeric polymer polydimethylsiloxane (PDMS) has facilitated rapid and inexpensive prototyping [48].

Microfluidics made its entry into the field of CTC isolation technologies with the CTC chip in 2007 [7]. Containing 78,000 microposts etched in silicon, the device captured CTCs from whole blood flowing through the device at 1 mL/h using antibodies against EpCAM conjugated to the silicon features. The device demonstrated capture in cancer cell lines from prostate, bladder, breast, and lung cancer as well as from cancer patients in 115 out of 116 samples. Immunofluorescence staining was used to verify that captured cells were

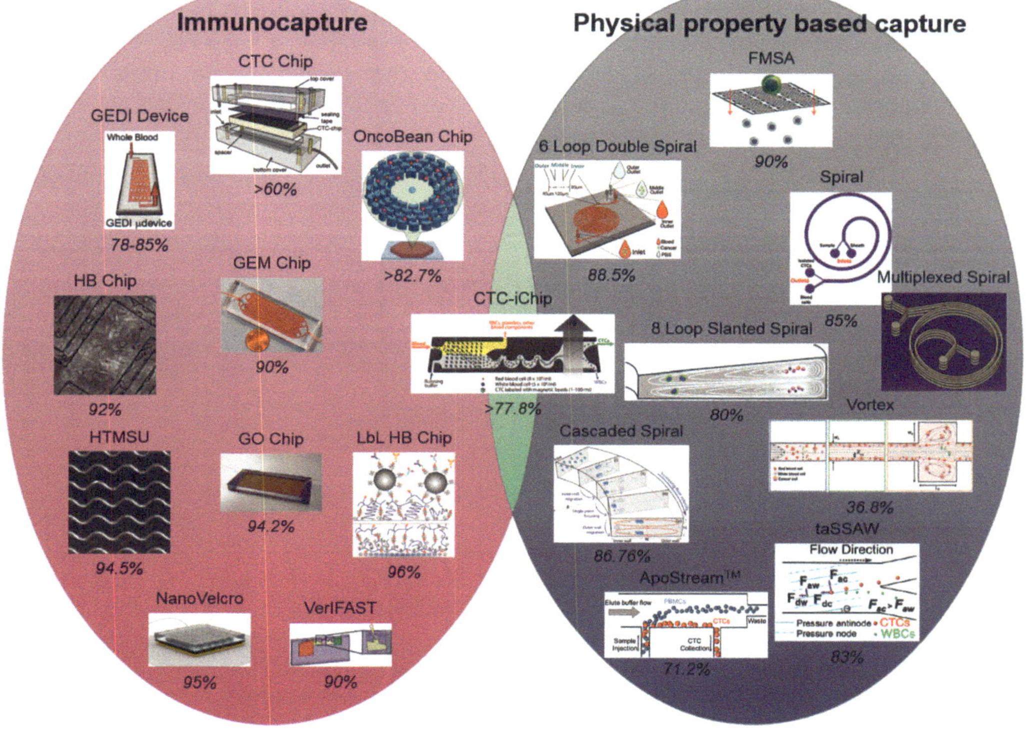

FIGURE 19.2
Microfluidic circulating tumor cell (CTC) isolation technologies. Examples of antigen-based (immunocapture) and physical property-based separation in microfluidic systems. Percentages denote capture efficiency as determined through cell line optimization. *GEDI*, geometrically enhanced differential immunocapture; *GEM*, geometrically enhanced mixing; *HB*, herringbone; *HTMSU*, high-throughput microsampling unit; *GO*, graphene oxide; *LbL*, layer-by-layer; *taSSAW*, tilted-angle standing surface acoustic waves; *FMSA*, flexible micro spring array *Figures reprinted with permission from: Macmillan Publishers Ltd, Copyright 2007 [7] [51]; John Wiley and Sons, Copyright 2014 [53] [54]; Royal Society of Chemistry, Copyright 2014 [56]; American Chemical Society, Copyright 2008 [57]; Macmillan Publishers Ltd, Copyright 2013 [60]; John Wiley and Sons, Copyright 2015 [61]; American Chemical Society, Copyright 2014 [62]; Royal Society of Chemistry, Copyright 2014 [63]; AAAS, Copyright 2013 [64]; Royal Society of Chemistry, Copyright 2012 [67]; Macmillan Publishers Ltd, Copyright 2013 [68] [69]; [70]; AIP Publishing LLC, Copyright 2014 [71]; Royal Society of Chemistry, Copyright 2014 [72]; AIP Publishing LLC, Copyright 2012 [75] [74].*

DAPI+/CK+/CD45−, while white blood cells showed a DAPI+/CD45+phenotype. Additional staining was performed for the PSA and thyroid transcription factor-1 (TTF-1) in patients with the relevant cancers. The device was also used to track patient response to treatment in one colon, two lung, and three pancreatic cancer patients. The initial work with this device was followed up with a study to investigate a specific mutation in the epidermal growth factor receptor (EGFR) gene in CTCs isolated from lung cancer patients [49]. Serial

Table 19.1 CTC Isolation Technologies. Relevant Metrics for CTC Isolation Technologies. Generally, immunocapture features high capture efficiency, while size based capture benefits from high throughput. Some numbers were recalculated from values given in the text to facilitate comparison. Values not included in the Table were not explicitly mentioned in the source text

	Technology	Cell Line Capture Efficiency	Purity	Throughput	References
Immunocapture	CellSearch	85.50%	Low		[6]
	CTC chip	>60%	50%	1 mL/h	[7]
	GEDI	78–85%	68%	1 mL/h	[50]
	Oncobean	82.7–100%	Increases with higher flow rates	Up to 10 mL/h	[53]
	HB chip	92%	14%	1.2 mL/h	[54]
	GEM chip	90%	84%	3.6 mL/h	[56]
	HTMSU	94.50%		1.6 mL/h	[57]
	GO chip	94.20%	High	1 mL/h	[60]
	LbL HB chip	96%	High		[61]
	NanoVelcro	95%		0.5 mL/h	[62]
	VerIFAST	90%			[63]
	CTC-iChip	77.8–98.6%	2.5–3.5 log depletion	8 mL/h	[64]
Physical property-based capture	FMSA	92.6%	1.4×10^4	Around 45 mL/h	[65]
	6 Loop double spiral	88.5%	Enrichment factor of 19	20 mL/h	[67]
	Spiral	>85%	3 log depletion	3 mL/h	[68]
	Multiplex spiral		Next generation spiral		[69]
	8 Loop slanted spiral	>80%	4 log depletion	56.2 mL/h	[70]
	Cascaded spiral	86.76%	97.91% depletion	33 mL/h	[71]
	Vortex technology	10–20%	57–95% for clinical samples		[72]
	taSSAW	>83%	1 log depletion	1.2 mL/h	[74]
	ApoStream (DEP)	70%	2–3 log depletion	1 mL/h	[75]

monitoring showed correlations between CTC counts and radiographic tumor responses. Following treatment, subsequent CTCs isolated showed increased mutational status. The multiple examples of clinical relevance demonstrated by the CTC chip show the ability of this type of technology to be incorporated into medical research, paving the way for subsequent technologies.

Advances in Microfluidic Immunocapture

While the CTC chip represented a positive first step toward the inclusion of microfluidics into CTC research, there remained room for improvement in metrics relevant to further analysis. Increased specificity, selectivity, purity, and throughput were common objectives for devices developed going forward. Many of these devices shared the CTC chip's use of capture antibodies conjugated to three dimensional structures as the initial basis for their design.

The etched silicon microposts themselves were subject to design optimization in the geometrically enhanced differential immunocapture (GEDI) device [50]. Computer simulations guided the shape and layout of the microposts. The specifics of the antibody choice, a monoclonal antibody against the prostate-specific membrane antigen (PSMA), and the cell line used for characterization, LNCaP, suggested prostate cancer–specific applications, and some follow-up work with this device has examined CTC counts, ERG rearrangement status, and on-chip drug testing in castrate resistant prostate cancer [51] and copy number variation in prostate cancer cell lines [52].

Further optimization of the micropost shape to reduce flow separation and consequently increase the surface of the micropost available for capture was conducted in the development of the OncoBean chip [53]. While the CTC chip and GEDI device were operated at 1 mL/h, the OncoBean chip was designed in a radial flow configuration to allow processing at higher flow rates while maintaining low shear stresses to allow viable cell capture. Initial validation of the device was conducted with two patient samples each from pancreatic, lung, and breast cancers, with parallel devices run at 1 and 10 mL/h yielding comparable results. An additional advantage of the OncoBean chip is its fabrication in PDMS, which is less expensive and chemically demanding than silicon etching.

Prior work in PDMS features had been conducted to increase collisions between cells in laminar flow and anti-EpCAM functionalized surfaces. Chaotic micromixers were fabricated using soft lithography to form the ceiling of a channel that was subsequently multiplexed in the Herringbone chip (HB chip) [54]. Like the GEDI device, much of the early work and optimization was performed in prostate cancer, with the ability to perform downstream analysis demonstrated through on-chip FISH for androgen receptor amplification in a cell line and RNA sequencing for ERG rearrangement in two patient samples. This device was also used to evaluate androgen signaling based on the expression of PSA and PSMA via immunofluorescence staining [55]. Expression phenotypes changed over time and were correlated with response to treatment. The concept of a PDMS herringbone chamber was also applied in the geometrically enhanced mixing (GEM) chip [56], which was characterized with pancreatic cancer cell lines, used to serially monitor pancreatic cancer patients.

Polymers other than PDMS have also been used for further CTC isolation technologies, with hot-embossing as an option for patterning and fabrication. Poly(methyl methacrylate) (PMMA) was embossed with 51 high-aspect ratio microchannels functionalized with anti-EpCAM for cell capture in the high-throughput microsampling unit (HTMSU) [57]. The original device featured a conductivity sensor to automatically enumerate cells released by trypsin, and subsequent generations of this technology have incorporated different polymers (cyclic olefin copolymer (COC)) and additional staining and imaging modules [58].

The materials serving as the basis or functional component of CTC isolation technologies has led the field to consider nanomaterials given their advantageous properties of high surface area to volume ratio and their length scale on par with that of extracellular features [59]. One example of the application of a nanomaterial to increase specificity and sensitivity through enhanced antibody presentation is the graphene oxide (GO) chip [60]. In contrast with many of the previous devices, the GO chip featured effectively 2D flower-shaped gold patterns to promote the self-assembly of functionalized GO on the substrate surface. Characterized using a high EpCAM expressing breast cancer cell line (MCF-7) and a low EpCAM expressing prostate cancer cell line (PC-3) with as few as 3–5 cells/mL whole blood, this device demonstrated the advantages of incorporating a nanomaterial into a capture platform. The GO chip was then used to isolate CTCs from breast, pancreatic, and early lung cancer patients. Additional staining for ZEB1 in pancreatic cancer patient samples and HER2 in breast cancer patient samples showed the utility afforded by this device in analyzing EMT and tissue-specific markers, respectively. At the transcriptomic level, RNA was extracted and relative HER2 expression was determined using qRT-PCR.

While the previously mentioned microfluidic devices feature improvements upon early CTC separation technologies, they also feature the drawback of cells remaining on the chip or necessitating trypsin for release. Two recently developed devices took advantage of nanomaterials and nanostructures within microfluidic devices for effective CTC capture and release. A next-generation HB chip consisted of a layer-by-layer (LbL) film within a PDMS herringbone chamber, alternating between biotinylated gelatin and streptavidin [61]. The surface was ultimately coated with streptavidin-coated nanoparticles to serve as a site to bind a cocktail of biotinylated antibodies including anti-EpCAM, anti-HER2, and anti-EGFR. Release could be triggered in two different modes: individual through microtip vibration or bulk through an increase in temperature to dissolve the gelatin. The device was used to analyze breast and lung samples, with single cells released for genomic analysis for point mutations in relevant oncogenes.

Bulk release was also achieved in the thermoresponsive NanoVelcro CTC purification system using a thermoresponsive polymer coated on an array of silicon nanopillars [62]. Cells were captured on the surface of a microfluidic device also enclosed in a PDMS herringbone chamber while the system was maintained at 37 °C with the polymer with the capture antibody exposed. Reducing the temperature to 4 °C (below the lower critical solution temperature or

LCST) caused the polymer to assume a conformation that renders the antibody hidden from the surface, releasing the cells. Purity was improved with additional capture/release cycles. Nonsmall cell lung cancer (NSCLC) patient samples were analyzed with NanoVelcro, with one patient undergoing serial profiling of released CTCs to check for changing mutational status in EGFR.

The issue of cell release is rendered moot in the case of microfluidic immunomagnetic separation wherein the capture antibodies are conjugated to magnetic beads. The beads are preincubated with the sample, which may have undergone some preprocessing and then flowed or moved through a channel to ultimately yield the cells of interest. Two prominent examples in this genre have different focuses with regard to application. The VerIFAST system [63], the next generation of the immiscible filtration assisted by surface tension (IFAST) system, was designed with an emphasis on practical operation. Following incubation with antibody conjugated paramagnetic particles, magnetically labeled cells could then be dragged through oil pinning wells using a handheld magnet, bringing target cells through to the next stage while blood cells remained in the input well. Following incubation in the staining well, the cells could be moved to a sieve well where magnetic particles were removed and any intracellular staining was conducted. This technique was applied to blood and lavage samples, allowing staining for TTF-1 and EGFR. While limited to staining applications, the VerIFAST used immunomagnetic separation and the lack of interaction between oil and aqueous solutions to streamline the capture and staining process.

Another example of immunomagnetic isolation prioritized versatility, resulting in a multiregion microfluidic device called the CTC-iChip that took advantage of multiple modes of separation [64]. Samples were labeled with magnetic beads before introduction to the chip, at which point running buffer diluted the sample while performing initial separation based on cell size. Remaining cells were then aligned using a focusing channel for ultimate magnetic separation. The device could be operated in either positive or negative selection mode based on the choice of antibody conjugated to the magnetic bead. Positive selection was performed using beads with antibodies to select CTCs, while negative selection selected for white blood cells, leaving all unlabeled cells for further analysis. Isolated cells could undergo a wide spectrum of analysis including immunofluorescence, hematoxylin and eosin, and Pap staining; FISH; and gene expression analysis. Over the course of the study, patient samples were processed from breast, lung, prostate, pancreatic, colorectal, and melanoma cancers.

Microfluidic Separation Based on Physical Properties

The initial separation performed in the CTC-iChip takes advantage of another difference between CTCs and white blood cells differing physical characteristics, such as size. Early CTC separation occurred using filtration, and there have been many updates to this particular technique. A flexible micro spring array (FMSA) fabricated in parylene-C using photolithography allowed the gentle processing of whole blood [65]. Breast cancer and melanoma cell lines were used to optimize gap width with respect to capture efficiency and enrichment against leukocytes. The pressure-driven setup was used to process 21 patient samples from breast, lung, and colorectal cancers and the results positively compared with parallel samples run using CellSearch.

Size also affects how cells experience the multiple forces acting in a microchannel. Consequently, a subfield has arisen that exploits the relative magnitudes these forces to focus cells along different streamlines. These streamlines are then separated into multiple outlets, with one outlet containing purified CTCs. Known as inertial focusing, a large body of work in this area describes cells traveling in curved channels where the cells undergo both inertial lift and Dean forces to influence their position in the channel [66].

The curved microchannels used in cell separation often take the form of spirals formed in PDMS. One example of such a device was a six-loop double spiral developed following numerical simulations conducted using the computational fluid dynamics software Fluent [67]. The device had three outlets with cancer cells focused at a position that enabled isolation from the middle outlet. An advantage of this device common across inertial focusing technologies was the ability to operate at substantially higher throughput than immunocapture technologies, in this case 20 mL/h. However, a common disadvantage of inertial sorting also displayed by this technology is the need for preprocessing. During device characterization with HeLa and MCF-7 cell lines spiked into whole blood, 50× dilution was required prior to running the sample.

Another spiral design consisted of a single spiral with two outlets [68]. This device was also characterized with MCF-7 cell line, and cells taken from outlets were subsequently cultured, with this protocol being facilitated by virtue of the cells not being tethered to the device surface. Beyond cell line studies, 20 metastatic lung cancer patient samples were processed and analyzed with staining for the traditional CTC markers. In addition to canonical staining, CTCs from some advanced stage patients were stained for CD133, showing the potential application of this technology in hunting for circulating CSCs. To further increase throughput, this spiral design was multiplexed by stacking three spiral channels, this time with two inlets to allow provide sheath flow [69]. Samples to be analyzed using this technology underwent extensive preprocessing: plasma was first separated to undergo additional genetic analysis after which red blood cells (RBCs) were lysed. Fifty-eight patients from metastatic breast or NSCLC were processed to yield CTCs in a format that enabled downstream analysis. CTCs from NSCLC patients underwent FISH for the ALK rearrangement or single allele base extension reaction (SABER), which showed EGFR mutations.

This level of downstream analysis was also displayed on cells isolated using an eight loop slanted spiral that featured a trapezoidal cross section [70]. The design change was prompted by efforts to increase separation resolution, which was achieved by introducing an asymmetry which skewed the Dean vortex formation. RBC lysis and dilution was required for sample processing. The device was optimized with cell spikes with MCF-7, T24, and MDA-MB-231 cell lines, and cell culture following isolation was demonstrated. Five metastatic breast cancer and five NSCLC patient samples were processed. Subsequent analysis consisted of FISH for HER2 or staining for the stemness markers CD44 and CD24. Staining results showed distinct subpopulations of CD44+/CD24− cells as well as CD44−/CD24+. Interestingly, it was observed that cells expressing the typical CSC phenotype (CD44+/CD24−) were larger than the CD44−/CD24 + cells, once more suggesting the role of this technology in CSC isolation and study.

Other spiral microdevices focused on the study of the optimization protocol. A cascaded spiral was used to study the effects of individually considered design parameters including flow velocity, channel height, width, and radius of curvature on the migration patterns of 10 and 20 μm beads [71]. While the channel width did not significantly affect migration patterns, increases in channel height resulted in a left shift of both sizes of particles. Inner wall migration was shown to increase with increasing curvature. The results were then used to design a spiral device with intermediate and final sets of outlets that successfully separated MCF-7 cells from leukocytes.

Similarly, the balance between shear-gradient lift forces and wall effect lift forces can be used to focus cells in a straight channel. Such a design was used to focus cancer cells toward channel walls where they could be trapped in vortices formed in expansion–contraction reservoirs [72]. Small cells, ideally nontarget blood cells, were washed out using high flow rates while larger cells caught in the reservoir vortices were subsequently washed out using low flow rates. While this technology required preprocessing through RBC lysis and dilution, it did feature high throughput and sufficient viability to perform culture of processed cell line cells. Four breast cancer and four lung cancer patients were processed, showing clinical utility.

In addition to the way different cells are influenced by different forces in a microchannel, different types of cells focus differently based on the balance of drag forces and acoustic radiation forces. Initial attempts at acoustophoresis using tilted-angle standing surface acoustic waves (taSSAW) followed up numerical simulations with cell line optimization with and MCF-7/isolated white blood cell solution to determine the angle of inclination and power that maximized separation [73]. While successful, this was hampered by the low throughput of 120 μL/h. Follow-up included characterization with additional cell lines, including for cell line culture, to optimize additional design parameters (tilt angle and length of interdigitated transducers) to increase the throughput to 1.2 mL/h [74]. A protocol of RBC lysis and centrifugation enabled the processing of two breast cancer patient samples.

Analogously, cells can be separated using dielectrophoresis (DEP) forces. ApoStream, a commercialized DEP-based technology, used a continuous flow setup in which positive DEP forces attracted cancer cells to the base of the device which contained one outlet while white blood cells were elevated by negative DEP forces to a different outlet [75]. Cell lines were used to optimize the frequency, voltage, flow rate, and buffer components. Operated at up to 1.5 mL/h, the device featured good viability and the potential for subsequent culture.

Relative Merits of Different Isolation Methods

Surface marker expression and physical property differences are two broad categories by which the various microfluidic methods of isolating CTCs can be classified, each having inherent advantages and disadvantages. Immunocapture promises high specificity. While downstream analysis of cells captured on antibody coated structures and substrates is often limited as they are affixed to the chip, many new technologies are placing an emphasis on release strategies. However, positive selection antibody-based capture techniques are also limited by the selected antibody. As not all CTCs will express any given marker, especially the commonly assumed EpCAM, immunocapture may miss those cells of strongest interest—such as CSCs. The downregulation of epithelial markers such as EpCAM during EMT make this the greatest concern about the application of immunocapture.

Physical property-based isolation techniques feature high throughput and antigen agnostic separation but are based on an assumption that all CTCs fall within a size range that does not overlap with blood cells. While the mutability of surface marker expression does not affect these techniques, making them promising for use for less well characterized or highly variable populations such as CSCs, more information about size distribution of CSCs will likely be necessary to ensure the adequacy of these technologies. Additionally, cumbersome preprocessing steps including RBC lysis, multiple centrifugations, and dilution affect the sample and can lead to the loss of CTCs. DEP techniques require buffer optimization, and the throughput of both DEP and acoustophoresis are areas for improvement.

As shown in the next section, many current CTC isolation techniques have proven their utility in the clinic, but the ideal device that encompasses a highly heterogeneous population in a logistically convenient way that permits downstream analysis is still a goal and not yet a reality.

MICROFLUIDIC TECHNOLOGIES ENABLING FURTHER CIRCULATING TUMOR CELL CHARACTERIZATION AND DOWNSTREAM ANALYSIS

Nearly all the technologies cataloged previously feature patient sample processing and subsequent CTC enumeration, the most basic benchmark of clinical potential. Beginning with CellSearch's use in breast and other cancers, CTC counts have been shown to correlate with various clinical endpoints and characteristics [43], making it a natural starting point for analysis, particularly given the ability to stain cells on-chip.

One potential application of CTC enumeration is in early disease detection. The GEDI platform has been used to sample for CTCs in three cohorts with the aim of assessing the potential for early detection, including patients with precancerous cystic lesions, patients with pancreatic ductal adenocarcinoma (PDAC), and a control group (Fig. 19.3A) [76]. Using a cutoff of 3 CTCs/mL, there was no detection in the control, while at least this level of CTCs was detected in 33% of patients with cystic lesions and 73% of patients with PDAC. Based on the disease trajectory of these patients with precancerous lesions, future studies could show the utility of this device for use in a cancer that desperately needs new techniques for early detection.

The relative ease of serial blood sampling compared with multiple invasive biopsies points to the potential value of CTC isolation technologies to serially monitor patient progression and response to treatment. The HTMSU has been used to try to correlate CTC burden with disease burden using a patient-derived xenograft (PDX) mouse model of PDAC [77]. CTCs were enumerated pre and post treatment, and while there was not a statistically significant correlation between CTC number and tumor volume, there was a correlation between the change in CTC counts and the change in tumor burden. Additionally, those mice undergoing treatment with BKM120, a phosphatidylinositol-3-kinase inhibitor, showed a decrease in CTC burden. In humans, the FMSA device was used to follow the number of CTCs present at various points during metastasectomy in colorectal cancer patients, obtaining counts at the incision, during the resection, 30 min after the resection, and one day follow-up (Fig. 19.3B) [78]. Significantly higher numbers of CTCs were detected at all time points using the FMSA device compared with CellSearch, and the FMSA device showed significantly higher numbers of CTCs during resection compared to other time points analyzed with the FMSA device. The GEM chip was not only verified with pancreatic patient samples but was also used to serially monitor CTC burden over the course of multiple treatment cycles in three patients, showing correlation tumor size as measured by CT scans [56]. Response to gefitinib treatment in one lung cancer patient was monitored using the NanoVelcro technology, not only by assessing the CTC burden and radiographically assessing the tumor

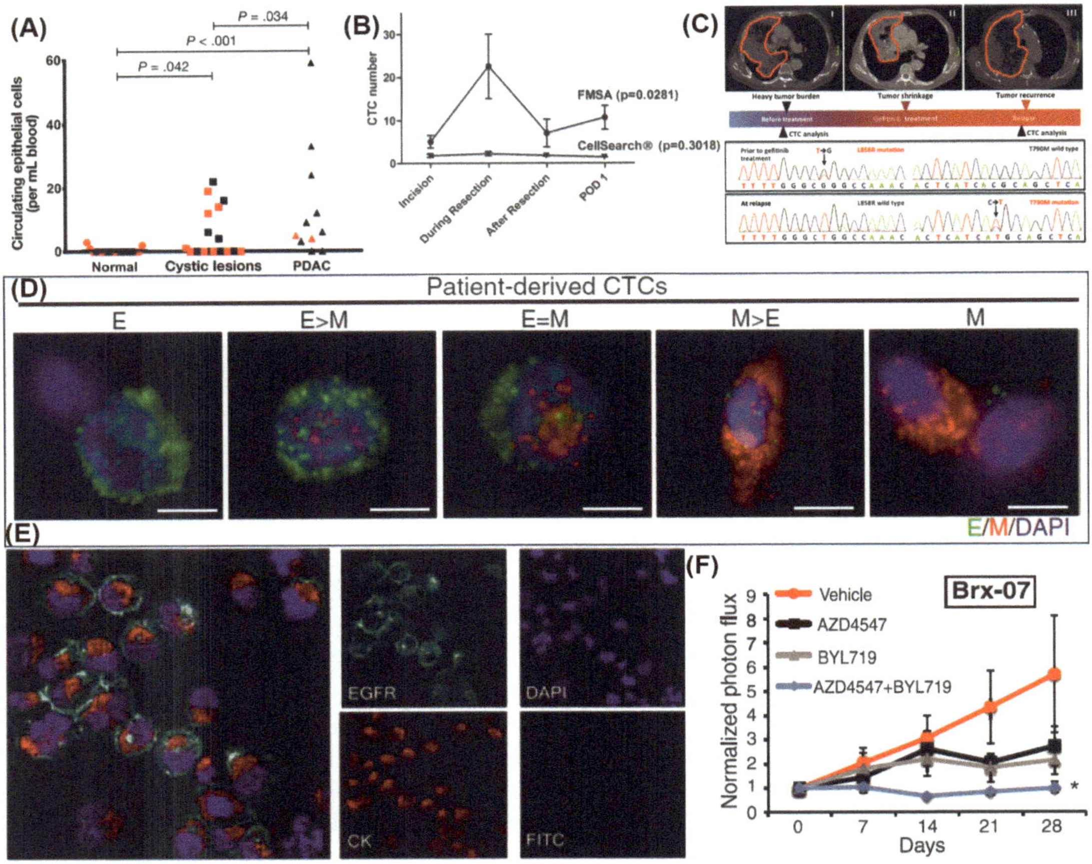

FIGURE 19.3

Clinical applications of circulating tumor cell (CTC) isolation technologies. (A) The application of the geometrically enhanced differential immunocapture microdevice for early detection in pancreatic ductal adenocarcinoma (PDAC). *(Reprinted with permission from Elsevier, Copyright 2014 [76].)* (B) The flexible micro spring array (FMSA) has been used to track CTC enumeration throughout metastasectomy in colorectal cancer patients. *(Reprinted with permission from Taylor & Francis, Copyright 2015 [78].)* (C) In addition to radiographic imaging of the primary tumor, CTCs isolated from a lung cancer patient sample using the NanoVelcro system underwent Sanger sequencing to detect mutations in epidermal growth factor receptor (EGFR). *(Reprinted with permission from American Chemical Society, Copyright 2014 [62].)* (D) Breast cancer CTCs fall along an epithelial spectrum as visualized through RNA in situ hybridization on patient-derived CTCs isolated using the herringbone chip. *(Reprinted with permission from AAAS, Copyright 2013 [12].)* (E) EGFR and cytokeratin (CK) staining on expanded CTCs isolated from lung cancer patients using a modified CTC chip [80]. (F) Drug testing on mouse xenograft model formed by expanded breast cancer CTCs isolated using the CTC-iChip. *(Reprinted with permission from AAAS, Copyright 2014 [81].)*

size, but also by conducting whole genome amplification and Sanger sequencing to detect relevant mutations in EGFR (Fig. 19.3C) [62].

As shown in the last example, enumerating CTCs may only be the tip of the informational iceberg, with genomic, expression, and proteomic analysis likely being able to provide insights into the fundamental mechanisms of how cancer spreads. The HB chip functionalized with an antibody cocktail including EpCAM, EGFR, and HER2 has been used to study RNA levels of transcripts associated with various points on the spectrum of EMT in breast cancer patient CTCs (Fig. 19.3D) [12]. RNA in situ hybridization (ISH) was used to visualize epithelial transcripts (keratins 5, 7, 8, 18, 19 as well as EpCAM and cadherin 1) and mesenchymal transcripts (fibronectin, cadherin 2, serpin peptidase inhibitor clade E). The varying levels were tracked in patients as they underwent treatment and noted increased mesenchymal marker expression as associated with clusters of CTCs. Additionally, RNA was sequenced yielding the discovery of enriched transcripts for ECM proteins, therapy resistance signatures, and transforming growth factor β (TGFβ) found in mesenchymal and clustering CTCs. CTCs isolated using the CTC-iChip from a mouse model of pancreatic cancer have undergone single cell RNA sequencing, also showing high expression of ECM proteins such as SPARC [79].

CTCs isolated using microfluidic platforms can yield information through a number of different staining techniques and a variety of "omics" analyses. However, a major limitation to the study of CTCs is their rarity. In addition to their low frequency in the bloodstream, 100% efficient separation would still yield few cells. Consequently, an area of emerging investigation is to culture these cells ex vivo. Culture of isolated CTCs from three colorectal patients was achieved using the FMSA device, and preliminary staining showed a macrophage-like phenotype in addition to the epithelial phenotype [78]. A technique to culture isolated cells on-chip has been optimizing using a modified CTC chip, a mixture of collagen and Matrigel, and fibroblasts for coculture (Fig. 19.3E) [80]. Expansion was achieved in 14 out of 19 lung cancer patients assessed for this purpose. Cultured CTCs were released with trypsin for further analysis such as transwell invasion assays and next-generation sequencing. CTCs isolated using the CTC-iChip have also been cultured as spheres under hypoxic conditions in a serum-free media containing growth factors, although this technique only showed success in 6 out of 36 attempts [81]. These cells could then be stained using immunofluorescence and immunohistochemistry (IHC) and could undergo next-generation sequencing, revealing several relevant mutations. These mutations were used to inform the analysis of the subsequent drug testing of various therapeutic regimens and targeted therapies on expanded CTCs. Response to the tested drugs was used to select two drugs to try on a mouse xenograft model formed using expanded CTCs. Each individual drug suppressed tumor growth while the combination prevented any additional growth of the tumor (Fig. 19.3F). The use of cultured CTCs for functional testing and therapeutic guidance represents the crossing into the next frontier of CTC research, where the results of CTC analysis can be used to characterize an individual's disease to optimize treatment.

From enumeration to genetic analysis to culture for even further analysis, the host of microfluidic technologies available for use in a clinical setting is enabling the study of individuals' diseases. These studies show the potential of microfluidic CTC isolation technologies in the clinic, though the applications with respect to CSCs remain largely unexplored. However, work with genetic and functional studies as well as the study of EMT shows the promise of these technologies in CSC research.

SUMMARY OF CIRCULATING TUMOR CELL CHARACTERIZATION APPROACHES

Given the mutable definition of a CTC, much of the study of CTCs has involved their characterization. CTCs are largely defined by their protein expression as determined by immunostaining, and the vast majority of studies mentioned thus far have been employing this technique. However, CTCs have also been defined by their size, morphology, or nuclear to cytoplasmic ratio [82]. Moreover, once these CTCs are indeed identified, it is their thorough characterization that will likely provide the most clinical insight.

Beyond the standard immunocytochemistry or immunohistochemistry, a number of techniques have been used to characterize CTCs at various levels of biology. At the DNA level, both Sanger and next-generation sequencing have been used to search for relevant mutations, although preliminary amplification was required either through whole genome amplification [62] or CTC expansion [81]. FISH has also been used to visualize androgen receptor amplification [54]. Transcriptomic analysis of CTCs has yielded RNA sequencing results [12,54,79–81] and expression levels [80]. Visualization of EMT markers in CTCs at the RNA level has been achieved through RNA ISH [12]. The CTC secretome has also been investigated via the ELISPOT and EPISPOT assays [83]. In addition to providing sufficient DNA for genomic analysis, expansion of viable CTCs through culture enables additional characterization, such as the EPISPOT assay as shown in the successful expansion of CTCs enriched through negative selection [84]. Cultured CTCs have enabled the evaluation of invasiveness [80] and drug susceptibility [81]. Additionally, these expanded cells have generated tumors when implanted as xenografts immunodeficient mice both orthotopically [81] and subcutaneously [84]. Xenografts have also been formed by CTCs from small cell lung cancer patients that were enriched but not expanded; the response of the mice to cisplatin and etoposide was tested, and the xenografts underwent genomic analysis [85]. While a number of techniques have been used to characterize CTCs isolated through microfluidic methods, most of them have been performed at the proof-of-concept scale. Incorporation of this host of analytical methods into the clinic will require further validation, the dedication of resources, and the refinement of such protocols. However, this plethora of tools can also allow CTCs to be further examined for a stem cell subpopulation.

CONCLUSION

Since their discovery in the late 19th century, CTCs have piqued the interest of researchers due to their critical role in the metastatic cascade. Given their clinical implications and accessible location, there has been major interest in the development of technologies to isolate and study CTCs. Through the use of antigen-based and size-based capture, a number of microfluidic devices have been put forth in attempts to improve upon initial macroscale separation methods. The objective remains to isolate these rare cells efficiently while maximizing specificity, purity, viability, and the ability to conduct downstream analysis, although great strides have been made in these areas as shown by applications in medical settings.

As there is a hypothesized stem population with the subpopulation of tumor cells that comprise CTCs, there is also an interest in ensuring that current technologies are able to address CTC heterogeneity. While a large number of clinical applications have entailed the use of immunocapture, to use this technique in CSC research will require clear surface marker-based definitions of CSCs that do not overlap with those markers expressed by blood cells. Similarly, size-based methods will only be effective so long as the CSC population is larger than the blood cell population. Challenges to the study of CTCs and CSCs are largely a function of this lack of consistent definition that is inherent to these populations. Consequently, research going forward will necessitate the incorporation of creative solutions in the development of new technologies to further approach the aim of studying these potentially valuable yet elusive cells.

Future directions beyond improvement to detection and isolation technologies include the further incorporation of CTC enumeration and study into clinical trials. The ability to study CTCs beyond enumeration is also an area of interest, with single cell genomics being of particular emphasis given the heterogeneity of the population. Additionally, the recovery and expansion of viable CTCs opens the door to functional studies such as invasion assays as well as the development of individualized treatment regimens through drug testing of patient CTCs. Future work flows could involve the identification of relevant mutations to select drugs targeting those mutations, or to first test panels of drugs on expanded CTCs and refining these choices based on additional testing in CTC-derived xenograft mice. Key issues still faced by the field are those that prevent such downstream analysis from occurring, such as purity and sensitivity considerations, low yield of captured CTCs, and in some instances, retrieval of CTCs from the capture system itself. With isolation and identification representing the first wave of CTC studies, the future holds their further interrogation for maximum utility in the clinical setting.

List of Acronyms and Abbreviations

ALDH Aldehyde dehydrogenase
ALK Anaplastic lymphoma kinase
ASCO American Society of Clinical Oncology
CK Cytokeratin
COC Cyclic olefin copolymer
CSCs Cancer stem cells
CTCs Circulating tumor cells
DAPI 4′,6-diamidino-2-phenylindole
DEP Dielectrophoresis
ELISA Enzyme-linked immunosorbent assay
ELISPOT Enzyme-linked immunospot
EMT Epithelial to mesenchymal transition
EpCAM Epithelial cellular adhesion molecule
EPISPOT Epithelial immunospot
FISH Fluorescence in situ hybridization
FMSA Flexible micro spring array
GEDI Geometrically enhanced differential immunocapture
GEM Geometrically enhanced mixing
GO Graphene oxide
HB Herringbone
HTMSU High-throughput microsampling unit
IFAST Immiscible filtration assisted by surface tension
IHC Immunohistochemistry
ISET Isolation by size of epithelial tumor cells
LbL Layer-by-layer
LCST Lower critical solution temperature
LOC Lab-on-a-chip
NSCLC Nonsmall cell lung cancer
OS Overall survival
PCR Polymerase chain reaction
PDAC Pancreatic ductal adenocarcinoma
PDMS Polydimethylsiloxane
PDX Patient-derived xenograft
PFS Progression-free survival
PMMA Poly(methyl methacrylate)
PSA Prostate-specific antigen
PSMA Prostate-specific membrane antigen
RBC Red blood cell
SABER Single allele base extension reaction
SCID Severe combined immunodeficiency
taSSAW Tilted-angle standing surface acoustic waves
TGFβ Transforming growth factor β

TTF-1 Thyroid transcription factor-1
WBC White blood cell
μTAS Micro total analysis systems

References

[1] Gupta GP, Massagué J. Cancer metastasis: building a framework. Cell 2006;127(4):679–95.

[2] Ashworth TR. A case of cancer in which cells similar to those in the tumours were seen in the blood after death. Aust Med J 1869;14(3):146–9.

[3] Paterlini-Brechot P, Benali NL. Circulating tumor cells (CTC) detection: clinical impact and future directions. Cancer Lett 2007;253(2):180–204.

[4] Seal SH. A sieve for the isolation of cancer cells and other large cells from the blood. Cancer 1964;17(5):637–42.

[5] Vona G, Sabile A, Louha M, Sitruk V, Romana S, Schutze K, et al. Isolation by size of epithelial tumor cells – a new method for the immunomorphological and molecular characterization of circulating tumor cells. Am J Pathol 2000;156(1):57–63.

[6] Allard WJ, Matera J, Miller MC, Repollet M, Connelly MC, Rao C, et al. Tumor cells circulate in the peripheral blood of all major carcinomas but not in healthy subjects or patients with nonmalignant diseases. Clin Cancer Res 2004;10(20):6897–904.

[7] Nagrath S, Sequist LV, Maheswaran S, Bell DW, Irimia D, Ulkus L, et al. Isolation of rare circulating tumour cells in cancer patients by microchip technology. Nature 2007;450(7173):1235–9.

[8] Talasaz AH, Powell AA, Huber DE, Berbee JG, Roh KH, Yu W, et al. Isolating highly enriched populations of circulating epithelial cells and other rare cells from blood using a magnetic sweeper device. Proc Natl Acad Sci USA 2009;106(10):3970–5.

[9] Peters C, Landsorp P, Eaves A, Thomas T, editors. Onestep non-magnetic prep highly purified cell populations unprocessed Hum whole blood. Exp Hematol 1999. Elsevier Science Inc. 655 Avenue of the Americas, New York, NU 10010 USA.

[10] Saucedo-Zeni N, Mewes S, Niestroj R, Gasiorowski L, Murawa D, Nowaczyk P, et al. A novel method for the in vivo isolation of circulating tumor cells from peripheral blood of cancer patients using a functionalized and structured medical wire. Int J Oncol 2012;41(4):1241–50.

[11] Wong SY, Hynes RO. Lymphatic or hematogenous dissemination: how does a metastatic tumor cell decide? Cell Cycle 2006;5(8):812–7.

[12] Yu M, Bardia A, Wittner BS, Stott SL, Smas ME, Ting DT, et al. Circulating breast tumor cells exhibit dynamic changes in epithelial and mesenchymal composition. Science 2013;339(6119):580–4.

[13] Gupta PB, Chaffer CL, Weinberg RA. Cancer stem cells: mirage or reality? Nat Med 2009;15(9):1010–2.

[14] Visvader JE, Lindeman GJ. Cancer stem cells in solid tumours: accumulating evidence and unresolved questions. Nat Rev Cancer 2008;8(10):755–68.

[15] Al-Hajj M, Wicha MS, Benito-Hernandez A, Morrison SJ, Clarke MF. Prospective identification of tumorigenic breast cancer cells. Proc Natl Acad Sci USA 2003;100(7):3983–8.

[16] Singh SK, Clarke ID, Terasaki M, Bonn VE, Hawkins C, Squire J, et al. Identification of a cancer stem cell in human brain tumors. Cancer Res 2003;63(18):5821–8.

[17] Collins AT, Berry PA, Hyde C, Stower MJ, Maitland NJ. Prospective identification of tumorigenic prostate cancer stem cells. Cancer Res 2005;65(23):10946–51.

[18] Ma I, Allan AL. The role of human aldehyde dehydrogenase in normal and cancer stem cells. Stem Cell Rev Rep 2011;7(2):292–306.

[19] Li C, Heidt DG, Dalerba P, Burant CF, Zhang L, Adsay V, et al. Identification of pancreatic cancer stem cells. Cancer Res 2007;67(3):1030–7.

[20] Nguyen LV, Vanner R, Dirks P, Eaves CJ. Cancer stem cells: an evolving concept. Nat Rev Cancer 2012;12(2):133–43.

[21] Wicha MS, Liu S, Dontu G. Cancer stem cells: an old idea—a paradigm shift. Cancer Res 2006;66(4):1883–90.

[22] Chaffer CL, Weinberg RA. A perspective on cancer cell metastasis. Science 2011;331(6024): 1559–64.

[23] Pantel K, Alix-Panabieres C. Circulating tumour cells in cancer patients: challenges and perspectives. Trends Mol Med 2010;16(9):398–406.

[24] Kalluri R, Weinberg RA. The basics of epithelial-mesenchymal transition. J Clin Invest 2009;119(6):1420.

[25] Friedl P, Wolf K. Tumour-cell invasion and migration: diversity and escape mechanisms. Nat Rev Cancer 2003;3(5):362–74.

[26] Alix-Panabières C, Pantel K. Challenges in circulating tumour cell research. Nat Rev Cancer 2014;14(9):623–31.

[27] Joyce JA, Pollard JW. Microenvironmental regulation of metastasis. Nat Rev Cancer 2009;9(4):239–52.

[28] Labelle M, Begum S, Hynes RO. Direct signaling between platelets and cancer cells induces an epithelial-mesenchymal-like transition and promotes metastasis. Cancer cell 2011;20(5):576–90.

[29] Pattabiraman DR, Weinberg RA. Tackling the cancer stem cells—what challenges do they pose? Nat Rev Drug Discov 2014;13(7):497–512.

[30] Brabletz T. To differentiate or not—routes towards metastasis. Nat Rev Cancer 2012;12(6): 425–436.

[31] Maheswaran S, Haber DA. Circulating tumor cells: a window into cancer biology and metastasis. Curr Opin Genet Dev 2010;20(1):96–9.

[32] Cristofanilli M, editor. Circulating tumor cells, disease progression, and survival in metastatic breast cancer. Seminars in Oncology-Supplements. New York: Grune & Stratton; 2006.

[33] Cristofanilli M, Hayes DF, Budd GT, Ellis MJ, Stopeck A, Reuben JM, et al. Circulating tumor cells: a novel prognostic factor for newly diagnosed metastatic breast cancer. J Clin Oncol 2005;23(7):1420–30.

[34] Hayes DF, Cristofanilli M, Budd GT, Ellis MJ, Stopeck A, Miller MC, et al. Circulating tumor cells at each follow-up time point during therapy of metastatic breast cancer patients predict progression-free and overall survival. Clin Cancer Res 2006;12(14):4218–24.

[35] Riethdorf S, Muller V, Zhang L, Rau T, Loibl S, Komor M, et al. Detection and HER2 expression of circulating tumor cells: prospective monitoring in breast cancer patients treated in the neoadjuvant GeparQuattro trial. Clin Cancer Res 2010;16(9):2634–45.

[36] Paoletti C, Muniz MC, Thomas DG, Griffith KA, Kidwell KM, Tokudome N, et al. Development of circulating tumor cell-endocrine therapy index in patients with hormone receptor-positive breast cancer. Clin Cancer Res 2015;21(11):2487–98.

[37] de Bono JS, Scher HI, Montgomery RB, Parker C, Miller MC, Tissing H, et al. Circulating tumor cells predict survival benefit from treatment in metastatic castration-resistant prostate cancer. Clin Cancer Res 2008;14(19):6302–9.

[38] Danila DC, Heller G, Gignac GA, Gonzalez-Espinoza R, Anand A, Tanaka E, et al. Circulating tumor cell number and prognosis in progressive castration-resistant prostate cancer. Clin Cancer Res 2007;13(23):7053–8.

[39] Attard G, Swennenhuis JF, Olmos D, Reid AH, Vickers E, A'Hern R, et al. Characterization of ERG, AR and PTEN gene status in circulating tumor cells from patients with castration-resistant prostate cancer. Cancer Res 2009;69(7):2912–8.

[40] Cohen SJ, Punt CJA, Iannotti N, Saidman BH, Sabbath KD, Gabrail NY, et al. Relationship of circulating tumor cells to tumor response, progression-free survival, and overall survival in patients with metastatic colorectal cancer. J Clin Oncol 2008;26(19):3213–21.

[41] Cohen SJ, Punt CJ, Iannotti N, Saidman BH, Sabbath KD, Gabrail NY, et al. Prognostic significance of circulating tumor cells in patients with metastatic colorectal cancer. Ann Oncol 2009;20(7):1223–9.

[42] Sastre J, Maestro ML, Puente J, Veganzones S, Alfonso R, Rafael S, et al. Circulating tumor cells in colorectal cancer: correlation with clinical and pathological variables. Ann Oncol 2008;19(5):935–8.

[43] Pantel K, Brakenhoff RH, Brandt B. Detection, clinical relevance and specific biological properties of disseminating tumour cells. Nat Rev Cancer 2008;8(5):329–40.

[44] Whitesides GM. The origins and the future of microfluidics. Nature 2006;442(7101):368–73.

[45] Sackmann EK, Fulton AL, Beebe DJ. The present and future role of microfluidics in biomedical research. Nature 2014;507(7491):181–9.

[46] Wang F, Burns MA. Performance of nanoliter-sized droplet-based microfluidic PCR. Biomed microdevices 2009;11(5):1071–80.

[47] Chin CD, Laksanasopin T, Cheung YK, Steinmiller D, Linder V, Parsa H, et al. Microfluidics-based diagnostics of infectious diseases in the developing world. Nat Med 2011;17(8):1015–9.

[48] McDonald JC, Whitesides GM. Poly (dimethylsiloxane) as a material for fabricating microfluidic devices. Accounts Chem Res 2002;35(7):491–9.

[49] Maheswaran S, Sequist LV, Nagrath S, Ulkus L, Brannigan B, Collura CV, et al. Detection of mutations in EGFR in circulating lung-cancer cells. N. Engl J Med 2008;359(4):366–77.

[50] Gleghorn JP, Pratt ED, Denning D, Liu H, Bander NH, Tagawa ST, et al. Capture of circulating tumor cells from whole blood of prostate cancer patients using geometrically enhanced differential immunocapture (GEDI) and a prostate-specific antibody. Lab chip 2010;10(1):27–9.

[51] Kirby BJ, Jodari M, Loftus MS, Gakhar G, Pratt ED, Chanel-Vos C, et al. Functional characterization of circulating tumor cells with a prostate-cancer-specific microfluidic device. PLoS One 2012;7(4):e35976.

[52] Pratt ED, Stepansky A, Hicks J, Kirby BJ. Single-cell copy number analysis of prostate cancer cells captured with geometrically enhanced differential immunocapture microdevices. Anal Chem 2014;86(22):11013–7.

[53] Murlidhar V, Zeinali M, Grabauskiene S, Ghannad-Rezaie M, Wicha MS, Simeone DM, et al. A radial flow microfluidic device for ultra-high-throughput affinity-based isolation of circulating tumor cells. Small 2014;10(23):4895–904.

[54] Stott SL, Hsu CH, Tsukrov DI, Yu M, Miyamoto DT, Waltman BA, et al. Isolation of circulating tumor cells using a microvortex-generating herringbone-chip. Proc Natl Acad Sci USA 2010;107(43):18392–7.

[55] Miyamoto DT, Lee RJ, Stott SL, Ting DT, Wittner BS, Ulman M, et al. Androgen receptor signaling in circulating tumor cells as a marker of hormonally responsive prostate cancer. Cancer Discov 2012;2(11):995–1003.

[56] Sheng W, Ogunwobi OO, Chen T, Zhang J, George TJ, Liu C, et al. Capture, release and culture of circulating tumor cells from pancreatic cancer patients using an enhanced mixing chip. Lab Chip 2014;14(1):89–98.

[57] Adams AA, Okagbare PI, Feng J, Hupert ML, Patterson D, Gottert J, et al. Highly efficient circulating tumor cell isolation from whole blood and label-free enumeration using polymer-based microfluidics with an integrated conductivity sensor. J Am Chem Soc 2008;130(27):8633–41.

[58] Kamande JW, Hupert ML, Witek MA, Wang H, Torphy RJ, Dharmasiri U, et al. Modular microsystem for the isolation, enumeration, and phenotyping of circulating tumor cells in patients with pancreatic cancer. Anal Chem 2013;85(19):9092–100.

[59] Yoon HJ, Kozminsky M, Nagrath S. Emerging role of nanomaterials in circulating tumor cell isolation and analysis. ACS Nano 2014;8(3):1995–2017.

[60] Yoon HJ, Kim TH, Zhang Z, Azizi E, Pham TM, Paoletti C, et al. Sensitive capture of circulating tumour cells by functionalized graphene oxide nanosheets. Nat Nanotechnol 2013;8(10):735–41.

[61] Reategui E, Aceto N, Lim EJ, Sullivan JP, Jensen AE, Zeinali M, et al. Tunable nanostructured coating for the capture and selective release of viable circulating tumor cells. Adv Mater 2015;27(9):1593–9.

[62] Ke Z, Lin M, Chen JF, Choi JS, Zhang Y, Fong A, et al. Programming thermoresponsiveness of NanoVelcro substrates enables effective purification of circulating tumor cells in lung cancer patients. ACS Nano 2015;9(1):62–70.

[63] Casavant BP, Strotman LN, Tokar JJ, Thiede SM, Traynor AM, Ferguson JS, et al. Paired diagnostic and pharmacodynamic analysis of rare non-small cell lung cancer cells enabled by the VerIFAST platform. Lab Chip 2014;14(1):99–105.

[64] Ozkumur E, Shah AM, Ciciliano JC, Emmink BL, Miyamoto DT, Brachtel E, et al. Inertial focusing for tumor antigen-dependent and -independent sorting of rare circulating tumor cells. Sci Transl Med 2013;5(179):179–247.

[65] Harouaka RA, Zhou MD, Yeh YT, Khan WJ, Das A, Liu X, et al. Flexible micro spring array device for high-throughput enrichment of viable circulating tumor cells. Clin Chem 2014;60(2):323–33.

[66] Di Carlo D. Inertial microfluidics. Lab Chip 2009;9(21):3038–46.

[67] Sun J, Li M, Liu C, Zhang Y, Liu D, Liu W, et al. Double spiral microchannel for label-free tumor cell separation and enrichment. Lab Chip 2012;12(20):3952–60.

[68] Hou HW, Warkiani ME, Khoo BL, Li ZR, Soo RA, Tan DS-W, et al. Isolation and retrieval of circulating tumor cells using centrifugal forces. Sci Rep 2013;3.

[69] Khoo BL, Warkiani ME, Tan DS, Bhagat AA, Irwin D, Lau DP, et al. Clinical validation of an ultra high-throughput spiral microfluidics for the detection and enrichment of viable circulating tumor cells. PLoS One 2014;9(7):e99409.

[70] Warkiani ME, Guan G, Luan KB, Lee WC, Bhagat AA, Chaudhuri PK, et al. Slanted spiral microfluidics for the ultra-fast, label-free isolation of circulating tumor cells. Lab Chip 2014;14(1):128–37.

[71] Kim TH, Yoon HJ, Stella P, Nagrath S. Cascaded spiral microfluidic device for deterministic and high purity continuous separation of circulating tumor cells. Biomicrofluidics 2014;8(6):064117.

[72] Sollier E, Go DE, Che J, Gossett DR, O'Byrne S, Weaver WM, et al. Size-selective collection of circulating tumor cells using Vortex technology. Lab Chip 2014;14(1):63–77.

[73] Ding X, Peng Z, Lin S-CS, Geri M, Li S, Li P, et al. Cell separation using tilted-angle standing surface acoustic waves. Proc Natl Acad Sci USA 2014;111(36):12992–7.

[74] Li P, Mao Z, Peng Z, Zhou L, Chen Y, Huang PH, et al. Acoustic separation of circulating tumor cells. Proc Natl Acad Sci USA 2015;112(16):4970–5.

[75] Gupta V, Jafferji I, Garza M, Melnikova VO, Hasegawa DK, Pethig R, et al. ApoStream™, a new dielectrophoretic device for antibody independent isolation and recovery of viable cancer cells from blood. Biomicrofluidics 2012;6(2):024133.

[76] Rhim AD, Thege FI, Santana SM, Lannin TB, Saha TN, Tsai S, et al. Detection of circulating pancreas epithelial cells in patients with pancreatic cystic lesions. Gastroenterology 2014;146(3):647–51.

[77] Torphy RJ, Tignanelli CJ, Kamande JW, Moffitt RA, Herrera Loeza SG, Soper SA, et al. Circulating tumor cells as a biomarker of response to treatment in patient-derived xenograft mouse models of pancreatic adenocarcinoma. PLoS One 2014;9(2):e89474.

[78] Kaifi JT, Kunkel M, Das A, Harouaka RA, Dicker DT, Li G, et al. Circulating tumor cell isolation during resection of colorectal cancer lung and liver metastases: a prospective trial with different detection techniques. Cancer Biol Ther 2015;16(5):699–708.

[79] Ting DT, Wittner BS, Ligorio M, Vincent Jordan N, Shah AM, Miyamoto DT, et al. Single-cell RNA sequencing identifies extracellular matrix gene expression by pancreatic circulating tumor cells. Cell Rep 2014;8(6):1905–18.

[80] Zhang Z, Shiratsuchi H, Lin J, Chen GA, Reddy RM, Azizi E, et al. Expansion of CTCs from early stage lung cancer patients using a microfluidic co-culture model. Oncotarget 2014;5(23):12383–97.

[81] Yu M, Bardia A, Aceto N, Bersani F, Madden MW, Donaldson MC, et al. Cancer therapy. Ex vivo culture of circulating breast tumor cells for individualized testing of drug susceptibility. Science (New York, NY) 2014;345(6193):216–20.

[82] Krebs MG, Hou JM, Sloane R, Lancashire L, Priest L, Nonaka D, et al. Analysis of circulating tumor cells in patients with non-small cell lung Cancer using epithelial marker-dependent and -independent approaches. J Thorac Oncol 2012;7(2):306–15.

[83] Alix-Panabières C. EPISPOT assay: detection of viable DTCs/CTCs in solid tumor patients. Minimal residual disease and circulating tumor cells in breast cancer. Springer; 2012. p. 69–76.

[84] Cayrefourcq L, Mazard T, Joosse S, Solassol J, Ramos J, Assenat E, et al. Establishment and characterization of a cell line from human circulating colon cancer cells. Cancer Res 2015;75(5):892–901.

[85] Hodgkinson CL, Morrow CJ, Li Y, Metcalf RL, Rothwell DG, Trapani F, et al. Tumorigenicity and genetic profiling of circulating tumor cells in small-cell lung cancer. Nat Med 2014;20(8):897–903.

Index

'*Note*: Page numbers followed by "f" indicate figures, "t" indicate tables.'

E

I

J

K

L

M

R

S

T

Kolibri 360 GmbH
Ferdinand-Jühlke-Straße 7
99095 Erfurt, Deutschland
produktsicherheit@kolibri360.de